Microbial Quality of Water Supply in Distribution Systems

Edwin E. Geldreich
Senior Research Microbiologist
United States Environmental Protection Agency
Drinking Water Research Division
Cincinnati, Ohio

CRC Press
Taylor & Francis Group
Boca Raton London New York

CRC Press is an imprint of the
Taylor & Francis Group, an **informa** business

CRC Press
Taylor & Francis Group
6000 Broken Sound Parkway NW, Suite 300
Boca Raton, FL 33487-2742

First issued in paperback 2019

© 1996 by Taylor & Francis Group, LLC
CRC Press is an imprint of Taylor & Francis Group, an Informa business

No claim to original U.S. Government works

ISBN-13: 978-1-56670-194-5 (hbk)
ISBN-13: 978-0-367-40141-2 (pbk)
Library of Congress Card Number 95-37894

Library of Congress Cataloging-in-Publication Data

Geldreich, Edwin E.
    Microbial quality of water supply in distribution systems / Edwin
E. Geldreich.
       p.   cm.
    Includes bibliographical references and index.
    ISBN 1-56670-194-5
    1. Water--Microbiology. 2. Water quality--Management. I. Title.
TD384.G45 1996
628.1'62--dc20   95-37894
                                                  CIP

Visit the Taylor & Francis Web site at
http://www.taylorandfrancis.com

and the CRC Press Web site at
http://www.crcpress.com

Preface

Why a book on the microbial quality of water in distribution systems? Simply because there is a need to better understand the problems hidden in the pipe network buried under streets that may become a utility nightmare when there is noncompliance with drinking water regulations or when a waterborne disease crisis occurs.

These problems are not new, having been well hidden for years in disbelief by management of some utilities that water quality could degrade upon passage through the network of pipes, storage reservoirs, and standpipes. Engineering textbooks and operator confidence in the ultimate power of a chlorine residual to inactivate organisms of public health significance was unshaken. When coliforms were reported from samples taken on the distribution system, the immediate reaction by some operators was to discredit the sample as being contaminated during collection or to reject the analysis as being a laboratory error.

This attitude began to change during the 1970s with the intensification of laboratory certification at the state level by the federal mandate to evaluate laboratory performance and adherence to appropriate methods specified in current editions of *Standard Methods for Examination of Water and Wastewater*. In turn, state programs were created to take this evaluation process to the local utility, county, and private laboratories involved in water supply monitoring. More intense appraisal of sample collection techniques, expanded verification procedures for coliform-positive samples, and regulation requirements to repeat sample analyses at sites of coliform occurrence had done much to strengthen the realization that coliform bacteria may co-exist with chlorine residuals under a specific set of circumstances.

As these summer time coliform regrowth occurrences became more obvious, utilities were forced to try drastic measures in the hope of extinguishing the microbial nightmare. One utility actually drained the entire supply of public water from the system over a weekend in an effort to clean the pipe network of these colonizations. This drastic action provided only a temporary measure of success. Other systems elevated the chlorine residuals to as much as 5 mg/l of free chlorine for brief periods of time. This action brought the system back into compliance, but as soon as stepwise drops in applied chlorine were begun, coliform-positive samples began to appear again. Some treatment operators began to apply various corrosion inhibitor additives only to find that somehow things got worse, not better.

All of these events led the EPA Drinking Water Research Division to begin an extensive research program to study the problem, identify the factors that were key to microbial growth, and work with universities and the water utility industry to study various factors in the field. Over the past 20 years many of the issues have been identified, including composition of pipe biofilms, their habitats in pipe sediments and tubercles, their nutrient base for survival, the protective factors that shield them from disinfection contact, and some treatment measures that appear to have a degree of success in specific instances. Much of this state of knowledge appears in this book, yet much still remains to be learned about the public health significance of biofilms, ecological progression of a biofilm community, detecting areas of biofilm activity, approach to successful preventive maintenance, and alternative treatment regimes for an aggressive biofilm progression in the distribution system. Hopefully, the book will stimulate other scientists, engineers, and operators to work on this aspect of drinking water microbiology and resolve the issue in the next decade.

Edwin E. Geldreich

The Author

 **Edwin E. Geldreich**, with more than 46 years of research experience, has directed investigations into the microbiology of water supply sources, treatment processes, methods development, criteria, and standards. He has also been a frequent microbiology consultant for World Health on projects in Europe and South America involving water supply, fresh water pollution, and recreational water and coastal water quality. A member of American Water Works Association (AWWA), American Society for Microbiology, and the International Association on Water Pollution Research, Geldreich has received the Kimble Methodology Research Award, the U.S. Environmental Protection Agency silver and bronze medals, AWWA's Research Award, and the 1989 Abel Wolman Award of Excellence and was the 1991 Allen Hazen Lecturer to the New England Water Works Association. He has also been a Fellow in the American Academy of Microbiology for many years. As author of approximately 100 research papers and two books, his work has been published in many peer review journals, including the *Journal of the American Water Works Association, Applied and Environmental Microbiology,* and *Water Pollution Research.*

Current interest during the past 20 years has focused on treatment barriers and microbial problems in water supply distribution systems, their cause and control. More than 30 utilities have invited Geldreich's participation in investigating coliform problems in their distribution systems. Activities have included an investigation into the waterborne outbreaks at Cabool and Gideon, Missouri, an investigation, at the request of the Peruvian Ministry of Health, into their cholera outbreak for a water supply connection, and involvement in an international conference urging water supply disinfection within the Latin American and Caribbean community of nations to combat further waterborne pathogen invasion.

Special Acknowledgments

In the development of the text it quickly became evident that visual aids would be essential in demonstrating problems from the hidden environment of the pipe network. A search was made to gather photographs and drawings that could illustrate the issues on water quality degradation. I am grateful for some of the pipe photographs supplied by colleagues in the field. Many of the drawings, figures, and tables were the creation of Richard C. Findsen, artist, Steve C. Waltrip, graphics expert, and Nancy T. Frazier, graphics assistant, who volunteered their professional services and provided suggestions on the selection of enhanced graphics for this book.

Table of Contents

*This book is dedicated to my wife Detta,
who provided much encouragement to
complete my improbable dream
of a book on this subject.*

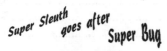

Searching for cause/effect.

CHAPTER 1

Characterizing the Distribution System: Microbial Issues

CONTENTS

INTRODUCTION

The purpose of a water supply distribution system is to deliver to each consumer safe drinking water that is also adequate in quantity and acceptable in terms of taste, odor, and appearance. Distribution of drinking water to the community has been a global major concern for many centuries. In ancient times, the primary objective was to provide delivery of adequate amounts of water to centralized fountains and other locations designated for public gathering plus restricted areas of official residences. Aqueducts to carry spring water or divert remote mountain streams were built by the Greeks and Romans during their rise to dominance as centers of ancient civilizations. The Romans also covered some of their aqueducts to protect the waters from the hot Mediterranean sun or willful contamination by enemy action.[1] Pipe materials in these early water transport systems were of natural stone, clay, or lead. Distribution lines were simple, generally limited to a main trunk line with a few dead-end branches. Water flow was dependent on gravity and the continuous discharge from springs and mountain stream diversions.[2]

Historically, the initial distribution network of pipes was a response to existing community needs that eventually created a legacy of problems of inadequate supply and low pressure as the population density increased. To resolve the problem of increased water demand along the distribution route, reservoir storage was created. Pressure pumping to move water to far reaches of the supply lines and standpipes was incorporated to afford relief from surges of pressure in pipelines. In some areas, population growth exceeded the capacity of a water resource so that other sources of water were incorporated and additional treatment plants were built to feed into the distribution network. In other situations, neighboring water systems were consolidated by interconnection.

To satisfy the delivery of a safe water to the consumer, management of public water supply systems today must still be continually vigilant for any intrusions of contamination in the distribution network and occurrences of microbial degradation. This job is complicated by the very nature of a distribution system, which is a network of mains, fire hydrants, valves, auxiliary pumping or booster chlorination substations, storage reservoirs, standpipes, and service lines. Added to these complications are the plumbing systems in large housing projects, highrise office buildings, hospitals, and public buildings. These building supply lines introduce many dead ends and a variety of attachment devices for special water supply applications.

Water supply released into the distribution system becomes altered during its passage through pipes, open reservoirs, standpipes, and storage tanks. These changes are caused to a large extent by a variety of microbiological activities that occur as water passes through the distribution network. Field experiences have revealed these quality changes most often cause taste, odor, and color complaints. More serious is a problem of noncompliance with federal regulations when there is a sudden occurrence of coliform bacteria that could lead to public notification or an order to boil water. Because coliform bacteria are easily controlled by treatment, the concern is, why do these organisms occasionally appear in the public water supply, what is their public health significance, and how can pipeline biofilms be suppressed? In an effort to better understand the reasons for these water quality changes, it is important to have a basic understanding of the hydraulic arrangements designed to deliver a constant supply of water to the consumer. Some of these network components may become excellent areas for microbial habitation. Any long-term persistence and biofilm development in this pipe network environment will be influenced by a variety of factors including distribution system age, soil stability and its corrosivity, and type of pipe material used. Other influences include water flow reversals, static water conditions, chemistry of the water, seasonal warm-water temperatures, and operational practices.

DISTRIBUTION SYSTEM HABITATS

There is a variety of situations in the distribution system where microbial activity may develop. It is amazing how tenacious bacteria are in establishing a foothold in some areas of the distribution network and using these sites as a base for colonization. Equally astounding is the fact that colonization occurs in an environment of minimal nutrients, disinfectant residuals and against the shearing forces of water flow. Key to habitat development are areas for sediment deposition, materials that are degradable, static water zones, and warm water. The following sections will identify areas of habitat potential and measures to consider to minimize the potential for microbial invasion.

PIPE, VALVES, AND FIRE HYDRANTS

By far, the pipe network is the most extensive part of the distribution system. Factors that provide an important contribution to organism colonization include length of the pipe network, predominant pipe material, age of pipe, number of breaks per year, corrosion, sediment accumulation, and zones of static water in the pipe network. Information obtained from various utilities (Table 1.1) suggests that while there is a general correlation between population served and size of the pipe network, some distribution systems apparently have a greater population density per square mile (Wilmette, IL, Rochester,

Table 1.1 Characterizing Distribution Systems for Selected Utilities

Water treatment system	Pop. served	Network size (mi)	Total storage (MG)	Type of pipe	Age of oldest pipe (years)	System pressure range (psi)	Retention time (d)	Routine flushing program
Brockport, NY	35,000	100	4.8	C	75	<200	1–2	Yes
Kennebunk, ME	12–50,000	181	4.0	C	90	30–110	Variable	Yes
Terre Haute, IN	66,700	320	3.05	C,D,AC	125	60–80	1–2	Yes
Muncie, IN	85,000	300	4.00	AC,C,D,S	80	50–90	2–5	Yes
Wilmette, IL	100,000	95	12.00	C,D	100	30–80	1	Yes
Springfield, IL	142,000	420	12.00	D	90	50–60	2	Yes
Lexington, KY	225,000	1,162	16.93	C,AC,P	100	50–80	2	Yes
Rochester, NY	245,000	625	245	C	>80	90	2	Yes
Monmouth, NJ	260,000	1,048	13.45	C,D	—	—	—	No
Grand Rapids, MI	263,000	864	73.75	C,D	70	<120	2	Yes
New Haven, CT	370,000	1,391	51.00	C,D	80	40–60	5–6	Yes
Fairfax, VA	820,000	2,376	25.00	C,D,CP,S	50	30–80	<1	Yes

Note: C, cast iron; D, ductile iron; AC, asbestos cement; S, steel; CP, concrete pressure; P, plastic.

Data from Geldreich, Goodrich, and Clark.[3]

NY) than others because the miles of pipe are much less than in other cities of similar population.

In a typical medium-sized city, the distribution system may contain approximately 18,000 fire hydrants and over 160,000 service lines connected to the pipe network. Most often this pipe network consists of water mains along each street with cross-connections to other mains at street intersections so that there is a grid pattern of pipelines to reach every building in the city. Large transmission lines of 60-in. diameter are used to supply water-storage tanks and finished water reservoirs. Pipe size in major trunk lines to different areas may be of 12- or 16-in. diameter, while those in the cross-grids are often of 6- or 8-in. size. From these latter water pipes, service lines of smaller size bring water to homes, apartment complexes, and businesses. This arrangement would in theory provide the most efficient way to distribute water at a uniform pressure; however, there are many factors that have impact on maintaining a uniform pressure. Most of their resistance to flow is caused by numerous cross-line connections, pipe tuberculation and sediment formation, restrictions in gate valve passage, and differences in ground elevation. Pressure drops are also due to summertime water demand to fill pools and sprinkle lawns. Major fire events also create pressure drop in water lines because of the need for water in fire control. As a consequence of sudden restoration of normal water demand, back pressures in the pipeline have been known to release biofilm into the water supply. In fact, in one instance following a major fire event, water turbidity became evident and for 24 h there was a release of over 100 *Pseudomonas aeruginosa* per milliliter in that district water lines.

Gate valves are set at nearly every street intersection for the purpose of shutting off the water supply in case of emergency. Some water systems augment their supply of water through a gate valve connection to an neighboring community water supply. This arrangement is done to meet summertime water demand, often because of tourism that may seasonally more than double the population served. Blending of two water supplies of different water qualities has on occasion introduced contamination into the receiving supply and failure to completely close off the gate valve provided leakages of the poorer quality purchased water supply. At the service connection to each customer, smaller gate valves are positioned near the street curb to provide utility control of water service at the premise for a variety of purposes.

Fire hydrants are generally located at each street intersection and at some intermediate point with spacing that ranges from 350 to 600 ft. These spacings should not be within 10 ft of sanitary sewers or storm drains as a precaution against some surface contamination entering the system.

From this generalized description of the pipe network it is apparent that there are many locations where pipe bends, joint interconnections, gate valves, fire hydrant attachments, and metered service lines will impede the flow of water and provide opportunities for particles to settle out and deposit a pipe sediment. Continuous runs of pipe for several miles with few active withdrawals of water may create a static water zone where bacteria can colonize the line and chlorine residuals disappear in the conversion of available organic carbon. This situation is of particular concern in rural water consolidation schemes that join two or more area distribution systems together. Frequent intermixing of water pipes should be discouraged because these mismatches of pipe size increase pumpage cost and set up areas of reduced water velocities that may become sites for pipe sediment accumulation.

Pipe Materials

Various materials have been used for water supply pipes over the centuries. Some of the wood, clay, stone, and lead pipes developed in antiquity are still functional delivery systems to gardens, fountains, and palaces of the Mediterranean region. In this part of the world, some eastern and western city utilities have evidence that wooden pipes (bored logs or wood staved) were part of their distribution system in the 18th and 19th centuries. In fact, Butte, MT is still using wood stave pipe (Figure 1.1) as a part of its transmission line from raw water impondments. Cast-iron pipe was introduced to the industry about 280 years ago. Now the choice of pipe materials include ductile iron, cast iron lined with cement, steel, reinforced concrete, asbestos combined with Portland cement, and three plastic materials: polyvinylchloride (PVC), polyethylene, and polybutylene. Pipe materials for service lines and building pipe networks are most often copper or plastic, but occasionally black and galvanized iron may be encountered.[4]

Figure 1.1   Wood stave water transmission pipe. (Courtesy of the Montana Health Department.)

Selection of pipe sizes and material involves consideration of cost, velocities at peak flow, pressure, climatic conditions, soil contaminants, transition fittings and jointing methods.[5,6] Permeation of plastic water pipe by organics is of special consideration in areas where the soil is contaminated with organics from gasoline spills (leaking storage tanks) abandoned industrial sites, bulk chemical storage, electroplaters, and dry cleaners.[7] Over time, permeation of the plastic will cause undesirable organic leaching into the water supply and possible hasten pipeline failures.

Service life of these pipe materials will vary, being related to water chemistry, microbial activity, soil corrosivity, climatic conditions, and vibrations around the street bedding site. Cast-iron and ductile-iron pipe appear to have a service life of approximately 100 years; reinforced concrete, 50 years; and asbestos cement pipe, 30 years. Little can be said with any degree of certainty about the service life for plastic pipe because it has been slow to gain wide acceptance by most utilities.

The continual presence of a microbial population in drinking water passing through the distribution system will eventually result in localized development of a biofilm somewhere in the pipe network. In iron pipe, chlorine doses as high as 4 mg/l have little impact on biofilm reduction because iron corrosion products interfere with free chlorine disinfection. In fact, chlorine demand was found to be as much as ten times higher in iron pipes than in pipes of other composition.[8] Various coatings have been developed to deter microbial activity that can lead to pipe corrosion or taste and odor problems. Some of these additives may be effective, but they should be evaluated in a pilot operation by the utility before making any extensive commitment because effectiveness may vary with water chemistry and microbial flora.

As an example of induced microbial activity, a 7-km-long iron water pipe was lined with cement mortar and after some time, substantial densities of heterotrophic bacteria were detected in water passing through the pipe section.[9] Repeated flushing and chlorination treatments only brought a temporary improvement in the microbial quality. Inspection of the line after 3.5 years in service revealed a slimy coating in the portion of cement mortar lining that had been applied to the pipe by hand. In the manual application, workmen had added a plastic additive to the cement mortar mix to improve quick adhesion. Unfortunately, this additive became the nutrient support for the development of a localized (2 m) slime deposit. The factory-produced cement mortar lining used in the remainder of the pipe did not contain a biodegradable substance and was not the site for bacterial growth.

Sand used as an aggregate in the making of concrete for lining new ductile iron pipe can also be a contaminating source of bacteria. Such was the situation experienced by a midwestern utility after accepting a new pipeline from the contractor. Suddenly there was a large increase in the heterotrophic bacterial population in the water supply passing through the new pipe section. The laboratory reported the predominant organisms were *Pseudomonas fluorescens, Ps. maltophilia* and *Ps. putida*. Tracing back through the pipeline construction project provided circumstantial evidence that the contaminate was probably introduced in the sand (the same strains of *Pseudomonas* were recovered in the sand). The subcontractor supplying the sand resolved the problem by increasing the wash cycle for better removal of silt in the raw material and elevating the sand drying temperature to a range between 180 and 200°F.

Gaskets, Sealants, and Lubricants

Materials used to join water pipes in a tight fit, seal threaded plastic pipe, and lubricate gate valves in the pipe network must also be considered as possible sites for microbial colonizations. Old water lines, prior to 1950, incorporated gasket and packing materials made of jute, leather, cork, string, or natural rubber.[10] Lubrication around these fittings was done with a variety of products such as linseed oil, soaps, or tallow. In this situation it was not surprising to note the appearance of localized biofilm because of the biodegradability of such materials.[11-13] Since then new products based entirely on synthetic materials have appeared on the market. Their appeal is based on stated claims for better leak protection, ease of application, and reduced costs. Unfortunately, in the selection of newer materials, little attention was given to concerns for microbial colonizations. Using a microbiological growth test (Table 1.2) it is easy to see that use of some available materials could be the reason for bacterial colonization. Use of organic accelerators in on-site application of epoxy resins in water main renovations is another contributing factor. In this case, benzyl alcohol, added as the accelerator, has been found to be the nutrient support that leads to bacterial multiplication on the epoxy resin deposit.[15,16] This growth-promoting effect may continue in the pipe for years, until the material is completely degraded or leached out of the water. Many of the chemicals used as accelerators, retarders, or fillers in formulation will support microbial growth when present in concentrations in excess of 1% of the total compound.[14] Mold release agents used in forming gaskets is another factor supporting bacterial growth. As a consequence, manufacturers are now using silicone as the mold release compound.

While some lubricants have supported microbial growth, one product incorporates a quaternary ammonium compound to help alleviate the contamination often encountered in new main installations.[17] During field trials of this lubricant, 12 different water systems used it for a period of 1 year. Taking the absence of coliform bacteria in 100 ml of sample as the criterion of successful disinfection, results were compared with control series obtained with conventional lubricants (Table 1.3). In mains using the formulation, only 19.8% of new main installations required repeat chlorination, while 47.9% of the mains laid using another lubricant (vegetable-based lubricant on rubber gaskets at pipe couplings) required repeat disinfection. The bactericidal gasket lubricant is not a replacement for routine disinfection nor does it provide a relaxation in the quality of installation practices. The lubricant is principally

Table 1.2 Performance of Coatings, Sealants, Jointings, Packings, and Other Materials in the Microbiological Growth Test

Material type	No. of samples[a]	Pass rate (%)
Coatings		
Acrylic	4	100
Bituminous	20	35
Epoxy	64	84
Polyester	4	100
Polyurethane	24	67
Unknowns	4	50
Sealants		
Polysulfide	9	11
Polyurethane	17	29
Silicone	11	73
Jointings		
Asbestos	6	100
Fiber	14	67
Packings	15	26
Adhesives	13	100
Greases	21	38
Lubricants	14	57
Graphite	2	100
Cementitious products	13	85
Ion-exchange resins	5	100
Oil sorbents	4	25
Total	264	

[a] Combined results from Thames Water and Albury Laboratories Ltd.
From Colbourne, J. and Brown, D., *J. Appl. Bacteriol.*, 47, 1–9, 1979. With permission.

effective in controlling contamination within the joint space and, therefore, the incidence of repeated failures. When used in conjunction with proper main-laying practices, adequate pipe cleaning, and chlorination procedures, significant reductions in unsuccessful disinfection attempts should be expected.

WATER SUPPLY STORAGE

Water supply in the community varies hourly as a reflection of the activities of the general public and local industries. While industrial uses of potable water are more constant and predictable, it is unrealistic to expect water treatment operations to gear production to those frequent and sudden changes in water demand from all consumers, particularly during heat wave conditions. For this reason storage reservoirs must be an essential component of the distribution network whose purpose is to serve as a buffer against a variety of situations related to treatment capacity and water supply demand. These water supply reserves supplement water flows in distribution during periods of fluctuating

Table 1.3   Mains Disinfection: Comparison of Results
for Lubricants

Item	Bactericidal gasket lubricant[a]	Other lubricants
Total no. of mains treated	364	2091
Mains in which the initial chlorination failed		
No.	55	490
Percent of total no. of mains chlorinated	15.1	23.4
Mains requiring more than a second chlorination		
No.	10	260
Percent of total no. of mains treated	2.8	12.4
Total no. of repeated chlorinations		
No.	72	1002
Percent of total no. of mains treated	19.8	47.9

[a] Medlube.
Data from Buelow et al.[18]

demand on the system, provide storage of water reservoirs during off-peak periods, equalize operational water pressures, and augment water supply from production wells that must be pumped at a constant rate best suited for the well field. Water storage also provides a protected reserve of drinking water to guard against shutdown of water treatment during oil spills in the raw source water, flooding of well fields, and destruction of transmission lines feeding the distribution system. An important secondary consideration is sufficient water storage capacity calculated to be adequate for fire emergencies.

Finished water reservoirs may be located near the beginning of a distribution system but most often are situated on high ground in suburban areas. Local topography plays an important part in determining the use of ground-level or high-level reservoirs. Underground storage basins are usually formed by excavation while ground-level reservoirs are generally constructed by earth embankments. Such reservoirs are lined with concrete, Gunite, asphalt, or a plastic sheet over the basin to prevent water loss.[19] In earthquake zones, reinforced cement or a series of flatbed steel compartments are mandatory. All things considered, reinforced concrete is the preferred material because it has a minimal rate of deterioration from water contact. Elevated storage tanks and standpipes are of steel design and may have bituminous coal tar, epoxy resin, PVC, silicon, acrylic, or other materials applied to the interior as a sealant.[9,10,20]

Microbial Considerations

Some of these coatings have been shown to support microbial growth (Table 1.2). Organic polymer solvents in bituminous coating materials do not entirely evaporate even after several weeks of ventilation. As a consequence, the water supply in storage becomes contaminated from the solvent-charged air and from contact at the side wall air–water interfaces. These nutrients are assimilable organics that support regrowth of heterotrophic bacteria during warm-water periods.[9,21-23] Liner materials, also used to prevent water loss, may contain bitumen, chlorinated rubber, epoxy resin, or a tar–epoxy resin combination that will eventually be colonized by microbial growth and slime development.[9] PVC film and PVC coating materials are other sources of microbial activity. Nonhardening sealants (containing polyamide and silicone) used in expansion joints should not be overlooked as a source of microbial habitation.

Water volumes in large storage tanks mix and interchange slowly with water that is actually distributed to service lines. In fact, water stratification may occur in some instances because of the design placement of the inlet–outlet port. Standpipes, in contrast, provide a fluctuating storage of water during a down surge, thereby preventing or minimizing the effects of rapid changes in hydraulic pressure, usually described as water hammer.[24] These occasional abrupt changes in water flow can cause the steady-state nature of sediment depositions to become unstable, releasing viable bacteria from biofilm sites into the main flow of water. Sediment depositions in storage tanks contain a variety of heterotrophic bacteria including ammonia-oxidizing bacteria. Incomplete nitrification by these organisms results in off-tastes and odors and the loss of chlorine residuals over time. The growth of ammonia-oxidizing bacteria is stimulated by warm-water temperatures and long detention times common to storage tanks. Such a problem is futher aggrevated in chloraminated systems when the chlorine-to-ammonia nitrogen ratio is less than 4 to 1.

Air pollution contaminants and surface-water runoff can also impact on the quality of finished water in open storage reservoirs. Accidental spills of chemicals in truck transports using adjacent highways are always a threat. Such spills are very serious because of the concentrated form of the chemical and its toxicological properties upon ingestion. Surface-water runoff into open reservoirs may introduce pesticides, herbicides, fertilizers, silt, and humic materials from decaying vegetation. These contaminants may be toxic, form trihalomethanes in contact with chlorine residuals, or stimulate algal blooms. All of these events degrade the treated water quality and may call for emergency measures to protect the public water supply.

Even covered water storage tanks are subject to contamination because there is always air movement to equalize air pressure in the tank as the water columns change in height. During these air transfers, the tank water is exposed to dust fallout and air pollution contaminants from inflowing air. Vent ports, overflow pipe, and the roof access hatch are candidates for entry of airborne contaminants and bird passage. In arid regions, vent ports should be equipped with suitable air filters to safeguard the water from dustborne particles. In other regions use of hardware cloth to cover all tank vents and attention to closing roof access hatches are essential to keep birds from gaining access to the tank interior. The concern is that bird excrement may fall into the water supply and become transported into the distribution system before dilution and residual disinfection are able to dissipate and inactivate the associated organisms.

The importance of effective screen protection on water tank air vents and closure of the access hatch should not be ignored. In Gideon, MO, a waterborne outbreak of salmonellosis was traced to pigeons colonizing three water storage tanks on the untreated groundwater system. This waterborne outbreak is further discussed in Chapter 8.

DISTRIBUTION WATER TEMPERATURE TRENDS

Water temperature is a primary factor in the bacterial regrowth that occurs in the distribution system and building plumbing networks. In temperate regions, water pipes are buried below the frost line to protect the water supply from freezing and the pipe network from bursting due to water expansion. The practice of pipe burial also insulates the passage of water supply from seasonal climatic air temperatures (Table 1.4). While groundwater temperatures are

Table 1.4   Temperature Variations of Cincinnati Tap Water[a]
from EPA Building Supply

Sample period	Temp (°C) of finished water at treatment plant (surface source water)	Temp (°C) of tap water in building supply system	
		Start of flush	After 1 min flush
January 12–February 17, 1981	6.9	5.9	13.0
April 1–May 15, 1981	14.0	16.2	12.4
June 15–July 15, 1981	22.5	19.1	16.4
August 3–September 1, 1981	25.0	21.3	17.0
September 21–October 6, 1981	22.0	15.8	18.2
November 18–December 18, 1981	12.8	13.6	9.1

[a] Average value from daily samples over test period.

relatively uniform throughout the year, surface waters used for raw source waters will introduce seasonal changes in the treated water that may range from 3 to 25°C or more in Sunbelt areas. Building plumbing networks also provide exposure to ambient air temperature fluctuations unless special precautions are taken to insulate pipes, particularly for those running through an outer wall chase or channel for utility lines. When water temperature rises above 15°C, accelerated growth begins for many heterotrophic bacteria colonizing the pipe environment. Data available from water systems located in geographical areas of pronounced seasonal temperature changes suggest that regrowth of heterotrophic plate count organisms is more pronounced than among coliforms and abrupt surges in density may occur in summer.[3,13,14] Regrowth was not correlated with temperature in southern California water systems,[15] possibly because water temperatures common to those groundwater systems are continually above 10°C and the assimilable organic carbon concentrations were below the 50 µg/l threshold level. Perhaps nutrient accumulation in pipeline sediments during periods of minimal microbial activity in winter may be the key to summer regrowth occurrences.[25]

Static water in supply lines may lead to water expansion just prior to freezing, which increases the risk of pipeline breakage. A solution to this problem is to bury pipelines at greater depth to provide more insulation from extended periods of subfreezing air temperatures. Draining the water from shallow lines that are used only in the summer at resort locations is not desirable because of the concern for ground movement that may cause more pipeline breakage in the dry pipe sections. See Chapter 7 for a case history on antifreeze use in water supply lines.

WATER SUPPLY TRANSIENT TIME

The optimum situation would be to utilize treated water within 24 h of production. In so doing, water quality deterioration from microbial activity in the distribution system would be negligible. Unfortunately, this does not happen for several reasons: water supply reserves for high-demand situations (fire control, peak demands in summer, and line breaks) and to achieve system pressure stability. Furthermore, distribution systems are often complex in their configurations, which in itself can create problems of slow flow in some areas and static water in dead ends. Static water areas are undesirable at any time because this condition provides opportunity for suspended particulates to settle into pipe sediments, biofilm development to proceed without the shearing action of water hydraulic changes, and corrosion to accelerate, particularly during warm-water periods. Redesigning the distribution network to create continuous loops from numerous dead-end sections has been helpful in reducing microbial degradation and improving the efficacy of disinfectant residuals in outreach areas.

SUPPLY AND DEMAND

Planning for future growth of the water supply system to keep pace with metropolitan population growth into the next decade is necessary, but may lead to overproduction of treated water and excessive water supply retention time in the distribution system. Based on information from various systems, the estimated retention time for water supply in the system would appear to range from 1 to 6 d for systems with populations of 35,000 to 820,000 people served (Table 1.1). However, the capacity of storage tanks in the system often extends water supply retention into weeks, depending on water demand. A case history on static water retention in a storage tank is discussed in Chapter 7.

OPERATIONAL FACTORS

Operation, maintenance, and expansion of the distribution network requires careful attention to protocols for repair of line breaks, corrosion control, new line acceptance, water storage, and avoidance of static water. Consolidation of water systems may unite distribution systems of dissimilar infrastructures and differing qualities of water supply. Rural water systems are often confronted with long supply lines that serve few customers per mile, creating extended periods for static water retention in the pipe network.

PROTECTIVE PROTOCOL FOR PIPE WORK

A rigorous protective protocol must be followed during the repair or replacement of existing mains and the installation of new mains in order to avoid bacteriological contamination of the distribution network.[18] There are six areas of concern: (1) protection of new or replacement pipe sections at the construction site, (2) restriction on the type of joint packing material used, (3) preliminary flushing of repaired or new pipe sections, (4) application of pipe disinfectant, (5) final flushing of new construction or repaired line, and (6) bacteriological testing to verify the absence of contamination.

Key to prevention of contamination is physical cleanliness of new sections being installed so that there will be opportunity for successful disinfection. A survey of five water systems in Britain indicated the failure rate of different types of materials to meet the coliform standard in new pipe installation was very similar (Table 1.5) and averaged 18.5% for all materials. Pipe diameter was not a factor in frequency of failures (Table 1.6) but more failures for all new pipelines were greater during the summer (Table 1.7). Whether this seasonal difference was caused by greater road traffic, animal activity, storm events, or soil friability is not known. The problem of continued bacterial contamination following initial disinfection could be the result of a variety of factors including the production of slower acting chloramines, high-pH water, inadequate flushing of sediment deposits in pipe joints, degradable gasket lubricants, and winter water temperatures that reduce the rate of microbial inactivation by chlorine.[26]

In Halifax, Nova Scotia, new line construction presented an unusual pathway for coliform entry into the distribution system.[27] This utility had not previously had a coliform problem in the distribution system, but after acceptance of a new line of approximately 5 km in length, there was a sudden occurrence of *Klebsiella pneumoniae*. The use of a video camera to scan the inside of the large pipe revealed residual pieces of wood construction material imbedded in the new section. An environmental *Klebsiella* strain associated with the wood forms adjusted to the pipe environment and colonized the exposed construction material. Shearing forces created by water flow velocity introduced this coliform into the bulk water, resulting in consistently unsatisfactory test results. The problem was solved by the addition of more than

Table 1.5   New Pipeline Disinfection: Recurrence of Failures
with Respect to Material

Material	No. laid	Pipeline passed first time	No. failing repeat chlorination					Failure (%)
			1×	2×	3×	4×	5+	
Asbestos cement	391	329	48	11	2	1	0	16
Iron	70	54	11	3	1	0	1	23
PVC	282	222	23	12	8	1	5	21
All materials	743	605	93	26	11	2	6	18.5

Data from Buelow et al.[18]

Table 1.6   New Pipeline Disinfection:
Failures with Respect to Pipe Diameter

Pipe diameter (mm)	Pipes passed after first chlorination	Pipes failing after initial chlorination	Failure (%)
50	71	10	12
75	99	25	20
100	285	61	18
150	93	29	24
200	17	1	a
225	17	6	a
300	23	6	a

[a] The total failure for 200, 225, and 300 pipe sizes is 19%.

Data from Buelow et al.[18]

Table 1.7   New Pipeline Disinfection:
Failures (for Five Water Utilities),
Expressed on a Monthly Basis

Month	New pipeline construction	Pipes failing initial chlorination	Failure (%)
January	78	18	23
February	80	4	5
March	63	7	11
April	70	5	7
May	81	9	11
June	52	14	27
July	49	17	35
August	48	17	35
September	82	26	32
October	42	9	21
November	53	5	9
December	45	7	16

Data from Buelow et al.[18]

5 mg/l lime to the raw water, which elevated the treated water pH to 9.1 in the distribution system. At this pH, *Klebsiella* was either inactivated or became entrapped in carbonate scale on the newly formed pipe sediment and could no longer be detected in the public water supply.

CROSS-CONNECTIONS AND BACK-SIPHONAGE

Cross-connections and back-siphonage are always a threat to water quality in the distribution system and dictate the need for maintaining a positive pressure in the system at all times. Authorized cross-connections involve a water supply pipe connected to a second pipeline carrying another water that is not obtained solely from the public supply. In industry this may be a process water, or water stored for use in a fire control situation. All of these situations call for adequate backflow preventer devices to serve as an effective barrier to water supply contamination. What is more dangerous is the illegal tap-in on water supply lines in the home or the interconnecting of a private well or cistern supply with the public water supply system.

Back-siphonage is the siphonage of water back against the direction of normal flow or pressure. Reduction in pressure can occur because of line breaks, pump failure, or sudden high demand for water used in fire fighting. These are situations that may lead to back flows from nearby leaking sewer lines and a variety of attachment devices that have water reservoirs. These are the situations involving garden, agricultural, and industrial hoses submerged in sinks, tanks, and ponds that suck contaminants into the water supply. A study in Britain has suggested that while the potential risk of back-siphonage in the home may reach 85%, the probability of this being a serious concern is really very low.[28]

A disinfection residual in the distribution system can be very effective in the inactivation of pathogens associated with contaminants seeping into large volumes of high-quality potable water. Obviously, limitations to the protective barrier exist. Chlorine residuals in distribution water often range from 0.1 to 0.3 mg/l and at these concentrations will not provide adequate protection against massive intrusions of gross contamination characterized by odors, color, and milky turbidities. Recent studies conducted in a small abandoned distribution system on a military base indicate that at tap water pH 8, with an initial free chlorine residual of 0.7 mg/l and wastewater added to levels of up to 1% by volume, 3 log (99.9%) or greater bacterial inactivation were obtained within 60 min. Viral inactivation under these conditions was less than 2 logs (<99%). In laboratory reservoir experiments, when the residual chlorine is replenished by inflow of fresh uncontaminated chlorinated tap water, greater inactivation was observed at the higher wastewater concentrations tested. Furthermore, a free chlorine residual was more effective than a combined chlorine residual in the rapid inactivation of microorganisms contained in the contaminant.[29] Maintaining free chlorine or chlorine dioxide residuals under low-flow conditions in some parts of the distribution system will be difficult because of disinfection demand and disinfectant instability. However, any sudden loss of free chlorine residual should be interpreted as an indication of a contamination event that may be followed quickly by elevated densities of heterotrophic bacteria, the occurrence of fecal coliforms, and possible entry of some pathogenic agent.

FLUSHING AND CLEANING PIPE NETWORKS

Perhaps nothing is more misunderstood in the technical operation of the distribution system than flushing the pipe network. Some utilities have never flushed the pipe network on a scheduled basis because of water conservation in arid regions. Others are reluctant to flush for fear flushing will stir up the release of coliforms into the distribution samples analyzed for compliance purposes. There are also a few utilities that assume that the fire hydrant pressure testing program on their system is sufficient to satisfy a system-wide flushing requirement. Systems in these categories eventually experience a noticeable loss in water quality in terms of taste, odor, loss of chlorine residual, or increased frequency of coliform occurrences.

The purpose of flushing is to remove significant accumulations of organic nutrients, inorganic corrosion particles, and biofilm and to restore chlorine residuals. To achieve a cleaner pipe environment will necessitate the rapid movement of water out of each fire hydrant (Figure 1.2) so that water pressure and the shearing action of water velocity will dislodge sediments and attached biofilms. As a consequence, there will be instability in water quality during this period that could last a few hours at a given site. After water quiescence is restored, residual sediment pockets will reform, anchoring any biofilm fragments in the process.

System-wide flushing requires a careful study of the pipe network and supply reservoirs, location of all street valves and their operational status, fire hydrants, pipe diameters, water pressure zones, and a general knowledge of the outward flow of water to each area. Records of customer complaints on taste and odor, as well as historical records of site locations with coliform occurrences are important to the action plan. Other important considerations[30] include time of year (spring or autumn are usually the best times for flushing because of lower water demand than characteristic of summer months), time of day (nighttime will minimize traffic congestion and customer inconvenience), and disposal of flushed wastewater (tanker truck or storm drain after dechlorination to protect a receiving stream or lake). Once this information is gathered, flushing can begin. Flush all main lines starting at the source and moving outward in the direction of flow. Open the hydrants slowly to full flow for 15 min or more, then slowly valve off flow to prevent contaminated water entering the pipe network from weep holes in the fire hydrant. Move on to the next hydrant on line and repeat the process. After the main lines are flushed, secondary lines are then opened in each quadrant to complete the pipe network flushing for different pressure zones.

Figure 1.2 Flushing a distribution main to release sediments and attached biofilms.

Localized flushing is done in response to taste and odor complaints in a neighborhood, reaction to a site with coliform occurrences, or as an effort to restore a disinfectant residual in dead ends of static water areas. The same careful planning of the flushing event for a localized area follows those established for system-wide flushing, but on a reduced scale. Once the schedule has been established for the field crew, it is essential to alert the public to the planned flushing operation, which may cause a temporary reduction in water pressure and a brief occurrence of dirty water at the faucet.

The strategy for a vigorous flushing is to systematically open each fire hydrant in sequence on the mains then on branch lines so that clean water is introduced through the network. The purpose of sequential flushing is to completely refresh the pipe environment, reducing all pockets of sediment and biofilm. In this operation it is important to avoid reducing water pressure below 20 psi because of the concerns for back-siphonage.

Sediment formation and tuberculation that have accumulated over the years in pipe networks either through neglect or because of aggressive corrosion are not going to be effectively removed by a rigorous flushing program alone. When this situation arises, it becomes necessary to use pipeline cleaning devices made of polyurethane rubber foam to push the soft deposits out of the exit fire hydrant, modified to pass the device. Low-density swabs (1 to 1.4 lb/ft^3), foam plugs (pigs), are able to be compressed in the pipe readily and ejected through the recovery hydrant easily, making them the method of choice for removal of soft deposits only. Optimum swabbing distance through 4- to 8-in. mains that have been drained is about 4000 ft at a maximum speed of 2.5 ft/s.[31-33] If the swab does not arrive at the exit hydrant at the estimated time, reverse flow is applied to recover at the point of entry. Following passage of the first compressed swab, similar compressible swabs are run through at about 5-min intervals until the water exiting behind the swab runs clear. This procedure may require 15 to 20 passages to achieve a clean pipe section. After the last swab is recovered, the pipe section is flushed for about 15 min to remove any residual loose particles and "dirty water," then put back in service.

High-density plugs are more rigid, some with spirals of hardened steel-wire brushes strapped to the device. When water pressure is applied to propel the plug through the pipe, the material expands to form a tight fit. The progressive cleaning method in this situation involves using a pig that tightly fits the encrusted pipe, then followed by pigs of successive increased size (1-in. increments) until the final pig is of the same diameter as the pipe interior. Ideal pigging speed is between 7 and 22 ft/s. It is important to flush after each pig passage for 20 to 30 min to remove fractured tubercle debris.[34]

A more drastic procedure for removing hard incrustations and tubercles involves hydroblasting, air scouring, or mechanical scrapers. Hydroblasting involves insertion of a jet head, 2 to 4 inches in diameter and containing 6 to 13 removable water jets into the pipe. Water pressure of 1000 to 10,000 psi is forced through the jets against the hard deposits encrusting the pipeline, shattering the material lining the pipe walls. Air scouring is another technique that has been used in England to remove loose deposits and appears to be particularly effective in small-diameter pipe. The technique involves the injection of filtered air into a water main at high pressure that is generated by a mobile compressor at the site. Injecting compressed air causes the formation of discrete slugs of water in the pipeline that are driven along by air pressure in a scouring action. There is no need to turn the water or air on and off to achieve the scouring effect over a few hundred feet of pipe length. Air releases at the customer's tap may be startling but they will help to loosen sediment in the building plumbing lines as well.

Mechanical scrapers and power rodding use friction and mechanical force to release deposits on the pipe walls. While these approaches are effective, there can be several serious side effects that must be recognized. Following this type of cleaning and after the pipe is placed back in service, pipe rust releases from tuberculation sites often bring dirty water complaints from customers who are irate over stains on garments washed in this water. Complaints are also made about water rings in toilet bowls and sinks, in addition to unpleasant-tasting drinking water. A localized flushing of lines over several days will generally be effective in restoring the aesthetic qualities of the water supply.

The most serious side effect may be from the violent vibrations on weaker pipe walls and the exposure of corrosion pinholes in the metal. These conditions ultimately lead to water pipe leaks and breaks in the water mains. For these reasons, relining the pipe after mechanical scraping is desirable to extend the service life of metal pipe and avoid loss of pipeline integrity. Pipelining will also seal off persisting sites of microbial habitation.

CORROSION CONTROL

Pipeline corrosion (Figure 1.3) is a major problem associated with distribution network aging, aggressiveness of water, microbial activity, and soil chemistry. Old, unlined cast-iron pipe sections are subject to more internal corrosion while ductile-iron pipe is more susceptible to external attack for conditions not yet completely understood.[35] The effects of corrosion on a water distribution system can be significant: water-carrying capacity is reduced (pumping costs increase), discolored water causes more customer complaints, microbial quality of treated water deteriorates, detection of disinfection residuals decline, and pipeline breaks occur at a rising frequency.

Water pipe corrosion can be sorted into two categories: internal and external corrosion.

Internal Corrosion

Internal corrosion in the distribution system is the result of galvanic actions in unlined metal pipe that is often aided by the microbial activity of both sulfate-reducing organisms and other aerobic heterotrophic bacteria.[36-38] Factors that cause water to be aggressive include low pH, low alkalinity and hardness, high chloride, or high sulfate in some interrelationships. For example, a ratio above 0.3 for chloride to bicarbonate may stimulate corrosion.[39]

Various corrosion indices have been suggested based on the water's chemical characteristics and used to manipulate the water chemistry in an effort to cause formation of a scale lining on pipe walls.[40,41] Adverse side effects can occur because irregular, porous deposits are more suitable for microbial habitation. Adjusting the calcium carbonate saturation index (Langelier index) for a slight oversaturation does not insure deposition of a uniform $CaCO_3$ scale.[42-44] A uniform calcium carbonate scale does not form under conditions of low velocity, high pH values, or high sulfate content.[45] At low velocities, scale is too porous and nonuniform. Thus, in low-flow and dead-end portions of the system, where a minimum flow velocity of 1 ft/s (0.3 m/s) cannot be achieved, scale will be more porous and corrosion protection will be inadequate. Tuberculation is increased at high pH values because scale formation in less uniform (insufficient calcium and alkalinity present) and less effective in coating the pipe walls.[46,47] Above pH 9, orthosilic acid becomes a factor and changes the chemical nature of the deposits. The presence of high sulfate content in the water adversely affects the calculation of a calcium carbonate saturation index, because of the formation of $CaSO_4$. This results in producing false high index (Langelier) values and misjudgment of the treatment application. In water that is slightly oversaturated with calcium carbonate, selective formation of inorganic precipitate may exist instead of inorganic scale on the pipe walls. The result is a light precipitate that can entrap and protect bacteria. This material will then settle in dead ends or low-flow sections and produce sediment accumulations in finished water reservoirs.

Phosphate addition in water treatment is another approach to pipe corrosion control. In theory, phosphates in solution react with iron corrosion products to produce colloidal particles of hydrous ferric oxides and phosphates that form a film over the inner wall of pipe. The advantage of using phosphates is that the formation of this film is not normally affected by water pH and is enhanced by the presence of calcium, magnesium, and zinc in the water. Orthophosphates used alone or with a zinc salt $(ZnSO_4)$ are the most commonly used corrosion inhibitors and are generally applied initially at a passivation dose (large dosage) for a few days to a month, then reduced in concentration to maintenance levels. Care must be exercised in the use of certain phosphate complexes because phosphate is a critical nutrient for microorganisms to synthesize membrane phospholipids, nucleic acids, and coenzymes. As a result, orthophosphates could be supportive rather than deterrent to colonization by heterotrophic bacteria in the pipe environment. Perhaps the reason zinc orthophosphates have been reported to be successful in avoiding coliform bacteria regrowth is due to metal ion (zinc) toxicity. Why other heterotrophic organisms will grow in the presence of some commercial grades of zinc orthophosphates is not known but could be a response to some uncontrolled physiological parameter such as nitrogen.

External Corrosion

External corrosion of pipe networks results from the intimate contact with aggressive soil environments. Some of the important factors include soil resistivity, soil pH, presence of sulfate-reducing bacteria, and differences in soil type along the pipeline. This variability in soils is particularly noticeable in urban areas at sites of redevelopment where a variety of backfill materials may be used to bring the ground up to a specified grade level. Pipes laid in areas of old landfills or across industrial sites may also be exposed to accelerated corrosion. Differences in the degree of compaction of the soil around the pipe produces heterogeneities in resistivity and soil pH that can stimulate localized corrosion. For soils with resistivities below 3000 Ω/cm, some form of external coating or sleeve is needed for a protective barrier.

Pipe surface imperfections can act as initiating sites for corrosion. These may result from the method by which cast-iron pipe is manufactured, producing shrinkage porosities, foreign inclusions, and the pronounced peen pattern found on some ductile-iron pipe. Installation damage to the factory-applied coating or the semiprotective metal oxide skin beneath it, as well as score marks from dragging the pipe along the trench, may become exposed areas for corrosion to begin.

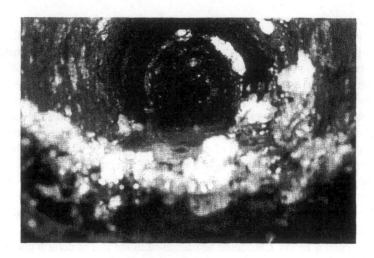

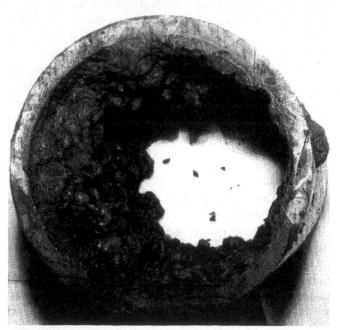

Figure 1.3   (top) Water main showing tuberculation sites. (bottom) Advanced tuberculation in pipe causing flow restriction. (opposite top) Lateral cut of pipe showing effect of corrosion on pipe wall; sand blasted in upper section. (opposite bottom) External pinhole corrosion of water pipe.

Biologically Mediated Corrosion

Biologically mediated corrosion of cast-iron pipe causes a substantial deterioration in water quality as it is transported through the distribution system to the consumer. Most obvious to the consumer are observations of dark

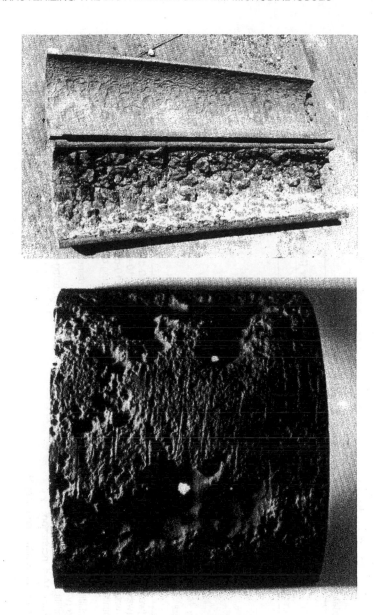

Figure 1.3 (continued)

staining of laundry and plumbing fixtures, unpleasant taste and odors, and the presence of sediments in the water supply. From the utility perspective, these changes are seen as a loss of chlorine residuals, depletion of dissolved oxygen, reduction of sulfate to hydrogen sulfide, and a buildup of iron precipitates in distributed water supply. All of these effects have been associated with the corrosion activities of microorganisms.[48,49]

Table 1.8 Microbial Composition of Water Samples and Pipe Corrosion Deposits

	Untreated water (August 1989)		Untreated water (March 1990)		Treated water (March 1990)		Corrosion tubercles	
	20°C	8°C	20°C	8°C	20°C	8°C	20°C	8°C
Aerobic SPC[a]	2.2×10^6	1.5×10^5	3.2×10^4	9×10^3	20	ND	2.9×10^7	2.0×10^6
Anaerobic SPC[a]	2.0×10^1	ND	1.0×10^1	ND	<1	ND	3.0×10^6	ND
Total coliforms[b]	570	350	200	75	<1	ND	5.0×10^4	ND
Fungal SPC[c]	4.8×10^4	2.1×10^2	3.5×10^3	2.0×10^2	3	ND	3.0×10^3	7.8×10^2
Iron reducers[d]	540	240	70	49	<0.3	<0.3	>24,000	430
Sulfate reducers[e]	280	130	<3	<3	<0.3	<0.3	280	110
Sulfite reducers[e]	120	93	4	4	<0.3	<0.3	460	210
Thiosulfate reducers[e]	540	170	240	79	9.3	1.5	920	540
Iron oxidizers[f]	54	24	7.9	7	<0.3	<0.3	75	64

a Cfu per milliliter (water) or cfu per gram (corrosion tubercle), 7-d incubation at 20°C, 10-d incubation at 8°C.
b Cfu per milliliter (water) or cfu per gram (corrosion tubercle), 48-h incubation at 35°C, 10-d incubation at 8°C.
c Cfu per milliliter (water) or, cfu per gram (corrosion tubercle), 7-d incubation at 20°C, 10-d incubation at 8°C.
d Organisms per milliliter (water) or cfu per gram (corrosion tubercle), by 5-tube MPN, using dilutions of 1.0–0.0001 ml in B_{10} broth, 14-d incubation.
e Organisms per milliliter (water) or cfu per gram (corrosion tubercle), by 5-tube MPN, using dilutions of 1.0–0.0001 ml in Buttlin's broth + either 1% (v/v) Na_2SO_4, Na_2SO_3, or $Na_2S_2O_3$, 14-d incubation.
f Organisms per milliliter (water) or cfu per gram (corrosion tubercle), by 5-tube MPN, using dilutions of 1.0–0.0001 ml in modified Winogradsky's broth, 14-d incubation.

Table adapted from Emde, Smith, and Facey.[54]

There is a wide variety of organisms involved in pipe corrosion: aerobic and facultative anaerobic heterotrophs, autotrophic nitrifiers, denitrifiers, nitrogen-fixing organisms, iron-precipitating organisms, sulfate reducers, sulfur oxidizers, and actinomycetes. Many of these organisms are bacteria while others are fungi living in a biofilm community.[50-52] The number of organisms growing in the attached biofilm has been estimated to be approximately 200 times more per square centimeter of attached growth in corrosion tubercles than found in 1 ml of flowing water in the pipe.[53] This relationship also includes total coliforms found in a tubercled water pipe section during low water temperature conditions.[54] Some perspective of the diversity of this biofilm community that play a role in corrosion of iron pipe can be seen in Table 1.8.

Microbial activity from this type of biofilm community may enhance the accumulation of a matrix of iron and aluminum oxides, silica, calcium carbonate, and other inorganic debris in areas beyond the bulk flow of water. Sporadically, these accumulations are dislodged and suspended in the bulk water causing water to be cloudy and take on a brown, orange, red, or black color. A variety of factors contribute to this problem, which may be a reflection of water quality changes in source water, loss of disinfection residual, or extended periods of static water and low levels of dissolved oxygen in the water supply. Two case histories follow to illustrate some of the factors that stimulate biologically mediated water quality changes in the water distribution system.[49]

The Moberly, MO public water supply serves a population of 13,400 customers with water from a surface impoundment that receives conventional treatment (coagulation, sedimentation, filtration, and chlorination). During summer, and following lake impoundment turnovers during 1976 to 1977, there were numerous consumer complaints of "red" water at the tap. Three areas on the distribution system (Figure 1.4) were identified as major complaint areas. Water samples taken at various points in the distribution system revealed the most obvious change in water quality in these areas was a decline in chlorine residual. Standard plate counts increased by two orders of magnitude compared to finished water. Sulfate reducers, iron-precipitating organisms, and nitrifiers were found in most samples but not in the finished water. Samples from many of the complaint locations had total iron ranging from 0.1 to 6 mg/l while other areas had <0.01 mg/l. It was concluded that changes in source water quality, warm-water temperatures, and slow water movement had provided opportunities for accelerated growth of these nuisance organisms. Flushing to reduce the sediments and restoration of a free chlorine were essential to reduce this microbial activity and adverse water quality.

A rural water district surrounding Hallsville, MO supplies 363 customers with groundwater extracted at three interconnected wells. No treatment is provided to the water beyond chlorination. This distribution pipe network (Figure 1.5) is almost entirely composed of 2- and 4-in. diameter PVC pipes.

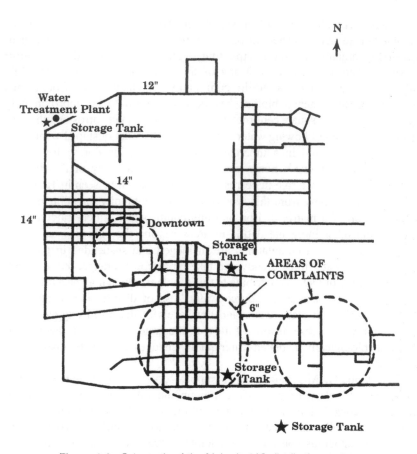

Figure 1.4   Schematic of the Moberly, MO distribution system.

During 1976 to 1977, there were numerous complaints of "black" water problems from sites in the distribution network near well no. 3. Chemical analysis of these black sediments indicated the presence of iron sulfide in these precipitates. Chemical and microbiological analyses were done on samples collected at various places in the distribution system. Again, where there were water quality problems, chlorine residuals were absent. Also of note was the somewhat higher concentrations of iron and sulfide in well no. 3 than detected at wells no. 1 and 2. The presence of sulfide and ferrous iron from the biologically mediated oxidation reduction reactions in that localized area may have caused the formation of iron sulfide precipitates. Feeding the reaction were iron pipe or galvanized iron pipe sections used as service lines to the homes or for road crossings to fire hydrant locations. Microbiological analyses of water from these iron pipe locations indicated the presence of sulfate-reducing bacteria. Keeping sufficient water moving through long PVC lines to a few customers per mile was a contributing factor; however, the use of 2-in.-diameter galvanized service lines that are prone to early corrosion attack

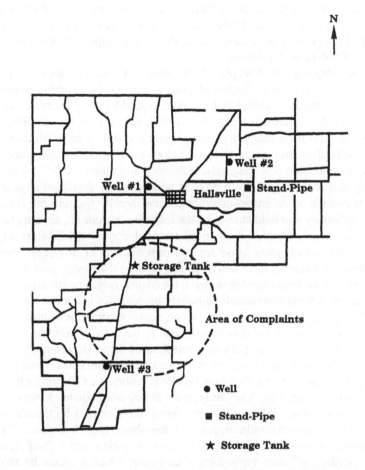

Figure 1.5   Schematic of Water District No. 4 (near Hallsville, MO) distribution system.

in groundwater, increased iron precipitates, and total dissolved sulfides were the major cause of deteriorating water quality.

Corrosion Treatment

Corrosion treatment for internal pipe surfaces is difficult to control because only certain nontoxic substances at low dosage levels can be employed for health reasons. With this in mind, corrosion treatment has taken several approaches: raising the pH and alkalinity or using corrosion inhibitors (carbonates, polyphosphates, and silicates). Lime feeding for water softening is being used satisfactorily but some utilities have found the chemical addition difficult to control so calcium carbonate deposition in the transmission lines may become excessive. Indirectly, fluoride additions may have an adverse effect through an acid pH shift and when sodium hydroxide is added to make a pH correction in this situation, there can be wide variances in the result that also have impact on chlorine disinfection effectiveness. In this situation, pH instability may become a major factor that causes pipe sediments to have a lack of firmness in deposition.

Polyphosphate and orthophosphates are often applied after the pipe network has been thoroughly flushed of loose sediments. These additions produce their effect either by stabilizing the corrosion product or by forming chemical complexes that reduce its deleterious effect, rather than by inhibiting iron corrosion. Corrosion inhibitors often require a conditioning period of up to one month of continuous application before there is the establishment of a thin film deposit on pipe materials, sediments, and tubercles.[55,56] The purpose of this film is to restrict the continued interfacing of dissolved oxygen and carbon dioxide at the metal–liquid boundary, thereby limiting the corrosion rate. Laboratory and field studies on the influence of sodium, zinc, metaphosphate, and zinc orthophosphate in commercial corrosion inhibitors suggest there is no growth stimulation for coliform bacteria (*Citrobacter freundii, Enterobacter cloacae*, and *Klebsiella pneumoniae*). However, growth of other organisms in the heterotrophic bacterial population might be stimulated by the phosphates in these compounds, although the evidence is not conclusive.[57] At Wilkinsburg, PA application of metaphosphate for corrosion control resulted in an increase in bacterial densities thought to have come from the sloughing off of old deposits after the pH was lowered from 9.0 to 7.0. No similar increase in bacterial deposits was noted from sections of new asbestos cement pipeline during this event.[58] In another case history (Cambridge, OH), metaphosphate application appeared to cause an increase in bacterial counts; however, later experiments suggest this increase was more likely caused by the dispersion of bacterial clumps in biofilm fragments rather than a growth response.[58] These case histories and others in recent years provide evidence that abrupt changes in pH during the initial application of corrosion inhibitors must be avoided. Such drastic action will cause the pipe sediments to become flaky, detach, and get into the bulk water, carrying along fragments of associated biofilm. As a consequence, there may be a sudden surge in coliform bacteria and other heterotrophic organisms released into the bulk water. In several instances, coliform detection at monitoring sites on the distribution system rose to near the 5% occurrence frequency limit permitted in federal drinking water regulations.

In application, polyphosphates and orthophosphates are often applied in the range of 1 to 5 mg/l to the finished water after the pipe network has been thoroughly flushed of loose sediments. No standardized concentration of these additives can be designated because each water supply will have different chemical constituents, different pipe materials, and varying effectiveness with a system-wide flushing program. When zinc orthrophosphate is used in corrosion control, the zinc concentration must be limited to 0.5 mg/l because of regulations on wastewater effluent limits for this metal discharged to receiving streams. In highly alkaline waters above pH 8.5, it may be necessary to add 0.5

to 1.0 mg/l of sodium hexametaphosphate[55] to minimize excessive deposition of calcium carbonate in the treated water transmission lines near the water plant. Any inadvertent precipitation of calcium carbonate at this point could impair the development of the desired zinc phosphate film in primary water mains. As the polyphosphate reverts to orthophosphate during transit, small amounts of calcium carbonate will coat secondary transmission lines and laterals throughout the system. Carbonate deposition is achieved by slightly raising pH, alkalinity, and total hardness levels at the water treatment plant. If the pH of the water must be adjusted upward, careful control must be maintained to avoid the formation of a soft, nonadhering carbonate coating.[59] Monitoring progress with the corrosion treatment is best accomplished by the placement of a few mild steel coupon inserts into the pipe network at various locations representative of the entire distribution system.[60] A few coupons placed near dead ends will help to determine whether the treatment is penetrating the slow-flow areas in the system.

Cleaning and lining cast-iron pipe with a cement mortar will increase the service life, combat corrosion and associated problems of red water, reduce some water loss from pinhole corrosion sites, and improve water movement through slow-flow areas in the project. All of these advantages contributed to improving the microbiological quality of water by removing or sealing off habitats and biofilm development.[61] This process, however, can do nothing to further the service life of pipe that is not structurally sound or has severe external corrosion.

Cleaning the pipe section to the bare metal is essential to provide an effective bonding of cement to the internal pipe surfaces. This procedure may involve the use of plastic swabs, the use of mechanical scrapers, or eroding with a cable to which a series of conically shaped scrapers incorporating steel blades are attached. This assemblage is then dragged through the pipe section to break loose tuberculations so that they can be moved in a water flush to the exit site. After the breakup of deposits has been accomplished, a series of circular rubber squeegees is passed through to remove any residual rust and to dry the pipe surface in preparation for cement lining. A cement mortar (cement–sand ratio of 1:1 to 1:5) is then applied (to a thickness of 1/8 to 1/4 in. [3.2 to 6.4 mm], depending on pipe diameter) by centrifugal force using a compressed air hose to rotate the lining spray head, as the machine moves through the pipe. House service lines are then back-fed with water to blow out any obstructions created by cement lining the main. Curing the cement-lined pipe may take approximately 2 to 7 d, after which the pipe section is flushed and sanitized by chlorination.[62,63]

External protection of cast iron from corrosion may be done by applying some form of a zinc-sprayed coating that eventually converts to an impervious, adherent product. Other approaches are to use a selective backfill material in the trench while laying new pipe in an aggressive soil area or by wrapping the pipe with a polyethylene sleeving. While the plastic sleeve is not heat-shrunk to the pipe, it does discourage a rapid pitting of the wall.[64] Use of PVC pipe material in areas of saltwater contact or insulating joints near subways and power mains to provide cathodic protection from stray direct currents are other options to minimize specific corrosion problems.[65]

PASS-THROUGH HEAT PUMPS

Water mains have received much attention in recent years as potential heat sink/sources for thermal exchange through pass-through heat pumps. These devices tap the temperature-stable water in the distribution system in a transfer of heat to cut the cost of heating and cooling public buildings and suburban homes. Pass-back of the water into the distribution system after use in heat transfer is a cross-connection concern for many utilities and public health authorities. Few states regulate such systems directly, rather relying on local plumbing codes and utility discretion.

The heat pump concept represents an innovative approach to energy conservation. There are several ways that a thermal exchange can be achieved: water from a private well may be passed through the heat exchanger and then discharged to a sewer; water may be recirculated in a closed pipe loop buried in the ground; or water from the distribution system may be circulated past the heat exchanger, then returned to the water supply lines for blending with the potable water supply. It is the pass-back of water by such devices to the distribution system that creates potential uncertainties for the public utility, which may lose control of distribution water quality. Major areas of unknown consequences are the clustering of units in a residential subdivision and placement of units in slow-flow areas, which might produce a stepwise temperature elevation in distribution water. Elevation of water temperatures during periods of peak heat transfer may stimulate growth to higher densities of heterotrophic bacteria in adjacent portions of the pipe network. Higher water temperatures also accelerate corrosion rates in the pipe network, cause the dissipation of disinfectant residuals, and increase total trihalomethanes (TTHMs) formation potential in the immediate vicinity of these pass-through devices.[66,67] For every 5% rise in water temperature there may be an increase in TTHMs by as much as 53%.

These threats to the integrity of the distribution system and any potential water quality changes need to be resolved through pilot testing in a Sunbelt area of the country where a cluster of units are proposed for a suburban real estate development. The study would define minimum spacing of units on-line related to water flow characteristics. Also important in such an investigation would be information gathered on corrosion associated with heat pump materials in contact with water supply and pipe network. Key factors of water temperature elevation and circulating nutrients will stimulate biofilm growth and detachment during turbulent flow conditions in the heat pump pipe loop and water supply return zone. Biofilm organisms of concern include *Klebsiella* (coliform), *Aeromonas hydrophila* (stimulated by warm, nutritive waters), *Microbacterium* strains (chlorine resistant), and *Legionella* (heat tolerant).

Because disinfectant residuals could be either free chlorine or chloramines, information on accelerated chlorine demand from pass-back of used water should be evaluated for effect on system residuals further out in the pipe network.

JOINT RESPONSIBILITY FOR WATER QUALITY

Fulfilling the obligation for the production, delivery, and maintenance of high-quality water supply to the consumer is the joint responsibility of the water utility and the user community of hospitals, highrise building complex management, and the individual family. Water quality is created at the water plant and is a reflection of treatment operations and distribution system maintenance and management. While treatment technology will provide a water free of health risks associated with primary pathogens, treatment processes were never intended to produce a sterile water supply. Opportunistic organisms will pass through or circumvent treatment barriers as now defined. Some of the pathways include passage of dust contaminants into open-air process basins, organism protection in cell clumping, viable organism transport in aquatic invertebrates, movement with carbon fines or unsettled particulates, and infiltration through fractured pipes, line breaks, and repair practices.

WATER UTILITIES

Opportunistic organisms are very adaptable in the establishment of a biofilm colonization that promotes the amplification of cell densities to levels that may be several logs higher than originally. In treatment basins, colonization occurs at the solid surface-to-water interfaces of process basins and connecting flumes and agitator paddles in the coagulation basin. This problem is best controlled by scheduling application of high-pressure (jet stream) washing of compartment walls and mechanical scraping of paddle surfaces. Colonization of the pipe network occurs in corrosion tubercles and pipe sediments plus the accumulated sediments in water storage tanks. Static water locations in the pipe system and stratified water in storage tanks promotes this colonization by a variety of bacteria during warm-water periods.

The key to suppressing colonization is to keep the water moving throughout the system and to remove accumulating pipe sediments. Flushing that is done at least every spring in a systematic fashion from the site of first connections to the end of the pipe network often contribute to the control of biofilm incursions. Draining and cleaning of all water storage tanks and standpipes may be more difficult to manage but, nevertheless, should be done every few years to suppress biofilm growth at these sites. For those water utilities that use chloramination as the postdisinfectant, it may be desirable to change to free chlorine for a 2-week period each year (after flushing the system) to effectively reduce chloramine-resistant heterotrophic bacterial populations growing in slow-flow sections and dead ends.

HOSPITALS

Hospitals also have a responsibility in the management of water supply quality. The entire pipe network needs to be flushed every 6 months because this water system is always in a warm ambient environment regardless of seasonal weather changes. Flushing in this case needs to be done at each faucet throughout the facility for a 15-min period or until a measurable disinfectant residual is detected. Attachment devices must be disassembled and cleaned of incrustations and sediments on an annual basis. Flexible hose connections should be replaced at this time as well as any gaskets and washers that have visible slime growth or incrustations.

Prior to the activation of a new hospital wing or the reactivation of closed patient wards and operating theaters, all faucets should be opened for 15 h to flush out the stagnant water and sediments in an effort to achieve water supply that is representative of the quality in the municipal water supply.

The problem of water quality deterioration in this situation was experienced by a Boston hospital in the activation of a new facility to their building complex.[66] Water entering the hospital supply lines was of high quality (no coliforms per 100 ml and heterotrophic plate counts averaging three organisms per milliliter). As the staff began to phase in the new building addition, there were numerous complaints of malaise by the personnel and the water supply became suspect. Laboratory analysis of the water taken from taps without any flushing revealed no detectable coliforms, but the heterotrophic bacterial count ranged from 3000 to 4000 organisms per milliliter. Turning on all faucets throughout the new building complex overnight for a minimum of 15 h was successful in flushing out the contaminants in the plumbing system. The bacterial densities decreased to 15 organisms per milliliter and by the following day averaged seven heterotrophic bacteria per milliliter, thereby achieving a water supply of the quality similar to that of the municipal water system. The ill-defined health complaints of the hospital staff declined following this action response; however, incrimination of the microbial quality of the water supply remained circumstantial.

BUILDING PLUMBING NETWORKS

Building plumbing systems designed for highrise office complexes, apartment buildings, hospitals, schools, and hotels are miniature versions of a public distribution system. Source water is the public water supply and the pipe network has several pipe sizes, fire control stations, water storage reservoirs to assist adequate pressure throughout the structure, many faucet connections, and attached fixtures for toilets and bathing facilities. While these plumbing networks are beyond the water service meter and responsibility of public water utilities, they are the sites for customer exposure to water supply and possible water quality problems from neglected maintenance.

Maintenance of the building plumbing system is often given little serious consideration until there are taste and odor problems or water pressure has had a significant reduction at some locations. The microbial quality of water in new buildings is influenced by the cleanliness of materials used in the construction of a pipe network (connectors, faucets, building storage tanks, and attachment of water fixtures). The other factor in new plumbing network construction is with pipe assembly techniques that leave dirt and excessive solder flux in the lines.

Water quality deterioration in building plumbing develops over the years from low-pH water, high mineral concentrations, elevated dissolved oxygen, corrosion of dissimilar materials used in plumbing attachments, faulty system design that creates numerous dead ends, and inherent slow-flow sections. Added to this is the failure to recognize the need for systematic flushing of the entire network or inability to apply appropriate corrosion control measures. Unoccupied apartments, office space, and hotel rooms vacant for 3 months or more may have a significant deterioration in the water quality because of static service lines. Flushing each faucet in the vacant rooms for 15 min would restore water quality that is characteristic of the building supply.

Consumer dissatisfaction with the building water supply is often perceived to originate in the municipal water supply rather than as a result of localized deterioration. Complaints of unpleasant tastes, color, odor, and sediment in the water that lead to stains on plumbing fixtures and laundry are most common. Other consumer complaints include low water pressure, leaks from the effects of water hydraulics (water hammer), and interrupted service resulting from pipeline freezing. While these undesirable characteristics are easily detected by consumers, the unseen health hazards of pathogen occurrences or toxic chemical contamination introduced through cross-connection and back-siphonage of waste waters are generally only recognized after there has been a waterborne outbreak.

Taste and odor control in building plumbing systems should be based on a regularly scheduled flushing of all water pipes because the pipe network is generally of a configuration that includes dead-end branch lines containing static water of varying age. A routine schedule period for storage tank drainage and sediment removal is also an important control measure for minimizing microbial growth that cause taste and odors. Storage tank covers must be in place to prevent airborne dust, birds, or algal growth from contaminating the building water supply. Several years ago, a localized waterborne outbreak among tenants of a building in Chicago was found to be caused by the growth of algae and release of a toxigenic metabolic by-product. Algal growth was possible because the tank cover was not in place to seal off light and nutrients in airborne dust particles from reaching the water supply.

Complete softening of the building water supply should be avoided because this practice may be quite corrosive to the water pipes. In retrospect, expert design of building plumbing networks to minimize dead ends, careful selection of plumbing network materials in new construction work, and proper care to avoid introduction of foreign substances during pipe assembly are very important for avoidance of a variety of water quality deterioration issues, immediately or years later.

INDIVIDUAL CONSUMERS

Consumers must also be aware of their responsibility to protect the public water supply quality in their home. The most common home problem with static water quality is in the first draw of water supply in the morning. Microbial growth will occur in the water during overnight periods of no flow due to warm ambient temperatures associated with proximity to furnace pipes, hot-water lines and locations under sinks. As a general practice, it is a good habit to flush the tap water line for 30 s each morning before ingesting that first glass of water. If the family has been away on an extended vacation, again flushing water lines from the bathroom and kitchen taps for several minutes will do much to remove elevated bacterial densities in static water lines. For families that attach point-of-use devices to the water supply for additional treatment, there should be a scheduled effort to change carbon filters every 4 to 6 weeks, depending on usage, or as recommended by the manufacturer. Morning flush of these devices is very important because of the microbial buildup in the unit overnight that often exceeds what occurs in static water at other home faucets.

REFERENCES

1. Frontinus, S.J. 97 A.D. *The Water Supply of the City of Rome*. Transl., C. Herschel, New England Water Works Association, Boston, MA, 1973.
2. Baker, M.N. 1981. *The Quest for Pure Water*. American Water Works Association, Denver, CO.
3. Geldreich, E.E., J.A. Goodrich, and R. Clark. 1989. Strategies for Monitoring the Bacteriological Quality of Water Supply in Distribution Systems. AWWA Annual Conference. Los Angeles, CA. June 18-22, 1989.
4. Davis, A.R. 1951. The Distribution System. In: *Manual for Water Works Operators*. Texas State Dept. Health and Texas Water and Sanitation Research Foundation Austin, TX. pp. 342-363.
5. Walski, T., J. Gessler, and J.W. Sjostrom. 1988. Selecting Optimal Pipe Sizes for Water Distribution Systems. *Jour. Amer. Water Works Assoc.*, 80:35-39.
6. Committee Report. 1979. The Use of Plastics in Distribution Systems. *Jour. Amer. Water Works Assoc.*, 71:373-375.
7. Lee, R.G. (moderator). 1987. Roundtable on Plastic Pipe Permeation. Water Quality Technology Conference, Houston, TX. AWWA Research Foundation Report 88, Jan. 1987. Denver, CO.
8. LeChevallier, M.W., C.D. Lowry, and R.G. Lee. 1990. Disinfecting Biofilms in a Model Distribution System. *Jour. Amer. Water Works Assoc.*, 82,7:87-99.

9. Schoenen, D. 1986. Microbial Growth due to Materials Used in Drinking Water Systems. In: *Biotechnology*, Vol. 8, H.J. Rehm and G. Reed, eds. VCH Verlagsgesellchaft, Weinheim, Germany.

10. Colbourne, J.S. 1985. Materials Usage and Their Effects on the Microbiological Quality of Water Supply. *Jour. Appl. Bacteriol. Symp. Suppl.*, 47S-59S.

11. Burman, N. and J. Colbourne. 1976. The Effect of Plumbing Materials on Water Quality. *Jour. Inst. Plumb.*, 3:12-13.

12. Burman, N. and J. Colbourne. 1977. Techniques for the Assessment of Growth of Microorganisms on Plumbing Materials Used in Contact with Potable Water. *Jour. Appl. Bacteriol.*, 43:137-144.

13. Schoenen, D. and H.F. Scholer. 1985. *Drinking Water Materials: Field Observations and Methods of Investigation.* Ellis Horwood Ltd., Chichester, England. 195 p.

14. Colbourne, J. and D. Brown. 1979. Dissolved Oxygen Utilization as an Indicator of Total Microbial Activity on Non-Metallic Materials Used in Contact with Potable Water. *Jour. Appl. Bacteriol.*, 47:1-9.

15. Schoenen, D. and M. Hotter. 1981. Mikrobielle Besiedlung von Auskleidungsmaterialien und Baustoffen in Trinkwasserbereich. 8. Mitteilung: Experimentelle Untersuchung von Epoxidharzenstrichen inter Laboratoriumsbedingungen. *Zentralbl. Bacteriol. Hyg.*, 173:356-364.

16. Parkinson, R., I. Warren, and K. North. 1983. Recent Investigations of the Epoxy Resin Lining of Water Mains. *Jour. Inst. Water Eng. Sci.*, 37:257-271.

17. Hutchinson, M. 1974. WRA Medlube: An Aid to Mains Disinfection. *Proc. Soc. Water Treat. Exam.*, 23:2:174.

18. Buelow, R.W., R.H. Taylor, E.E. Geldreich, A. Goodenkauf, L. Wilwerding, F. Holdren, M. Hutchinson, and I.H. Nelson. 1976. Disinfection of New Water Mains. *Jour. Amer. Water Works Assoc.*, 68: 283-288.

19. Harem, F.E., K.D. Bielman, and J.E. Worth. 1976. Reservoir Linings. *Jour. Amer. Water Works Assoc.*, 68:238-242.

20. Wade, J.A., Jr. 1974. Design Guidelines for Distribution Systems as Developed and Used by an Investor-Owned Utility. *Jour. Amer. Water Works Assoc.*, 66:346-348.

21. Thofern, E., D. Schoenen, and G.J. Tuschewitzki. 1987. Mikrobielle Wandbesiedlung und Desinfektionsprobleme. *Off. Gesundh.-Wes.*, 49:14-20.

22. Mackle, H. et al. 1988. Koloniezahlerhobung sowie. Geruchs-und Geschmachsbiemtrachtigungen des Trinkwassers durch Losemettelhaltige auskleidematerialien. *GWF, Gas-Wasserfach:Wasser/abwasser*, 129:22-27.

23. Bernhardt, H. and H.V. Liesen. 1988. Bacterial Growth in Drinking Water Supply Systems Following Bituminous Corrosion Protection Coatings. *GWF, Gas-Wasserfach:Wasser/abwasser*, 129:28-32.

24. Kroon, J.R., M.A. Stoner, and W.A. Hunt. 1984. Water Hammer: Causes and Effect. *Jour. Amer. Water Works Assoc.*, 76:39-45.

25. Geldreich, E.E. 1990. Microbiological Quality Control in Distribution Systems. In: *Water Quality Treatment.* American Water Works Assoc., Denver, CO.

26. Geldreich, E.E. 1988. Coliform Non-Compliance Nightmares in Water Supply Distribution Systems. In: *Water Quality: A Realistic Perspective.* University of Michigan Press, Ann Arbor, MI.

27. Martin, R.S., W.H. Gates, R.S. Tobin, et al. 1982. Factors Affecting Coliform Bacteria Growth in Distribution Systems. *Jour. Amer. Water Works Assoc.*, 74:34-37.

28. Gilfillan, D.J. 1971. Report on the Investigation of Back-Siphonage Risks in Domestic Properties, Winter 1970-71. Report T.P. 82, Water Research Association, Medmenham, England.

29. Snead, M.C., V.P. Olivieri, K. Kawata, and C.W. Kruse. 1980. The Effectiveness of Chlorine Residuals in Inactivation of Bacteria and Viruses Introduced by Post-Treatment Contamination. *Water Res.*, 14:403-408.

30. California-Nevada Section AWWA. 1981. Distribution Main Flushing and Cleaning. Manual No. 3. American Water Works Association.

31. Stearn, H.M. Cleaning Water Mains with Foam Plugs. 1971. *Jour. Amer. Water Works Assoc.*, 63:7;414-415.

32. Jenkins, C.A. 1968. Foam Swabs for Water Main Cleaning. *Jour. Amer. Water Works Assoc.*, 60:8;899-908.

33. Denton, F.G. 1969. Foam Swabbing of Water Lines. *Pub. Works,* 100:118.

34. Landes, M. 1981. An Effective Method for Cleaning Pipelines. *Opflow,* 7:3.

35. Stokes, R.F. 1983. Research on Materials for Use in Water Distribution Systems. *Chemistry and Industry,* 17:659-663.

36. O'Connor, J.T., S.K. Banerji, and B.J. Brazos. 1988. Water Quality Deterioration: Recognizing the Symptoms. Pub. Works, 119:(13);44-46.

37. Allen, M.J., R.H. Taylor, and E.E. Geldreich. 1980. The Occurrence of Microorganisms in Water Main Encrustations. *Jour. Amer. Water Works Assoc.*, 72:614-625.

38. Tuovinen, O.H., K.S. Button, A. Vuorinen, L. Carison, et al. 1980. Bacterial, Chemical, and Mineralogical Characteristics of Tubercles in Distribution Pipelines. *Jour. Amer. Water Works Assoc.*, 72:626-635.

39. Larson, T.E. 1967. Loss in Water Main Carrying Capacity. *Jour. Amer. Water Works Assoc.*, 59:1565-1572.

40. Singley, J.E., B.A. Beaudet, and P.H. Markey. 1984. Corrosion Manual for Internal Corrosion of Water Distribution Systems, U.S. Environmental Protection Agency, Rep. 570/9-84-001, Office of Drinking Water, Washington, D.C.

41. DVGW-Forchungstolle. 1985. AWWA Research Foundation, Internal Corrosion of Water Distribution Systems, AWWA Research Foundation, Denver, CO.

42. Langelier, W.F. 1936. The Analytical Control of Anti-Corrosion Treatment. *Jour. Amer. Water Works Assoc.*, 28:10, 1400-1521.

43. McCauley, R. 1960. Controlled Deposition of Protection Calcite Coating in Water Mains. *Jour. Amer. Water Works Assoc.*, 52:11;1386-1396.

44. Tillmans, J. and O. Heublein. 1913. Investigation of the Carbon Dioxide Which Attacks Calcium Carbonate in Natural Waters. *Gesundh. Ing.*, 35:669-677.

45. Stumm, W. 1960. Investigation on the Corrosive Behavior of Waters. *Amer. Soc. Civ. Eng., Jour. San. Eng. Div.*, 86(6):27-45.

46. Larson, T.E. and R.V. Skold. 1958. Current Research on Corrosion and Tuberculation of Cast Iron. *Jour. Amer. Water Works Assoc.*, 50:11;1429-1432.

47. Larson, T.E. 1966. Deterioration of Water Quality in Distribution Systems. *Jour. Amer. Water Works Assoc.*, 58:10;1307.

48. O'Connor, J.T., L. Hash, and A.B. Edwards. 1975. Deterioration of Water Quality in Distribution Systems. *Jour. Amer. Water Works Assoc.*, 67:113-116.

49. Banerji, S.K., W.R. Knocke, S.H. Lee, and J.T. O'Connor. 1978. Biologically Mediated Water Quality Changes in Water Distribution Systems. U.S. EPA Project Report, Cincinnati, OH.

50. Burman, N.P. 1965. Taste and Odour Due to Stagnation and Local Warming in Long Lengths of Piping. *Proc. Soc. Water Treat. Exam.*, 14:125-131.

51. Emde, K.M.E., D.W. Smith, and R. Facey. 1992. Initial Investigation of Microbially Influenced Corrosion (MIC) in a Low Temperature Water Distribution System. *Water Res.*, 26:169-175.

52. Dott, W. and D. Washko-Dransmann. 1981. Occurrence and Significance of Actinomycetes in Drinking Water. *Zbl. Bakt. Hyg. Abt. 1 Orig. B*, 173:217-232.

53. Costoston, J.W. and G.C. Geesey. 1979. Microbial Contamination of Surfaces. In: *Surface Contamination*, K.L. Mittal, ed. Plenum Press, New York. pp. 211-221.

54. Facey, R.M., D.E. Smith, and K.M.E. Emde. 1991. Case Study: Water Supply Distribution Corrosion, Yellowknife, N.W.T. In: *Proc. Microbially Influenced Corrosion*, N.J. Dowling, M.W. Mittleman, and J.C. Danko, eds. MIC Consortium, Knoxville, TN. pp. 4-45-4-49.

55. Murray, W.B. 1970. A Corrosion Inhibitor Process for Domestic Water. *Jour. Amer. Water Works Assoc.*, 62:659-662.

56. Rice, O. 1947. Corrosion Control with Calgon. *Jour. Amer. Water Works Assoc.*, 39:552-560.

57. Rosenzweig, W.D. 1987. Influence of Phosphate Corrosion Control Compounds on Bacterial Growth. Project Summary. EPA/600/S2-87/045. Water Eng. Research Lab., U.S. Environmental Protection Agency, Cincinnati, OH.

58. AWWA Committee Report. 1942. The Value of Sodium Hexametaphosphate in the Control of Difficulties Due to Corrosion in Water Systems. *Jour. Amer. Water Works Assoc.*, 34:1807-1830.

59. Powell, S.T., H.E. Bacon, and J.R. Till. 1946. Recent Developments in Corrosion Control. *Jour. Amer. Water Works Assoc.*, 38:808-825.

60. Harms, L.L., B.C. Kwan, and D.S. Sarai. 1994. How to Tame an Aggressive Softened Water. *Opflow*, 20:1,4-5.

61. Martin, R.S., W.H. Gates, R.S. Tobin, D. Granthan, et al. 1982. Factors Affecting Coliform Bacterial Growth in Distribution Systems. *Jour. Amer. Water Works Assoc.*, 74:34-37.

62. Dahl, R. 1974. Cleaning and Lining Small-Diameter Pipe. *Jour. Amer. Water Works Assoc.*, 66:483-484.

63. Scanga, P. and H. Guttman. 1992. Restore Old Pipe with Cleaning, Lining Program. *Opflow*, 18:1,6-7.

64. Triners, R.E., J.F. Armacost, and J.A. Lee. 1974. Water Distribution System Corrosion Control. *Jour. Amer. Water Works Assoc.*, 66:453-455.

65. von Baeckmann, W.G. 1974. Cathodic Protection of Underground Pipelines with Special Reference to Urban Areas. *Jour. Amer. Water Works Assoc.*, 66:466-470.

66. Amy, G.T., Chadik, P.A., and Chowdhury, Z.K. 1987. Developing Models for Predicting Trihalomethane Formation Potential and Kinetics. *Jour. Amer. Water Works Assoc.*, 79(7):89-97.

67. Singer, P.C. 1993. Formation and Characterization of Disinfection By-Products. In: *Safety of Water Disinfection: Balancing Chemical and Microbial Risks*, G.F. Craun, ed. ILSI Press, Washington, D.C.

Beware of microbial-mediated corrosion.

CHAPTER 2

Creating Microbial Quality in Drinking Water

CONTENTS

INTRODUCTION

Drinking water is not sterile. Public water supply has never been intended to provide a sterile product. The objective has always been to produce a water supply of minimal public health risk with secondary considerations being given to taste and odor. Specialized uses of public water supply (hospitals, industries, and space stations) need to recognize their obligation to further treat the water at their site if desired quality requires enhancement beyond consumer health protection.

The microbial quality of potable water is a reflection of those organisms introduced in source water, modified in composition through treatment processes, colonized in the distribution system, and selectively amplified in various attachment devices. While there may be significant differences in the microbial flora of some water supplies that impact on taste and odor, all safe water supplies either contain no organisms derived from the intestinal tract of warm-blooded animals (fecal organisms) or are treated to minimize this risk to public health.

Those organisms that are indigenous to water, the true water bacteria or environmental strains, have, in the past, received minimal attention because they often were considered to have little significance. This attitude is beginning to change because of concern with biofilm development in pipe networks and the increasing evidence that some opportunities pathogens, such as *Legionella*, may be ubiquitous in the water environment.

RAW SOURCE WATERS

A number of factors have entered into the choice of a best available raw source water for potable supply. These considerations include adequate quantity during any seasonal variations in flow, water quality that is amenable to cost-effective treatment, and some measure of watershed protection from domestic, industrial, and agricultural pollution.[1] Increasing urban populations have often placed a greater burden on water supply utilities, both in terms of the quantity demanded and in terms of contamination potential through the use of less desirable raw waters derived from polluted rivers, lakes, and groundwater.[2] Wastewater treatment plants, urban and rural runoff, agricultural activities, and industrial chemical spills often discharge to a watercourse that may be the source for a public water supply. Many of these wastes contain pathogens and a myriad of other waterborne organisms. Waste minimization management, waste treatment of point source discharges, and controlled treatment of urban stormwater are essential ingredients to protecting downstream use by water supply. There must be a concerted effort to maximize public health protection through a multiple treatment barrier concept and not force the water purveyor to either demonstrate a public health threat to make upstream treatment of waste mandatory or have utilities assume the full burden of treatment for poor-quality water resources.

SURFACE WATERS

Raw surface water is subject to a variety of bacterial contaminants introduced in stormwater runoff over the watershed and the upstream discharges of domestic and industrial wastes. These sources of pollution introduce a wide range of infectious agents to waters that may ultimately be used in water supply.[3-6] The more prevalent infectious agents may be classified within four broad groups: bacteria, viruses, protozoa, and helminths (Table 2.1). That these infectious agents, derived principally from infected persons and other warm-blooded animals, are found in polluted water is unquestionable.

Receiving waters between the point of discharge of sewage effluents and water intakes may not only serve as a buffer to accidental spills and treatment bypasses but can contribute to water quality improvements through natural self-purification. Every stream, lake, and groundwater aquifer has some limited capacity to assimilate waste effluents and stormwater runoff entering the drainage basin. This self-purification process is a complex and ill-defined mechanism that involves bacterial adsorption with sedimentation, predation, dilution, hydrologic tributary effects, water temperature, and solar radiation.[7] For example, sedimentation can rapidly move pathogens from the overlying water into the bottom deposits. Over a 1-year period, approximately 90% of *Salmonella* species isolated from the North Ocone River were recovered in bottom sediments downstream. Adsorption of the organisms to sand and clay and the action of sedimentation resulted in their concentration in the river bottom sediments.[8]

Table 2.1   Major Infectious Agents Found in Contaminated Drinking Waters Worldwide

Bacteria	Viruses	Protozoa
Campylobacter jejuni	Adenovirus (31 types)	*Balantidium coli*
Enteropathogenic *E. coli*	Enteroviruses (71 types)	*Entamoeba histolytica*
Salmonella (1700 spp.)	Hepatitis A	*Giardia lamblia*
Shigella (4 spp.)	Norwalk agent	*Cryptosporidium*
Vibrio cholerae	Reovirus	
Yersinia enterocolitica	Rotavirus	
Helminths	Coxsackie virus	
Ancylostoma duodenale		
Ascaris lumbricoides		
Echinococcus granulosis		
Enterobius vermicularis		
Necator americanus		
Strongyloides stercaralis		
Taenia (spp.)		
Trichuris trichiura		

Data from Geldreich.[3]

The rate at which these natural processes proceed is dependent upon time, generally referred to as flow time downstream to achieve 99% reduction in the microbial population. Mitigating circumstances that may nullify these beneficial effects include discharges of industrial wastes that alter the water pH to above 9 or below 4, food processing and paper mill wastes that provide excessive bacterial nutrients, stormwater runoff characteristics of the watershed, magnitude of pollutional loading from all discharges, and their degree of treatment. While natural self-purification may be effective in a given water course during periods of warm temperatures and limited conditions of precipitation, results will be less impressive following wet seasonal periods and be particularly poor during winter environmental conditions.

Rivers

Rivers are subject to big fluctuations in water quality as a result of stormwater runoff and spills of municipal and industrial wastes. While these source waters are not necessarily of the best quality, there is assurance that quantity will be available to meet the needs of the growing community. Locating river intakes for water supply many miles upstream of the community domestic and industrial waste discharges was done to achieve some quality improvement in source water prior to treatment. This degree of separation provided a buffer zone in which some measure of stream self-purification could produce a limited water quality improvement before reaching the intake. In recent years, there has been both a rapid expansion of cities and the development of satellite residential communities beyond the political boundaries. As a consequence, the river miles of buffer zone have diminished, because there are new sources of sewage and urban stormwater runoff to further degrade raw surface-water quality. For example, there are 56 water supply intakes along 977.8 mi of the Ohio River and 43 of these intakes are within 5 mi of an upstream wastewater treatment plant. While low-flow conditions are minimized by navigational dams on this river, numerous tributary streams are not so protected from flash flooding and dramatic burdens of pollution. In a study of surface-water supplies used by 20 cities, serving a total population of seven million people, it was estimated that the minimum wastewater component of the raw source water ranged from 2.3 to 16% and increased to predominantly wastewater for several municipal intakes during low-flow periods.[9]

Several states have proposed suspension of the requirement for sewage effluent disinfection at waste treatment plants that are greater than 20 to 40 mi upstream of a water supply intake.[10] The argument is that chlorination is not cost-effective for controlling fecal coliform densities in sewage effluents. Furthermore, based on bacteriological studies, it was argued that the water quality at downstream water supply intakes has not improved substantially as a direct result of sewage effluent disinfection practices. The crux of the problem is that chlorination of sewage effluents would be more effective if disinfection demand was reduced through effective nutrient removal (TOC or BOD), total suspended solids reduction, and minimization of ammonia content. Because many plants have difficulty in consistently achieving limits below 20 mg/l BOD and 30 mg/l suspended solids, sewage treatment plant operators have generally applied excessive chlorine dosages to meet a specific bacterial limit. Efforts to further minimize organic releases in sewage effluent releases and process water in water supply treatment are of critical importance. The concern here is that these organic residuals may combine with disinfectant in water supply treatment to create by-products that may be carcinogens.

Retention of Surface Water

Retention of surface water in lakes and impoundments of streams is an important contribution to water quality enhancement. Storage and sedimentation of surface water often improves the chemical quality.[11] Many organisms are inactivated by natural self-purification processes while others are removed from the water column through natural siltation.[12-14] These water quality improvements are variable, being directly proportional to holding time, water dilution, and the magnitude of pollutional inputs.[15] Large lakes and impoundments are often very effective in dampening additions of contaminants introduced from feeder streams and stormwater runoff so that water quality conditions should be very uniform over time (Table 2.2).

While impoundments and lakes provide volume and buffering capacity to dilute bacterial contamination and thereby reduce density fluctuations, counterproductive factors must be considered. Lake turnovers, decaying algal blooms, and bacterial nutrient conditions deteriorate water quality and introduce a wide range of organisms (some of which may be pathogenic) to the raw water intake

Table 2.2   Characteristics of Raw Water Reservoirs Related to Size

Reservoir	Storage capacity	Retention time	Ranges		
			Coliforms per 100 ml	Turbidity (NTU)[a]	Color (ACU)[b]
Cruises	82 million gal	—	96–3,800	—	—
Whitney	886 million gal	19 d	1–10,000	0.2–16	<1–45
Hansen Res.	8.47 billion gal	8 months	<1–198	<0.2–2.0	8–18
Sebago 1	9.4 billion gal	5.4 years	<1–170	0.16–0.62	6–27
Sebago 2	9.4 billion gal	5.4 years	<1–78	<0.19–0.55	8–13
Cannonsville	9.75 billion gal	208 d	<1–2,400	0.6–4.6	1–18
Bull Run	16 billion gal	Variable	<1–125	0.15–2.0	7.5–17
Gaillard	16 billion gal	266 d	<1–280	<1.0–9.0	1–40
Tolt Res.	18.24 billion gal	6 months	<1–80	—	—
Hemlock	46 billion gal	3–3.5 years	2–2,400	0.6–10.4	—
Wachusett	65 billion gal	4 months	<1–900	0.25–0.74	6–12
Casitas	83 billion gal	Variable	2–220	—	—

[a] NTU = nephelometric turbidity units.
[b] ACU = apparent color unit.
Data from Geldreich, Goodrich, and Clark.[15]

that proceed to pass through marginal treatment processes or improperly operated treatment trains.

Eutrophication is another water quality concern. These nutrient-rich waters will support the periodic growth of substantial populations of blue-green algae (*Anabaena flos-aquae, Microcystis aeruginosa, Oscillatoria* spp., and *Aphanizomenon flos aquae*) which release exotoxins that, if ingested by any mammal, bird, or fish, can cause illness or death. Ingestion of these toxins by humans causes gastroenteritis and body contact through water recreation sports may induce skin irritations.[16] Secondary effects of algal mass development include oxygen depletion, release of methane, formation of hydrogen sulfite and reductive compounds of iron and manganese, plus increased concentration of ammonia ions.[17] Furthermore, intakes of algae into water treatment systems using only disinfection will result in increase of assimilable organic carbon from the dead cells reaching the distribution system and becoming a potential nutrient for bacterial regrowth.

Not only is the blue-green algae density stimulated by eutrophic qualities of these waters, but there is also greater activity among phytoplankton, zooplankton, bacteria, and fungi. These diversified organisms quickly reach higher population levels and their metabolism and cellular decomposition release a variety of taste and odor compounds. It must also be realized that some of these organisms are disinfectant resistant and will pass through water treatment and become a part of the microbial flora of potable water.

In temperate zones with pronounced seasonal temperature changes, many lakes and impoundments are subject to periods of thermal stratification in the summer and winter and destratification (water turnover) periods in spring and autumn. During the interval of stratification, water movement in the deeper portions of raw water reservoirs becomes restricted, generally creating a zone of maximum bacterial occurrence in the water layer above thermocline.[18-21] With destratification, water from the bottom and top layers mix and a more uniform dispersion of water quality develops throughout. This mixing process causes decaying vegetation and entrapped organisms in settled particulates to reenter the water column.[20,22] As a result, heterotrophic bacterial populations (including coliform bacteria), turbidity, and humics from partially decomposed vegetation will temporarily increase in the water. Upon entering the water plant intake, this water quality change many create a significant but temporary increase in disinfectant demand and variety of organisms that may survive treatment.

GROUNDWATER

As surface water percolates through the soil, changes occur in both chemical and microbiological characteristics. These changes are the result of ion exchange, adsorption, precipitation, and biodegradation occurring in soil–water interfaces during passage. Thus is the origins of groundwater. Other factors that influence groundwater quality are climate, depth of aquifer, geochemistry of the unsaturated zone where percolation occurs, and land use at the surface.[24]

Percolation through soil normally provides a groundwater that is virtually free of organisms due to filtering action or particle adsorption. Provided there is sufficient depth, texture, and structure to the soil, many bacteria and protozoan cysts will be removed in water passage. However, viruses may penetrate further into the soil barrier. Apparently, adsorption–desorption phenomenon and ion exchanges are more crucial to the fate of virus movement through soil.[25]

Natural self-purification in aquifers proceeds at a very slow pace (more often measured in years); therefore, once groundwater is contaminated, it remains deteriorated in quality for periods longer than observed in surface water. Attempts to accelerate purification in deep aquifers have met with limited success. Most treatment approaches are applied after the water is pumped to the surface for immediate use. Some efforts have been made to purify aquifers by use of recharge water of high quality in an effort to blend waters and improve the quality in the contaminated aquifer. Because of these difficulties, any increased reliance on land disposal of wastes must evaluate the impact this practice may have on the future quality of groundwater resources.

Groundwaters are used extensively as a source of public water in arid regions and in areas where surface supplies are insufficient in volume, poor in quality, and require extensive purification or treatment. Because groundwater has long been considered to be of unquestionably excellent quality, treatment is frequently nonexistent or is limited to water hardness reduction or taste and odor removal. In contrast, the waters from shallow wells are frequently grossly polluted with a variety of pollutants from surface runoff. These supplies must be disinfected to provide protection from waterborne agents.

As greater areas of watersheds are used to accommodate increasing populations and industrial growth, there is an increased risk that the soil barrier will become inadequate to continually protect high-quality groundwater resources from pollution. Pollutants enter groundwater supplies by various routes including direct injection of wastes through wells, percolation of liquids sprayed over the land or leached from soluble solids at the surface, leaking or broken sewer lines, seepage from waste lagoons, infiltration of polluted surface streams, interaquifer leakage, irrigation return waters, leachates from landfills, septic tank effluents, and upwelling of salt water into freshwater aquifers. As a consequence of improper source protection and inadequate treatment, it is not surprising that many small public water systems using groundwater have the poorest records for compliance with the drinking water standards.

MICROBIAL COMPOSITION

Study of the microbial flora of raw source waters used in water supply treatment reveals a wide spectrum of diverse organisms in an ever-changing kaleidoscope, which is a reflection of many influences. Many of these organisms are of no known significance in water supply while others may be pathogens, fecal organisms, and opportunists. It is these latter three groups that are of significance; they cause health problems, interfere with indicator detection, and cause taste and odor problems in potable water. Fortunately, waterborne agents can be inactivated by application of appropriate treatment processes but other organisms may form biofilms in sand filters, resist disinfectant applications, colonize in the distribution system, or amplify in plumbing attachment devices.

Survival of these organisms in raw source water is determined largely by available nutrient, water temperature, water pH, UV light penetration, clumping in particulates, microbial predation, parasitism (bacteriophage), and antagonism (actinomycetes) in the flora. While microbial antagonisms are a major deterrent to some bacteria, other organisms, including viruses, are not affected.[26] Active feeding by various bacterial predators and lysing by bacteriophages are limiting influences on maximum densities achieved by enteric bacteria. Even so, microbial populations may adjust to these adversities by accelerating generation times and still present a formidable challenge to be resolved in water treatment.

PATHOGENS

Fecal discharges from infected humans, animal pets, farm animals, and wildlife are continually being released directly to the water environment or indirectly through stormwater runoff over the watershed. General estimates of the number of infected individuals, domestic animals, and wildlife range from

<1 to 25% of the total population for a given warm-blooded animal type
(Table 2.3). While pathogenic bacteria, protozoans, and parasitic worms may
be found in a wide range of warm-blooded animal hosts, human viral pathogens are shed only by infected people. The profile of organisms and their
densities in raw waste water from two cities in South Africa (Table 2.4)
illustrate what may be expected. These findings could be repeated in any raw
domestic sewage from a city, worldwide.

Table 2.3 Enteric Pathogen Excreters Worldwide

		Individual excreters (%)				
Pathogen	Animal	North America	Europe	Asia	Australia	Africa
Salmonella	Human	1.0	1.0	3.9	3.0	—
	Cattle	13.0	14.0	—	—	—
	Sheep	3.7	15.0	—	—	—
	Pig	—	7–22.0	—	—	—
Shigella	Human	0.46	0.33	2.4	—	—
Leptospira	Human	<1.0–3.0	<1.0–3.0	<1.0–3.0	—	—
	Cattle	2.0	—	—	—	—
	Pig	—	—	—	—	—
	Mice	33.0	—	—	—	—
	Dogs	26.6	—	—	—	—
Enteropath. E. coli	Human	1.2–15.5	2.4–3.3	2.0	—	—
	Pig	9.0	—	—	—	—
Vibrio cholera	Human	—	—	1.9–9.0	—	—
V. cholera, biotype El Tor	Human	—	—	9.5–25.0	—	—
M. tuberculosis	Human	—	3.1–5.6	—	—	—
Enteroviruses	Children	0.88[a]	—	—	—	—
Entamoeba histolytica	Human	10	10	16	—	17
Giardia lamblia[b]	Human	7.4	—	—	—	—
	Domestic cat	2.5	—	—	—	—
	Dog	13	—	—	—	—
	Coyote	6	—	—	—	—
	Cattle	10	—	—	—	—
Cryptosporidium[c]	Human	0.6–4.3%	1–2%	3–20%	3–20%	3–20%

[a] Seasonal occurrence in young children.
[b] Data from Burke;[28] Davies and Hibler.[29]
[c] Data from Fayer and Ungar.[30]
Data from Geldreich.[27]

Several other pathogens such as *Legionella* and *Naegleria* do not have an
animal host reservoir but appear to be ubiquitous in the environment and are
comprised of both nonpathogenic and pathogenic strains.[32,33] Why some strains
in the environment are predisposed to be nonpathogenic, while others of the

Table 2.4   Microbial Densities in Municipal Raw Sewage
from Two Cities in South Africa

	Average count per 100 ml	
Organism or microbial group	Worcester sewage	Pietermaritzburg sewage
Aerobic plate count (37°C; 48 h)	1,110,000,000	1,370,000,000
Total coliforms	10,000,000	—
E. coli type 1	930,000	1,470,000
Fecal streptococci	2,080,000	—
C. perfringens	89,000	—
Staphylococci (coagulase positive)	41,400	28,100
Ps. aeruginosa	800,000	400,000
Salmonella	31	32
Acid-fast bacteria	410	530
Ascaris ova	16	12
Taenia ova	2	9
Trichuris ova	2	1
Enteroviruses and reoviruses (TCID$_{50}$)	2,890	9,500

Data from Grabow and Nupen.[31]

same genus are pathogenic, is unclear. These latter strains can quickly become opportunistic, invading the human body under conditions of stress, weakening immunity systems, or general physical degeneration associated with advancing age.

The degree and frequency of pathogen exposure is exacerbated by expanding human populations worldwide that intensify human activities in metropolitan areas and by the trend toward concentrated domestic animal husbandry. Wildlife contributions to pathogen releases are also significant, being more noticeable in remote watersheds where man's activities are minimal or where seasonal colonizations occur.

Bacteria

Bacterial pathogens will have a variable persistence upon discharge into a water source, which is predicated on many factors. Sewage treatment serving a residential community of 4000 persons was regularly found to release various *Salmonella* strains into receiving waters.[34] It has been hypothesized that a sewage collection network of 50 to 100 homes is a minimal size expected before there is a reasonable chance of successful detection of salmonellae.[35] Prevalent *Salmonella* serotypes will vary in wastewater, being a reflection of the common illness in the community at a given time. A total of 32 *Salmonella* serotypes were found to be common during one survey of sewage effluent discharges and the downstream sections of the Oker River in Germany.[36] How far downstream *Salmonella* persist is also variable. For example, *Salmonella* were detected only 250 m downstream of a wastewater treatment plant with regularity but never at stream sites 1.5 to 4.0 km downstream.[37] Excessive BOD or TOC and low stream temperatures provide support to longer *Salmonella* persistence downstream. As an illustration, salmonellae were isolated in the Red River of the North, 22 mi downstream of sewage discharge from Fargo, ND and Moorhead, MN during September and prior to the sugar beet-processing season.[38] By November, salmonellae were found 62 mi downstream of these two municipal sewage effluent discharge points. With increased sugar beet processing, waste reaching the stream in January, under cover of ice, brought higher concentrations of degradable nutrients. In this situation, *Salmonella* strains were able to persist longer; they were detected 73 mi downstream or 4 d flow time from their source in the sewage effluent.

Water utilities using major rivers as a source of raw water are confronted with widely fluctuating water qualities that often contain a variety of pathogens. Salmonellae densities alone may be substantial. Kampelmacher and Jansen[39] calculated that the Rhine and Mause Rivers carried approximately 50 million and 7 million salmonellae per second, respectively. The Missouri River is a source of raw water for various water treatment plants supplying drinking water to some major cities in the Midwest. This river system also transports a fecal and nutrient pollution load of treated sewage effluents, cattle feedlot runoffs, paper and pulp waste effluents, and partially treated wastes from meat and poultry processing plants. Analysis of food, beverage, meat processing, and wood pulp and paper wastes[40] reveals some *Salmonella* occurrences because of direct or indirect contact with feces in sanitary wastes, irrigation water, and polluted processing waters (Tables 2.5 and 2.6). Bacteriological examination of raw water quality at the water treatment plant intakes of Omaha, NE, St. Joseph, MO, and Kansas City, MO on the Missouri River (Table 2.7) frequently revealed fecal coliform densities in excess of 2000 organisms per 100 ml.[5] This fecal pollution load also contained various *Salmonella* serotypes that had their origins in the wastes released in effluents and stormwater runoff upstream.

Table 2.5   Enteric Bacterial Profiles in Raw Industrial Wastes

	Occurrences (%)			
	Wood pulp and paper	Beverage	Food processing	Meat processing
Coliform species				
E. coli	0.4	5.6	35.0	56.9
Klebsiella pneumoniae	92.3	68.0	55.0	21.5
Enterobacter species	6.7	15.0	3.3	13.8
Pectobacterium	0.6	7.0	6.0	0.5
Salmonella strains	0.008	4.4	0.7	7.3
Densities (per 100 ml)				
Klebsiella	10^5–10^6	10^4–10^7	10^3–10^5	10^5–10^8
Fecal coliform	10^1–10^4	10^3–10^5	10^2–10^4	10^6–10^9

Data from Herman.[40]

Table 2.6 Enteric Bacterial Profiles in Waste Effluents

Waste source	Occurrences (%)					
	BOD (mg/l)	Kleb-siella	E. coli	Entero-bacter	Pecto-bacterium	Salmon-ella
Sugarcane	ND	86.0	ND	ND	ND	ND
Wood pulp and paper	85–4,500	85.0	4.4	9.5	0.8	0.3
Potato	1,500–45,000	81.1	0.9	9.4	6.9	1.6
Textile	ND	80.0	ND	ND	ND	ND
Forest vegetation	ND	71.0	ND	ND	ND	ND
Beverage industry	800–18,000	68.9	5.6	15.0	7.0	4.4
Food processing	160–2,400	55.0	35.0	3.3	6.0	0.7
Meat processing	2,600–45,000	21.5	56.9	13.8	0.5	7.3
Municipal sewage	ND	18.0	62.0	14.3	3.6	2.1

Note: ND, not determined.
Data adapted from Herman.[32]

Untreated water supplies are at greatest risk from pathogen contamination, particularly surface supplies. In one instance, an untreated surface supply in an Alaskan community was contaminated with *Salmonella* from sea gull pollution, resulting in several cases of salmonellosis.[41,42] Wildlife was also believed to be the reservoir of *Campylobacter* that contaminated untreated or poorly treated surface water (streams and reservoirs) of low turbidity in small communities located in Vermont[43] and British Columbia.[44] Vacationers in national parks in Wyoming became ill from drinking water from mountain streams and were the reason for a 25% increase statewide in campylobacter enteritis.[45]

A variety of waterborne disease outbreaks have been attributed to unprotected groundwater supplies. Specific bacterial pathogens that have been isolated from well waters include enteropathogenic *Escherichia coli, Vibrio cholera, Shigella flexneri, S. sonnei, Salmonella typhimurium, Yersinia enterocolitica,* and *Campylobacter jejuni.*[46-58] In these situations surface pollution has either penetrated the soil barrier or well construction has failed to seal off stormwater runoff from entering the well.

Viruses

Enteric viruses can also be found in polluted surface waters and unprotected groundwater supplies that are used for public and private potable water supplies.[59-63] In the latter case, poliovirus and other enteroviruses have been isolated from well water used to provide drinking water to restaurant patrons.[64]

Referring back to the Missouri River study (Table 2.7), virus examination performed on raw water from the intake at St. Joseph resulted in the detection of poliovirus types 2 and 3 and ECHO virus types 7 and 33. In another study of Missouri River water quality, total coliforms, fecal coliforms, virus and turbidity levels were quantified at the Lexington, MO intake during 1976 to

Table 2.7 Fecal Coliform Densities and Pathogen Occurrence at the Missouri River Public Water Supply Intakes

Raw water intake	River mile	Date	Fecal coliforms per 100 ml[a]	Pathogen occurrence
Omaha, NE	626.2	October 7–18, 1968	8,300	N.T.
		January 20–February 2, 1969	4,900	N.T.
		September 8–12, 1969	2,000	*Salmonella enteritides*
		October 9–14, 1969	3,500	*Salmonella anatum*
		November 3–7, 1969	1,950	N.T.
St. Joseph, MO	452.3	October 7–18, 1968	6,500	N.T.
		January 20–February 2, 1969	2,800	N.T.
		September 18–22, 1969	4,300	N.T.
		October 9–14, 1969	N.T.	*Salmonella montevideo*
		November 3–7, 1969	6,500	N.T.
		January 22, 1970	N.T.	19 virus PFU; polio types 2,3; echo types 7,33
		April 23, 1970	N.T.	3 virus PFU, not typed
Kansas City, MO	370.5	October 28–November 8, 1968	6,500	N.T.
		January 20–February 2, 1969	8,300	N.T.
		September 18–22, 1969	3,800	*Salmonella newport* *Salmonella give* *Salmonella infantis* *Salmonella poona*

Note: N.T., no test for pathogens done.

[a] Geometric means.

Data from U.S. Environmental Protection Agency report.[5]

1978.[65] Data in Table 2.8 were grouped by results obtained during water temperature periods that closely corresponded to seasonal periods. During the winter period, the resulting fecal contamination was derived largely from municipal wastewater effluent discharges and meat processing plants upstream of the water supply intake. Spring thaws brought increasing flood conditions and associated increases in turbidity and movement of fecal contamination in runoff to the mainstream portions of the river. Occasional high turbidities in the summer were related to major storm periods that increased river flow and

Table 2.8   Raw Water Quality at Lexington, MO Water Treatment Plant

Temp range[a]	No. of samples	Turbidity (NTU) range	Fecal coliforms per 100 ml[b]	Virus (PFU) per 100 gal	
				Range	Occurrence (%)
0–10°C	68	0.05–200	93,000	0–38	79.4
10.5–20°C[c]	32	32–600	113,000	0–9.1	53.1
20.5–28°C	42	40–1,000	162,000	0–10	42.9

[a] Sampling period December 3, 1976 to December 27, 1978.
[b] Geometric mean values.
[c] Spring and autumn periods combined.
Data from O'Connor, Hemphill, and Reach, Jr.[65]

turbidities periodically to maximum values. During wet weather periods, stormwater runoff included bypasses of raw sewage in treatment facilities and movements of fecal wastes from numerous feedlot operations. Autumn periods showed a decline in turbidities, river flows, and fecal coliforms as a reflection of reduced rainfall events. The occurrences of enterovirus in these samples suggest that virus levels correlated inversely with coliform levels, temperature, and turbidity. This unexpected finding could be due to more rapid virus die-off at higher water temperatures or, more likely, the decreased recovery efficiency for virus because of interference with virus detection methodology from the higher stream turbidities created by stormwater runoff silts. An average of 2.5 viruses were recovered in 216 samples. Polioviruses were the most abundant (69%), followed by Coxsackie (11%) and ECHO viruses (5%). Only 10% of the total virus particles detected were recovered on 3- to 5-μm-pore size prefilters, suggesting that most of the recovered virus particles were not associated with the suspended sediment in the Missouri River, being more likely a part of fecal cell debris or viral aggregates in the water. The evidence presented in Tables 2.7 and 2.8 illustrate some of the potential health hazards that become more prevalent and more challenging to a water treatment system processing the widely fluctuating water qualities in a major river.

Protozoans

Detection of protozoan pathogens in pristine waters over the past 30 years has changed forever the time-honored concept that restricted-use watersheds surrounding a lake or impoundment insure freedom from pathogen risk.[66-70] A total of 22,897 people became ill with giardiasis acquired from 84 waterborne outbreaks in the United States during 1961 to 1983. The source of the pathogen was quickly revealed to be in the wildlife population living in the "protected" watershed environment. Beavers were most often considered to be the primary animal reservoir because of the high incidence of infection in beaver colonies and their habitat association with near-shore water environments. However, coyotes, muskrats, voles, cattle, and pets may also be involved in the perpetuation of this pathogen in the natural environment.[71,72] Many of these animals are either part of the natural fauna or are introduced onto the watershed as domestic farm animals.

Cryptosporidium is another pathogenic, enteric protozoan that can be transmitted by contaminated water supply. Waterborne outbreaks have been reported from inadequate treatment of sewage-contaminated groundwater[73] and from water systems using poorly operated filtration backwashing procedures.[74,75] The largest outbreak to date occurred in Carrollton, GA and involved 13,000 illnesses. Other waterborne outbreaks have been cited in England and Scotland.[76] The largest *Cryptosporidium* outbreak to date occurred in Milwaukee in 1992 causing 400,000 illnesses and approximately 80 deaths.

Human reservoirs of this pathogen are found in children (particularly at day-care centers), among people exposed to substandard public health sanitation practices, travelers returning from some Third World nations and some patients with immunodeficiencies due to AIDS or organ transplants.[77-86] Therefore, it is not surprising to find *Cryptosporidium* concentrations as high as 52,000 per gallon of raw sewage and 15,000 per gallon of treated effluent.[87] Farm animals (cattle, sheep, swine, turkeys, and goats), animal pets (dogs and cats), and wild animals (deer, raccoons, foxes, coyotes, beavers, muskrats, rabbits, and squirrels) are also important reservoirs of *Cryptosporidium*[30,88-93] living on the watershed that may be transmitted via the stormwater route. Thus, *Cryptosporidium* densities in surface waters may range from 8 to 20,000 oocysts per gallon,[87] so it is not surprising that these oocysts can be found in filter backwash water at the water treatment plant.

COLIFORMS

Origins of coliforms in water may often be traced to a variety of warm-blooded animal feces; however, some strains occur naturally in the aquatic environment, being common to soil and vegetation. While these diverse origins can obscure the significance of a few coliforms in untreated groundwater, coliforms in waters of lesser quality are more often of great magnitude.[94] Furthermore, these latter waters most likely contain 10 to 20% or more strains that are of fecal origin. For this reason, it is important to have some vision of the potential coliform densities to be encountered in raw water.

Origins and Densities

Coliforms are generally found in large densities in the microbial flora of various animal feces (Table 2.9). As might be expected, these coliform densities magnify many times more as the population of farm animals, wildlife, pets, and humans increases per square mile. Further increases will occur in the collection of raw sewage (Table 2.10) but are dramatically reduced through sewage treatment processes (Table 2.11) and dilution in receiving waters before they reach a downstream water intake. Some mitigating circumstances are nutrient pollution and chlorination of poor-quality sewage (Table 2.12) that

Table 2.9 Microbial Flora of Animal Feces

Animal group	Average density per gram				
	Fecal coliforms	Fecal streptococci	C. perfringens	Bacteroides	Lactobacilli
Farm animals					
Cow	230,000	1,300,000	200	<1	250
Pig	3,300,000	84,000,000	3,980	500,000	251,000,000
Sheep	16,000,000	38,000,000	199,000	<1	79,000
Horse	12,600	6,300,000	<1	<1	10,000,000
Duck	33,000,000	54,000,000	—	—	—
Chicken	1,300,000	3,400,000	250	<1	316,000,000
Turkey	290,000	2,800,000	—	—	—
Animal pets					
Cat	7,900,000	27,000,000	25,100,000	795,000,000	630,000,000
Dog	23,000,000	980,000,000	251,000,000	500,000,000	39,600
Wild animals					
Mouse	330,000	7,700,000	<1	795,000,000	1,260,000,000
Rabbit	20	47,000	<1	396,000,000	<1
Chipmunk	148,000	6,000,000	—	—	—
Human	13,000,000	3,000,000	1,580	5,000,000,000	630,000,000

Data from Geldreich.[94]

Table 2.10 Indicator Composition in Raw Sewages from Various Communities

Sewer source	Estimated sewered population	Total coliforms	Fecal coliforms	Fecal streptococci	FC (%)	Ratio FC/FS
		Densities per 100 ml				
Esparto, CA	—	23,500,000	6,200,000	—	26.4	—
Shastina, CA	—	9,600,000	2,300,000	—	24	—
Los Banos, CA	6,090	62,000,000	23,000,000	—	37	—
Anaka, MN	9,500	47,400,000	10,200,000	—	21.5	—
Newport, MN	800	13,600,000	3,580,000	—	26.3	—
Red Wing, MN	11,000	17,700,000	4,050,000	—	22.9	—
Mankato, MN	81,490	5,525,000	2,630,000	—	47.6	—
Oakwood Beach, NJ	45,000	13,250,000	4,240,000	—	32.0	—
Perth Amboy, NJ	38,000	1,600,000	387,000	—	24.2	—
Middlesex, NJ	300,000	12,900,000	1,070,000	—	8.3	—
Keyport, NJ	5,600	2,210,000	641,000	—	29.0	—
Omaha, NE	180,000	45,800,000	5,360,000	—	11.7	—
Anderson Township, OH	11,000	17,200,000	4,600,000	—	26.7	—
Mt. Washington (Cincinnati), OH	20,000	34,800,000	4,900,000	—	14.1	—
Linwood (Cincinnati), OH	22,000	—	10,900,000	2,470,000	—	4.4
Preston, ID	3,640	—	340,000	64,000	—	5.3
Fargo, ND	50,500	—	1,300,000	290,000	—	4.5
Moorhead, MN	22,934	—	1,600,000	—	—	4.9
Lawrence, MA	67,000	—	17,900,000	4,500,000	—	4.0
Monroe, MI	22,968	—	19,200,000	700,000	—	27.9
Denver, CO	520,000	—	49,000,000	2,900,000	—	16.9
Average value		21,900,000	8,260,000	1,610,000		
Calculated %					37.7	
Ratio FC/FS						5.1

Data from Geldreich.[94]

Table 2.11   Effect of Sewage Treatment Processes
on Indicator and Pathogenic Microbial
Populations in Raw Sewage

Type of sewage treatment	Removal range for various organisms (%)
Primary	5–40
Septic tanks	25–75
Trickling filters	18–99
Activated sludge	25–99
Anaerobic digestion	25–92
Waste stabilization ponds	60–99
Tertiary (flocculation, sand filtration, etc.)	93–99.99

Data from Geldreich.[94]

Table 2.12   Aftergrowth of Coliforms in Dilute Chlorinated
Primary Effluent

Time (d)	Chlorinated primary effluent	Coliform densities (MPN per 100 ml) Chlorinated effluent diluted with river water		
		25%	10%	5%
0.1 mg/l Total Chlorine Residual, 30-min Contact				
0	250	250	250	250
1	6,000	350	2,500	250
2	2,500,000	200,000	60,000	60
3	2,000,000	115,000	6,000	250
4	2,500,000	95,000	115,000	250
5	1,150,000	25,000	2,500	250
7	2,500,000	9,500	2,500	250
0.5 mg/l Total Chlorine Residual, 30-min Contact				
0	20	20	20	20
1	140,000	130	20	20
2	>1,600,000	160,000	20	20
3	>16,000,000	35,000	20	20
4	>16,000,000	35,000	20	250
5	>16,000,000	7,000,000	20	20
7	>16,000,000	1,100	200	250

Data from Heukelekian.[95]

can result in some persistence and aftergrowth of several coliform species (*Enterobacter, Klebsiella,* and *Citrobacter*) miles downstream of the discharge point.[94-95]

Many of the bacteria common to sewage contribute to the overall biological process that degrades the waste material through species selection and adaptation.[96] Conversion of some portion of the complex organic matter to simpler chemical end products by a specialized group of microorganisms creates the basic building blocks for new growth by other organisms in the sewage micro-flora. Thus, a succession of differing bacterial populations, coupled with phys-ical and chemical changes, are the phenomena that make biological treatment of sewage a reality. Organisms most active in biological waste treatment include zoogleal bacteria, *Pseudomonas, Flavobacterium, Achromobacter,* nitrifying bacteria, methane producers, and other organisms whose roles are of special importance in the utilization of unique waste components.

Separate and combined stormwater sewer systems that bypass treatment during intense storms are a major contributor of coliforms (Table 2.13) during wet weather periods. Although the fluctuating quality of stormwater runoff has been recognized for many years, the health hazard implications of these intermittent discharges into raw water impoundments have largely been ignored.[94,97] Major contributors of microbial pollution in the urban community are derived from the fecal material deposited on soil, asphalt, and cement by cats, dogs, and rodents.[98] The release of fecal waste from pets in urban areas is substantial. There are about 40 million owned dogs in the United States and 46% of all American households have at least one dog.[99] It has been estimated that the 500,000 owned dogs in New York City alone deposit about 150,000 lb of feces and 90,000 gal of urine each day on the streets, sidewalks, and park areas. Added to this pet waste are the fecal droppings and urine discharge by other pets and by the rodents in the city environs. It is therefore realistic to consider urban stormwater runoff as a serious cause of surface-water quality deterioration following every storm event.

Table 2.13   Comparison of Bacteriological Data from Separate and Combined Systems during Stormwater Runoff Periods

Period	Organisms per 100 ml			Ratio FC/FS	(FC/TC) × 100
	Total coliforms	Fecal coliforms	Fecal streptococci		
Separate System — Ann Arbor					
April	340,000	10,000	20,000	0.5	2.9
May	510,000	51,000	200,000	0.3	10.0
June	4,000,000	78,000	120,000	0.7	2.0
July	4,000,000	120,000	390,000	0.3	3.0
August	1,700,000	350,000	310,000	1.1	20.5
Combined System — Overflows — Detroit					
April	2,400,000	890,000	—	—	37.1
May	4,400,000	1,500,000	320,000	4.7	34.1
June	12,000,000	2,700,000	740,000	3.6	22.5
July	37,000,000	7,600,000	350,000	21.7	20.5
August	26,000,000	4,400,000	530,000	8.3	16.9

From Burns, R. J. and Vaughan, R. D., *J. Water Pollut. Control Fed.*, 38, 400–409, 1966. With permission.

Stormwater runoff over watersheds in rural and wilderness areas flush fecal contamination released to soil from the fecal droppings of farm animals and wildlife.[100-106] There is, of course, some contribution from inadequate home sewage disposal systems, but because of the low human population densities in rural areas, compared to domestic animals and wildlife, such household sewage contributions are generally insignificant. In geographical locations where farm animals and wildlife roam freely (large cattle ranches, wilderness areas, and national parks), fecal contamination that reaches streams in stormwater runoff is generally restricted to adjacent stream-side areas while fecal deposits in upland areas are largely limited by ground cover, sunlight exposure, frequency and intensity of rainfall, soil temperature, soil pH, moisture retention, and the antagonistic action of the soil microbial flora.[106-114]

The possible contribution of coliforms and pathogens from the burial of solid wastes should not be overlooked, particularly as a source of groundwater contamination. Natural drainage of landfill sites releases a variety of microorganisms that may be present in substantial numbers even after 2 months of confinement (Table 2.14). Reductions of pollution indicator densities to one or two logs below the densities found in fresh garbage are probably related to rises in temperatures within landfills caused by the biological decomposition of organic materials during the first week of landfill operation.[115] Many of the organisms appearing in seepages (leachates) are from the higher-density bacterial populations in peripheral areas of the landfill. When leachates are a serious threat to surface waters, lagooning these seepages will improve the microbial quality prior to discharge into receiving waters. Where test wells in the groundwater table indicate declining water quality caused by these seepages penetrating the soil barrier, landfill operations at the site of intrusion should be suspended immediately until corrective landfill practices can be devised.

Table 2.14   Fecal Coliform and Fecal Streptococcus
Survivals in Leachates from Sanitary
Landfills (6 Weeks after Placement)

Days after placement	Densities per 100 ml	
	Fecal coliforms	Fecal streptococci
42	2,600,000	240,000,000
43	4,900,000	790,000
56	2,000	79,000
63	9,000	33,000
70	33,000	170,000

Data from Blannon and Peterson.[115]

Intake Raw Water Quality

Intake raw water quality for public water supply treatment is quite variable. Among the various water systems evaluated in "Survey of Community Water Supply Systems,"[116] groundwater was by far the major raw water source with

820 well or spring sources supplying 570 water plants. During this study, 774 (94.4%) of these groundwater supplies contained no detectable fecal coliforms per 100 ml (Table 2.15). Other groundwater sources were of a lesser quality. One supply in Vermont processed contaminated groundwater containing a fecal coliform density in excess of 2000 organisms per 100 ml.

Table 2.15   Groundwater Quality for 820 Public Water Intakes[a]

| Fecal coliforms per 100 ml | Public supplies | | | |
	Total	Percent	Cumulative total	Cumulative percent
<1	774	94.4	774	94.4
1–20	29	3.5	803	97.9
21–200	14	1.7	817	99.6
201–2,000	2	0.2	819	99.9
2,001–4,000	1	0.1	820	100.0

[a] These intakes included wells and springs serving as source water for 570 public water systems.

Data from McCabe et al.[116]

Surface water was the source for 160 water plants in this nationwide survey. The results (Table 2.16) showed that 99.4% of these surface-water sources had fecal coliform densities of less than 2000 organisms per 100 ml. Possibly less than 0.1% of all utilities throughout the nation are treating raw water that periodically exceeds 2000 fecal coliforms per 100 ml.

Table 2.16   Surface Water Quality for 160 Public Water Intakes[a]

| Fecal coliforms per 100 ml | Public supplies | | | |
	Total	Percent	Cumulative total	Cumulative percent
<1	70	43.8	70	43.8
1–20	30	18.8	100	62.5
21–200	37	23.1	137	85.6
201–2,000	22	13.8	159	99.4
2,001–4,000	0	0.0	159	99.4
4,001–8,000	1	0.6	160	100.0

[a] These intakes included streams, lakes, and reservoirs serving as source water for 160 public water systems.

Data from a survey of community water systems in 9 metropolitan areas; McCabe et al.[116]

The fecal coliform component in raw surface waters can be expected to approximate 15 to 30% of the less definitive total coliform population (Table 2.17). The reason for such differences in total coliform–fecal coliform densities lies in the more rapid inactivation of *E. coli* and other fecal coliforms as compared to the longer survival of environmental coliform strains of no sanitary significance. In untreated but protected groundwater, total coliform

Table 2.17   Total Coliforms to Fecal Coliform Relationships at Intakes

Water source	No. of samples	Coliforms per 100 ml[a]		Fecal coliforms (%)
		Total	Fecal	
Lakes and impoundments				
Lake Seymor	49	7.0	0.2	2.9
Hemlock Lake	363	64	4.1	6.4
Lake Coquintan	50	3.9	0.5	12.8
Lake Guillard	58	16.8	2.2	13.1
Casitas	113	8.7	2.3	26.4
Bull Run	78	29	8.5	29.3
Missouri River (Lexington, MO intake) Temp. range				
0–10°C	68	93,000	4,600	4.9
10.5–20°C	32	114,000	7,900	6.9
20.5–28°C	42	162,000	8,000	5.3

[a] Mean values.

densities are often less than 10 organisms per 100 ml and fecal coliforms are not detectable in 100-ml sample portions.

HETEROTROPHIC BACTERIA

The heterotrophic bacterial population in source waters consists of numerous organisms whose origins are aquatic by nature — the true indigenous bacteria — and other bacteria introduced in stormwater runoff, waste effluent discharges, wildlife sanctuaries, and airborne dust particles. Their continued survival or potential for colonization of the aquatic environment is qualified by availability of nutrients, favorable water temperatures and pH, and the status of dissolved oxygen concentration. While these factors, among others, will determine a positive growth response by heterotrophic bacteria, there are counteracting forces in the aquatic environment, such as microbial antagonists, predators, sunlight exposure, and sedimentation, that limit the magnitude of population growth. Seasonal changes in water temperature and pollution spills into the aquatic environment are powerful influences on the aquatic ecology and will cause distortions in bacterial profiles, creating changes in the dominance of organisms at any given time.

Because there is a myriad of heterotrophic organisms that have been isolated from water, perhaps it is more meaningful to examine the bacterial population in terms of profiling groups with common environmental characteristics that may be significant to drinking water quality. The most important of these groups to water supply treatment and distribution quality are the wide range of organisms that utilize organic contaminants for rapid growth in warm source waters and other organisms that can create taste and odor problems. In so doing, some sense may be made in characterizing raw water quality for water supply purposes.

Survival and growth of this large group of heterotrophic bacteria depends on dissolved or suspended organic matter originating in vegetation decay and organic releases in waste discharges from domestic sewage, food processing, beverage production, pharmaceutical manufacture, and agricultural activities (Table 2.6). While a few fastidious organisms may require 10 to 100 mg of organic nutrients per liter,[117] most of these bacteria (including coliforms and some bacterial pathogens) can grow in polluted waters containing only a few milligrams of degradable organic material per liter.

It should be noted that chlorinated effluent releases with poor nutrient reductions will create a sudden sag in the expected population density for heterotrophic bacteria in the vicinity of a discharge point. Some 24 to 48 h flow time downstream of this discharge point, there may be a dramatic growth of these organisms, particularly coliform bacteria, to densities that are approximately 1000 times or more above the residual population at the effluent discharge point (Table 2.12). This phenomenon, known as coliform aftergrowth, results from an initial decline of a variety of competitive organisms that are sensitive to chlorination, thereby providing a brief opportunity for coliforms to grow unchallenged before the ecosystem balance is restored.

It has been estimated that the community of heterotrophic bacteria may comprise up to 90% of all aquatic bacteria detected in surface waters[118,119] and are largely responsible for the wide fluctuations in bacterial populations noted in surface waters. Because groundwaters are generally protected from organic contaminants by a soil barrier, heterotrophic bacterial species requiring organic nutrients are few in number, comprise a limited species diversity and show minimal density fluctuations.

Predominant genera of heterotrophic bacteria include *Flavobacterium, Cytophaga, Achromobacter, Acinetobacter, Alcaligenes, Moraxella, Vibrio, Aeromonas,* and *Pseudomonas.* Environmental coliforms (*Klebsiella, Aerobacter,* and *Citrobacter*) and fecal coliforms (*Escherichia* and fecal *Klebsiella*) are also a part of this community of heterotrophic bacteria that is found in raw water intakes to water supply treatment facilities. Density magnitudes that may be expected for these organisms in rivers of varying qualities can be seen in Table 2.18. These relative numbers will, of course, fluctuate widely during periods of stormwater runoff. Typical densities for heterotrophic bacteria in a stratified lake are given in Table 2.19. Note the bacterial population response to copper sulfate additions to the lake for algae bloom control.

NUISANCE ORGANISMS

This assorted group of bacteria, fungi, and algae can cause a variety of problems in water treatment, water supply storage, and distribution. Many of these organisms obtain their energy resources from the oxidation of inorganics (ammonia, sulfur, and reduced iron and/or manganese). Others are stimulated by inorganic nitrogen (N), inorganic phosphorous (P), sunlight for photosynthesis,

Table 2.18   Heterotrophic Bacterial Densities
in Various River Water Qualities

Water quality	Heterotrophic bacteria per ml[a]
Pristine river water	<500
Relatively clean river water	5,000–10,000
Moderately polluted river water	25,000–50,000
Polluted river water	≤100,000
Grossly polluted river water	≤1,000,000

[a] Plates incubated at 22°C.
Data adapted from Daubner.[120]

and warm-water conditions. Dominance in the microbial profile is achieved in some environmental niche of water supply because of chemical characteristics of the source water or the surrounding area of the habitat site.

Significant concentrations of essential inorganics often play a key role. These chemical constituents may occur naturally in the source water but more often are introduced in stormwater runoff that brings agricultural fertilizers and soil particles into the source water while industrial wastes supply a steady release of sulfides, phosphates, and other inorganics. Microbial activity for some of these organisms also becomes optimized in the bottom waters of stratified lakes with transformation of sulfides[121-123] and ferrous[121,122,124] and manganous compounds[121,122,124-126] occurring in the oxygen-depleted bottom sediments of stratified lakes.

Massive blooms of algae (green and blue-green species, predominately) often occur in summer, being stimulated by high concentrations of inorganic N and P. Most algal blooms occur when N and P concentrations reach 0.8 and 0.4 mg/l, respectively. A bloom condition has been arbitrarily defined as an algal density of 500 cells or more per milliliter of raw source water.[127] When this condition occurs, organisms clog treatment filters, algal cell clumps provide bacterial protection from disinfection exposure, and disinfectant demand increases. Some of the blue-green algae (*Cyanobacteria*) can produce toxins that, if ingested in sufficient concentration, may cause gastrointestinal upsets.[120,128,129] Blue-green algal blooms also create a problem of taste and odor as a consequence of geosmin production in concentrations that exceed a threshold point.[130] Geosmin is also produced by actinomycetes,[131] particularly by members of the genus *Streptomyces*, growing in stagnated, warm, surface waters used in water supply treatment.

Sulfur-oxidizing bacteria (*Beggiatoa* and *Thiobacillus*) become active participants in surface waters receiving substantial inputs of wastewater containing elemental sulfur, thiosulfate, or sulfides. Substantial growth of some filamentous species causes fouling of treatment processes, while other strains produce corrosion problems of concrete structures used for water storage and distribution.[132] Of immediate concern is the formation of various undesirable sulfurous compounds that lead to taste and odor problems.

Table 2.19 Effect of Copper Sulfate Additions on Bacterial Concentrations in Cruises Creek Reservoir

	Surface				Mid-depth[a]		
Date, 1974	Fecal coliforms per 100 ml	SPC per ml	Temperature (°C)	Date, 1974	Fecal coliforms per 100 ml	SPC per ml	Temperature (°C)
June 4	41	330	24.0	June 4	680	2,700	15.1
11[b]	53	730	24.6	11	350	3,300	17.0
13	3	22,000	22.0	13	180	7,300	15.0
18	1	1,200	23.7	18	87	1,300	18.6
20	2	2,700	24.8	20	5	2,400	18.0
July 1	3	500	28.5	July 1	280	13,000	17.8
9[c]	110	5,100	27.3	9	86	46,000	15.3
18	<1	50,000	26.2	18	144	31,000	16.5
22	2	27,000	26.5	22	48	25,000	16.0
24	12	4,700	27.0	24	50	19,000	16.0
29	<1	2,600	26.0	29	18	11,000	17.0
31	1	470	24.5	31	11	5,100	15.7
August 5[d]	3	6,200	24.9	August 5	19	4,000	17.1
7	<1	13,000	25.8	7	16	4,800	16.5
14	1	15,000	25.5	14	25	2,000	18.7
19	35	25,000		19	2,400	14,000	
September 16	103	3,900	21.0	September 16	580	5,700	18.3
18	31	2,700	18.5	18	240	7,200	18.2
23	18	2,400	18.0	23	49	4,100	18.0
25	8	900		25	11	1,300	

[a] 3.0 to 3.6 m (10 to 12 ft).
[b] Copper sulfate treatment, June 12.
[c] Copper sulfate treatment, July 9.
[d] Copper sulfate treatment, August 6.

Data from Geldreich et al.[13]

Sulfate-reducing bacteria (*Desulfovibrio*) can be a major nuisance in oxygen-depleted bottom waters of reservoirs and impoundments. These organisms are strict anaerobes with sulfate replacing oxygen in the oxidation of vegetation debris accumulating in bottom sediments. Hydrogen sulfide produced in this microbial activity can cause odor problems, produce suspended back particles, and be a corrosive agent on steel and other metals exposed to water in the distribution network.[133]

The application of ammonia to water containing chlorine to provide a persistent combined chlorine residual or applied to control the production of trihalomethanes may stimulate various nitrifying bacteria in the water flora to oxidize excess ammonia to nitrite. In so doing, these organisms develop a slime on the walls of pipe that periodically sloughs off and appears at hydrants or household taps.[134] *Nitrosomonas* strains are the predominant nitrifying bacteria involved, but they may be accompanied by other organisms such as *Bacillus, Micrococcus, Pseudomonas,* and *Achromobacter* in the growth of slime at the far ends of the distribution system. Origins of these organisms may be either source water or soil that enters the pipe network during line construction and repairs.

Nuisance bacteria can create an especially serious problem in some groundwater supplies, affecting not only aesthetics (taste, color, and odor), but also producing a biofilm that restricts water movement through the well casing or in the distribution pipe network.[135-136] Factors that contribute to biologically mediated pipeline corrosion or interfere with groundwater extraction include elevated concentrations of ferrous or manganous compounds, sulfides, or other incompletely oxidized minerals. These inorganics may be naturally occurring in the soil barrier and leach into the aquifer. Other sources are introduced through pollution migration or are found in corrosion of iron pipe.

Because groundwater in the aquifer is generally depleted of dissolved oxygen, microbial activity occurs either in the well casing, on the reservoir walls, or at other sites in the system where some air contact is achieved with the anaerobic water. At these sites, fouling bacteria (*Sphaerotilus, Leptothrix, Crenothrix,* and *Gallionella*) become active, utilizing the available inorganic nutrient resources, thereby causing the precipitation of ferric hydrate. These deposits of iron oxide or iron hydroxide on bacterial cell surfaces form a protective layer that can impair disinfection effectiveness. Other undesirable changes include giving the water a rusty brown or black appearance and creation of tastes that make the water supply unpalatable. In view of the problems caused worldwide by these nuisance bacteria, it is surprising that so little is known about their density occurrences in raw source waters.[136-138]

MICROBIAL CONTROL THROUGH TREATMENT

Freedom from gross suspended matter, odors, and bad tastes were considered by early man to be critical characteristics sought in an adequate supply of safe drinking water. Although today's water quality concerns are more complex, reductions of these undesirable properties are frequently achieved by minimizing particulates in the water supply. Waterborne particulates vary in size, shape, and chemical composition. They enter the water environment as silt, finely divided inorganic or organic matter, including fecal cell debris, any of which can carry bacteria, virus, algae, or macroinvertebrates in aggregates or microcolonies.[139-143]

Rivers are particularly prone to wide fluctuations in suspended matter as a result of stormwater runoffs in the watershed plus seasonal floods that transport massive pollution loads and are accompanied by high flow velocities that resuspend stream bottom materials. As a consequence, intake towers at extraction points on a river (Figure 2.1) must be cleaned periodically to remove heavy siltation and debris that obstructs the passage of raw water to the treatment plant. Raw water reservoir and impoundment water quality are less sensitive to these events but more impacted by massive algal blooms and water turnover if there are periods of water stratification in spring and autumn. As a consequence, raw surface-water quality varies considerably, experiencing shifts in the profile of microorganisms and pathogenic agents present over time. These inconsistencies in water quality, therefore, dictate treatment of some magnitude to modify the microbial quality for all surface waters. Where groundwater is protected from surface contamination by an effective soil barrier, water quality is uniform and microbial problems minimal.

RAW WATER STORAGE

Storage of water in a quiescent state will permit the settling of many waterborne particles provided sufficient time is available. For many waters, however, complete natural sedimentation is difficult to achieve because various colloidal materials remain suspended almost indefinitely (Table 2.20). In practice, raw water is often held for only a few hours to a few days because of space limitations for storage. Even with limited storage time, there may be significant removal of heavy particles with some uneven reductions in bacterial densities. Coliform reductions by raw water storage (<1 to 2 d) involves settling and other natural self-purification processes. Some coliform die-off in storage has been reported to range from 13 to 97% at water treatment facilities using water along the Ohio River basin.[145] *Salmonella typhi* contamination of a storage reservoir be roosting sea gulls was observed to be reduced by 90% after 2 weeks storage. However, within the period of <1 to 2 d, storage normally practiced in water treatment operations, most of the viruses in associated fecal cell debris will still remain suspended along with the fine particles that often exert the greatest influence on the chemical and microbial quality of drinking water.

Figure 2.1   Water intake station on the Missouri River.

Table 2.20   Settling Rate for Waterborne Particles

Diameter of particle (mm)	Type of material	Time required to settle[a]
10	Gravel	0.3 s
1	Coarse sand	3 s
0.1	Fine sand	38 s
0.01	Silt	33 min
0.001	Bacteria	55 h
0.0001	Colloidal particles	230 d
0.00001	Colloidal particles	6.3 years

[a] Calculations based on time required for particles with a specific gravity of 2.65 to settle 0.3 m.

Data adapted from Powell.[144]

AERATION

Aeration is often used as a treatment for groundwaters that need improvement in some physical or chemical characteristic before use as a drinking water supply. In the past, most of these concerns were with iron removal, taste and odors produced by hydrogen sulfide, or with removal of substances that interfere with water softening or disinfection. More recently, the concern has been expanded to include a need to reduce volatile organic pollutants that become the precursors for trihalomethane production when disinfectant is applied to safeguard the water from microbial contamination.

Aeration treatment or air-stripping for removal of volatile organics could be accomplished simply by open sprays of water in fountains located in a treatment basin. Unfortunately, when this approach to air-stripping is used, there is a risk of airborne contamination in dust particles and from defecation of birds that frequent such aquatic fountains. For example, coliform organisms were reported in the effluent from one water spray aeration process, but were not detected in the influent, which was filtered water.[146] Such sources of contamination may be minimized by enclosing the aerator in a tower that is properly vented for exchange of gases in the water–air interfaces. These enclosures also restrict available light, which is beneficial in suppressing algal growth.

Aeration towers may use a waterfall technique, cascading a film of water over wood slats (redwood or cypress) or a packed column of PVC ladder rings. In this design, water flows downward over multiple surfaces in a tower filled with ceramic or plastic packing materials. Use of an air diffuser to produce upward air movement through the aeration tower is another technique. A spiraling water turbulence is created by injection of compressed air from one side of the column. This agitation at a selected air-to-water ratio is done in an effort to obtain maximum removal of volatile organics from the water.

Some of these process designs, particularly those using wood slats as the contact surface may provide an ecological niche for the proliferation of microbial slimes. The problem is complicated by the fact that environmental strains of *Klebsiella* have been demonstrated to be common in redwood lumber (often used as baffle or slat material in aeration towers). New redwood contains cyclitols, which are an excellent source of bacterial nutrients, and may require several years of leaching before becoming exhausted. Furthermore, redwood pores form excellent sites for colonization by various heterotrophic bacteria including *Klebsiella*.[147] Ceramic or plastic packing material (also used in aeration towers) will become sites of nutrient entrapment (trace organics and salts) and favored locations for microbial colonization. Because groundwater temperatures are relatively uniform year-round, microbial growth on the multiple surfaces of the aeration packing material will not exhibit periods of seasonal acceleration when the process is in continuous use. However, microbial quality degradation may be seen in effluent water from an aeration tower (Table 2.21) that was used on an intermittent basis. During nonoperating periods, increased ambient temperatures may stimulate regrowth of colonized organisms in the entrapped water. Then when the system is brought on-line again, water flowing through the columns will release this increased growth into the effluent. Aeration systems that exhibit problems with changing water pressure or air pressure may also create water pulses that shear off some of the microbial growth at irregular intervals. All of these situations cause the microbial quality of the process effluent to vary significantly at times from that of the original groundwater source.

Table 2.21   Seasonal Microbial Quality of Groundwater Treated by Aeration

Date of sample	Water temp (°C)		Heterotrophic bacteria[a] (CFU per ml)		Total coliform (CFU per 100 ml)	
	Raw	Aerated	Raw	Aerated	Raw	Aerated
October 27, 1981	11.5	11.6	40	84	<1	<1
November 3, 1981	11.5	11.5	390	880	<1	<1
December 2, 1981	11.5	11.6	128	2,200	<1	<1
December 8, 1981	11.9	12.0	27	200	<1	<1
December 15, 1981	11.3	11.3	42	1,150	1[b]	1[b]
January 5, 1982	9.1	8.8	74	700	<1	<1
January 20, 1982	9.9	10.8	115	14,800	<1	<1
May 11, 1982	9.2	9.7	128	168	<1	1[c]
July 21, 1982	11.3	12.3	119	1,800	<1	<10[d]
December 21, 1982	11.4	11.4	10	178	<1	<1
January 11, 1983	11.3	11.3	9	34	<1	<1
January 25, 1983	11.3	11.4	53	120	—	—
February 9, 1983	11.3	11.3	186	190	—	—
February 23, 1983	11.2	11.2	1,300	316	<1	<1
February 25, 1983	11.4	11.4	53	120	1[e]	<1
March 29, 1983	6.0	6.0	57	108	<1	<1

[a] R-2A medium (27°C for 7 d).
[b] *Enterobacter agglomerans* in influent and aerated effluent.
[c] *Citrobacter freundii.*
[d] *Klebsiella pneumonia, K. oxytoca, C. freundii,* and *E. coli.*
[e] *Enterobacter agglomerans.*
Data from Spino et al.[148]

HARDNESS CONTROL

Hardness in source waters is caused by dissolved minerals, primarily calcium and magnesium compounds (carbonates, chlorides, and sulfates). This condition is more likely to occur with groundwater, although it can be a serious problem with some surface source supplies. Without effective treatment, hardness may result in excessive scale formation in water pipes and water heaters and generate consumer complaints about increased soap consumption and reduced service life for hot-water heaters. Treatment usually involves a lime-soda process, although cation exchange may be used. The lime-soda process requires water pH to be adjusted to above 10 for optimum hardness reduction. An incidental benefit of this pH adjustment is often a marked reduction in microorganisms either by removal through entrapment with the insoluble carbonates produced or by inactivation when the water pH is approximately 10.0.[149-151] For instance, 99% removal of virus was reported when a high-lime process was applied and water was adjusted to pH 11.0.[152] In water reclamation, lime clarification at pH 11.3 resulted in a 5-log removal of total and fecal coliforms and a 98 to 99.9% reduction of viruses. Regardless of these secondary benefits, the lime-soda process using elevated water pH values should not be considered a disinfection substitute in water supply treatment, because greater reduction levels of surviving bacteria (6-log reduction) and virus (4-log reduction) must be achieved as a public health safeguard.

CONVENTIONAL TREATMENT

Conventional treatment is here defined to consist of a combination of processes (coagulation, sedimentation, filtration, and disinfection) to provide a safe water supply. Each of these treatment processes contributes to the reduction of various organisms of public health concern in raw source water, and in so doing significantly reshapes the microbial flora of drinking water.

Coagulation

Coagulation of source water is a primary processing step used to hasten the agglomeration of fine particles in turbidity. Water coagulants commonly used include aluminum sulfate (alum), sodium aluminate, ferric iron salts (ferric sulfate and ferric chloride), ferrous sulfate, and polyelectrolytes, which may be used as primary coagulants or as filter aids to remove colloidal particles. With proper dosage and efficient mixing to rapidly and intimately blend the coagulant chemicals with every bit of raw water, floc particles begin to form, causing organisms, fecal cell debris, and colloids to adhere together with subsequent growth into larger masses that either settle rapidly or become entrapped in subsequent filtrations. Murray and Parks[153] demonstrated that poliovirus adsorption preference was

$$\text{metals (strong adsorption)} > \text{sulfides} \geq$$
$$\text{transition metal oxides} > SiO_2 > \text{organics}$$

Coagulation–flocculation processes are generally optimized for removal of hydrophobic colloids (clay and various inorganic particles) because this material is the major component of turbidity in water. Optimizing the treatment process for removal of hydrophilic colloids (organic particles, including fecal cell debris) would not only result in more efficient virus and bacterial removal, but also in reductions in some dissolved organics that may be of toxicological significance.[154]

Coagulation and sedimentation pilot-plant studies performed on Ohio River water (Table 2.22) demonstrated that suspensions of coxsackie virus and coliform bacteria, added prior to coagulation with 25 mg/l aluminum sulfate or ferric chloride, were reduced in the clarified water to less than 1% of their original densities.[155] Proof that this effect was associated with coagulation was shown in parallel experiments in which the microorganisms were added 5 min or more after introduction of alum coagulant. In these latter experiments, there was no apparent organism reduction. Thus, it could be concluded that a metal–organism complex must form within seconds following the addition of coagulant to water.

Table 2.22 Survival of Coxsackie Virus and Native Bacteria in Ohio River Water at 25°C, Treated by Coagulation and Sedimentation

Test series	Initial turbidity (J.U.)	$Al_2(SO_4)_3$ dosage (mg/l)	Final pH units	Final turbidity (J.U.)	Percent surviving	
					Coxsackie virus	Coliform bact.
1	60–100	15	7.1–7.4	5–10	4.3	36.2
2	16–240	25	6.7–7.3	1–5	1.4	0.2
3	160–240	25:25[a]	7.3–7.8	0.1	0.1	0.01

[a] Two-state coagulation and sedimentation: second-stage coagulant — $FeCl_3$.
Data from Chang et al.[155]

In full-scale operations, complete mixing to achieve optimized microorganism removals may not reach the levels found in pilot-plant studies. Total coliform removals have been reported to range from 42 to 83%.[145,156,157] For larger organisms such as *Giardia muris* and *Entamoeba histolytica*, removal by coagulation and sedimentation may be expected to exceed 99%.[158] Based on these observations, perhaps turbidity measurements should be replaced by particle counts to optimize virus and bacterial removal in coagulation–flocculation processes.[159,160]

Filtration

Filtration of the water supply was originally done to improve the aesthetic qualities of drinking water. In retrospect, following a cholera outbreak in 1892, it was noted that illness cases were largely confined to Hamburg (unfiltered supply) while the adjacent German city of Altona (filtered supply) was relatively unaffected. This event demonstrated in a dramatic fashion the value of a sand filtration process. Slow sand filters had been used in water supply processing at Paisley, Scotland as early as 1804.[161]

Slow Sand Filters

Slow sand filtration involves the passage of raw source water through a filtering medium consisting of a layer of sand from 24 to 40 in. (0.6 to 1 m) deep. Rate of passage is commonly 3 to 6 mgd (million gallons per day) per acre of filter area. Much of the effectiveness of this process depends on the biofilm (schmutzdecke) that forms in the top few millimeters of sand. The filter is cleaned by carefully scraping off the clogged sand at the surface so as to minimize loss of the biofilm activity.

Studies on slow sand filters indicate that bacterial reductions may range from 98 to 99%[162] in a properly operated unit. Many of the important operating conditions (fine-grain sand, deep beds, and lower loading rates to improve effluent bacterial quality) were first identified by Hazen.[163] Recent studies on the microbial barrier suggest that 99.9% or more of the heterotrophic bacterial population (including *E. coli*) could be removed.[158] At a flow rate of 0.12 m/h for low-turbidity source water containing 1000 total coliforms per 100 ml, coliform reductions of greater than 99% could be achieved in pilot-plant filter operation. Total coliforms and 7- to 12-μm size particle reductions were found to be somewhat inferior during the 48-h start-up period of each filter run. Therefore, filtration to waste during this initial phase is appropriate in those surface supplies where *Giardia* cysts are of concern.[164] Furthermore, Burman,[165] reported removals of fecal and total coliforms were only 41 and 88%, respectively, during cold weather. Low temperatures slow the development of the biological slimy, gelatinous layer on the sand surface layer (schmutzdecke), which is the key to effective removal of a variety of microorganisms in a slow sand filter. Another contributing factor is flow rate; a slow filtration rate effectively entraps more organisms. For these reasons, poliovirus removal also varied and has been reported to range from 22 to 99.999%.[168-171] The lower removal rates reported often reflect application of filters that have not achieved an adequate "schmutzdecke".

Recent research suggests that a properly operated slow sand filter will remove *Giardia* cysts.[170-171] The efficiency with which *Giardia* cysts are removed through slow sand filtration correlates with the process effectiveness on turbidity removal. Although *Giardia* cysts do not contribute significantly to water turbidity, turbidity reductions may serve as a surrogate marker to indicate filtration process effectiveness in cyst removal. Increased cyst occurrences in suspect water supplies have been found to correspond to subtle changes in finished water turbidities. Maintenance of a turbidity limit at 0.1 NTU in finished water processed from surface waters with a past history of *Giardia* occurrences appears to assure absence of cysts in the drinking water supply.[172,173]

Rapid Sand Filters

Rapid sand filtration involves the downflow of water through a 50- to 80-cm (20- to 31-in.) bed of coarse sand (0.35- to 0.55-mm particle size) supported over a gravel or anthracite base (0.4- to 1-mm particle size). The stratified mixed media bed permits water passage at a rate 40 to 50 times that of slow sand filters. However, without preconditioning the raw water by coagulation and flocculation, removal of organisms may be very erratic.[174-177] For example, percent removal of total coliform bacteria in the rapid sand filter range from 86 to 96% in all seasons, being affected by raw water turbidities, bacterial densities, and algal blooms.[164]

In a worst-case situation, rapid filtration of a water with added poliovirus through sand alone or through sand plus anthracite removed only 1 to 50% of the virus when no coagulants were used.[171] When the same water was first coagulated, settled, and then passed through these pilot-plant filter beds, virus removal rates exceeded 99%. Carefully applied coagulation prior to rapid sand filtration also resulted in better than 95% reductions in *Giardia muris* cysts[172] and 99.99% removals of *E. histolytica*.[178] These observations demonstrate the traditional rapid sand filter is not effective in producing a reliable reduction in microorganisms or in clarifying raw waters without prior source water conditioning. Therefore, pretreatment by coagulation and flocculation should be considered as essential components of a single treatment process involving rapid sand filtration.[166]

Diatomaceous Earth Filters

Diatomaceous earth (DE) filtration is sometimes used by small water treatment systems where source water is generally low in turbidity and bacteriological quality is good, because this treatment provides a lower total operational cost. As applied, a slurry of DE is hydraulically deposited on a filter septum, in a precoat approximately 0.16 to 0.32 cm (1/16 to 1/2 in.) thick. As raw water is introduced in treatment, an additional quantity of DE is added to maintain a uniformly porous filter cake. Stopping the flow of water through the filter will cause the filter cake to immediately drop from the septum and the treatment barrier to be breached, so DE filters should be backwashed and recoated when flow interruptions occur. Because commercial diatomaceous filter materials have substantially different permeabilities, it is not surprising that reports on coliform removal vary with the type of filter medium used. The most permeable (coarsest) filter media reduce coliform densities by up to 90% while the least permeable (finest) grade is capable of coliform removal in excess of 99.8%.[179] The mean particle size of various commercial grade DE filter materials ranges from 9.5 to 50 μm. Because DE particles at the upper size limit are less effective for virus removal, use of a coated DE filter aid or a cationic polyelectrolyte applied to the raw source water prior to filtration is recommended.[171,180-182] Virus removal was also found to be significantly better in DE filtration when the source water pH was not reduced from pH 9.5 to 6.5. DE filters were also found to consistently remove greater than 99.5% of *Giardia muris* cysts.[172,173] For DE filtration to be an effective microbial barrier with optimal removal rates requires the use of a filter aid precoat, uniform body feed, and continuous filtration at a rate between 0.1 and 2 gal/min/ft² (0.38 and 7.6 l/min/0.09 m²) of filter area.

GAC Filters

Granulated activated carbon (GAC) filtration/adsorption has often been used for taste and odor control. Undesirable tastes and odors are often associated with waters from impoundments and frequently result from algal blooms and thermal destratification of bottom waters. In other instances, decaying vegetation (humics) and industrial pollutants are important contributors that may be precursors to the formation of trihalomethanes (suspected carcinogens) when chlorine is applied for control of microbial hazards in drinking water. Existing trihalomethane concentrations have led some water plant operations to incorporate GAC filtration in the existing treatment scheme in hopes of minimizing a variety of organic residuals that might be involved in disinfection by-product formation.[183] By so doing, GAC particles become focal points for the concentration of bacterial nutrients and provide suitable attachment sites for microbial habitation and subsequent colonization. This process can thereby be of microbiological concern if it provides opportunities for regrowth of indicator organisms, disinfectant-resistant or -tolerant strains, opportunistic pathogens, antibiotic-resistant transfers among biofilm residences, and amplifications of antagonistic bacteria to coliform detection.

When the Beaver Falls, PA municipal authority replaced three of its eight sand filters with granular activated carbon for full-scale taste and odor control, a concurrent evaluation of trihalomethane control and bacteriologic conditions was conducted.[145] Source water from the Beaver River was treated with coagulant, settled, then mixed with lime and chlorine and settled again before filtration. These treatment processes were sufficient to reduce the total coliform densities of 6000 to 220,000 organisms per 100 ml in the source water to densities in the settled GAC influent that were generally undetectable in 100 ml of sample. Coliform densities in the activated carbon filter/adsorber effluents (Table 2.23) were in excess of influent densities during the first 12 weeks of filter/adsorber operations. Warm water temperatures appeared to be the critical factor, because by week 13, when the source water temperature had declined to 4°C (39°F), total coliforms became undetectable in 100-ml sample volumes. A significant drop in the standard plate count densities of the GAC filter/adsorber effluent from all beds occurred as the source water declined to below 10°C (50°F).

Because of the apparent correlation between bacterial densities in activated carbon filter/adsorber effluents and source water temperature, additional bacteriological sampling was done when the source water temperature again rose above 10°C (50°F). Data obtained verified the initial observations that total coliform densities in effluents from activated carbon filter/adsorber beds exceeded the influent densities of less than one organism per 100 ml when temperatures were above 10°C. When the temperature again dropped below 10°C, effluent total coliform densities returned to below detectable levels per 100 ml of sample. High initial total coliform occurrences may also be attributed to the difficulty of disinfecting adsorption beds when putting them into service. These field data suggest occasional coliform penetration past the early stages of treatment and before filtration can occur. Furthermore these organisms *may* become temporarily established in the carbon filter/adsorber and release some coliforms in the process effluent.

Table 2.23 Granular Activated Carbon Study of Beaver River Source Water

		GAC influent			GAC effluent[a]		
Week	Temp. (°C)	Turbidity (NTU)	Total coliform per 100 ml	Standard plate count per ml	Turbidity (NTU)	Total coliform per 100 ml	Standard plate count per ml
1	21	5.6	<1	NR	0.44	64	NR
2	21	4.8	<1	NR	0.36	75	NR
3	15	2.3	<1	NR	0.31	98	1,000
4	11	2.9	<1	100	0.33	45	1,400
5	16	2.5	<1	800	0.36	34	25,000
6	16	3.3	<1	350	0.60	42	2,000
7	16	3.6	<1	10	0.34	28	20,000
8	10	3.2	<1	42	0.36	22	29,000
9	10	4.6	2	110	0.48	13	6,500
10	8	4.5	<1	33	0.34	12	1,600
11	6	3.7	<1	95	0.58	2	960
12	3	5.9	1	360	0.64	1	270
13	4	4.6	<1	660	0.38	<1	480
14	2	6.6	1	200	0.13	1	440
15	1	4.8	<1	120	0.36	<1	44
17	1	5.9	<1	150	0.40	<1	50
18	1	5.5	<1	33	0.52	<1	21
22	1	6.4	<1	30	0.64	<1	30

Note: NR, not run.

[a] Filtrasorb 400®.

From Beaver Falls, PA municipal authority (Ohio River Valley Sanitation Commission;[145] Symons et al.[183]).

The microbial population that develops in activated carbon filter/adsorbers includes both a specialized group of organisms capable of biodegrading organics adsorbed from the source water and a residual population of heterotrophic bacteria in source water that survived passage in early treatment processes. Among the coliform bacteria that are occasionally found in activated carbon filter/adsorbers (*Citrobacter freundii, Enterobacter cloacae, Klebsiella pneumoniae,* and *K. oxytoca*) none of them appear to include *Escherichia* species. All of these coliform species have been found to also colonize pipe sediments and become active participants in biofilm development within the distribution system.

GAC filter/adsorbers receiving ozonated influent water were observed to have an approximate 10- to 100-fold increase in the general bacterial population, compared with control GAC filters not receiving ozonated influent waters. This stimulation of bacterial growth on GAC adsorbers is presumably caused by the oxidative breakdown of some organics by ozone treatment, which results in more usable nutrients for bacterial metabolism. As a consequence, a different profile of dominant organisms emerges in the carbon filter/adsorber-fed environment. The size and composition of this microbial flora will change more as a response to seasonal water temperatures and the organic nature of the raw source water than with system operations. In so doing, the biological activity may lead to slime development in the carbon bed that may interfere with filter operation and provide a periodic release of large clusters of organisms into the process effluent. Biofilm sloughing into the effluent can interfere with effective disinfection contact for all cells in the aggregate thereby providing protective passage through the barrier.

Bacterial population studies of activated carbon particles taken from different sites in the filter suggest that bottom samples or areas near the effluent discharge in the filter bed are often the locations of greatest densities that may approach 10^7 to 10^8 organisms per gram of carbon wet weight.[183,184] However, sites for intense bacterial colonization in GAC adsorbers appear to vary with the adsorber bed age (Table 2.24), bacterial species dominance, influent disinfection, and, perhaps, approach (flow-through) velocity. Furthermore, these populations may pulse widely in densities because they are a reflection of numerous other variables in the adsorber column ecosystem. Bacterial densities in GAC effluent were found to range from 10^4 to 10^5 organisms per milliliter.

While commonality as to the dominant organisms (*Pseudomonas, Moraxella, Acinetobacter, Alcaligenes,* and gram-positive bacilli) was encountered in profile studies of bacteria released from activated carbon adsorbers and sand filters, bacterial survivors of ozonation (*Alcaligenes* spp., 70%) appeared greatly restricted in species diversity.[185] This change in microbial flora composition in turn stimulated a significant increase in the bacterial density of ozonated effluent samples. Among the infrequently occurring strains (i.e., a broad spectrum of bacteria with less than 5% occurrence) were a variety

Table 2.24   Bacterial Counts from Top, Midpoint, and Bottom of an
Activated Carbon Bed and from Its Effluent[a]

| Column age[b] (d) | Organisms per 0.5 g dry wt.[c] | | | Effluent (counts per ml) |
| | GAC section | | | |
	Top	Midpoint	Bottom	
6	—	58,000	55,000	250,000
11	550,000	45,000	28,000	135,000
17	130,000	4,400,000	2,700,000	30,000
20	2,790,000	460,000	320,000	44,000
25	7,700	90,000,000	50,000,000	520,000

[a] R2-A pour plates (35°C incubation for 6 d).
[b] Ambient room temperatures.
[c] Activated carbon particles sonicated for 4 min.

Data from Symons et al.[183]

of pigmented organisms that became established in the adsorbers and found their way into the effluent. Although the significance of these organisms is uncertain, they appear to colonize carbon adsorbers and sand filters, and to be resistant to control through the final water treatment barrier, i.e., disinfection.[185]

Virus adsorption in GAC filters does occur[186,187] and may be as high as 90% initially, then decreasing as operational time increases.[188-192] When carbon filtration is used for organics removal, monitoring for DNA and RNA coliphages as surrogate virus indicators[193] may be desirable to gain baseline information on the magnitude of virus penetration through the carbon filter adsorber into the effluent. No information is currently available on the effectiveness of carbon filtration to stop the passage of protozoan cysts and macroinvertabrates.

Disinfectants

Disinfectants are primarily used in the treatment of drinking water to ensure the destruction of pathogenic organisms that may be present in surface waters or unprotected groundwater sources.[188,192] Many chemical and physical agents have been studied with regard to their potential as drinking water disinfectants. Chlorine, in one form or another, continues to be the dominant primary disinfectant used in water supply treatment. The reasons for this predominance in the past have been related mainly to availability, cost, and ease of handling and measurement. Of the disinfectant options, only ozone, chlorine dioxide, and chloramines are considered to be viable alternatives to free chlorine for use as primary disinfectants.[193] Other potentially useful disinfectant candidates include iodine species, bromine species, permanganate, hydrogen peroxide, ferrate, silver, UV light, ionizing radiation, high pH, and elevated temperatures in desalination. Certain combinations of these oxidants (ozone with UV radiation, hydrogen peroxide with UV radiation, and hydrogen peroxide catalyzed ozonation) being investigated for reduction of organic substances in water may also have beneficial application in disinfection.[194]

Some of these oxidants may also have application in specialized water treatment technology (small water treatment systems, point-of-use devices, bottled water); however, lack of sufficient data on their biocidal activity for various waterborne pathogens, toxicity risk, or excessive costs associated with installation and operation have limited their practicability in public water supply treatment.

Disinfection effectiveness is controlled by the concentration of disinfectant (C) measured in milligrams per liter and contact time (T) determined in minutes to achieve a 6-log reduction of bacteria, 4-log reduction of virus, and 3-log reduction of protozoans. These C·T values are further influenced by other variables that include temperature, pH, organics, disinfectant demand, cell aggregates and disinfection mixing rates. The order of organism resistance to inactivation by disinfection is generally considered to be

coliforms < virus < protozoan cysts

The general order of disinfection efficiency, based on coliform inactivation is

ozone > chlorine dioxide > hypochlorous acid >
hypochlorite ion > chloramines

Based on data obtained from studies with "clean systems" (no extraneous disinfectant-demand materials and use of pure cultures), it has been estimated that 3 to 100 times more chlorine is required to inactivate enteric viruses than is needed to kill coliform bacteria, *Salmonella typhosa,* and *Shigella dysenteriae* when factors such as contact time, water temperature, and water pH are held constant.[195] For example, hypochlorous acid at a concentration of 0.01 mg/l, water pH 7.0, and temperature of 0 to 6°C required the following times to achieve 99% destruction: adenovirus type 3, 15 s; *E. coli,* 90 s; poliovirus type I, 9 min; and Coxsackie virus, type A2, 40 min. The contact times would be extended approximately 100 times for monochloramines used at the same concentration. For this reason, a free chlorine residual of at least 0.3 mg/l, with a 30-min contact time in low-turbidity water (1 NTU) is frequently used. Unfortunately, this C·T value of 9 (0.3 × 30) is not adequate in surface waters contaminated with *Endamoeba histolytica, Giardia lamblia,* or parasitic helminth ova. These pathogens require free chlorine concentrations and contact times that are neither cost-effective nor achievable in disinfection contact chambers normally used.[196-200] In these situations effective coagulation and filtration are the essential control processes for cysts and ova while disinfection is necessary to inactivate any bacterial and viral pathogens that may also be in the same raw source water.

Under field conditions, the raw water being treated does not consist of a "clean system" with only one strain of microorganisms. The menstrum is not free of extraneous materials that might react with disinfectant used. Water quality fluctuations may so change the biocidal capability of a disinfectant that inactivation of organisms is hindered or made completely ineffective. A variety of microorganisms are present in their "natural" state, which may include encapsulation or clumping, suspended in water that contains a variety of other solids, sediments, and dissolved materials, some of which can have pronounced effects on disinfectant activity and effective concentration. Because of these environmental characteristics, it can be assumed that disinfection does not operate as a constant-rate process in the field as it does in laboratory experiments.

For a number of years, scientists have suspected that microorganisms associated with particulate matter could be shielded from disinfectant action. This surmise has had some influence on the establishment of a turbidity limit in drinking water. Now it is known, for instance, that washed cell debris associated with poliovirus offers substantial protection to the associated virus against inactivation by HOCl, compared to freely suspended virus.[142] Coliforms associated with washed primary effluent solids are inactivated by HOCl much more slowly than clean suspensions of laboratory-grown E. coli. Cell debris-associated virus are also protected from inactivation by ozone, the most efficient biocide among the primary disinfectants.[201] Ozone levels in excess of 2 mg/l failed to completely inactivate viruses associated with cell debris in 30 s. In longer-term experiments, cell debris and associated viruses could be detected even after exposure for 75 min to an initial ozone level of 2.5 mg/l. Comparable information for chlorine dioxide and chloramines is not yet available, but it seems probable that similar limitations in inactivating viruses associated with cell debris will be noted.

From these observations on the influence of environmental factors, precise ranking of the three alternative disinfectants (ozone, chlorine dioxide, and chloramines) cannot be made.[192] Different bacteria and viruses vary in their resistance to a given disinfectant under specific conditions of water pH, temperature, disinfectant concentration, and contact time. Changing any one specific condition for disinfection treatment will also produce different rates of inactivation for the same organism. The major features regarding disinfection efficiency are (1) viruses, as a group, are substantially more resistant to disinfectants than are enteric bacteria and (2) these differences for inactivation are especially evident for combined chlorine. The relatively high resistance of enterovirus to combined chlorine has been responsible for the concern about the adequacy of chloramines as primary drinking water disinfectants. Despite this problem, some water treatment operations have had good success over a number of years using chloramine disinfection, probably because contact times often range from 1 to 2 h.

WATER SUPPLY STORAGE AND DISTRIBUTION

Treatment to reduce the microbial risk associated with polluted raw water does have an influence on the composition of organisms found in drinking water. For the most part, those organisms associated with intestinal diseases are either killed by disinfectant exposure or physically removed in filtration; however, many environmental strains survive water treatment and become part of the microbial flora as the water enters storage reservoirs and moves throughout the distribution system.

WATER SUPPLY STORAGE

Storage of treated drinking water is essential because there should be at least a 24 h reserve available to meet the water supply peak demand and to provide a buffer during emergency stoppage of water treatment processes. These structures cannot be excluded as sources of post-treatment contamination, introducing coliform bacteria and possible pathogens into the water supply.

Finished water reservoirs include storage underground, at ground level or in elevated tanks plus various standpipes on the system. Underground reservoirs are generally constructed of concrete while older basins may be of masonry, stone, or native rock caverns. In an effort to prevent surface leakage and groundwater contamination, underground reservoirs are lined with concrete, asphalt, or butyl rubber. Steel, cement, and wood may be used in the construction of reservoirs and standpipes that are elevated above ground level.

Reservoirs of potable water supply should be covered whenever possible to avoid contamination of the supply with bird excrement[41,42] and surface runoff. The microbiological concern with birds, which may frequent finished water reservoirs, is that 5 to 20% of this wildlife population is periodically infected with *Salmonella* and other intestinal pathogens that are also pathogenic to man. For example, sea gulls are scavengers and are often found at landfill areas and waste discharge sites searching for food that is obviously very contaminated with a variety of pathogens. At night, these marine birds return inland to roost on impoundments and open finished water reservoirs, thereby introducing pathogens to the water through their excrement.[202]

Occasional coliform occurrences may be found associated with birds roosting in the vent ports of covered water reservoirs. In one instance, inspection of these venting ports revealed several vents to be unscreened and birds had built their nests in these locations. Fecal droppings around the nest sites were thought to have entered the stored water supply and thereby transported microbial contamination into the distribution system before water dilution and residual disinfection were able to dissipate and inactivate the associated fecal clump of organisms.

As a measure to avoid microbial quality deterioration originating in water supply storage, all reservoir structures should be placed on a regularly scheduled cleaning program to remove sediments and slime development in the water storage compartments. Sediments accumulating can be sites for bacterial colonization and slime development over the inner surfaces, a cause for taste and odor problems in nearby service areas.

STATIC WATER CONDITIONS

Microbial activity in drinking water is a dynamic process that may slow down as a response to temperature declines and nutrient limitations, but it never ceases completely. In static water situations, available nutrients are minimal and continually recycle through the growth and death of various organisms. With nutrients being the limiting factor, cell generation times become extended, individual daughter cells develop into smaller replicates, and population dominance changes with a succession of species that adapts to the spartan environment.

Static water does occur in distribution systems, being a short-term phenomenon in dead-end lines and some service lines or a long-term phenomenon in overdesigned water storage reservoirs. This condition is undesirable at any time because this situation provides opportunity for suspended particulates to settle into pipe sediments, biofilm development to proceed without the shearing action of water hydraulic changes, and biologically mediated corrosion to accelerate. Redesigning the distribution network to create continuous loops from numerous dead-end sections has been helpful in reducing microbial degradation and improving the efficacy of disinfectant residuals in outreach areas. In essence, it is important to keep the water supply moving throughout the distribution system.

Long-term water storage in reservoirs is not desirable from a water quality viewpoint. During these quiescent water periods, lasting weeks or several months, persisting waterborne heterotrophic bacteria begin to grow in sediments, attach to inner walls, and spread a biofilm over the surfaces. Obviously, these organisms enter the water column and cause significant changes in the microbial flora of the water supply.

Some appreciation for this bacterial adjustment in static water can be seen from a study of stored public water supplies prepared for emergency civil defense.[203] In this investigation, water supply taken from the public water system in each city was collected in 17.5-gal (66.2-l) drums with polyethylene liners, disinfected, and sealed for long-term storage at room temperature in basement areas. Careful sampling of these stored water supplies from different cities on a monthly basis revealed a significant decline in the overall quality of water over the first year of storage (Table 2.25). While no total coliforms, fecal coliforms, and *Pseudomonas aeruginosa* were detected, heterotrophic plate counts indicated increased occurrences of pigmented bacteria and fungi with length of water supply storage. Long-term storage of water to meet seasonal peak demand during the summer without efforts made to periodically utilize this water with scheduled replenishment will lead to creation of an inferior water quality accompanied by taste and odor problems. In one instance (Chapter 7), purchased water inadvertently held for several months led to the release of persisting coliforms and increased heterotrophic bacterial densities into the distribution system.

Table 2.25 Changes in Bacterial Quality for Civil Defense
Emergency Stored Water

Density per ml[b]	Occurrence (%)[a] at storage time (months)					
	1	3	4	6	8	12
<1	45.1	39.2	29.2	29.6	36.4	29.4
1–500	29.7	34.2	31.9	27.0	30.2	35.2
501–1,000	3.1	6.3	4.2	2.7	3.2	4.4
1,001–10,000	8.0	15.2	22.2	27.1	15.9	19.2
>10,000	14.1	5.1	12.5	13.6	14.3	11.8

[a] Thirty emergency water supplies examined.
[b] SPC agar, 5-d incubation at 35°C.
Data from Geldreich et al.[203]

CORROSION

The service life of pipe, regardless of composition, is finite. Pipe deterioration is affected not only by composition but also by treatment practices, corrosivity of the water, biological activity, soil reactivity, water hydraulics, and ground movement. As pipe sections age, corrosion accumulation or heavy scale development may become a serious problem that restricts water passage, creates taste and odor complaints, produces frequent line breaks, and provides potential sites for bacterial colonization and biofilm development. Many of these quality changes are brought about by the activity of microorganisms.

There are a wide variety of organisms involved in pipe corrosion: aerobic and facultative anaerobic heterotrophs, autotrophic nitrifiers, denitrifiers, nitrogen fixers, iron-precipitating bacteria, sulfate reducers, and sulfur-oxidizing organisms, among other bacteria in the microbial community. These organisms belong to groups that can transform carbon, nitrogen, sulfur, and iron as part of their microbial activities. A national survey of the extent and nature of water quality problems in distribution systems indicated that taste and odor, red water, and cloudy and black water were the most frequently cited water quality complaints from customers.[204]

Taste and Odor

Taste and odor in water supply result from chemical action on iron, copper, and lead, including a significant contribution from the effects of biologically mediated corrosion. Creation of these disagreeable tastes and odors in water supply is often associated with an environment of locally warmed static water or water subject to slow flows over a long period.[205] Organisms involved directly or indirectly in corrosion may be algae, bacteria (including actinomycetes), and fungi. Green and blue-green algae growing on the surface of submerged concrete have been observed to cause this material to become pitted and friable. The corrosive action of these organisms (which give water a grassy, musty odor) was most pronounced when the content of SO_3 in the mortar was 0.6% or higher.[206]

Among other organisms fungi, actinomycetes, and iron- and sulfate-reducing bacteria are frequently involved in growth on distribution pipes, fittings, and mastic packing materials used in concrete expansion joints. Their metabolic end products may include substances released into the water that are odoriferous or musty tasting. Fungal counts of 10 to 100 per 100 ml and <10 actinomycetes per 100 ml of water are often associated with taste and odor complaints.[205] Unpleasant tastes and odors from iron bacteria (*Gallionella, Crenothrix,* and *Leptothrin*) are produced either directly or indirectly as the dead cells are decomposed either by disinfection residuals or by other microorganisms. In other situations, sulfate-reducing bacteria (*Desulfovibrio*) convert sulfates in pipe sediments and tubercles to hydrogen sulfide, thereby creating a foul-smelling water and a corrosive agent to steel and other metals exposed to water in the distribution network.[133] Finally, under the most severe situations, microbial activity in the distribution system transporting groundwaters with organic residuals can produce an unpleasant methane odor.

Red Water

Red water observed in line flushing and by consumers in areas of slow flow is caused by the microbial activity of iron-oxidizing bacteria. These filamentous bacteria may either form slimes in pipe sediments, hard deposits that tend to restrict water flow, or scums on the water surface in reservoirs. As a result of their induced corrosion of pipe materials, iron bacteria may be a cause of turbidity and discoloration in addition to unpleasant tastes and odors. While clean water may reveal only a few iron bacteria per milliliter, red water conditions are generally found to have these bacteria in excess of 10 million per millimiliter.[207] The occurrence of iron bacteria in the distribution systems for 76 municipal wells in Wisconsin revealed *Gallionella* more frequently than *Leptothrix* and in higher densities. Generally, these densities were under 100 organisms per milliliter, but in two samples taken from a fire hydrant and an infrequently used tap, the densities were as much as 10 million organisms per milliliter. In these instances, the high density of iron bacteria created a red sediment that, upon settling, comprised between one quarter and one third of the total "water" volume, consisting of iron particles and iron bacteria.

Black (Dirty) Water

Black (dirty) water conditions are caused by other microorganisms colonizing the biofilm in corrosion sites. The source of the dissolved manganese to be oxidized within the distribution system may be salts in the rock formations that define the aquifer, leaching from cement catchment areas and sediments in stratified lakes or impoundments.[208] While various water treatment measures (softening, sequestering, aeration, permanganate addition, and filtration) can effectively minimize dissolved manganese in finished water, there will be occasional inadvertent passage out into the distribution system at concentrations above 0.3 mg/l. These inorganic complexes become adsorbed in pipe sediments, tubercles, and biofilm at corrosion sites where a variety of bacteria (*Pseudomonas, Corynebacterium, Pedomicrobium,* and aerobic *Actinomycetes*) and various fungi imperfects (*Mycelia sterilia*) begin the oxidizing process.[209] With changes in water velocities, the extensive biofilm growth that has incorporated dissolved manganese sloughs and, upon further oxidation by available chlorine residuals or by laundry detergent bleach products, causes a dirty water problem to suddenly emerge in the home.

BUILDING PLUMBING NETWORKS

Building plumbing systems can be characterized as a complex network of small-diameter pipe with numerous dead ends and attachment devices. In this situation, water on demand is irregular, creating a variety of flow patterns with many opportunities for stagnant water conditions to develop in continual warm air surroundings.

Under this situation water quality deterioration from chemical and microbial action will occur over time. Deterioration will be accelerated by numerous factors including warm-water temperatures, low-pH water, high mineral concentrations, depletion in dissolved oxygen, pipe material composition, and faulty system design. Failure to recognize the need for system flushing or application of water softening in special cases will exacerbate the problem and in time lead to system failure.

ENVIRONMENTAL IMPACT ON WATER QUALITY

Reduction in user satisfaction with the water supply is often a result of the development of unpleasant tastes, color, odor, and sediment. Other complaints include reduction in water flow rate, low water pressure, leaks from changes in water hydraulics, and interrupted service resulting from pipeline freezing. While these undesirable characteristics are easily detected by building occupants, the unseen health hazards of pathogen occurrences or toxic chemicals introduced through cross-connections and back-siphonage of wastewater are generally only recognized after there has been a waterborne outbreak.

Water supply lines located in areas of vacant offices and closed sections of hospitals are prime sites for static water accumulations. In this environment, some bacteria in the microflora become established and proceed to produce biofilms or create metabolic by-products responsible for taste and odors. From these sites colonization may spread to more active areas of water supply lines in the building unless a regular scheduled flushing program is performed to move pockets of stagnant water out of the system. Flushing of lines for about 5 min should be scheduled every 6 to 8 weeks in all inactive building areas and a flush of the entire water system should be performed at least once per year.

As part of the effort to prevent further deterioration in building water supply, storage tanks should be drained annually so that sediment and biofilm may be removed by cleaning. It has been estimated that Rio de Janeiro (Brazil) has approximately 10,000 highrise buildings with rooftop water storage tanks and the quality of water in the structures is a source of repeated complaints expressed by the residents. In these situations building management should be required by law or ordinance to flush the storage tank each year, followed by a sample collection to be examined by the local health department or certified private laboratory for coliform and HPC analysis. The annual record of storage tank flushing and results of the follow-up bacteriological analysis should be forwarded to the local health agency for review and approval. A fee charged for an annual permit would help the agency budget the manpower necessary to review permits, make sanitary surveys within the building, and provide guidance in resolving contamination events that could have impact on the quality of the public water supply.

ATTACHMENT DEVICES

Attachment devices connected to plumbing networks may introduce deterioration in water quality because of device design or poor maintenance. In both situations there can be an amplification of bacteria introduced in low densities from the public water supply followed by dissemination in the water use application. For this reason, there must be isolation barriers (backflow preventers) established to avoid microbial contamination entering the water supply to all users in the building.

HOT-WATER STORAGE TANKS

Amplification of *Legionella* in attachment devices, particularly hot-water storage tanks is an important consideration in the formation of the microbial flora of water supply to which the consumer is exposed through inhalation of water vapor. *Legionella* is an opportunistic pathogen that is ubiquitous in the aquatic environment and passes through conventional water treatment processes with little reduction in densities that may be only a few organisms per liter. The mere presence of these few *Legionella* in distribution water does not necessarily represent a health threat; rather, it is the threat of amplification of this opportunistic pathogen in some environmental niche associated with building plumbing or attachment devices that is the public health concern.[210]

Upon entry into building plumbing systems this status may change if colonization occurs on pipe joint packing materials, valve stem seals, vacuum breakers (used in backflow prevention), and faucet aerators.[211,212] Static warm-water zones in building plumbing networks are also areas where *Legionella* may multiply. A major amplifier of this organism is the hot-water supply tank. Hot-water storage tanks, particularly those units that are operated at a temperature below 55°C, are the amplifiers of *Legionella* growing in the supply sediments and thereby become the greatest dissimulator of *Legionella*.[213] The route of exposure risk is not from ingesting water but in the inhalation of waterborne *Legionella* in vapor droplets generated during shower baths or during use of whirlpools, hot tubs, or exposure to air coolers and room humidifiers.[214] Careful cleaning of these devices to remove scums and wall deposits will minimize risk of *Legionella* and fungal spore dissemination in vapor droplets.[215]

Control measures for legionellae occurrences in water supply should focus on building plumbing networks and attachment devices because these are the locations where colonization, amplification, and dissemination are most severe. It must be recognized that water treatment processes provide an ecological niche that also needs to be minimized. Efforts to eliminate the few organisms per liter in process waters are not as cost-effective as control through plumbing networks and their attachment devices. Careful redesign of pipe networks to avoid static water conditions, adequate flushing of all lines on a scheduled basis, and application of corrosion control measures are as important as are the removal of all sediment accumulations in hot-water storage tanks, showerheads, and faucet aerators. When possible, elevate the temperature in hot-water tanks to 55°C and use a periodic heat-shock treatment (70°C for 1 h) to eradicate legionellae that may be persisting in the protective nature of sediment accumulations. Local plumbing codes, manufacturer modifications in attachment devices, and consumer awareness should be the main thrust of providing adequate protection of water supply beyond the service connection to the water main.

POINT-OF-USE TREATMENT

On-site treatment of water supply in buildings involves devices that have been specifically designed to soften water, remove organic residuals, or serve as a microbial barrier to various pathogens (bacterial, viral, and protozoan). A growing number of special users (hospitals, dialysis clinics, food and pharmaceutical industries, and laboratories) and some of the general public, who are dissatisfied with the taste, odor, and potential health risks associated with their municipal water supply, are attempting to further refine the water supply at the tap. While these treatment devices may be very effective initially for some specific contaminant, their usefulness over time may become limited. For this reason, carbon filter cartridges need to be replaced, water softeners recharged, and microbial filter barriers backwashed at intervals determined by service life measured in liters of water procured and quality of water being treated.

The microbial quality of the product water from these devices is extremely variable.[216-219] Bacterial densities in water from POU carbon filters can be expected to increase by one to two logs over the number detected in public water supplies at the building tap (Figure 2.2). This proliferation relates to the species of organisms passing through the filter devices, seasonal changes in water temperatures, the presence or absence of a free chlorine residual, ambient air temperatures around the device, and the service duration for a given carbon cartridge (Figure 2.3). Nonuse periods overnight or for longer intervals will also provide an opportunity for continued growth of organisms colonizing carbon filters. Under no-flow conditions, water temperatures can increase to ambient temperatures quickly and there is no chlorine residual brought to the carbon sites that might suppress bacterial growth. Examinations of the bacterial quality changes that might result from accumulated overnight growth in the device revealed that product water in these test units frequently had higher bacterial densities in morning samples (Table 2.26) than in the untreated tap water source.

Another aspect to the colonization concerns with point-of-use devices is the nature of the organisms that either pass through the device or are amplified in these units with release at higher density levels than in the building tap water. Challenge of carbon filter devices with coliforms, opportunistic pathogens, and primary pathogens (bacteria found in cross-connections, line breaks, or back-siphonage) revealed that such treatment units do not provide an effective barrier. Such organisms as *K. pneumoniae, Ps. aeruginosa, Serratia marcesens, A. hydrophila,* and *L. pneumophila* can colonize these devices. As a consequence, carbon filters without an associated microbial barrier to protect product water quality should not be used on untreated water supplies of questionable quality.[216-219] Along with these concerns is the problem of GAC bed materials unloading absorbed organic compounds into the product water at higher concentrations as selective adsorption shifts from one group of organic compounds to another group or as the total adsorption capacity is reached for the mass of GAC material available.

Product engineering of POU treatment devices must take into consideration that certain design configurations may result in microbial problems. For example, cellulosic paper filters must be avoided because the fibers are quickly degraded by bacteria. Filter units should also be designed to avoid sumps that could be the source of perpetual contamination. Insulation of POU devices installed below the kitchen sink, where ambient air temperatures are often elevated above supply water, should be insulated to minimize water temperature changes. Each unit should have a built-in indicator that signals the need for treatment cartridge replacement. The warning indicator could be related to increased pressure from filter clogging or total volume of water processed. Backflow preventers should be considered essential on those building plumbing networks that have chronic problems with low line pressure, because effluents from old POU treatment units (carbon filters, deionizing columns,

Table 2.26 Daily Fluctuations in Heterotrophic Bacterial Densities (per ml) from Five Point-of-Use Home Treatment Devices

	Nonchlorinated source				Chlorinated tap
	Acrylic fiber-wound carbon	Carbon-impregnated paper	Carbon filter no. 1	Carbon filter no. 2	Carbon filter no. 3 (control)
Winter[a]					
A.M.	84–480	2,600–6,700	1,200–2,300	3,000–23,000	210–550
P.M.	106–220	4,200–7,300	680–860	4,000–19,000	89–110
Summer[b]					
A.M.	5,400–11,000	180,000–220,000	4,600–10,000	7,400–7,500	49–17,000
P.M.	5,000–7,300	150,000–530,000	2,300–5,300	7,000–15,000	150–9,800

[a] Test period, January 1981 to May 1981; average water temperature, 10.5°C; average, 26 samples.
[b] Test period, June 1981 to September 1981; average water temperature, 23.8°C; average, 25 samples.

Data revised from Geldreich and Reasoner.[219]

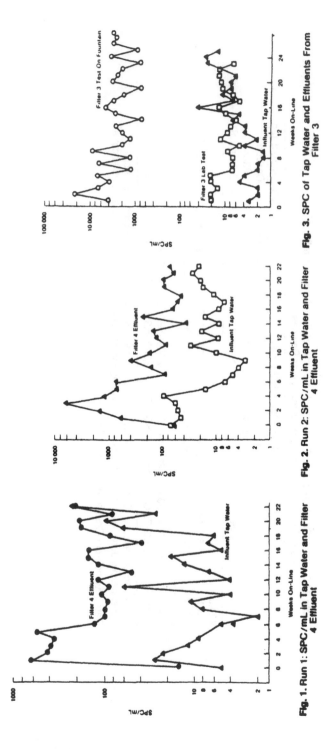

Figure 2.2 Influent and effluent SPC per milliliter for tap water and GAC filter units. (Data from Taylor, Allen, and Geldreich.[216])

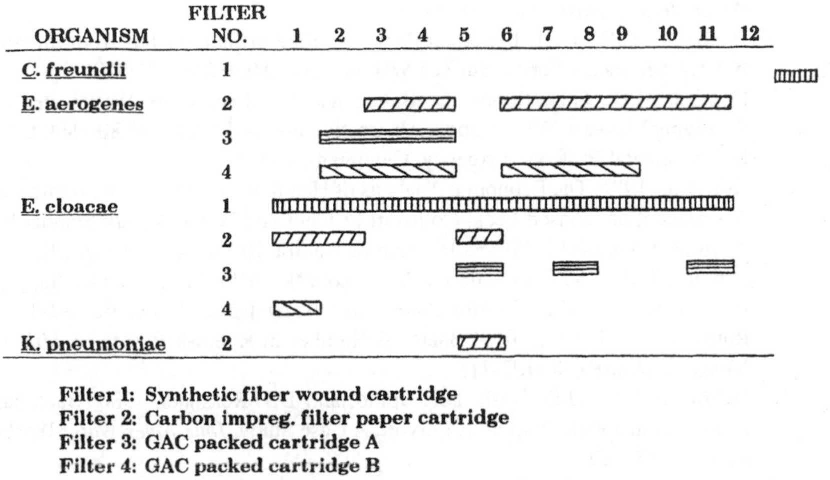

ORGANISM	FILTER NO.	1	2	3	4	5	6	7	8	9	10	11	12
C. freundii	1											▥	
E. aerogenes	2				▨▨			▨▨▨▨▨▨					
	3			▭▭▭									
	4		◪◪◪◪				◪◪◪◪						
E. cloacae	1	▥▥▥▥▥▥▥▥▥▥▥▥▥▥▥▥▥▥▥▥▥▥											
	2	▨▨▨			▨								
	3				▭		▭			▭			
	4	◪											
K. pneumoniae	2				▨								

Filter 1: **Synthetic fiber wound cartridge**
Filter 2: **Carbon impreg. filter paper cartridge**
Filter 3: **GAC packed cartridge A**
Filter 4: **GAC packed cartridge B**

Figure 2.3 Coliforms isolated from unchallenged point-of-use unit product water over a 12-month period. (Data from Geldreich and Reasoner.[219])

etc.) could discharge a variety of microbial contaminants back into the service lines for other users in the building. In the final analysis, building plumbing networks and devices attached to these lines should not be ignored as potential contributors to water quality deterioration in public water supply.

REFERENCES

1. Office of Drinking Water. 1978. Guidance for Planning the Location of Water Supply Intakes Downstream from Municipal Wastewater Treatment Facilities. Contract No. 68-01-4473. Office of Water Supply, U.S. Environmental Protection Agency, Washington, D.C.
2. AWWA Organisms in Water Committee. 1987. Committee Report: Microbiological Considerations for Drinking Water Regulation Revisions. *Jour. Amer. Water Works Assoc.,* 79:81-84,88.
3. Geldreich, E.E. 1990. Microbiological Quality of Source Waters for Water Supply. In: *Drinking Water Microbiology,* G.A. McFeters, Ed., Springer-Verlag, New York.
4. Smith, R.J. and R.M. Twedt. 1971. Natural Relationships of Indicator and Pathogenic Bacteria in Stream Waters. *Jour. Water Poll. Contr. Fed.,* 43:2200-2209.
5. U.S. Environmental Protection Agency, Office of Water Quality. 1971. Report on Missouri River Water Quality Studies. Regional Office, Kansas City, MO.
6. Olivieri, V.P., C.W. Kruse, and K. Kawata. 1977. Microorganisms in Urban Stormwater. Environmental Protection Technology Series, EPA- 600/2-77-087, MERL, U.S. Environmental Protection Agency, Cincinnati, OH.

7. Geldreich, E.E. 1986. Control of Microorganisms of Public Health Concern in Water. *Jour. Environ. Sci.,* 29:34-37.

8. Hendricks, C.W. 1971. Increased Recovery Rate of Salmonellae from Stream Bottom Sediments Versus Surface Waters. *Appl. Microbiol.,* 21:379-380.

9. Swayne, M.D., G.H. Boone, D. Bauer, and J.S. Lee. 1980. Wastewater in Receiving Waters at Water Supply Abstraction Points. EPA-600/2-80- 044, U.S. Environmental Protection Agency, Cincinnati, OH.

10. Huff, L.L. 1981. The Economic Analysis of Health Risk and the Environmental Assessment of Revised Fecal Coliform Effluent and Water Quality Standards, Document No. 81/15, Illinois Institute of Natural Resources, Chicago, IL.

11. Symons, J.M., J.K. Carswell, and G.G. Robeck. 1970. Mixing Water Supply Reservoirs for Quality Control. *Jour. Amer. Water Works Assoc.,* 62:322-334.

12. Romaninko, V.I. 1971. Total Bacterial Number in Rybinsk Reservoir. *Mikrobiologiya (USSR),* 40:707-713.

13. Geldreich, E.E., H.D. Nash, D.F. Spino, and D.J. Reasoner. 1980. Bacterial Dynamics in a Water Supply Reservoir: A Case Study. *Jour. Amer. Water Works Assoc.,* 72:31-40.

14. Dzyuban, A.N. 1975. The Number and Generation Time of Bacteria and Production of Bacterial Biomass in Water of the Saratov Reservoir. *Gidrobiol. Zh.,* 11:14-19.

15. Geldreich, E.E., J.A. Goodrich, and R.M. Clark. 1988. Characterizing Raw Surface Water Amenable to Minimal Water Supply Treatment. Proc. Amer. Water Works Assoc. Confr., June 19-23, 1988, Orlando, FL.

16. Carmichael, W.W. 1981. *The Water Environment.* Plenum Press, New York.

17. Bernhardt, H. 1981. General Impacts of Eutrophication on Public Water Preparation. In: Restoration of Lakes and Inland Waters, U.S. Environmental Protection Agency, EPA 440/5-81-010. Office of Water Regulations and Standards, Washington, D.C.

18. Niewolak, S. 1974. Distribution of Microorganisms in the Waters of the Kortowskie Lake. *Pol. Arch. Hydrobiol.,* 21:315-333.

19. Drury, D.D. and R.A. Gearheart. 1975. Bacterial Population Dynamics and Dissolved Oxygen Minimum. *Jour. Amer. Water Works,* 67:154-158.

20. Weiss, C.M. and R.T. Oglesby. 1960. Limnology and Water Quality of Raw Water in Impoundments. *Public Works,* 91:97-101.

21. Collins, V.G. 1963. The Distribution and Ecology of Bacteria. *Proc. Soc. Water Treat. Exam. (England),* 12:40-73.

22. Niewolak, S. 1987. Bacteriological Water Quality of an Artificially Destratified Lake. *Roczniki Nouk Rolniczyck.,* 101:115-154.

23. Niewolak, S. 1987. Microbiological Study of an Artificially Destratified Lake. *Roczniki Nouk Rolniczyck.,* 101:155-172.

24. Committee on the Challenges of Modern Society (NATO/CCMS) 1984. Drinking Water Microbiology, NATO/CCMS Drinking Water Pilot Project Series, EPA 570/9-84-006, U.S. Environmental Protection Agency, Office of Drinking Water, Washington, D.C.

25. Gerba, C.P. and G. Bitton. 1984. Microbial Pollutants, Their Survival and Transport Pattern to Ground Water. In: *Groundwater Pollution Microbiology,* G. Bitton and C.P. Gerba, Ed., John Wiley and Sons, New York. pp. 65-88.

26. Gerba, C.P. and J.B. Rose 1990. Viruses in Source and Drinking Water. In: *Drinking Water Microbiology,* G.A. McFeters, Ed., Springer-Verlag, New York. pp. 380-396.

27. Geldreich, E.E. 1972. Waterborne Pathogens. In: *Water Pollution Microbiology,* R. Mitchell, Ed., John Wiley & Sons, Inc., New York. pp. 207-241.

28. Burke, J.A. 1977. The clinical and laboratory diagnosis of giardiasis. *CRC Crit. Rev. Clin. Lab. Sci.,* 7:373-391.

29. Davies, R.B. and C.P. Hibler. 1979. Animal Reservoirs and Cross Species Transmission of *Giardia.* In: Waterborne Transmission of Giardiasis, W. Jakubowski and J.C. Hoff, Eds., Environmental Research Center, EPA-600/9-79-001, U.S. Environmental Protection Agency, Cincinnati, OH, pp. 104-126.

30. Fayer, R. and Ungar, B.L.P. 1986. *Cryptosporidium* and *Cryptosporidiosis.* Microbiol. Rev., 50:458-483.

31. Grabow, W.O.K. and E.M. Nupen. 1972. The Load of Infectious Microorganisms in the Waste Water of Two South African Hospitals. *Water Res.,* 6:1557-1563.

32. Thornsberry, C., A. Balows, J.C. Feeley, and W. Jakubowski. 1984. Legionella, *Proceedings of the 2nd International Symposium,* Amer. Soc. Microbiol., Washington, D.C. p. 369.

33. Chang, S.L. 1970. 10th International Congress of Microbiologists, Mexico City, Mexico.

34. Harvey, R.W.S. et al. 1969. Salmonellas in Sewage. A Study in Latent Human Infections. *Jour. Hyg. (Brit.),* 67:517-523.

35. Callaghan, P. and Brodie, J. 1969. Laboratory Investigation of Sewer Swabs following the Aberdeen Typhoid Outbreak of 1964. *Jour. Hyg.,* 66:489-497.

36. Popp, L. 1974. *Salmonella* and Natural Purification of Polluted Waters. *Zentralbl. Bakteriol. Hyg., Abt. I,* 158:432-445.

37. Claudon, D.G., D.I. Thompson, E. H. Christenson, G.W. Lawton, et al. 1971. Prolonged *Salmonella* Contamination of a Recreational Lake by Runoff Waters. *Appl. Microbiol.,* 21:875-877.

38. U.S. Department of Health, Education, and Welfare. 1965. Report on Pollution of Interstate Waters of the Red River of the North (Minnesota, North Dakota). Public Health Service, Field Investigation Branch, Cincinnati, OH.

39. Kampelmacher, E. H. and L.M. van Noorle Jansen. 1976. *Salmonella* Effluent from Sewage Treatment Plants, Wastepipes of Butcher's Shops and Surface Water in Walcherne. *Zentralbl. Bakteriol. Hyg., I. Abt. Orig. B.,* 159:307-319.

40. Herman, E. 1972. Experiences with Coliform and Enteric Organisms Isolated from Industrial Wastes. In: U.S. Environmental Protection Agency Seminar: The Significance of Fecal Coliforms in Industrial Wastes, Denver Field Investigations Center, U.S. Environmental Protection Agency, Denver, CO. pp. 26-40.

41. Alter, A.J. 1954. Appearance of Intestinal Wastes in Surface Water Supplies at Ketchikan, Alaska. Proc. 5th Alaska Sci. Confr. AAAS, Anchorage, AK.

42. Anon. 1954. Ketchikan Laboratory Studies Disclose Gulls Are Implicated in Disease Spread. *Alaska's Health,* 11:1.

43. Vogt, R.L., H.E. Sours, T. Barrett, et al. 1982. *Campylobacter Enteritis* Associated with Contaminated Water. *Ann. Intern. Med.,* 96:292-296.

44. Health and Welfare, Canada. 1981. Possible Waterborne *Campylobacter* Outbreak — British Columbia. *Can. Dis. Weekly Rpt.*, 7:223, 226-227.
45. Taylor, D.N., D.T. McDermott, J.R. Little, et al. 1983. *Campylobacter Enteritis* Associated with Drinking Untreated Water in Back Country Areas of the Rocky Mountains. *Ann. Intern. Med.*, 99:38-40.
46. Greensberg, A.E. and J. Jongerth. 1966. Salmonellosis in Riverside, California. *Jour. Amer. Water Works Assoc.*, 58:1145-1150.
47. Schroeder, S.A., J.R. Caldwell, T.M. Vernon, P.C. White, et al. 1968. A Waterborne Outbreak of Gastroenteritis in Adults Associated with *Escherichia coli*. *Lancet*, 6:737-740.
48. Lassen, J. 1972. *Yersinia entercolitica* in Drinking Water. *Scand. Jour. Infect. Dis.*, 4:125-127.
49. Evison, L.M. and A. James. 1973. A Comparison of the Distribution of Intestinal Bacteria in British and East African Water Sources. *Jour. Appl. Bacteriol.*, 36:109-118.
50. Lindel, S.S. and P. Quinn. 1973. *Shigella sonnei* Isolated from Well Water. *Jour. Bacteriol.*, 26:424-524.
51. Centers for Disease Control. 1973. Typhoid Fever — Florida. *Morbidity and Mortality Weekly Report*, 22:77-78 and 85.
52. Woodward, W.E., N. Hirschhorn, R.B. Sack, R.A. Cash, et al. 1974. Acute Diarrhea on an Apache Indian Reservation. *Amer. Jour. Epidemiol.*, 99:281-290.
53. Centers for Disease Control. 1974. Acute Gastrointestinal Illness — Florida. *Morbidity and Mortality Weekly Report*, 23:134.
54. Dragas, A-Z. and M. Tradnik. 1975. Is the Examination of Drinkable Water and Swimming Pools on Presence of Entero-pathogenic *E. coli* Necessary? *Zentralbl. Bakteriol. Hyg., I. Abt. Orig. B*, 160:60-64.
55. Highsmith, A.K., J.D. Feeley, P. Shalig, J.G. Wells, et al. 1977. Isolation of *Yersinia enterocolitica* from Well Water and Growth in Distilled Water. *Appl. Environ. Microbiol.*, 34:745-750.
56. Schiemann, D.A. 1978. Isolation of *Yersinia enterocolitica* from Surface and Well Waters in Ontario. *Can. Jour. Microbiol.*, 24:1048-1052.
57. Centers for Disease Control. 1980. Waterborne Disease Outbreaks in the United States-1978. *Morbidity and Mortality Weekly Report*, 29:46-48.
58. Mentzing, L.O. 1981. Waterborne Outbreak of *Campylobacter enteritis* in Central Sweden. *Lancet*, 2:(8242) 352-354.
59. Sekla, L., W. Stackeiv, C. Kay, and L. Van Buckentrout. 1980. Enteric Viruses in Renovated Water in Manitoba. *Can. Jour. Microbiol.*, 26:518-523.
60. Hejkal, T.W., B. Keswick, R.L. LaBelle, C.P. Gerba, et al. 1982. Viruses in a Community Water Supply Associated with an Outbreak of Gastro-enteritis and Infectious Hepatitis. *Jour. Amer. Water Works Assoc.*, 74:318-321.
61. Keswick, B.H., C.P. Gerba, H.L. DuPont, and J.B. Rose. 1984. Detection of Enteric Viruses in Treated Drinking Water. *Appl. Environ. Microbiol.*, 47:1290-1294.
62. Schwartzbrod, L., C. Finance, M. Aymard, M. Brigand, and F. Lucena. 1985. Recovery of Reoviruses from Tap Water. *Zbl. Bakt. Hyg., I. Abt. Orig. B*, 181:383-389.
63. Payment, P. Isolation of Viruses from Drinking Water at the Point-Viau Water Treatment Plant. 1981. *Can. Jour. Microbiol.*, 27:417-420.

64. Vander Velde, T. L. and W.M. Mack. 1973. Poliovirus in Water Supply. *Jour. Amer. Water Works Assoc.*, 65:345-348.

65. O'Connor, J.T., L. Hemphill, and C.D. Reach, Jr. 1982. Removal of Virus from Public Water Supplies. EPA-600/52-82-024, U.S. Environmental Protection Agency, Cincinnati, OH.

66. Mackenthun, K.M. and W.M. Ingram. 1967. Biological Associated Problems in Freshwater Environments. 287 pp. Federal Water Pollution Control Administration, U.S. Department of the Interior, Washington, D.C.

67. Hibler, C.P., K. MacLeod, and D.O. Lyman. 1975. Giardiasis — in Residents of Rome, NY and in J.S. Travelers to the Soviet Union. *Morbidity and Mortality Weekly Report*, 24:366 and 371.

68. Allard, J. et al. 1977. Waterborne Giardiasis Outbreaks — Washington, New Hampshire. *Morbidity and Mortality Weekly Report*, 26:169-170 and 175.

69. Kirner, J.C., J.D. Littler, and L. Angelo. 1978. A Waterborne Outbreak of Giardiasis in Camas, Washington. *Jour. Amer. Water Works Assoc.*, 70:35-40.

70. Pluntze, J.C. 1984. The Need for Filtration of Surface Water Supplies: Viewpoint. *Jour. Amer. Water Works Assoc.*, 76:11 and 84.

71. U.S. Environmental Protection Agency. 1979. Waterborne Transmission of Giardiasis. Jakubowski, W. and J.C. Hoff, Eds., EPA-600/9-79/001, Office of Research and Development, Cincinnati, OH. p. 306.

72. Jakubowski, W. et al. 1984. Waterborne *Giardia*: It's Enough to Make You Sick, Roundtable. *Jour. Amer. Water Works Assoc.*, 77:14, 19, 22, 26, and 84.

73. D'Antonio, R.G., R.E. Winn, J.P. Taylor, T.L. Gustafson, et al. 1985. A Waterborne Outbreak of Cryptosporidiosis in Normal Hosts. *Ann. Intern. Med.*, 103:886-888.

74. Mason, L. 1987. Experience with *Cryptosporidium* at Carrollton, Georgia. Amer. Water Works Assoc., Proceedings of Water Quality Technol. Confr. 889-898, Baltimore, MD.

75. Silverman, G.P. 1988. *Cryptosporidium* — The Industry's New Superbug. *Opflow*, 14:4-5.

76. Galbraith, N.S., N.J. Barrett, and R. Stanwell-Smith. 1987. Water and Disease after Croydon: A Review of Waterborne and Water Associated Disease in the UK 1937-86. *Jour. Inst. Water Environ. Manage.*, 1:7-21.

77. Taylor, J.P., J.N. Perdue, D. Dingley, M. Gustafson, et al. 1985. Cryptosporidiosis Outbreak in a Day-Care Center. *Amer. Jour. Dis. Child.*, 139:1023-1025.

78. Weikel, C.S., L.I. Johnston, M.A. DeSousa, and R.L. Guerrant. 1985. Cryptosporidiosis in Northern Brazil: Associated with Sporadic Diarrhea. *Jour. Infect. Dis.*, 151:963-965.

79. Alpert, G., L.M. Bell, C.E. Kirkpatrick, L. D. Budnick, et al. 1986. Outbreak of Cryptosporidiosis in a Day-Care Center. *Pediatrics*, 77:152-157.

80. Bogaerts, J., P. Lepage, D. Rouvroy, and J. Vandepitte. 1984. *Cryptosporidium* spp., a Frequent Cause of Diarrhea in Central Africa. *Jour. Clin. Microbiol.*, 20:874-876.

81. Centers for Disease Control. 1984. Cryptosporidiosis Among Children Attending Day-Care Centers — Georgia, Pennsylvania, Michigan, California, and New Mexico. *Morbidity and Mortality Weekly Report*, 33:599-601.

82. Perez-Schael, I., Y. Boher, L. Mata, M. Perez, and F.J. Tapia. 1985. Cryptosporidiosis in Venezuelan Children with Acute Diarrhea. *Amer. Jour. Trop. Med. Hyg.,* 34:721-722.

83. Current, W.L., N.C. Reese, J.V. Ernst, W.B. Bailey, et al. 1983. Human Cryptosporidiosis in Immunocompetent and Immunodeficient Persons. *New Engl. Jour. Med.,* 308:1252-1257.

84. Jokipii, L., S. Pohjola, and A.M.M. Jokippi. 1985. Cryptosporidiosis Associated with Traveling and Giardiasis. *Gastroenterology,* 89:838-842.

85. Ma, P., D.L. Kaufman, C.G. Helmick, A.J. D'Souza, and T.R. Navin. 1985. Cryptosporidiosis in Tourists Returning from the Caribbean. *New Engl. Jour. Med.,* 312:647-648.

86. Mata, L., H. Bolanos, D. Pizarro, and M. Vives. 1984. Cryptosporidiosis in Children from Some Highland Costa Rican Rural and Urban Areas. *Amer. Jour. Trop. Med. Hyg.,* 33:24-29.

87. Rose, J.B., A. Cifrino, M.S. Madore, C.P. Gerba, et al. 1986. Detection of *Cryptosporidium* from Wastewater and Freshwater Environments. *Water Sci. Tech.,* 18:233-239.

88. Angus, K.W. 1983. Cryptosporidiosis in Man, Domestic Animals and Birds: A Review. *Jour. Roy. Soc. Med.,* 76:62-69.

89. Anderson, B.C. 1978. Patterns of Shedding of Cryptosporidial Oocysts in Idaho Calves. *Jour. Amer. Vet. Med. Assoc.,* 178:982-984.

90. Dhillon, A.S., H.L. Thacker, V. Dietzel, and R.W. Winterfield. 1981. Respiratory Cryptosporidiosis in Broiler Chickens. *Avian Dis.,* 25:747-751.

91. Hoerr, F.J., F.M. Ranck, Jr., and T.F. Hastings. 1978. Respiratory Cryptosporidiosis in Turkeys. *Jour. Amer. Vet. Med. Assoc.,* 173:1591-1593.

92. Tzipori, S. 1983. Cryptosporidiosis in Animals and Humans. *Microbiol. Rev.,* 47:84-96.

93. Tzipori, S. and I. Campbell. 1981. Prevalence of *Cryptosporidium* Antibodies in 10 Animal Species. *Jour. Clin. Microbiol.,* 14:455-456.

94. Geldreich, E.E. 1978. Bacterial Populations and Indicator Concepts in Feces, Sewage, Stormwater and Solid Wastes. In: *Indicators of Viruses in Water and Food,* G. Berg, Ed., Ann Arbor Science, Ann Arbor, MI.

95. Heukelekian, H. 1951. Disinfection of Sewage with Chlorine. IV. Aftergrowth of Coliform Organisms in Streams Receiving Chlorinated Sewage. *Sewage Ind. Wastes,* 23:273-277.

96. Kabler, P.W. 1960. Selection and Adaptation of Microorganisms in Waste Treatment. *Amer. Jour. Pub. Health,* 50:215-219.

97. Burns, R.J. and R.D. Vaughan. 1966. Bacteriological Comparison between Combined and Separate Sewer Discharges in Southeastern Michigan. *Jour. Water Poll. Contr. Fed.,* 38:400-409.

98. Geldreich, E.E., L.C. Best, B.A. Kenner, and D.J. Van Donsel. 1968. The Bacteriological Aspects of Stormwater Pollution. *Jour. Water Poll. Contr. Fed.,* 40:1861-1872.

99. Feldman, B.M. 1974. The Problem of Urban Dogs. *Science,* 185:903.

100. Fair, J.F. and J.F. Morrison. 1967. Recovery of Bacterial Pathogens from High Quality Surface Water. *Water Resources Res.,* 3:799-803.

101. Feachem, R. 1974. Faecal Coliforms and Faecal Streptococci in Streams in the New Guinea Highlands. *Water Res.,* 8:367-374.

102. Glantz, P.J. 1973. *Escherichia coli* Serogroups Isolated from Streams in Pennsylvania, 1665 to 1972. *Appl. Microbiol.*, 26:741-743.

103. Goodrich, T.D., D.G. Stuart, and W.G. Walter. 1973. Enteric Bacteria Found in Elk Droppings in a Municipal Watershed in Montana. *Jour. Environ. Health*, 35:374-377.

104. Stuart, D.G., G.K. Bissonnette, T.D. Goodrich, and W.G. Walter. 1971. Effects of Multiple Use on Water Quality of High Mountain Watersheds: Bacteriological Investigations of Mountain Streams. *Appl. Microbiol.*, 22:1048-1054.

105. Weidner, R.B., A.G. Christianson, S.R. Weibel, and G.G. Robeck. 1969. Rural Runoff as a Factor in Stream Pollution. *Jour. Water Poll. Contr. Fed.*, 41:377-384.

106. Geldreich, E.E., C.B. Huff, R.H. Bordner, P.W. Kabler, and H.F. Clark. 1962. The Faecal Coli-Aerogenes Flora of Soils from Various Geographical Areas. *Jour. Appl. Bacteriol.*, 25:87-93.

107. Brasfield, H. 1972. Environmental Factors Correlated with Size of Bacterial Populations in a Polluted Stream. *Appl. Microbiol.*, 24:349-352.

108. Brown, T.J. 1969. The Significance of Agricultural Pollution in New Zealand. *Publ. Health*, 84:21 .

109. Evans, M.R. and J.D. Owens. 1972. Factors Affecting the Concentration of Faecal Bacteria in Land-Drainage Water. *Jour. Gen. Microbiol.*, 71: 477-485.

110. Kunkle, S.H. January, 1970. Concentrations and Cycles of Bacterial Indicators in Farm Surface Runoff. Presented at Cornell Agricultural Waste Management Conference, The Relationship of Agriculture to Soil and Water Pollution.

111. Kunkle, S.H. August, 1970. Sources and Transport of Bacterial Indicators in Rural Streams. Presented at Symp. on Interdisciplinary Aspects of Watershed Management, Montana State Univ., Bozeman, MT.

112. McFeters, G.A. and D.G. Stuart. 1972. Survival of Coliform Bacteria in Natural Waters: Field and Laboratory Studies with Membrane-Filter Chambers. *Appl. Microbiol.*, 24:805-811.

113. Van Donsel, D.J., E.E. Geldreich, and N.A. Clarke. 1967. Seasonal Variations of Indicator Bacteria in Soil and Their Contribution to Stormwater Pollution. *Appl. Microbiol.*, 15:1362-1370.

114. Ellis, J.R. and T.M. McCalla. 1978. Fate of Pathogens in Soil Receiving Animal Wastes — A Review. *Trans. ASAE*, 21:309-313.

115. Blannon, J.C. and M.L. Peterson. April 12, 1974. Survival of Fecal Coliforms and Fecal Streptococci in a Sanitary Landfill. U.S. Environmental Protection Agency, News of Environmental Research in Cincinnati.

116. McCabe, L.J., J.M. Symons, R.O. Lee, and G.G. Robeck. 1970. Survey of Community Water Supply Systems. *Jour. Amer. Water Works Assoc.*, 62:670-687.

117. ZoBell, C.E. and C.W. Grant. 1942. Bacterial Utilization of Low Concentrations of Organic Matter. *Jour. Bacteriol.*, 43:555-564.

118. Druce, R.G. and S.B. Thomas. 1970. An Ecological Study of the Psychrotrophic Bacteria of Soil, Water, Grass, and Hay. *Jour. Appl. Bacteriol.*, 33:420-435.

119. Yoshimizu, M., K. Kamiyana, T. Kimura, and M. Sakai. 1976. Studies on the Intestinal Microflora of Salmonids. IV. The Intestinal Microflora of Freshwater Salmon. *Bull. Jap. Soc. Sci. Fisch.*, 42:1281-1290.

120. Daubner, I. Drittes Kapitel. 1972. In: *Mikrobiologie des Wassers*. Inst. f. Limnologie, Slowakischen Academie der Wissenschaften, Bratislava. pp. 53-104.

121. Symons, J.M. 1969. *Water Quality Behavior in Reservoirs*. PHS Pub. 1930. Cincinnati, OH.

122. Purcell, L.T. and C.H. Capen. North Jersey District Water Supply Commission Report on Treatment Plant and Transmission Main for Water Supply from Round Valley–Spruce Run Project for Participating Municipalities. New Jersey District Water Supply Commission. Wanaque, NJ.

123. Ridley, J.E. and J.M. Symons. 1972. New Approaches to Water Quality Control in Impoundments. In: *Microbiology of Polluted Waters*. R. Mitchell, Ed., Wiley-Interscience, New York.

124. Walesh, S.G. 1967. Natural Processes and Their Influence on Reservoir Water Quality Control. *Jour. Amer. Water Works Assoc.*, 59:63-79.

125. Bernhardt, H. 1967. Aeration of Wahnbach Reservoir Without Changing the Temperature Profile. *Jour. Amer. Water Works Assoc.*, 59:943-961.

126. Tortoriella, R.C. 1971. Manganic Oxide Reduction by Microorganisms in Freshwater Environments. Dissertation Abs. 32:B:3406. *Water Poll. Abs. (G.B.)* 44:2602.

127. Lackey, J.B. 1949. Plankton as Related to Nuisance Conditions in Surface Water. In: *Limnological Aspects of Water Supply and Waste Disposal*. Amer. Assoc. Advance. Sci., Washington, D.C. pp. 56-63.

128. Mackenthun, K.M. and Ingram, W.M. 1967. Biological Associated Problems in Freshwater Environments. U.S. Dept. of the Interior, FWPCA, Washington, D.C.

129. Carmichael, W.W. 1981. Freshwater Blue-Green Algae (Cyanobacteria) Toxins — A Review. In: *The Water Environment: Algal Toxins and Health*. W.W. Carmichael, Ed., Plenum Press, New York. pp. 1-14.

130. Izaguirre, G., C.J. Hwang, S.W. Krasner, and M.J. McGuire. 1982. Geosmin and 2-Methylisoborneol from Cyanobacteria in Three Water Supply Systems. *Appl. Environ. Microbiol.*, 43:708-714.

131. Medsker, L.L., D. Jenkins, and J.F. Thomas. 1968. Odorous Compounds in Natural Waters. An Earthy-Smelling Compound Associated with Blue-Green Alae and Actinomycetes. *Environ. Sci. Technol.*, 2:461-464.

132. Montgomery, J.M. Consulting Engineers. 1985. *Water Treatment Principles and Design*. Wiley-Interscience, New York.

133. Mackenthun, K.M. 1969. The Practice of Water Pollution Biology. U.S. Department of the Interior. U.S. Government Printing Office, Washington, D.C.

134. Larson, T.E. Deterioration of Water Quality in Distribution Systems. *Jour. Amer. Water Works Assoc.*, 58:1307-1316.

135. Cullimore, D.R. and A.E. McCann. 1977. The Identification, Cultivation and Control of Iron Bacteria in Groundwater. In: *Aquatic Microbiology*, F.A. Skinner and J.M. Shewan, Eds., Academic Press, New York.

136. Schorler, B. 1906. Die Rostbilding ni den-Wasserleitungsrohren. *Centro. Bakt. II*, 15:564-568.

137. Harder, E.C. 1919. Iron-Depositing Bacteria and Their Geologic Relations. U.S. Geol. Survey, Prof. Paper 113.

138. Cliver, D.O. and R.A. Newman. 1987. Drinking Water Microbiology. *Jour. Environ. Pathol. Toxicol. Oncol.*, 7:1-366.

139. Chang, S.L., G. Berg, N.A. Clarke, and P.W. Kabler. 1960. Survival and Protection Against Chlorination of Human Enteric Pathogens in Free Living Nematodes Isolated from Water Supplies. *Amer. Jour. Trop. Med. Hyg.,* 9:136-142.

140. Clarke, N.A., G. Berg, P.W. Kabler, and S.L. Chang. 1962. Human Enteric Viruses in Water: Source, Survival and Removability. In: *International Conference on Water Pollution Research, London,* Vol. 2., Pergamon Press, Oxford. pp. 523-535.

141. Tracy, H.W., V.M. Camarena, and F. Wing. 1966. Coliform Persistence in Highly Chlorinated Waters. *Jour. Amer. Water Works Assoc.,* 58:1151-1159.

142. Hoff, J.C. 1978. The Relationships of Turbidity to Disinfection of Potable Water. In: Evaluation of the Microbiology Standards for Drinking Water, C.W. Hendricks, Ed., U.S. Environmental Protection Agency, EPA-570/9-78-00C, Washington, D.C. pp. 103-117.

143. Floyd, R. and D.G. Sharp. 1978. Viral Aggregation Quantitation and Kinetics of the Aggregation of Poliovirus and Riovirus. *Appl. Environ. Microbiol.,* 35:1079-1083.

144. Powell, S.T. 1954. *Water Conditioning for Industry.* McGraw-Hill, New York.

145. Ohio River Valley Sanitation Commission. 1980. Water Treatment Process Modification for Trihalomethane Control and Organic Substances in the Ohio River. EPA-600/2-8-028, U.S. Environmental Protection Agency, Cincinnati, OH.

146. American Water Works Association. 1971. *Water Quality and Treatment,* 3rd ed., McGraw-Hill, New York.

147. Seidler, R.J., J.E. Morrow, and S.T. Bagley. 1977. *Klebsiella* in Drinking Water Emanating from Redwood Tanks. *Appl. Environ. Microbiol.,* 33:893-900.

148. Spino, D.F., E.W. Rice, and E.E. Geldreich. 1984. Preliminary Investigation on the Occurrence of *Legionella* spp. and Other Aquatic Bacteria in Chemically Contaminated Ground Water Treated by Aeration. In: Legionella, *Proceedings of the 2nd International Symposium,* C. Thornsberry, A. Ballows, J.C. Feeley, and W. Jakubowski, Eds., Amer, Soc. Microbiol. Washington, D.C. pp. 318-320.

149. Wentworth, D.F., R.T. Thorup, and O.J. Sproul. 1968. Poliovirus Inactivation in Water-Softening Precipitation Processes. *Jour. Amer. Water Works Assoc.,* 60:939-946.

150. Thayer, S.E. and O.J. Sproul. 1966. Virus Inactivation in Water Softening Precipitation Processes. *Jour. Amer. Water Works Assoc.,* 58:1063-1074.

151. Sproul, O.J. 1971. Recent Research Results on Virus Inactivation by Water Treatment Processes. In: *Virus Water Quality: Occurrence and Control,* V. Snoeyink and V. Griffin, Eds., Proc. 13th Water Quality Confr., Univ. of Illinois, Pull. 69, Urbana, IL. pp. 159-169.

152. McCarty, P.L., M. Reinhard, J. Graydon, J. Schreiner, K. Southerland, T. Everhart, and D.G. Argo. 1980. Advanced Treatment for Wastewater Reclamation at Water Factory 21. Dept. Civil Engr. Tech. Report 236, Stanford Univ., Stanford, CA.

153. Murray, J.P. and G.A. Parks. 1980. Poliovirus Adsorption on Oxide Surfaces. Correspondence with the DLVO-Lifshitz Theory of Colloid Stability. In: *Particles in Water, Characterization, Fate, Effects and Removal,* M.C. Kavanaugh and J.O. Leckie, Eds., Amer. Chem. Soc. Adv. Chem. Ser. 189, Washington, D.C.

154. Hoff, J.C. and E.W. Akin. 1983. Removal of Viruses from Raw Waters by Treatment Processes. In: *Viral Pollution of the Environment*, G. Berg, Ed., CRC Press, Boca Raton, FL.

155. Chang, S.L., R.E. Stevenson, A.R. Bryant, R.L. Woodward, and P.W. Kabler. 1958. Removal of Coxsackie and Bacterial Viruses in Water by Flocculation, II. Removal of Coxsackie and Bacterial Viruses and the Native Bacteria in Raw Ohio River Water by Flocculation with Aluminum Sulfate and Ferric Chloride. *Amer. Jour. Public Health*, 48:159-169.

156. Streeter, H.W. 1927. Studies of the Efficiency of Water Purification Processes. U.S. Public Health Service, Bull. 172, Washington, D.C.

157. Streeter, H.W. 1929. Studies of the Efficiency of Water Purification Processes. U.S. Public Health Service, Bull. 193, Washington, D.C.

158. Logsdon, G.S. and K.R. Fox. 1981. Getting Your Money's Worth from Filtration. *Jour. Amer. Water Works Assoc.*, 74:249-256.

159. Hannah, S.A., J.M. Cohen, and G.G. Robeck. 1967. Control Techniques for Coagulation-Filtration. *Jour. Amer. Water Works Assoc.*, 59:1149- 1163.

160. Kavanaugh, M.C. C.H. Tate, A.R. Trussell, R.R. Trussell, and G. Treweek. 1980. Use of Particle Size Distribution Measurements for Selection and Control of Solid/Liquid Separation Processes. In: *Particulates in Water, Characterization, Fate, Effects and Removal*, M.C. Kavanaugh and J.O. Leckie, Eds., Amer. Chem. Soc., Adv. Chem. Ser. 189, Washington, D.C.

161. Hudson, H.E., Jr., 1981. *Water Clarification Processes: Practical Design and Evaluation*, Van Nostrand Reinhold, New York.

162. Huisman, L. and W.E. Wood. 1974. Slow Sand Filtration. World Health Organization, Geneva, Switzerland.

163. Hazen, A. 1913. *The Filtration of Public Water Supplies*, 3rd ed. John Wiley and Sons, New York.

164. Cleasby, J.L., D.L. Hilmoe, and C.J. Dimitracopoulos. 1984. Slow Sand Filtration and Direct In-Line Filtration of a Surface Water. *Jour. Amer. Water Works Assoc.*, 76:44-55.

165. Burman, N.P. 1962. Bacteriological Control of Slow Sand Filtration. *Eff. Water Treat. Jour.*, 2:674-677.

166. Robeck, G.G., N.A. Clarke, and K.A. Dostal. 1962. Effectiveness of Water Treatment in Virus Removal. *Jour. Amer. Water Works Assoc.*, 54:1275-1292.

167. Taylor, E.W. 1970. Forty-Fourth Report of the Results of the Bacteriological, Chemical and Biological Examination of London Waters for the Years 1969-1970. Metropolitan Water Board, London, England.

168. Taylor, E.W. 1973. Forty-Fifth Report on the Results of the Bacteriological, Chemical and Biological Examination of London Waters for the Years 1971-1973. Metropolitan Water Board, London, England.

169. Amirhor, P. and R.S. Englebrecht. 1975. Virus Removal by Polyelectrolyte Aided Filtration. *Jour. Amer. Water Works Assoc.*, 67:187-192.

170. Bellamy, W.D., G.P. Silverman, D.W. Hendricks, and G.S. Logsdon. 1985. Removing *Giardia* Cysts with Slow Sand Filtration. *Jour. Amer. Water Works Assoc.*, 77:52-60.

171. Bellamy, W.D., D.W. Hendricks, and G.S. Logsdon. 1985. Slow Sand Filtration: Influences of Selected Process Variables. *Jour. Amer. Water Works Assoc.*, 77:62-66.

172. Logsdon, G.S., J.M. Symons, R.L. Hoge, Jr., and M.M. Arozarena. 1981. Alternative Filtration Methods for Removal of *Giardia* Cysts and Cyst Models. *Jour. Amer. Water Works Assoc.*, 73:111-117.

173. Logsdon, G.S. 1982. Comparison of Some Filtration Processes Appropriate for *Giardia* Removal. In: *Advances in* Giardia *Research*, P.M. Wallis and B.R. Hammond, Eds., Univ. of Calgary Press, Calgary, Canada. pp. 95-102.

174. Berg, G., R.E. Dean, and D.R. Dahling. 1968. Removal of Poliovirus I from Secondary Effluent by Lime Flocculation and Rapid Sand Filtration. *Jour. Amer. Water Works Assoc.*, 60:193-198.

175. Guy, M.D., J.D. McIver, and M.J. Lewis. 1977. The Removal of Virus by a Pilot Treatment Plant. *Water Res.*, 11:421-428.

176. McCormick, R.F. and P.H. King. 1980. Direct Filtration of Virginia Surface Waters: Feasibility and Cost. Virginia Water Resources Res. Center, Bull. 129, Polytechnic Inst. and State Univ., Blacksburg, VA.

177. Logsdon, G.S. and E.C. Lippy. 1982. The Role of Filtration in Preventing Waterborne Disease. *Jour. Amer. Water Works Assoc.*, 74:649-655.

178. Baylis, J.R., O. Gullans, and B.K. Spector. 1936. The Efficiency of Rapid Sand Filters in Removing the Cysts of Amoebic Dysentery Organisms from Water. *Public Health Rep.*, 51:1567-1575.

179. Hunter, J.V., G.R. Bell, and C.N. Henderson. 1966. Coliform Organisms Removal by Diatomite Filtration. *Jour. Amer. Water Works Assoc.*, 58:1160-1169.

180. Chaudhuri, M., P. Amirhor, and R.S. Engelbrecht. 1974. Virus Removal by Diatomaceous Earth Filtration. *Jour. Environ. Eng. Div. Amer. Soc. Civ. Eng.*, 100:937-953.

181. Poynter, S.F.B. and J.S. Slade. 1977. The Removal of Viruses by Slow Sand Filtration. *Prog. Water Technol.*, 9:75-78.

182. Brown, T.S., J.F. Malina, Jr., and B.D. Moore. 1974. Virus Removal by Diatomaceous Earth Filtration, Part I. *Jour. Amer. Water Works Assoc.*, 66:98-102; Part 2, 66:735-738.

183. Symons, J.M., A.A. Stevens, R.M. Clark, E.E. Geldreich, O.T. Love, Jr., and J. DeMarco. 1981. Treatment Techniques for Controlling Trihalomethanes in Drinking Water. EPA-600/2-81-156, Office of Research and Development, U.S. Environmental Protection Agency, Cincinnati, OH.

184. Klotz, M. P. Werner and R. Schweisfurth. 1976. Investigations Concerning the Microbiology of Activated Carbon Filters. In: Translation of Report on Special Problems of Water Technology. Vol. 9. Adsorption. Conference at Karlsruhe FRG, 1975, EPA-600/9-76-030, U.S. Environmental Protection Agency, Cincinnati, OH. pp. 312-330.

185. Reasoner, D.J., J.C. Blannon, E.E. Geldreich, and J. Barnick. 1989. Nonphotosynthetic Pigmented Bacteria in a Potable Water Treatment and Distribution System. *Appl. Environ. Microbiol.*, 55:912-921.

186. Cookson, J.T., Jr. and W.J. North. 1967. Adsorption of Viruses on Activated Carbon — Equilibria and Kinetics of the Attachment of *Escherichia coli* bacteriophage T4 on Activated Carbon. *Environ. Sci. Technol.*, 1:46-50.

187. Cookson, J.T., Jr. 1969. Mechanism of Virus Adsorption on Activated Carbon. *Jour. Amer. Water Works Assoc.*, 61:52-59.

188. Sproul, O.J., L.R. Larochelle, D.F. Wentworth, and R.T. Thorup. 1967. Virus Removal in Water Reuse Treating Processes. *Chem. Eng. Symp. Ser.,* 63 (78):130-136.

189. Cliver, D.O. 1971. Viruses in Water and Wastewater — Effects of Some Treatment Methods. In: *Virus and Water Quality: Occurrence and Control,* V. Snoeyink and V. Griffin, Eds., Proc. 13th Water Quality Confr. Univ. IL, Urbana, IL. pp. 149-157.

190. IAWPRC Study Group on Water Virology. 1983. The Health Significance of Viruses in Water. *Water Res.,* 17:121-132.

191. Hoff, J.C. and E.E. Geldreich. 1978. Alternative Disinfectants for Drinking Water — Ozone, Chlorine Dioxide, Chloramines. In: Proc. 20th Ann. Pub. Water Supply Engr. Confr. Water Treat., Part III, Univ. IL, Urbana-Champaign, IL.

192. Hoff, J.C. and E.E. Geldreich. 1981. Comparison of the Biocidal Efficiency of Alternative Disinfectants. *Jour. Amer. Water Works Assoc.,* 73:40-44.

193. Akin, E.W., J.C. Hoff, and E.C. Lippy. 1982. Waterborne Outbreak Control: Which Disinfectant? *Environ. Health Perspectives,* 46:7-12.

194. Berger, B.B. 1983. Control of Organic Substances in Water and Wastewater. EPA-600/8-83-011. Office of Research and Development, U.S. Environmental Protection Agency, Washington, D.C.

195. Kabler, P.W., S.Y. Chang, N.H. Clarke, and H.F. Clark. 1963. Pathogenic Bacteria and Viruses in Water Supply. Proc. 5th Sanitary Engr. Confr., Jan 29-30, 1963, Univ. Illinois Engr. Exper. Sta. Circular No. 81, Urbana, IL.

196. Newton, W.L. and M.F. Jones. 1949. Effect of Ozone in Water on Cysts of *Endamoeba histolytica. Amer. Jour. Trop. Med.,* 29:669-681.

197. Stringer, R.P., W.M. Cramer, and C.W. Kruse. 1975. Comparison of Bromine, Chlorine and Iodine as Disinfectants for Amoebic Cysts. In: *Disinfection, Water and Wastewater,* J.D. Johnson, Ed., Ann Arbor Science, Ann Arbor, MI. pp. 193-210.

198. Jarroll, E.L., A.K. Bingham, and E.A. Meyer. 1981. Effect of Chlorine on *Giardia lamblia* Cyst Viability. *Appl. Environ. Microbiol.,* 41:483- 487.

199. Rice, E.W. and J.C. Hoff. 1981. Inactivation of *Giardia lamblia* Cysts by Ultraviolet Irradiation. *Appl. Environ. Microbiol.,* 52:546-547.

200. Rice, E.W., J.C. Hoff, and F.W. Schaefer. 1982. Inactivation of *Giardia* Cysts by Chlorine. *Appl. Environ. Microbiol.,* 43:250-251.

201. Foster, D. M., M.A. Emerson, C.E. Buck, D.S. Walsh, and O.J. Sproul. 1980. Ozone Inactivation of Cell- and Fecal-Associated Viruses and Bacteria. *Jour. Water Poll. Contr. Fed.,* 52:2174-2184.

202. Fennel, H., D.B. James, and J. Morris. 1974. Pollution of a Storage Reservoir by Roosting Gulls. *Water Treat. Exam.,* 23:5-24.

203. Geldreich, E.E., H.D. Nash, D.J. Reasoner, and R.H. Taylor. 1975. The Necessity of Controlling Bacterial Populations in Potable Waters — Bottled Water and Emergency Water Supplies. *Jour. Amer. Water Works Assoc.,* 67:117-124.

204. O'Connor, J.T., L. Hash, and A.B. Edwards. 1975. Deterioration of Water Quality in Distribution Systems. *Jour. Amer. Water Works Assoc.,* 67:113-116.

205. Burman, N.P. 1965. Taste and Odour Due to Stagnation and Local Warming in Long Lengths of Piping. *Soc. Water Treat. Exam.,* 14:125-131.

206. Oborn, E.T. and E.C. Higginson. 1954. Biological Corrosion of Concrete. Joint Report Field Crops Res. Branch, Agri. Res. Ser., U.S. Dept. of Agriculture and Bureau of Reclamation, U.S. Dept. of the Interior.

207. Lueschow, L.A. and K.M. Mackenthun. 1962. Detection and Enumeration of Iron Bacteria in Municipal Water Supplies. *Jour. Amer. Water Works Assoc.*, 54:751-756.

208. Loos, E.T. 1962. Experiences with Manganese in Queensland Water Supplies. *Water (Jour. Aust. Water Wastewater Assoc.)*, 14:751-756.

209. Schweissfurth, R. 1978. Manganese and Iron Oxidizing Microorganisms. *Landwirtsch. Forsch.*, 31:127-133.

210. Colbourne, J.S. and R.M. Trew. 1985. Presence of *Legionella* in London's Water Supplies. *Isr. Jour. Med. Sci.*, 22:633-639.

211. Colbourne, J.S., M.G. Smith, S.P. Fisher-Hock, and D. Harper. 1984. Source of *Legionella pneumophila* Infection in a Hospital Hot Water System: Materials Used in Water Fittings Capable of Supporting *L. pneumophila* Growth. In: Legionella, *Proceedings of the 2nd International Symposium*, C.T. Thornsberry, A. Balows, J.C. Feeley, and W. Jakubowski, Eds., Amer. Soc. Microbiol., Washington, D.C. pp. 305-307.

212. Ciesielski, C.A., M.J. Blason, F.M. LaForce, and W.L.L. Wang. 1984. Role of Stagnation and Obstruction of Water Flow in Isolation of *Legionella pneumophila* from Hospital Plumbing. In: Legionella, *Proceedings of the 2nd International Symposium*. C. Thornsberry, A. Balows, J.C. Feeley, and W. Jakubowski, Eds., Amer. Soc. Microbiol., Washington, D.C. pp. 307-309.

213. Bornstein, N., C. Yieilly, M. Nowicki, J.C. Paucod, and J. Fleurette. 1986. Epidemiological Evidence of Legionellosis Transmission and Possibilities of Control. *Isr. Jour. Med. Sci.*, 22:655-661.

214. Muraca, P.W., V.L. Yu, and J.E. Stout. 1988. Environmental Aspects of Legionnaires' Disease. *Jour. Amer. Water Works Assoc.*, 80:78-86.

215. Geldreich, E.E., A.K. Highsmith, and W.J. Martone. 1985. Public Whirlpools — The Epidemiology and Microbiology of Disease. *Infect. Control.*, 6:392-393.

216. Taylor, R.H., M.J. Allen, and E.E. Geldreich. 1979. Testing of Home Use Carbon Filters. *Jour. Amer. Water Works Assoc.*, 71:577-579.

217. Geldreich, E.E., R.H. Taylor, J.C. Blannon, and D.J. Reasoner. 1985. Bacterial Colonization of Point-of-Use Water Treatment Devices. *Jour. Amer. Water Works Assoc.*, 77:72-80.

218. Reasoner, D.J., J.C. Blannon, and E.E. Geldreich. 1987. Microbiological Characteristics of Third-Faucet Point-of-Use Devices. *Jour. Amer. Water Works Assoc.*, 79:60-66.

219. Geldreich, E.E. and D.J. Reasoner. 1989. Home Treatment Devices and Water Quality. In: *Advances in Drinking Water Microbiology*, G.A. McFeters, Ed., Springer-Verlag, New York.

CHAPTER 3

Biological Profiles in Drinking Water

CONTENTS

INTRODUCTION

Public health considerations in water supply have largely addressed the issue of minimizing the occurrence of coliform bacteria and thereby the probability that intestinal pathogens may be present. Thus it is assumed that potable water of good bacteriological quality is generally associated with attainment of less than one total coliform per 100 ml of water sample. This position places little concern with the wide variety of other organisms that may be present and their significance in water quality deterioration.

PRIMARY MICROBIAL GROUPS

The general population of other organisms includes many gram-negative and gram-positive bacteria, sporeformers, acid-fast bacilli, pigmented organisms, and free-living amoebas and nematodes. The amoebas and nematodes feed on bacteria in the sediments and graze on biofilms in the distribution system. Most of these organisms in drinking water are ubiquitous and may include some potentially pathogenic strains as well as a large number of saprophytic organisms. All of these microorganisms utilize a variety of organic compounds in their metabolism and are thereby often referred to as heterotrophs. Some idea of their varied influence in water supply microbiology can be observed in Table 3.1. In these instances, it can be seen that some heterotrophic bacteria such as *Pseudomonas, Flavobacterium*, and *Klebsiella*, may be placed in several different categories that involve interrelationships with other bacteria in the microflora, interferences with indicator detection, or represent potential health risks as opportunistic bacteria or as R factor genetic transfer hosts. Because this group of heterotrophic organisms is so large and diverse in their influence on water quality, it is logical to profile major subsets to study the range of concerns.

COLIFORM BACTERIA

Coliform bacteria may occasionally be found among the variety of gram-negative bacteria present in drinking water. Speciation of coliforms recovered from 111 water distribution systems reveals (Table 3.2) a wide listing of species that have been found at one time or another.[1,5-10] While coliform bacteria are chlorine sensitive, they may be protected from disinfectant exposure if associated with particulates in source water turbidity,[11] if colonized on carbon fines released from granular activated carbon filtration beds,[12] if imbedded in inorganic sediments in the contract basin, if engulfed but not inactivated by amoebas or ingested by nematodes migrating from filter process basins,[13-15] or if exposed to inadequate conditions for disinfection, i.e., contact time, water pH, and temperatures.[16] Coliforms may also enter via the distribution system through water line breaks, negative water pressure and cross-connections.[16,17] In these situations, coliform occurrence is a significant warning of a potential risk for pathogen invasion from a contaminated source water or wastewater backflow. All of these pathways must be blocked by application of effective treatment barriers and protected distribution of water supply to the consumer.

Coliform colonization of the distribution system should not be confused with coliform breakthrough due to treatment barrier failure or contamination penetration in the pipe or storage reservoir system for the community supply area. Correction of the latter occurrences is urgent because of public health risk. Colonization does represent a more remote contamination event in time and one that includes coliform species capable of populating the distribution system environment. In this situation, their presence in the water supply is more an indication of biofilm development in sediment accumulations in pipelines and storage reservoirs that needs to be removed through flushing and treatment refinements. These actions should reduce the availability of AOC (nutrient) concentrations in the water supply.

A profile of coliform organisms that colonize distribution networks reveals *Klebsiella pneumoniae, Enterobacter aerogenes, Enterobacter cloacae,* and *Citrobacter freundii* to be the most successful colonizers because they can grow at minimal nutrient concentrations and are capable of encapsulating in adverse environments. Encapsulation of these coliforms provides protection from the effects of chlorine or other disinfectants and the organisms proceed to adjust their metabolic activity to the availability of nutrients adsorbed by the pipe sediments. Once total coliforms become established in an appropriate pipe network habitat, growth can occur and result in occasional sloughing of cells into the flowing water. This condition of biofilm growth can persist until either the hydraulic shearing effects[19] limit further expansion of the biofilm or exposure to elevated disinfectant residuals penetrates the protective habitat and inactivates the microbial masses.

COLIFORM ANTAGONISTS

Excessive densities of organisms antagonistic to coliform detection in water supplies can desensitize both the multiple-tube test (including the P–A test) and the membrane filter (MF) total coliform procedure. Coliform bacteria co-exist with antagonistic organisms in the aquatic environment, but when the

Table 3.1 Heterotrophic Bacterial Group Cross-overs from Water Treatment to Distribution Supply

Water plant filter effluent and clear well		Distribution water	
Significant categories	Organisms isolated	Significant categories	Organisms isolated
Total coliforms	Klebsiella pneumoniae Enterobacter cloacae Erwinia herbicola	Total coliforms	Klebsiella pneumoniae Enterobacter cloacae Enterobacter aerogenes Erwinia herbicola Escherichia coli Aeromonas hydrophila Citrobacter freundii
Coliform antagonists	Pseudomonas fluorescens Pseudomonas fluorescens Flavobacterium sp.	Coliform antagonists	Pseudomonas aeruginosa Pseudomonas maltophila Pseudomonas fluorescens Pseudomonas cepacia Pseudomonas putida Bacillus sp. Actinomycetes sp.

Opportunistic pathogens	Opportunistic pathogens
Pseudomonas maltophila	*Pseudomonas maltophila*
Klebsiella pneumoniae	*Klebsiella pneumoniae*
Moraxella sp.	*Moraxella* sp.
Staphylococcus (coagulase+)	*Staphylococcus* (coagulase+)
Acinetobacter calcoaceticus	*Acinetobacter calcoaceticus*
	Pseudomonas aeruginosa
	Klebsiella rhinoscheromatis
	Serratia liquefaciens
	Serratia marcescens
	Mycobacterium gordonae
Pigmented bacteria	Pigmented bacteria
Serratia	*Serratia*
Mycobacterium	*Mycobacterium*
Flavobacterium	*Flavobacterium*
Corynebacterium	*Corynebacterium*
Pseudomonas	*Micrococcus*
	Chromobacterium

Data adapted from Geldreich, Nash, and Spino;[1] Reasoner and Geldreich;[2] and duMoulin et al.[3,4]

**Table 3.2   Coliform Species Identified in Various
Public Water Distribution Systems**

Citrobacter	Escherchia
C. freindii	E. coli
C. diversus	
Enterobacter	Klebsiella
Enter. aerogenes	K. pneumoniae
Enter. agglomerans	K. rhinoscleromatis
Enter. cloacae	K. oxytoca
	K. ozaenae

Published data[1,5-10] from 111 water distribution systems
in six states (U.S.) and Ontario Province (Canada).

mixed flora is introduced into lactose broth, competition for available nutrients begins. For some organisms, there is an immediate, accelerated growth that surpasses the growth of stressed coliforms in the same sample and thereby dominates the test culture with a heavy turbidity in broth or an abundance of colonies on the MF surface. This response can obscure gas production in the broth culture or coliform differentiation or discrete colony development on the MF. In other situations, the antagonistic bacteria may release toxic metabolic products that depress coliform growth, resulting in the inhibition of gas production in the fermentation tube or interference with aldehyde production necessary for total coliform detection on Endo type media used in the membrane filter procedure.

In a review of data from the national survey of 969 public water supplies,[20] it was noted that the frequency of detecting total and fecal coliforms by the MF procedure increased as the heterotrophic plate counts increased up to densities of 500 organisms per milliliter (Table 3.3) but decreased in detection frequency when the noncoliform bacterial population exceeded 1000 organisms per milliliter. In another study involving 613 water samples from 32 sampling sites on distribution dead-end lines in Cincinnati,[21] 12.4% of the samples were positive by the multiple-tube test (Table 3.4) while only 3.1% were positive when examined by the MF test. Apparent background interference from noncoliform colonies on the membrane filter surface were responsible for the difference.

Laboratory experiments using cell suspensions of *Pseudomonas, Sarcina,* and *Micrococcus* in a density range of 1000 to 2000 organisms per milliliter were added to lactose tubes simultaneously with ten *E. coli* per milliliter.[22] After incubation at 37°C for 24 h, there was a 28 to 97% loss of sensitivity to *E. coli* detection, depending on the combination of mixed strains in the experiment. *Arthrobacter* antagonism to *E. coli* in the same substrate occurs when there are density ratios of ten or more *Arthrobacter* to every *E. coli* present in the medium.[7]

Table 3.3 Bacterial Plate Count vs. Coliform Detection in Distribution
Water Networks for 969 Public Water Supplies

General bacterial population[a]

Density range per ml	Number of samples	Total coliform		Fecal coliform	
		Occurrences	Percent	Occurrences	Percent
<1–10	1013	47	4.6	22	2.2
11–30	317	28	7.5	12	3.2
31–100	396	72	18.2	28	7.1
101–300	272	48	17.6	20	7.4
301–500	120	30	25.0	11	9.2
501–1000	110	21	19.1	9	8.2
>1000	164	31	18.9	5	3.0
Total	2446	277	—	107	—

[a] Standard plate count (48 h incubation, 35°C).

Data from Geldreich, Allen, and Taylor.[21]

Table 3.4 Heterotrophic Bacterial Densities and Coliform
Detection in Distribution Water Samples

HPC per ml range[a]	Total samples examined	Coliform occurrences (%)	
		MPN[b]	MF[c]
<500	502	10	2.2
500–1000	34	22.9	5.7
>1000	51	29.4	9.8

[a] Of 613 samples, 588 had countable HPC.
[b] Three MPN-positive samples did not have countable SPC.
[c] One MF-positive sample did not have countable SPC.

Data from Geldreich, Allen, and Taylor.[21]

ANTIBIOTIC-RESISTANT BACTERIA

Heterotrophic bacteria in water supplies that are resistant to one or more antibiotics may pose a health threat if these strains are opportunistic pathogens or serve as donors of the resistant factor to other bacteria that could be pathogens. Antibiotic resistant (R factor) bacteria may originate in surface-water sources used for public water supplies.[23-26] Polluted waters acquire bacteria with R factors from the fecal wastes of man and domestic animals in wastewater effluents and stormwater runoff from farm pasture lands and feed-lots. Farm animals, in particular, may receive continuous doses of antibiotics in animal feed and become constant generators of a variety of antibiotic-resistant bacteria. Although treatment processes inactivate or remove introduced antibiotic resistant organisms (*Aeromonas, Hafnia,* and *Enterobacter*) in the source water, a shift of this transmissable factor to the *Pseudomonas/Alcaligenes* group, *Acinetobacter, Moraxella, Staphylococcus,* and *Micrococcus* may occur.[27]

Water supply treatment processes apparently act as a mixing chamber for resistant R factor transfers with some surviving organisms acquiring multiple resistances to different antibiotics. Many of the transformations occur in the biofilm established on activated carbon and sand filters.[23] The disinfection process may also have a major impact on the selection of drug-resistant bacteria. The reason for the common occurrence of streptomycin resistance among bacteria that survive chlorination is not known. Multiple-antibiotic-resistant bacteria passing through water treatment are more tolerant to metal salts (i.e., $CuCl_2$, $Pb(NO_3)_2$, and $ZnCl_2$).[28] Examination of bacteria for multiple-antibiotic resistance from two sites in a distribution system indicate a dynamic state of fluctuation (16.7% R factor organisms at one site and 52.4% at the other location). In a typical population of 100 herterotrophic bacteria per milliliter of water from the distribution system, 40 to 70 of these organisms could be expected to have some antibiotic resistance factors.[27] What health risk this represents, particularily when the heterotrophic bacterial population is above the 500 organisms or more per milliliter limit suggested for potable water, is not yet clearly understood.

DISINFECTANT-RESISTANT ORGANISMS

Chlorination of water introduces a strong selective pressure on the bacterial flora but does not insure a sterile water supply. Some appreciation of this fact can be observed in a study of bacteriological data and chlorine residuals obtained from the national survey of public water systems.[29] These data (Table 3.5) indicate that the heterotrophic bacterial population in distribution lines was being controlled to densities below 500 organisms per milliliter in numerous community water supplies by maintaining approximately 0.3 mg/l residual chlorine in these samples. Further increases in the residual chlorine concentration did not result in any significant decreases in the heterotrophic bacterial densities. Protective sediment habitats and selective survival of disinfectant resistant organisms were the reasons residual chlorine concentrations greater than 0.3 mg/l chlorine produced no further decreases in the heterotrophic plate counts.

From another perspective, bacteria isolated from a chlorinated surface-water supply were more resistant to both combined and free chlorine than strains of the same genera originating in an unchlorinated groundwater system.[30] Differences in the sensitivity of the two bacterial populations did not appear to be related to substances other than chlorine, because overall water chemistry of the two public water supplies was very similar. The most resistant microorganisms in both systems included gram-positive, spore-forming bacilli, actinomycetes, and some micrococci. These organisms were found to survive 2 min exposure to 10 mg/l free chlorine. In contrast, organisms most sensitive

Table 3.5 Effect of Varying Levels of Residual Chlorine on the Total Plate Count in Potable Water Distribution Systems

Standard plate count[a]	Standard plate count (%)							
	Residual chlorine (mg/l)							
	0.0	0.01	0.1	0.2	0.3	0.4	0.5	0.6
<1	8.1[b]	14.6	19.7	12.8	16.4	17.9	4.5	17.9
1–10	20.4	29.2	38.2	48.9	45.5	51.3	59.1	42.9
11–100	37.3	33.7	28.9	26.6	23.6	23.1	31.8	28.6
101–500	18.6	11.2	7.9	9.6	12.7	5.1	4.5	10.7
501–1000	5.6	6.7	1.3	2.1	1.8	0	0	0
>1000	10.0	4.5	3.9	0	0	0	0	0
Number of samples	520	89	76	94	55	39	22	28

[a] Standard plate count (48 h incubation, 35°C).
[b] All values are percent of samples that had the indicated standard plate count.

Data in percent from a survey of 923 community water systems in nine metropolitan areas; Geldreich et al.[29]

to chlorine contact that were recovered from these two water systems were *Corynebacterium/Arthrobacter, Klebsiella, Pseudomonas/Alcaligenes, Flavobacterium/Moraxella, Acinetobacter,* and *Micrococcus.* These water supply isolates were inactivated by 1.0 mg/l or less of free chlorine. The apparent contradiction in occurrences of genera that have strains in both categories reflects variations among species within the same genus, physical aggregation of cells, and chance association of organisms with particulate matter in a water sample.

PIGMENTED BACTERIA

A characteristic of some bacteria found in water supplies is the ability to form brightly colored nonphotosynthetic, nondiffusible pigments (Figure 3.1). This unique property is not often observed in the routine processing of water samples unless incubation time for heterotrophic plate cultures is extended beyond 3 d. While pigmented bacteria are frequently found to be *Flavobacterium* species, there are other genera (*Mycobacterium, Serratia, Corynebacterium,* and *Chromobacterium*) known to have some strains capable of pigmentation. These organisms appear in greatest abundance in high-quality waters (Table 3.6) so it is not surprising that they occur not only in distribution water, but also in bottled water and a variety of attachment devices including drinking water fountains, ice machines, point-of-use treated waters, laboratory high-quality water systems, humidifying units, and hemodialysis equipment.[2,4,6,30-44]

Pigmented bacterial occurrence was investigated at a raw water supply intake on the Ohio River in the vicinity of Cincinnati.[2] At this site, yellow

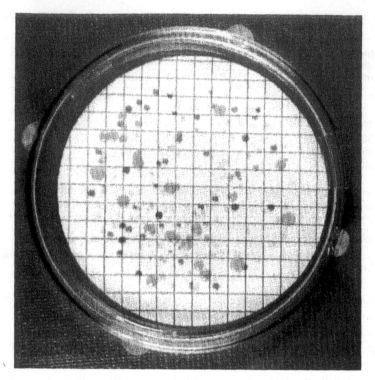

Figure 3.1   Pigmented bacteria in water supply. (Cultured on R-2A medium.)

**Table 3.6   Reported Occurrences of Pigmented Bacteria
in Potable Water**

Water source	Bacterial count (CFU per ml)	Pigmented (%)	Ref.
Bottled water	140–570,000	0–100	31
Distribution water	1,000	80–90	34
Distribution water (Cl$_2$)	200	62	6, 35
Distribution, well water	500	35	6, 35
Reservoir water (Cl$_2$)	5–150,000	55–90	36
Well water	30–690	10–14	37

Table revised from Reasoner et al.[2]

pigmented bacteria were the predominant color group with all other color groups (orange, pink, purple, brown, and black) accounting for less than 6% of the total percentage of pigmented bacteria in any season. The proportion of yellow pigmented bacteria in the raw water was least in the autumn and highest in the summer and the mean for the year was 26% of the heterotrophic population detected on R-2A agar with incubation for 7 d at 35°C.

Water supply treatment processes, particularly GAC filtration and disinfection, may be a selective factor for pigmented bacterial occurrence in some water supply floras. For example, the microbial population that develops in activated carbon adsorbers, both in the sand replacement and postfilter mode, includes a specialized group of organisms capable of biodegrading organics adsorbed from the source water and those organisms passing through early stages of the water treatment train. Among the recessive strains encountered, i.e., a broad spectrum of bacteria in percentages of less than 5%, are a variety of pigmented organisms that become established in the adsorbers and find their way into the effluent.

A study of pigmented organisms in the activated carbon adsorbers at Evansville, IN revealed a periodic colonization.[44] Both virgin GAC and reactivated carbon adsorber effluents contained some pigmented bacteria, even though the influent to the GAC adsorbers sometimes showed no significant pigmented bacterial population during periods when increased concentrations of chlorine dioxide were applied to the untreated river water. Apparently, disinfectant residuals during May–December were inadequate to be an effective, controlling force in the GAC adsorbers. No disinfectant residuals were detected in these GAC adsorber effluents because of specific oxidant/GAC reactions. Analysis during March–April of the following year, however, showed a few pigmented bacteria in the source water and essentially none from the GAC adsorber. This change may have been caused by a drastic seasonal decline in the occurrence of pigmented bacteria in the source water.

There is some evidence that pigmented bacteria may be more chlorine resistant than many nonpigmented organisms. For example, a *Flavobacterium* species was shown to survive after 10 min exposure to 10 mg of chlorine per liter.[34] *Flavobacterium* implicated in airway colonization was observed to survive exposure to 1.0 mg of chlorine per liter for 24 h.[31] The *Flavobacterium–Moraxella* group of strains isolated from a chlorinated distribution system showed smaller zones of inhibition when exposed to chlorine in a disk assay procedure than did bacterial isolates of the same groups obtained from a nonchlorinated distribution system. A red pigmented *Corynebacterium rubrum* strain isolated from tap water was not only chlorine resistant to 0.3 mg/l free chlorine per liter, but also thermotolerant at 80°C.[45]

Field investigation on the impact of disinfection indicates there are significant changes in the percent of pigmented bacteria as well as a significant reduction (one to ten organisms per milliliter) in the heterotrophic bacterial population as the plant effluent enters the distribution system. With densities of heterotrophic bacteria this low, the occurrence of one to a few pigmented colonies can account for a large percentage of the HPC. Nevertheless, these findings suggest that pigmented bacteria can pass through the treatment barrier and appear in the distribution system. The other route of passage into the pipe network is via soil contamination that may occur during repairs to line breaks.

Pigmented bacteria occurrence was also studied in the distribution system at six hydrant sampling points at or near dead-end areas of the distribution network plus one site that was approximately 25 mi distant from the treatment plant.[2] At these locations, yellow and orange pigmented organisms again became predominant on R-2A medium. Once again the percentage of yellow pigmented organisms showed peak values in summer and lowest values in winter, a reflection of similar changes occurring in the source water. Orange pigmented bacteria showed higher densities in winter and spring. Pink pigmented bacteria were present in very low numbers throughout the year.

The pigmented bacteria, in general, appear to adapt well to the distribution system environment, with at least two populations, the yellow and orange pigmented groups, being always present in slow-flow and dead-end sections of the pipe network, though occurring in different cyclic patterns of population growth. Little is known about the types and concentrations of nutrients (inorganic and organic) that stimulate growth of these organisms. While chromogenic bacterial occurrences of 25% or more in a natural water heterotrophic population may be indicative of a nonpolluted aquatic environment,[46] similar occurrences in treated water supplies are more often a reflection of treatment conditions that alter the microbial flora and habitat opportunities in the pipe network.

ULTRAMICROBACTERIA

Ultramicrobacteria are "normal"-sized organisms that have become very small (less than 0.3 μm in size) through gradual starvation[47,48] and thereby are capable of surviving for extremely long periods of time in low-nutrient waters.[48-50] In a water supply, these organisms may become a significant part of the microbial flora for water stored in reservoirs for many months prior to release during summer peak periods. They may also survive in emergency supplies of drinking water stockpiled for use during natural disasters.

Detection of these organisms may be difficult if attempts are made to immediately cultivate them on nutrient-rich media. Therefore, their occurrence is often overlooked unless very dilute media are used with extended incubation time or culture transfers are made through a series of media with gradually increasing concentrations of enrichments. Upon resuscitation in water, laboratory media or through host colonization, these organisms then return to normal size and typical metabolic functions.[51]

It has been suggested that many common aquatic bacteria are capable of survival as ultramicrobacteria in a spartan aquatic environment, some of them being strains of *Klebsiella, Escherichia, Bacillus, Micrococcus, Staphylococcus,* and *Vibrio*. The greatest concern is that some pathogenic bacteria can become ultramicrobacteria, given similar circumstances. Some pathogens known to survive in low-nutrient waters include *Pseudomonas cepacia,*[49]

Pseudomonas aeruginosa,[52] *Legionella pneumophila,*[53] *Salmonella typhimurium, Yersinia entercolitica, Shigella* sp., and enteropathogenic *Escherichia coli.*[54]

Survival in these aquatic environments is possible if the organisms can slow down generation times and metabolic rates and subsist in a near dormant state by a process of reduction division. In this situation, cells divide to create smaller daughter cells with this division continuing until the original store of nutrients in the mother cell is depleted and all subsequent daughter cells remain diminutive in size.[55] Another survival mechanism is for the bacterial cell to gradually condense all protoplasmic material into a compact sphere surrounded by a collapsed cell wall.[47,50,51]

The ability of *Klebsiella* strains to grow at low nutrient concentrations has been demonstrated by the routine cultivation of *K. oxytocia* on noble agar with no additional nutrients. In the process of adjusting to this situation the organism developed a mucoid mutant that became more dominant as storage time increased.[56] The ability to develop an enhanced adhesion property was also found to be part of the survival strategy of a marine vibrio.[57] With the advent of encapsulation and enhanced adhesion, ultramicrobacteria become more difficult to remove from protective sites on pipe walls and storage tank surfaces or from attachment sites in sediments and tuberculations.

FREE-LIVING MACROINVERTEBRATES:
PROTECTORS AND AMPLIFIERS OF BACTERIA

Drinking water may not only contain a variety of bacteria but also various free-living, microscopic animals that feed on the bacterial population.[58] These amoebas and macroinvertebrates (amphipods, copepods, and nematodes) are not normally encountered in groundwater unless the source is contaminated by surface runoff or is stored in uncovered reservoirs. Being common to surface waters, they are found colonizing various filter media and entrapped in settled coagulant particles searching for their food supply. Passage through treatment processes, including disinfection, is not impossible. In the distribution system, their subsistence may be found at biofilm sites or in areas of static water where bacterial populations are amplified.

Only two members of these free-living organisms are known to be pathogenic; these are *Naegleria gruberi* and species of *Acanthamoeba,* which are opportunistic pathogens associated with swimming activities, not water supply. Other pathogenic protozoans (*Entamoeba, Balantidium,* and *Cryptosporidium*) are host specific (warm-blooded animal tract) and not capable of adaptation to a free-living state in the aquatic environment.

Origins of macroinvertebrates may be traced to the raw source water used in water supply treatment or to open water reservoirs[14,59-67] degraded by dirt, lawn fertilizers, leaves, or other organic pollutants in the atmosphere. Such contaminants increase the reservoir water fertility and encourage the development of organic food chains similar to those in natural surface waters.

To cite an example, Indianapolis reported an outbreak of copepods in the distribution system during 1953 and 1954[60] caused by eggs penetrating the sand barrier and hatching in the finished water. As many as 20 eggs per liter were reported in the sand filter effluent. No mention of increased coliform occurrences was given in the report.

Water mains colonization begins with a breakthrough of macroinvertebrates infesting a filter bed or in the chance passage from an open reservoir environment. Once in the distribution network, their establishment depends on location of an attachment site, availability of sufficient food resources, and the ability to reproduce faster than the rate of die-off. Such sites include hydrants, dead ends, and water filter attachment devices.[14,61,64,67]

The chief complaints with these animal infestations in water supplies have focused either on interferences with filtration or other water treatment processes and customer complaints about water palatability and aesthetics. The role that biofilm plays in encouraging the colonization of various macroinvertebrates and the influence amoebas, crustaceans, and nematodes have in providing a vehicle of protective transport for bacteria into the distribution system is only beginning to be recognized.

PROTOZOANS

Amoebas are single-cellular microscopic animals that range from 10 to 30 μm or more in diameter. Their life cycle consists of an adult form that is a naked mass of oozing protoplasm that engulfs food and a resting stage that forms cysts that are resistant to adverse environmental conditions. Ciliates are another group of protozoans that are generally free-living but differ from amoebas in possessing cilia (hairlike feelers) for movement and rudimentary mouths for intake of solid food particles and have nuclei of two kinds. Both groups of protozoans ingest bacteria, algae, and other protozoans as food sources with digestion of these captured organisms occurring in vacuoles.

Free-living amoebas have, not unexpectedly, been isolated from a raw water reservoir in Nebraska used in water supply treatment.[68] A survey of municipal water supplies in the United States found live cysts of *Naegleria* and *Hartmannella* (few per liter) in 6 of 22 water supplies examined.[69] These types of amoebas (*Naegleria, Hartmanella,* and *Acanthamoeba*) were also isolated from the drinking water distribution system of Poiters, France.[70] While only the cyst stage of amoeba was detected (due to methodology limitations) in distribution water, there can be little doubt that active, feeding amoebas were also present in those water supplies at sites in the pipe network where bacteria colonize sediment and tubercle biofilms.

Bacteria engulfed by amoebas (Figure 3.2) and ciliates are generally digested within the food vacuole over a 1- to 4-h period.[71,72] This process may be interrupted by a physiological response to various environmental stresses, including free chlorine residuals.[73] While these bacterial feeders can withstand free chlorine residuals of 4 mg/l or more for periods of 30 to 60 min,[74] their metabolic activity may be so greatly altered that otherwise digestible bacteria survive intracellular digestion and amplify at the vacuole site before passage in the excretion of waste products or release through burst impact on the host cell, internally or externally.[75-79] This latter phenomenon may account for the safe passage of various bacteria through the treatment barrier and protected passage in the distribution system to sites of dissipated available chlorine.

Research has shown that ingested coliforms (*E. coli, Citrobacter freundii, Enterobacter agglomerous, Enterobacter cloacae, Klebsiella pneumonia,* and *Klebsiella oxytoca*), bacterial pathogens (*Salmonella typhinurium, Yersinia enterocolitica, Shigella sonnii, Campylobacter jejuni*), and an opportunistic pathogen (*Legionella gormanii*) showed increased resistance to free chlorine residuals when protected within predatory protozoans.[15] All of these strains in a free-living state (not associated with particulates, algae, or protozoans) were inactivated within 1 min by free chlorine residuals of 1 mg/l or more at pH 7.0 and 25°C. When the coliform and pathogenic bacteria were ingested by either the amoeba (*Acanthamoeba castellanii*) or ciliates (*Tetrahymena pyriformis* and *Cyclidium* sp.), these entrapped bacteria were shown to survive free chlorine residuals of 2 to 10 mg/l. Differences in chlorine resistance among the engulfed bacteria species may be in part due to different rates of chlorine diffusion into the protozoan cell as well as state of digestibility in the food vacuole. The important point is that protozoans may be an important mechanism by which various bacterial species survive disinfection in the treatment train and provide a vehicle of protected transport through a distribution network, safe from the adversities of disinfectant contact, osmotic stress, other predators, and antagonistic organisms.

CRUSTACEANS

These macroinvertebrates are hard-shelled multicellular animals with a flexuous body inside the shell, which is relatively large and can be seen with the naked eye. Included in this group of organisms are amphipods and copepods. Both groups are omnivorous, i.e., capable of browsing on algae, bacteria, and aquatic plants in the surface-water environment. The frequency of observing the juvenile stage and the mature size of adults in a sample are a direct indication of a stable aquatic environment.

Crustaceans were the predominant zooplankton in two English rivers used as raw source waters and during their seasonal growth period (June to August) were in such abundance as to cause interference with the water supply filtration

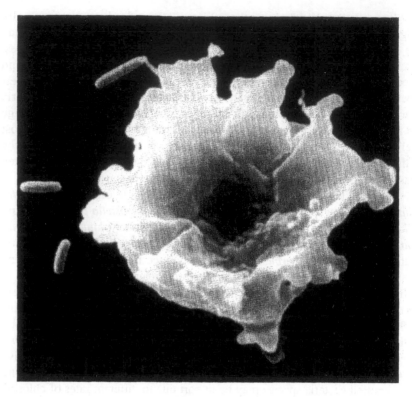

Figure 3.2   Micrographs showing: (above) amoeba psuedopod reaching out to entrap a bacterial cell; (opposite) amoeba engulfing a bacterial cell. (Original magnifications × 2000.) (Courtesy of Drs. Schotts and Steffens, University of Georgia.)

process.[80] During this period, densities of amphipods and copepods reached 20 to 50 organisms each per liter of raw water. Crustaceans passing into the San Francisco water distribution system from raw water receiving disinfection only were considered to be the most likely explanation for the survival of coliforms after hours of contact time with 1 mg/l or more of free chlorine.[62] Water supply systems lacking filtration or providing treated water from storage in open reservoirs foster the greatest potential for invasion of these organisms (Figure 3.3) and other larger invertebrates (bloodworms, hydras, snails, etc.).[66]

The Worcester, MA water system had a serious problem with invertebrate colonization within the water distribution system.[14] Infestations were mediated by flow rate within the pipe, hydraulic disturbances that caused a release of attached organisms into the flowing water, pipeline deterioration with buildup of tuberculations, and chlorine demand within the distribution network. Some characteristics of the situations where invertebrates were found in the pipe network can be seen in Table 3.7. Amphipods were collected in areas of low flow (<0.5 mgd), low chlorine residuals (<0.5 mg/l), and concurrent hydraulic

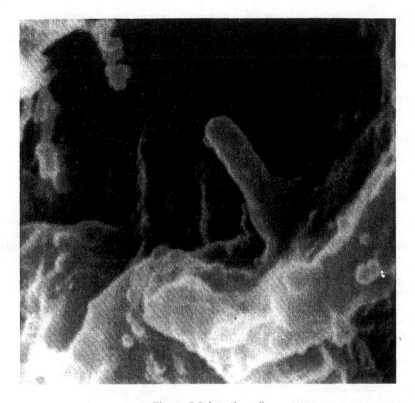

Figure 3.2 (continued)

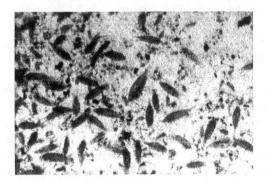

Figure 3.3 Macroinvertebrates (copepods) in distribution water pipe.

changes. The appearance of copepods, nematodes, and fly larvae were more directly associated with seasonal occurrences and increased rainfall events.

These circumstances provide the opportunity to explore invertebrate–bacterial associations and the data obtained (Table 3.8) did provide evidence that a variety of heterotrophic bacteria could be isolated from individual crustaceans living in

Table 3.7   Conditions Present when Invertebrates Were Found within the
Distribution System of Worcester, MA

Invertebrate type	Flow rate		Total chlorine (mg/l)	Pipeline characteristics
	mgd	Ml/d		
Amphipod	<0.5	<1.9	<0.5	<8 in. (200 mm), cast iron, older pipe
Copepod	>1.0	>3.8	>0.5	>8 in. (200 mm), all types, all ages
Fly larva	>1.0	>3.8	>0.5	>8 in. (200 mm), cast iron, all ages
Nematode	>1.0	>3.8	>0.5	>8 in. (200 mm), cast iron, all ages

Data from Levy, Hart, and Cheetham.[14]

Table 3.8   Bacteria Isolated from Invertebrates Collected from the Drinking
Water Distribution System of Worcester, MA, 1982 to 1984

Bacterium	Invertebrate type			
	Amphipod	Copepod	Fly larva	Nematode
Acinetobacter sp.	+			
Achromobacter xylooxidans	+			
Aeromonas hydrophila			+	
Bacillus sp.	+			
Chromobacter violaceum	+			
Flavobacterium meningosepticum		+		
Moraxella sp.	+	+		
Pasteurella sp.	+	+		
Pseudomonas diminuta		+		
Pseudomonas cepacia			+	+
Pseudomonas fluorescens	+		+	
Pseudomonas maltophilia	+			
Pseudomonas paucimobilis	+	+		
Pseudomonas vesicularis	+			
Serratia sp.	+			
Staphylococcus sp.	+			

Data from Levy, Hart, and Cheetham.[14]

the Worcester water supply lines.[14] While none of the organisms were coliforms, most genera could be considered to be opportunistic bacterial candidates. Scanning electron microscopy confirmed that bacteria associated with copepods were entrapped in the digestive system while the bacterial flora of other invertebrates examined was found either internal or external to the particular individuals. The density of bacteria encountered per animal (Table 3.9) appeared to be related to the size of an individual animal, to attachment sites on the animal skeletal structure, and passage of viable organisms in animal excreta. Apparently, not all entrapped bacteria are digested in every instance.

Disinfection experiments using the amphipod *Hyalella azteca* (which was fed *Enterobacter cloacae, Klebsiella pneumoniae,* and *Salmonella livingstone*) indicated that bacteria associated with *H. azteca* survived the disinfection

Table 3.9 Quantification of Bacteria Associated with
Invertebrates Collected from the Distribution
System of Worcester, MA, 1982 to 1984

Invertebrate type	Length in.	mm	Number of bacteria (cfu per animal)
Amphipod	0.12–0.39	3–10	10^2–10^1
Copepod	0.04–0.08	1–2	1–10
Nematode	0.02–0.08	0.5–2	10–10^2
Fly larva	0.20–0.39	5–10	10–10^2

Data from Levy, Hart, and Cheetham.[14]

process better (Table 3.10) than the same bacteria in an unassociated aquatic state (Table 3.11). Also noted was that free available chlorine was always more effective for inactivating associated bacteria than monochloramine at equivalent contact times and concentrations. This was true at both study temperatures (4 and 10°C). However, it must be realized that these data are from bench experiments and may not include additional protective factors common to pipe biofilm and porous tubercle formations.

NEMATODES

Nematodes are unsegmented roundworms or threadworms (Figure 3.4). Many of these macroinvertebrates are free-living in soil and fresh water. Some species are plant pathogens while others live in animal tissue as parasites. Those that parasitize man and other animals are visible to the naked eye. Free-living species are microscopic (10 to 40 μm in width and 100 to 500 μm in length). Their life cycle includes mating, egg laying, and the development of larvae and adults.

Some idea of the occurrence of nematodes in the aquatic environment may be deduced from information on genera diversity in different surface-water conditions. For example, the Saluda River (South Carolina) provided a wide variety of nematodes with 13 genera being detected.[81] Impoundment of the water into Lake Saluda yielded only five genera in the water column. Perhaps settling in these quiet waters caused more nematode types to move to the bottom sediment. Several investigations on the origins of nematodes entering water purification plants concluded that the major contribution in densities released is from sewage effluents, with lesser numbers being contributed by subsurface drainage and surface runoff.[82-84] Effluents from trickling filters contained 2000 to 2500 nematodes per gallon; primary settling, 200 to 500 nematodes per gallon; and stabilization pond effluents, only 20 nematodes per gallon. More diversity in genera were found in effluents from trickling filters and each nematode from this source carried about 100 viable bacteria. Of the nematode carrying bacteria, 5 to 10% were coliforms. No *Salmonella, Shigella*, or enteroviruses were recovered from these nematodes.[85]

Table 3.10 Kinetics of Disinfection of Bacteria Associated with *Hyalella azteca* with Free Available Chlorine (pH 6) and Monochloramine (pH 8)

Bacterium	Disinfectant	Concentration (mg/l)	Survival (%)									
			20°C					40°C				
			0 min	20 min	40 min	60 min	120 min	0 min	20 min	40 min	60 min	120 min
Enterobacter cloacae	Free available chlorine	0.5	100	96	78	68	a	100	94	88	85	a
		1.0	100	26	24	20	a	100	70	58	47	a
		2.0	100	28	10	4	a	100	61	51	30	a
	Monochloramine	0.5	100	96	81	78	88	100	96	95	93	90
		1.0	100	65	50	22	29	100	75	62	49	47
		2.0	100	60	48	18	24	100	74	51	35	42
Klebsiella pneumoniae	Free available chlorine	0.5	100	89	80	67	58	100	96	90	83	86
		1.0	100	36	30	21	15	100	54	40	31	26
		2.0	100	31	26	9	2	100	48	41	36	14
	Monochloramine	0.5	100	83	62	50	47	100	86	78	69	66
		1.0	100	76	43	37	30	100	80	44	40	39
		2.0	100	58	36	30	23	100	75	50	36	29
Salmonella livingstone	Free available chlorine	0.5	100	93	82	71	56	100	95	85	80	61
		1.0	100	60	42	27	19	100	69	43	35	21
		2.0	100	36	26	19	6	100	39	36	24	10
	Monochloramine	0.5	100	80	58	51	59	100	85	80	75	70
		1.0	100	70	52	34	30	100	86	56	54	35
		2.0	100	61	45	29	20	100	69	54	36	33

a No data available.

Data from Levy, Hart, and Cheetham.[14]

Table 3.11   Kinetics of Disinfection of Unassociated Bacteria with 1 mg/l of Free
Available Chlorine (pH 6) and Monochloramine (pH 8)

Bacterium	Disinfectant	Temperature (°C)	Time to 99% inactivation (min)
Enterobacter cloacae	Free available chlorine	4	4
	Free available chlorine	20	4
	Monochloramine	4	25
	Monochloramine	20	15
Klebsiella pneumoniae	Free available chlorine	4	2
	Free available chlorine	20	1
	Monochloramine	4	21
	Monochloramine	20	12
Salmonella livingstone	Free available chlorine	4	2.5
	Free available chlorine	20	1
	Monochloramine	4	16
	Monochloramine	20	11

Data from Levy, Hart, and Cheetham.[14]

Figure 3.4   Nematodes in distribution water.

Sixty-six nematode genera were isolated in surface waters used by several water treatment plants in Ontario.[82] Estimated mean densities ranged from 0.6 nematodes per liter in winter (under cover of ice) to 10.6 organisms per liter during the spring thaw. Fluctuations in nematode densities throughout the year are a reflection of activities on the watershed, major rainfall events, lake (impoundment) destratifications, turbidity fluctuations, and methodology configurations used in the survey.

In water treatment processes, free-living nematodes have been found to extensively inhabit slow sand filter beds.[86] It was calculated that 10^7 nematodes might be present in the top 3-in. layer of sand in an acre of surface area. This should not be surprising because the biological surface activity of the slow sand filter involves many organisms in the biofilm community referred to as schmutzdecke. Rapid sand filters provide a different picture; no nematodes were found in sand samples taken at various depths, although nematodes were demonstrated to be present in the finished water.[87] Water treatment processes that include coagulation, sedimentation, and filtration remove between 90 to 98% of the nematodes from raw source water.[81]

Because the disinfection process as generally applied to inactive bacterial and viral organisms of public health concern will not kill nematodes, an occasional few nematodes per liter may pass the treatment barriers and appear in the water supply. During an 11-month period, 16 of 22 water supplies were found to contain nematodes in the finished water assigned to eight genera.[69,88] While many of the water samples contained 10 or less nematodes per gallon, the two highest densities were 30 and 38 nematodes per gallon, showing no reduction after the water was processed. While this passage of macroinvertebrates in itself does not present a health risk, it might be a route for protected transport of bacteria, given favorable circumstances.

Laboratory experimentation has demonstrated that there could be a 5 to 6% survival of *Salmonella typhosa* and *Shigella sonnei* in nematodes within 24 h or 0.1% survival 48 h after ingestion. Furthermore, *S. paratyphi* and the Coxsackie A-9 virus were able to survive at higher percentages (12 to 16% after 24 h and 1% after 48 h).[89] Preferential digestion may also account for the report of no *S. typhi* survivors in nematode defecation in tap water while there was a 11.7% recovery of *S. wichita* in tap water stored at room temperature for 10 to 24 h.[90] These inconsistencies may be a reflection of the effectiveness of nematode digestive enzymes in attacking specific cell wall compositions of different bacterial species.[91]

Protection of enteric bacteria from the impact of chlorine exposure by passage in carrier nematodes is a possibility. Again, laboratory experimentation has demonstrated this aspect. First, unassociated *Salmonella* and *Shigella* organisms in water were completely destroyed in 1 min with as little as 2.5 mg/l chlorine residual while Coxsackie A-9 and Echo 7 virus inactivation in the same unassociated state required 1 to 10 min and 2.0 to 2.2 mg/l residual chlorine at 25°C. When ingested by nematodes, these organisms survived chlorination exposures ranging from 120 min at 11.8 to 12.6 mg/l or 30 min at 83.2 to 87.5 mg/l of residual chlorine. All of the resulting C·T values were above those used for disinfection by most water utilities.

This evidence suggests that while the release of viable bacterial cells or virus particles in defecation by nematodes is inconsistent, release of any ingested pathogen via rupture of nematodes in high-velocity movement through a water main should not be disregarded. For these reasons, finished water and water in distribution should be periodically examined during the summer months for nematode occurrence and density. When nematode densities exceed 10 per gallon, action should be taken to locate the source of this population and remedy the problem as a potential threat to water quality.

OPPORTUNISTIC ORGANISMS AND
THE WATER SUPPLY CONNECTION

The heterotrophic bacterial population in drinking water is composed of many transient organisms that never colonize the distribution system, while other associate organisms are more opportunistic, being capable of surviving on minimal nutrients, attachment to pipe sediments, and becoming participants in the developing biofilm. While these bacteria are generally of no public health significance, some opportunistic colonizers of the pipe network may, in addition, become colonizers of the human body through contact with the water supply.

Opportunistic pathogens are generally understood to include those organisms that may exist as part of the normal body microflora but under certain conditions cause disease in compromised hosts. Such organisms become particularly invasive to susceptible individuals (elderly, newborns, AIDS victims, cancer patients receiving chemotherapy, burn cases, dialysis patients, trauma patients, and individuals receiving organ transplants). The route of exposure may be ingestion, inhalation or body contact with the water supply during bathing (whirlpool use, dental equipment, etc.), and indoor air climate control devices (for humidification and air cooling).

COLONIZATION CONCERNS IN THE HOSPITAL

Opportunistic pathogen infections are a serious public health threat anywhere there are large numbers of people in close confinement, such as nurseries, preschools, summer camps, and, in particular, hospitals and senior-care facilities. At least 5% of patients admitted to hospitals acquire nosocomial infections and about 1% of the patients die as a direct result.[92] Many of these organisms occur in the diverse heterotrophic flora found in water supplies.[93-98] In general, these are the organisms that, when found in large enough numbers and in the wrong place at the right time, have the potential to cause an infection. Some examples of nosocomial outbreaks associated with contaminated potable water are given in Table 3.12.

Heterotrophic bacterial densities in most municipal water supplies are generally below 100 organisms per milliliter except at static water locations in buildings where densities are often one or two log higher because of warm ambient temperatures. By contrast, the infective dose levels for a 50% attack rate for the opportunistic pathogen subset of the heterotrophic bacterial population may range up to 1×10^{10} cells per dose. While the number of cells

Table 3.12   Documented Nosocomial
Outbreaks Associated with
Contaminated Potable Water

Etiologic agent	Illness	Ref.
Pseudomonas	Wound	94
	Wound	95
	Dermatitis	96
	Meningitis	97
	Respiratory	98
	Respiratory	99
	Cellulitis	100
Acinetobacter	Peritonitis	101
Mycobacterium	Septicemia	102
	Bacteremia	103
	Peritonitis	4
Flavobacterium	Septicemia	29
	Respiratory	26
Legionella	Respiratory	105
Klebsiella	Urinary, respiratory	106

Information reported from Highsmith et al.[99]

required to achieve an infective dose may seem unlikely to occur often, the volume of water used to take a shower or bath can easily supply this density during a given exposure period. By contrast, the density of such opportunistic organisms needed to establish infection in newborn babies, postoperative or immunosuppressed patients, the elderly, and the infirm is generally lower than for healthy children and adults.

An additional factor to consider is colonization by these organisms in water attachment devices used in hospitals and clinics. *Acinetobacter* infections have been associated with the use of a ventilator spirometer,[112] room humidifiers,[113,114] and moisturized Wright respirometers.[115] *Serratia marcescens* infec-. tions have been transmitted via medical solutions,[116,117] and peritoneal-dialysis effluents.[118] While such water-related systems and equipment may be amplifiers of opportunistic pathogens, the sources of these organisms may be the water supply, handling of the device, or airborne contaminants, to name a few. The contribution of the water supply to the problem has been the subject of two studies on water supply-associated bacteria and patient/consumer illness.[101,119] Both studies suggested that water supply organisms are part of the problem but not necessarily the major source of nosocomial infections in the hospital environment or in drinking water at the home tap.

BACTERIAL AMPLIFIERS IN BUILDING PLUMBING

Water supply systems in large housing projects, highrise office buildings, hotels, and large public buildings exacerbates the problem of deteriorating water quality as a consequence of static water or infrequent water demands.

Static water in building plumbing networks is often at warm-water tempera-tures that stimulate bacterial growth in the accumulated sediments. Hot-water tanks in homes and building water systems attachment devices should not be overlooked as a cause of water quality deterioration in home or care centers. The opportunistic pathogen *Legionella* illustrates this risk in relationship to hot-water generation for use in bathing. *Legionella* is remarkable in its ability to grow in hot-water environments such as hot-water tanks and associated attachment devices or materials (e.g., shower heads, gaskets). When thermo-stats on hot water tanks in homes or in senior citizen facilities have been set to below 55°C, as an energy conservation measure or to prevent scalding of elderly patients, amplification of this opportunistic pathogen may occur and be released in aerosols that will be inhaled.[120-128] Most residential hot-water tanks are heated from the bottom, near the cold-water entrance pipe, so the water supply can be quickly heated to above 55°C; however, accumulating sediments at the bottom of the tank provide a heat-buffered environment for *Legionella* colonization. Water in large institutions is often heated by internal steam coils located at mid-depth in the tank; thus, the cooler water in the bottom may not be heated sufficiently to kill *Legionella*. Recirculation of the hot water may spread the organisms to all parts of the system.

Legionella exposure through inhalation of water aerosols is a classic exam-ple of why risk in water supply must be viewed beyond the concern for ingestion of intestinal pathogens. Added to this risk is a concern for body contact during bathing with other opportunistic pathogens such as *Pseudomo-nas aeruginosa*, mycobacteria, and *Staphylococcus aerus* that may be present in some high-density occurrences of heterotrophic bacteria.

Cold-water storage tanks in highrise buildings must be covered to prevent introduction of contamination from nesting birds and atmospheric dust and to reduce or prevent algae growth. Various heterotrophic bacteria may enter via this route and colonize in the accumulating bottom sediments. Algae may also be introduced by the same routes and proceed to grow in the available light from the open storage tank. Growth in an open water supply tank and subse-quent release of algal toxin was the cause of one waterborne outbreak of diarrhea confined to a Chicago apartment building.[126]

Increasingly, water utility customers who are dissatisfied with taste and odor or fear the potential health risks alleged to be associated with their municipal water supply are attempting to further refine the water quality at the tap. While treatment devices may be very effective initially in providing aesthetic treatment of the water, their usefulness over time may diminish because of unpredictable service capacity to adsorb a varying mixture of trace contaminants, the quality characteristics of the water supply, and the volume of water processed over time. Poor design of attachment devices may provide recesses that do not drain. Such tiny pools of water become active sites for biofilm development that accelerates in the warm environment and periodically diffuses into the interrupted flows of product water. Cellulosic paper filters must be avoided because the fibers are quickly degraded by bacteria. Point-of-use devices installed below the kitchen sink (where ambient air temperatures are often warm) should be insulated to minimize water temperature elevations within the unit. Each unit should have a built-in flow meter that indicates when there is a need for treatment cartridge replacement based on water volume passage. Backflow preventers should be considered essential on those building plumbing networks that have chronic problems with low line pressure, because effluents from old filter units and deionizing columns could discharge a variety of contaminants back into the service line. Because monitoring of product water will rarely be considered by most users of water treatment devices, it is important that manufacturers provide careful guidance in their instruction manuals on proper installation and maintenance. These manuals should stress the importance of carbon filter replacements and routine early morning flushing (30 s) of the system prior to use each day.

Another aspect of bacterial proliferation is colonization in devices with carbon filters. While some organisms may pass through the device with little or no retention, others are amplified in these units and bacteria are released at densities higher than those found in the incoming public water supply. Challenge of carbon filter devices with coliforms, opportunistic organisms, and primary pathogens (bacteria anticipated in cross-connections, line breaks, or back-siphonage) revealed that such treatment units do not provide an effective barrier. While *Escherichia coli, Salmonella,* and other organisms pass through the filter, other organisms such as *Klebsiella pneumonia, Aeromonas hydrophila,* and *Legionella pneumophila* can colonize these devices. As a consequence, devices using carbon filters should not be used on an untreated water supply of questionable quality.[37,38,127]

While treatment devices may be very effective initially in processing the water supply, their usefulness over time may diminish because of capacity to adsorb, quality of source water, and volume of water processed. For these reasons, carbon filter cartridges need to be replaced, water softeners recharged, and microbial filter barriers backwashed periodically. The microbial quality of the product water produced by these devices is extremely variable.[37,38] Bacterial densities in water from point-of-use carbon filters can be expected to increase by one to two logs over the number detected in public water supplied at the service tap. This proliferation relates to the species of organisms passing through the filter device, seasonal changes in water temperatures (cold-water period, warm-water period), the presence or absence of a free chlorine residual, air temperatures around the unit, and the service duration for a given carbon cartridge. Nonuse periods overnight or for longer intervals will also provide an opportunity for continued growth of organisms colonizing carbon filters. Under no-flow conditions, water temperatures can increase to ambient temperatures, thereby stimulating bacterial multiplication. Chlorine residuals in this situation either form by-products over time in the warm static water or are effectively absorbed by carbon in the unit, thereby providing no deterrent to bacterial quality changes in the water. As a consequence, product water in these test units frequently had higher bacterial densities in morning samples (Table 3.13).

REPRESENTATIVE OPPORTUNISTIC ORGANISMS

There is a variety of heterotrophic organisms that can occur in any "safe water supply." They are, indeed, opportunistic in the broad sense, adjusting to a harsh environment and taking advantage of selected sites in the water supply system to colonize. Given a special set of circumstances, water supply contacts with breaks in the human body barriers against disease result in a similar pattern of colonization of selected tissues that lead to illness if natural defenses prove ineffective. While many heterotrophic organisms in the water supply may be capable of colonizing the pipe network, only a few have the potential to be significant opportunistic pathogens. Such is the nature of eight candidates, which are discussed here.

Acid-Fast (Nontuberculous) Bacteria

Acid-fast (nontuberculous) bacteria pass through water treatment barriers in very low densities. Speciation of water supply isolates reveals that this group of acid-fast bacteria includes *Mycobacterium fortuitum, M. phei, M. gordonae, M. xenopei, M. kansasii, M. avium,* and *M. chelonae.*[3,128-132] The pathological significance of these organisms is that colonization in humans may occur in the lungs and lymph nodes or they may cause skin lesions, septicemia, and postsurgery complications. Furthermore, nontuberculous mycobacterial disease is the third most common opportunistic fatal infection in patients with AIDS.[128] Waterborne mycobacterial infections present the greatest risk to patients in the hospital setting, particularly those susceptible older individuals who bathe in aerosolized water during the summer months.[3] Extensive searches for the cause of two nosocomial outbreaks of *Mycobacterium* infection (*M. fortuitum* and *M. gordonae*) in different hospitals revealed that these organisms were associated with ice and ice water taken from contaminated ice machines.[133,134]

Nontuberculous mycobacteria can be isolated from human fecal material.[135,136] These organisms were isolated in 40% of stool samples examined from health subjects (the mean density of 19 acid-fast organisms per gram of feces).[136] Wastes from pig farms also contained mycobacteria[137] and wastewater effluents were reported to contain approximately 10,000 organisms per 100 ml.[138]

Raw source waters at water supply intakes have been shown to contain acid-fast bacteria.[131,139,140] Acid-fast bacteria were found in the raw water to the Oakwood, IL water treatment plant at a geometric mean density of 271 organisms per liter, while at Decatur, IL, raw water densities were approximately

Table 3.13 Heterotrophic Bacterial Density (Monthly Mean), Comparisons for A.M. and P.M. Samples[a]

| | Organisms per ml | | | | | | | |
| Month | 1984[b] | | 1985[b] | | 1986 | | 1987[b] | |
	A.M.	P.M.	A.M.	P.M.	A.M.	P.M.	A.M.	P.M.
January	—	—	140,000	130,000	230,000	190,000	71,000	35,000
February	470,000	190,000	130,000	120,000	260,000	200,000	110,000	72,000
March	270,000	150,000	110,000	88,000	230,000	200,000	100,000	73,000
April	220,000	130,000	190,000	110,000	270,000	180,000	110,000	120,000
May	460,000	260,000	240,000	63,000	330,000	220,000	170,000	140,000
June	410,000	210,000	120,000	47,000	230,000	160,000	120,000	81,000
July	150,000	42,000	150,000	48,000	220,000	170,000	64,000	97,000
August	15,000	14,000	360,000	43,000	220,000	140,000	4,900	7,800
September	46,000	41,000	5,500	6,600	150,000	100,000	27,000	34,000
October	84,000	49,000	19,000	19,000	140,000	130,000	91,000	83,000
November	140,000	110,000	140,000	110,000	150,000	92,000	140,000	83,000
December	150,000	93,000	270,000	250,000	210,000	120,000	—	—
Overall mean	220,000	117,000	156,000	86,000	220,000	158,000	92,000	80,000
A.M. to P.M. decrease	−47%		−45%		−28%		−13%	

[a] Data for third faucet carbon filter treatment device.
[b] Filter replacements: July 6, 1984; September 9, 1985; August 8, 1987.
Data from Geldreich and Reasoner[127]

one order of magnitude less. Upon passage through treatment, the most significant reductions of acid-fast bacteria occurs during sand filtration. In an 18-month study of these two water systems, reductions in the concentrations of acid-fast bacteria by rapid sand filtration ranged from 59 to 74%.[141] In the finished water these organisms could be isolated in 36% of all 1-1 samples.

Reported findings of acid-fast bacteria in finished water demonstrate that these organisms are resistant to the usual chlorine disinfectant C·T values applied to inactivate coliforms and viruses.[142] Experiments using recent *Mycobacterium* isolates from chlorinated water supply (*M. fortuitum*, *M. gordonae*, and *M. avium*) and clinical isolates of *M. chelonae*, *M. Kansasii*, and *M. intracellulare* revealed that chlorine levels of less than 1.0 mg/l may not be adequate for effective inactivation of these opportunistic pathogens.[143] Even the presence of a free chlorine residual (<0.2 mg/l) at a low water pH (5.9 to 7.1) did little to reduce acid-fast bacteria in the distribution system. Mycobacteria were also reported to be more resistant than *E. coli* to inactivation by inorganic chloramines[144] and by ozone.[145]

While densities of mycobacteria entering the distribution system may be only a few organisms per liter, this density may change significantly during warm-water periods in static sections of the distribution network. Regrowth may also be intensified in older portions of the pipe network where corrosion is a problem and water pH is elevated to combat corrosivity. The trade-off is less effective disinfectant action of free chlorine at higher pH. Some regrowth was also noted in dead-end areas where chlorine residuals disappeared and total organic carbon concentration and turbidity were higher.

Devices attached to building plumbing systems may also be amplifiers of mycobacteria. Mycobacteria are among the first organisms to colonize reverse-osmosis membranes used in producing reagent grade water, reuse water systems, and medical devices. For example, nontuberculous mycobacteria were detected in water from 95 of 115 hemodialysis centers that reused disposable hemodialyzers (artificial kidneys) for the same patient.[146] Increased incidence or patient infections caused by acid-fast mycobacteria prompted an investigation that concluded water was the source of these opportunistic pathogens.

The water tap may also be a source of mycobacteria, with the organisms colonizing the sediment accumulations in the device itself. Apparently, the presence of these organisms can invariably be found in scrapings or swabbings from the cold- and hot-water taps.[132] In a study involving three hospitals, *M. xenopei* was recovered from 61 of 111 pairs of hot- and cold-water taps, 20 of 74 tap pairs in another hospital, but from only 3 of 61 pairs of taps in the third hospital. Positive findings were more often reported from the hot-water tap, an observation that is not surprising because the optimum growth temperature for acid-fast bacteria is 42 to 44°C.

Fecal Klebsiellae

Fecal klebsiellae have their origin in the intestinal tract of warm-blooded animals[50,142,143] while many other *Klebsiella* are of environmental origin. The genus includes *K. pneumoniae, K. oxytoca, K. ozaenae, K. planticola, K. terrigena,* and *K. rhinoscheromatis.* Most of these species have been detected in coliform-contaminated public water supplies.[5-7,10,95,149-154] *K. pneumoniae* and *K. oxytoca* have often been reported to be the predominant organisms in distribution biofilm occurrences. These occurrences in water supply pose the questions: Are these klebsiellae of fecal origin? Can they be a potential opportunistic pathogen to susceptible individuals in the community?

In response to the first question, approximately 30 to 40% of all warm-blooded animals, humans included, have *Klebsiella* in their intestinal tracts, with individual densities ranging up to 10^8 *Klebsiella* per gram of feces.[5-7,10,95,149-157] An estimated 60 to 85% of all *Klebsiella* isolated from feces and clinical specimens were positive in the fecal coliform test and were identified as *K. pneumoniae.*[158-161]

K. pneumoniae, particularly antibiotic-resistant serotypes, can cause human infections of the respiratory system, genitourinary tract, nose, and throat, and occasionally meningitis and septicemia.[162-163] *Klebsiella*-caused infection is sometimes of apparent primary etiology, but more often is found in mixed infection or as a secondary invader.[164] In the hospital environment, the nosocomial pathogenic infection rate was 16.7 infections per 10,000 patients from 94 hospitals.[165] *Klebsiella pneumoniae* was the cause of 1.1% of all nosocomial hospital deaths during the same period. Infections of the urinary system, lower respiratory tract, and surgical wounds were the most frequent cause of *Klebsiella*-associated illnesses or deaths. The lack of evidence of increased illness in a community during a coliform biofilm event may relate to difficulties in gathering reports of water-related illness cases among susceptible people at home or in the work environment vs. patients in the hospital setting.

Surface water and unprotected groundwater receive *Klebsiella* from both environmental and fecal sources. Environmental strains are introduced to the source of supply from urban and rural runoff and by discharges of wood-, paper-, and textile-processing wastes.[148,166-173] Fecal *Klebsiella* enters the water in point source discharges from municipal sewage and meat-processing works and in nonpoint source discharges from farm animals.[174-177]

Klebsiella in surface waters and groundwaters under the influence of surface-water contamination will vary considerably in density, which is related to stormwater, runoff events, and domestic and industrial discharges (paper, pulp, and textile mills) to receiving streams and lakes on the watershed. Total *Klebsiella* in these waters are often present in higher densities than fecal coliforms (Table 3.14). This observation supports the argument that more environmental strains of *Klebsiella* than fecal coliforms are to be found in

Table 3.14 Coliform Profiles in Wells, Springs, Lakes, Creeks, and Sewage Effluent

Water source	No. of samples	Total coliforms per 100 ml[a]	Fecal coliforms per 100 ml[a]	Klebsiella per 100 ml[a]	Occurrence (%) Fecal coliforms	Occurrence (%) Klebsiella
Wells						
Brown County	1	2,600	240	683	9.2	26.3
Indiana	1	330	<1	283	—	85.8
Tylersville	1	22	<1	8	—	36.4
Springs						
Clough	1	39	2	5	5.1	12.8
Ludlow	1	115	89	8	77.4	7.0
Lakes						
Burnet	3	680	330	67	48.5	9.9
Tepper	1	56,000	3,300	4,500	5.9	8.0
Creeks						
Mill	1	32,000	2,200	9,300	6.9	29.1
Sycamore	1	170,000	8,300	33,000	4.9	19.4
Muddy	1	2,600,000	68,000	70,000	2.6	26.9
Rivers						
Ohio	3	15,000	1,300	2,100	8.7	14.0
Little Miami	1	18,600	1,100	2,060	5.9	11.1
Licking	1	25,360	2,600	2,300	10.3	9.1
Sewage (Sycamore) primary effluent	3	21,380,000	1,947,000	8,860,000	9.1	41.4

[a] Mean of three replicates for each sample using appropriate MF procedures.
Data adapted from Geldreich and Rice.[56]

polluted waters and sewage effluents where high concentrations of nutrients provide regrowth potential to these organisms. The data also demonstrate that fecal *Klebsiella* are less numerous than environmental strains of the same genus.

Water quality at surface intakes can be quite variable if the source is a river, with total coliform counts ranging from 100 to 20,000 organisms per 100 ml. In these instances, the fecal coliform content may range from 1 to 10% and *Klebsiella* from 9 to 25%. Nondisinfection treatment may reduce the total coliform densities to between 20 and 200 organisms per 100 ml. Adequate disinfection C·T application to these process waters will further reduce the total coliform population to less than one organism per 100 ml in a properly operated treatment train. So the question is, how do *Klebsiella*, particularly fecal *Klebsiella,* enter the distribution system? The answer may be found in momentary breaks in the treatment barriers (unstable filtration after backwashing and inadequate C·T applications) or through cross-connections (low line pressure), water main breaks, line repairs, and finished water contamination during storage in open reservoirs.

Upon entering the distribution system *Klebsiella* may colonize biofilms located in pipe sediments and tubercles situated in slow-flow sections, dead ends, and on walls or sediments in storage tanks. Most of the *Klebsiella* waterborne occurrences are not of fecal origin. In those infrequent situations where the laboratory analyses reveal fecal *Klebsiella* in the distribution system, there should be a high-priority effort to destroy the colonization sites because of the concern for more frequent releases of this opportunistic pathogen at higher densities into the water supply. Infective dose (ID_{50}) values for environmental and clinical isolates of *Klebsiella* have been reported to be between 3.5×10^1 to 7.9×10^5 cells per milliliter.[178] Therefore, ingestion of 100 ml of drinking water (approximately one glass of water) containing greater than 3.5×10^1 *Klebsiella* per milliliter could present a risk to susceptible individuals. Inhalation of moisture from vaporizers using drinking water contaminated with opportunistic *Klebsiella* should also be considered a risk to some individuals.

Flavobacterium, *a Genus of Pigmented Bacteria*

Flavobacteria have often been identified in water supplies. Many of these chromogenic bacteria form bright-yellow colonies while others have been reported to produce red or green colonies (when grown at incubation temperatures below 35°C)[30,179,180] This group of organisms are generally considered to be gram-negative, aerobic, oxidase-positive, nonfermentative, and nonglucose oxidizers.[41] In addition, they generally do not demonstrate typical reactions (or any at all) in most biochemical identification media used to characterize to species. Thus, *Flavobacterium* turns out to be a difficult group to properly identify.

Flavobacteria can be opportunistic pathogens and have been implicated in human infections including meningitis (particularly in infants), pneumonia, endocarditis, and septicemia.[181-185] *F. meningosepticum* is the species most commonly involved in nosocomial infections. In the hospital, flavobacteria have been recovered from a variety of water system attachment devices including nebulizers, humidifiers, hemodialysis systems, distilled water systems, sink faucets, and drinking fountains.[30,32,184,186]

In the aquatic environment, flavobacteria are ubiquitous, so it is not unusual to detect these organisms to some degree in all water supplies. During a 5-month study of the Boston metropolitan water supply, *Flavobacterium* species were detected in the disinfected water from five of nine finished water reservoirs.[30] A study of the seasonal occurrence of various pigmented bacteria (including *Flavobacterium*) in the Cincinnati treated water system suggested that these organisms originated from the raw water supply at some previous point in time, or from line breaks and repairs to the distribution system.[187] In this study, the predominant pigmented bacteria at most sample locations, either in treatment stages or in the distribution network, were yellow and orange strains with a small incidence of pink organisms specific to a selected flowing distribution site. Changes in water temperature due to seasonal weather changes were a major influence on regrowth of these organisms in the distribution network. At one dead-end site, 80.5% of all pigmented bacteria detected during the summer were yellow pigmented organisms, while in spring 93.8% of the pigmented bacteria observed in samples collected at the same site were orange pigmented.

Little is known about the environmental conditions in the distribution system that favor the colonization of these heterotrophic bacteria but the available organic concentration may be below 1 μg of carbon per liter.[188] Current evidence suggests that the conditions that contribute to *Flavobacterium* regrowth include absence of free chlorine residual, water temperatures above 15°C, accumulations of bacterial nutrients (assimilable organic carbon and specific inorganics) in pipe sediments, and static water conditions.

Stagnation of building plumbing systems can provide opportunities for *Flavobacterium* colonization that may become public health concerns. In one situation there was an investigation in response to complaints about taste or color of water taken from drinking fountains and faucets in several modular office buildings.[184] Bacteriological analyses revealed no coliform bacteria, but heterotrophic bacterial populations that exceeded 6000 organisms per milliliter were found in many of the samples. Most of the water samples collected from stagnant lines contained a predominance of *Flavobacterium* that was both chlorine and copper resistant. Six employees complained about intense abdominal cramping 6 to 8 h after drinking water from fountains that were demonstrated to contain high counts of *Flavobacterium*. It was theorized that enough *Flavobacterium* was released through water ingestion to cause endotoxin gastroenteritis. Further investigation of the plumbing pipe network suggested that the bacterial colonization occurred in half-inch copper lines leading to drinking fountains. *Flavobacterium* was not isolated from the tubing or reservoirs within the refrigerated drinking fountains. No stagnation was evident in the public water supply lines entering the buildings.

Legionella pneumophila

Legionella pneumophila is an important waterborne opportunistic pathogen that causes Legionnaire's disease in susceptible individuals exposed to contaminated aerosols from shower baths and air-conditioner heat exchanges. The respiratory disease results in a complex colonization of the body that is responsible for pneumonia with significant mortality rates among senior citizens. Pontiac fever, another illness caused by legionellae, is a nonpneumonic, nonfatal, and self-limiting disease. Apparently, there is no human carrier state or reservoir for legionellae bacteria in warm-blooded animals.

While this group of small, gram-negative bacteria have an absolute nutritional requirement for L-cysteine,[189] it is somewhat surprising to find legionellae widespread in the aquatic environment. These organisms have been detected in freshwater streams and lakes in both North America and Europe, and also the tropical waters of Puerto Rico.[190-192] In one study of 793 water samples collected from 67 different lakes and rivers throughout the United States, virtually all sources were positive for *L. pneumophila*, using the direct fluorescent antibody technique for detection.[193] There is some indication that legionellae are very infrequently found in groundwater, unless there is some surface-water runoff seepage or poor soil barrier protection.[194,195]

Water treatment processes may play some role in the development of an ecological niche for *Legionella* through the release of assimilable organic nutrients, particulates, and some heterotrophic bacteria into the distribution system as a result of uneven, interrupted, or failed treatment processes.[196-201] Among the microorganisms in the raw source water that sequester *Legionella* and provide safe passage through the disinfection process are algae, amoebas, and ciliates.[106,107] Airborne legionellae in dust or particulate laden rain showers may find their way to the open-air treatment basins. Common pathways for their entry into the distribution system include reservoir air vents, main construction, pipeline repairs, cross-connections, and back-siphonage.[202]

Establishment of *Legionella* in the distribution system is most likely to occur in biofilm locations where symbiotic relationships with other heterotrophic bacteria (*Flavobacterium breve, Pseudomonas, Alcaligenes,* or *Acinetobacter*) provide the critical nutritive requirements necessary for the long-term persistence of this opportunistic pathogen.[199,203,204] Many of these sites will be found at the periphery of the system (long pipe runs into dead ends) and in little-used service connections throughout the pipe network where the water can stagnate. Densities of *Legionella* may be only a few cells per liter in water supplies[78,205] and the mere presence of these few legionellae in drinking water does not pose a direct health threat until there are opportunities for amplification (hot-water tanks, shower heads, water evaporator cooling devices, etc.).

Efforts to eliminate low levels of these organisms (a few per liter) in water supply treatment processes and in the distribution system network are not cost-effective. Water utility operations, however, can minimize regrowth potential through good housekeeping practices that include removal of scums and bio-film accumulations at air–water interfaces in treatment basins, connecting flumes, and attachments to agitator paddles in flocculation basins. System-wide flushing, with particular emphasis on dead-end sections during warm-water periods, will significantly suppress further development of biofilm and introduce detectable disinfectant residuals to those areas where *Legionella* and other heterotrophic bacteria may persist. The net effect desired is to suppress microbial symbiotic relationships that are essential to *Legionella* metabolism.

Pseudomonas aeruginosa

Pseudomonas aeruginosa is the most significant pseudomonad in drinking water. Pseudomonads are ubiquitous bacteria that are able to flourish in a wide variety of habitats (surface waters, aquifers, sea water, soil, and vegetation). Some pseudomonads are among the prominent denitrifiers while others grow prodigiously in and on tertiary treatment devices such as reverse-osmosis and electrodialysis membranes and in sand or carbon filtration beds. Pseudomonads reported in some drinking water supplies include *P. aeruginosa*, *P. cepacia*, *P. fluorescens*, *P. mallei*, *P. maltophila*, *P. putida*, and *P. testosteroni*.[206,207] To this list can be added *P. stutzeri*, *P. diminuta*, and *P. acidovorans*, which have been found in bottled waters at densities ranging from 1×10^3 to 1×10^5 organisms per milliliter.[208-210] These organisms meta-bolically adapt to survive on minimal nutrient concentrations typical of pro-tected aquifers and treated drinking water.

The ability of *P. aeruginosa* to rapidly colonize a variety of environments, including the susceptible human, makes it a major opportunistic pathogen, particularly *P. aeruginosa* serogroups 11 and possibly serogroup 9, which are the most frequently isolated pathogenic strains. Bacteremia attributable to *Pseudomonas* has become a major concern in the management of trauma as well as in the management of susceptible patients recovering from burns, intensive surgery, and others exposed to cancer therapy.[211-215] Other serious infections for susceptible individuals involve eye, ear, nose, and occasionally the gastrointestinal tract.[216,217]

The infrequent occurrence (3 to 19%) of *Pseudomonas* in the human intestinal tract[218] suggests that colonization of the gastrointestinal system rarely occurs in healthy adults, indicating that there are potent host-defense mecha-nisms against this group of gram-negative bacteria.[219,220] Because municipal sewage contaihs a mixture of domestic wastes, industrial discharges, and intermittent stormwater runoff, it is not unexpected to find *P. aeruginosa* in 90% of sewage samples.[221] Densities of *P. aeruginosa* in surface waters receiving waste and stormwater discharges may range from 1 to 10,000 cells per 100 ml, and are influenced by available nutrients and seasonal water temperatures.[222]

P. aeruginosa found in a contaminated water supply has been linked to one waterborne outbreak that occurred in a nursery for newborns.[223] In this case study, the groundwater supply was contaminated by seepage of sewage and infiltration of contaminated surface water. Because *P. aeruginosa* is the most prevalent *Pseudomonas* in human disease[189] its occurrence has been limited to less than one organism in 250 ml of bottled drinking water by the European community. Other *Pseudomonas* species found occasionally in water (*P. fluorescens, P. putida, P. multivorans, P. maltophilia,* and *P. stutzeri*) have not yet been linked to waterborne outbreaks, suggesting they are indigenous aquatic bacteria in every water environment.

Staphyloccus aureus

Staphylococcus aureus is a gram-positive coccus in the heterotrophic bacterial population. This organism and other *Staphylococcus* species are a major component of the normal human flora of the skin and to a lesser extent in fecal wastes. In the water environment, *Staphylococcus* may persist for an extended time compared to coliform bacteria. *S. aureus*, among several other species (*S. epidermidis* and *S. saprophyticus*) may be associated with human infections of the skin (cellulitis, pustules, boils, carbuncles, and impetigo), bacteremia, peritonitis associated with dialysis, genitourinary infections, and postoperative wound infections. Obviously, *S. aureus* may be involved in some nosocomial outbreaks in the hospital setting.

S. aureus concentrations in drinking water may be a health concern for individuals who are exposed to water contact for extended periods (dishwashing, whirlpool therapy, and dental hygiene). Densities of 200 to 400 cocci per millimeter have been shown to set up a carrier state in the nose of 50% of newborn infants[224] and an *S. aureus* density of a few hundred cells per millimeter in a water contact may induce infection in traumatized skin.[225]

Of additional concern in the production of thermostable enterotoxins during the colonization of *S. aureus* in food preparations, such as salads, baby formula, puddings, and custards, can provide *S. aureus* with an excellent source of nutrients for rapid growth at ambient temperatures. Such colonization within a few hours would produce sufficient enterotoxins to cause food poisoning. As an example, one study of a food poisoning incident was traced to water supply contaminated with 100 *S. aureus* per milliliter.[226] This tap water was used to cool hard-boiled eggs and *S. aureus* penetrated the broken egg shells, multiplying on the eggs in the holding water, and in the process produced enough endotoxin to cause illness.

Not much is known about the extent of an *S. aureus* problem in treated public water supplies. In a study of private water supplies in Oregon, *S. aureus* was isolated from over 6% of 320 rural water samples.[227] Inspection of heterotrophic plate count data on these same samples (Table 3.15) suggested a 63% *S. aureus* occurrence for HPCs above 300 organisms per milliliter. Analysis of untreated drinking water supplies of Port Harcourt, Nigeria suggest *S. aureus* was much higher, probably because the water supplies frequently had total and fecal coliforms.[228] Densities of *S. aureus* in both studies ranged from a few to several hundred cells per 100 milliliter.

Table 3.15 Numbers of *S. aureus* and Total Bacteria in Relation to Coliforms and Enterotoxin Production

Sample no.	Typical staphylococci per 100 ml	Average total plate	Confirmed coliforms	Coagulase reaction	Enterotoxin A production
17	1	1.0×10^2	−	+	+
48	4	2.0×10^2	−	+	−
49	12	<10	−	+(83)[a]	+
67	38	10	−	+	−
14	2	50	−	+	−
29	>600	1.9×10^4	−	+(25)	−
1152	1	NS	−	+	+
1162	3	5.2×10^2	−	+	+
1169	3	$>3.0 \times 10^3$	−	+	+
1195	205	2.6×10^3	−	+(77)	−
1199	159	3.1×10^2	+	+(25)	+
1203	6	7.5×10^2	−	+	−
1209	13	$>3.0 \times 10^3$	−	+	−
1223	>330	3.4×10^3	+	+(30)	+
1296	4	30	−	+	−
1316	4	$>3.0 \times 10^3$	+	+	+
1318	1	$>3.0 \times 10^3$	+	+	−
1353	1	<10	−	+	−
1559	27	$>3.0 \times 10^3$	−	+(48)	−
1365	>375	$>3.0 \times 10^3$	+	+(50)	−

[a] Numbers in brackets are the estimated percentage of coagulase-positive *S. aureus* in the sample based on the testing of at least ten colonies from each specimen.
Data from LeChevallier and Seidler.[227]

Colonization sites within water supply lines of private water supplies are most often aerator screens on faucets. Other important sites to consider are supply lines in the home and the water supply storage tank.[227] As previously noted, faucet aerators are a frequent site for opportunistic bacterial colonization so the public should be made aware of the problem and advised to periodically disassemble all faucet aerators for a cleaning to remove sediment and biofilm accumulation. In the case of private water system contamination, additional maintenance should include flushing of infrequently used water supply lines to the garden, drainage of storage tanks for sediment removal, and replacing point-of-use carbon filter cartridges (if used). In severe cases, the well may be the source of contamination, in which instance, intermittent chlorination of the groundwater supply may be essential.

Opportunistic Fungi

Waterborne fungi are other opportunistic organisms that are often of little public health significance, yet when present in certain situations may colonize the human body and possible lead to lethal consequences.[229] Often these situations arise from use of massive doses of antibiotics and immunosuppressive therapy. In nonfatal episodes, fungi may be a cause of allergenic reactions (sauna-takers disease and reactions to cool mist vapor inhalations) mycotoxins, and infections of the skin, nails, hair, and genitalia.[229-237] Very often the route of transmission is via air or water devices used in the home, hospital, or health spa.

Filamentous fungi have been found to be common in the aquatic environment,[238-241] but little attention has been given to their occurrences in drinking water. As a consequence of mycelia fragmentation and spore resistance to disinfection, various strains present in the raw source water may pass through sand filtration and disinfection treatment processes with wide variability in their reduction.[242-246] While raw water treatment by chemical coagulation and disinfection was more effective in removing fungi, these organisms could still be detected in 4 of 12 treated water samples. Failure of treatment barriers to completely remove fungi filaments and their spores provide a pathway into the distribution system. Additional inputs of fungi may also result from airborne dust contamination of finished water reservoirs or as an aftermath of soil particle contamination introduced in line breaks and subsequent repairs. Upon entering the distribution system, fungi may quickly establish themselves in pipe sediments and at sites of biofilm development.[235,244,245,247] Favorite colonization sites are pipe gaskets and joint sealant compounds.[242,247-249] Because conidia of filamentous fungi are resistant to chlorine residuals usually found in distribution lines (up to 0.3 mg/l free chlorine), colonization may proceed uninhibited by conventional disinfection practices.[246] In fact, prodigious growth of fungi at biofilm sites can also exert a chlorine demand that could further reduce available disinfectant residuals in the pipe network.

Frequency and profile of fungi genera in water supply occurrences may be variable as a reflection of seasonal temperature patterns, source water, and treatment patterns. For example, 17.6% of 978 samples collected from the transmission lines and distribution system for the Southern Nevada Water Utility were positive for fungi.[237] In another study involving six chlorinated groundwater systems in Pennsylvania and New Jersey, fungi were detected in 55.2% of 727 samples examined.[236] Warm water temperatures would seem to favor the growth of fungi in water systems and indeed was so noted in the

Omaha, NE water system during summer months (personal observation in the field evaluation of the delayed coliform incubation test during the 1960s). However, a recent report suggested growth might occur during periods when the water temperature was between 13 and 20°C rather than at higher temperatures (20 to 29°C). Perhaps this conflict in conclusions may be influenced by the regrowth competition from heterotrophic bacteria during the higher water temperature period that had an adverse impact on fungi methodology.

Fungal populations vary also with quality of raw source water. It has been observed that river water contains more thermophilia fungi than lake water.[245] Little is known about fungal species profiles as they relate to water treatment practices. It seems reasonable to surmise that there is a change in fungi flora while being exposed to the impact of pH shifts due to treatment (lime addition) and most certainly to chlorination practices.

Fungal densities reported in water supplies are variable, partly because of media selected and incubation time–temperature relationships. For a water system using alkaline lake source water and conventional treatment methods,[237] fungal densities ranged from 1 to 15 organisms per 100 ml. Several other studies have reported an average 5.5 or 18 fungal colonies per 100 ml in groundwaters[235,244] and 34 fungi per 100 ml in a chlorinated surface water.[244] Fungi have also been reported in the Salem and Beverly, MA,[95] Riverside, CA,[249] and Sioux Falls, SD[250] distribution systems.

The most common genera of filamentous fungi isolated from water supply will vary somewhat among water systems. Diversity among species may indicate a high-quality raw surface water supply. The number of different filamentous fungi and yeast detected in 978 samples collected from the Southern Nevada Water Utility during February to July 1988 classified into 20 genera. In this study the three most frequent fungi identified were *Cladosporium, Phoma*, and *Candida parapsilosis* (Table 3.16). The four most frequently occurring genera of filamentous fungi in two southern California distribution systems (one chlorinated and the other unchlorinated) were *Penicillium, Sporocybe, Acremonium*, and *Paeciolomyces*.[244] In the unchlorinated supply, *Penicillium* and *Acremonium* represented approximately 50% of 538 colonies identified among 14 genera. *Sporocybe* and *Penicillium* accounted for 56% of 923 fungal strains distributed within 19 genera isolated from the chlorinated supply. The mean density of fungi from the unchlorinated and chlorinated systems were 18 and 34 organisms per 100 ml, respectively. Conidia of *Aspergillus fumigatus, A. niger,* and *Penicllium oxalicum* isolated from distribution systems of three small water supplies (Pennsylvania) showed a greater resistance to chlorine inactivation than yeast, which in turn were more resistant than coliform bacteria.[246] A total of 15 fungi genera were detected in three small distribution systems.

In Europe, much the same pattern of filamentous fungi was reported in distribution water. For instance, *Aspergillus fumigatus* was the predominant species detected in the distribution systems of 15 water supplies in Finland.[230]

Table 3.16 Profile of Filamentous Fungi and Yeast in Water Supply Distribution

Microbial genera or species		Occurrence (%)[a]
Fungi	Yeast	
Cladosporium		19.4
Phoma		13.3
	Candida parapsilosis	11.7
	Rhodotorula rubra	5.6
Alternaria		5.1
Exophiala		5.1
Aspergillus		4.6
	Aureobasidium pullulans	3.6
Penicillium		3.6
	Cryptococcus albidus var. diffluens	3.1
Cephalosporium		2.0
Drechslera		2.0
Phialophora		2.0
	Cryptococcus albidus var. albidus	1.5
	Cryptococcus uniguttulatus	1.5
Geotrichum		1.5
Absidia		1.5
	Torulaspora rosei	1.0
	Rhodotorula minuta	1.0
	Rhodotorula pilimonae	1.0
Acremonium		1.0
Ulocladium		1.0
Trichophyton		1.0
Other filamentous fungi and yeast (0.5% each)	subtotal	5.0
Unidentified organisms	subtotal	1.5

[a] Percentages based on mean values.
Data revised from West.[237]

A variety of fungi (*Cephalosporium* sp., *Verticillium* sp., *Trichoderma sporulosum, Nectria veridesceus, Phoma* sp., and *Phialophora* sp.) were identified from water service mains in several water supplies in England.[242,243] Fungi densities in these drinking waters were usually less than 10 organisms per 100 ml. Water supplies with customer complaints of taste and odor often had fungal densities of 10 to 100 organisms per 100 ml.[242] *Penicillium, Aspergillus,* and *Rhizopus* occurred in almost half of the positive samples examined from the distribution networks for the cities of Nancy and Metz, France.[251] A total of seven genera of fungi were identified in these two chlorinated water supplies and densities ranged between 2 and 65 filamentous fungi per liter. As reported by other investigators, there were no correlations between total bacteria, yeast, filamentous fungi, and chlorine.

Because fungi are propagated usually by spores rather than fragments of mycelia, it might be argued that fungi occurrence in water supplies is accidental, i.e., a result of airborne contamination rather than a result of colonization of the distribution system. Available evidence suggests that the number of fungi found in some samples are too high to be explained by adventitious spores entering the system or sample collection.[252] Fungi colonization does occur in biofilm locations that may be in fire hydrants, elevated storage tanks, or attachment devices on building plumbing systems (faucets, humidifiers, and point-of-use devices). Slime growth appearing at the outlet of a slowly leaking faucet was reported to be caused by the fungal hyphae of two species of *Fusaria*.[253] In this case, it seems probable that the fungi colonization was the result of adventitious contamination from outside the dripping faucet rather than as a result of fungal passage from the water supply distribution system to the water tap.

Yeast

Yeast has been documented in water supplies, but information on frequency, density, and regrowth is limited. This situation is not because these organisms have no significance as opportunistic pathogens, but rather that additional research needs to be done on all environmental links that lead to the established niche for yeast in water and their contribution to water contact illness.

Candida albicans is an opportunistic pathogen that becomes dangerous to individuals stressed from diabetes, cancer, and immunological defects associated with AIDS, or through use of immunosuppressants in organ transplants and the application of broad-spectrum antibiotics.[254] Less severe but still very debilitating are yeast infections associated with diaper rash among infants and vaginitis in women from contaminated water contact or from sexually transmitted infections.[255]

Yeast is known to occur in various warm-blooded animals including man, dogs, pigeons, and sea gulls.[256-258] Densities in human feces may range up to 100,000 per gram, and in raw sewage these densities range from a few thousand to 25,000 colonies per liter.[142,244] Yeast densities of up to 400 colonies per liter have been reported in chlorinated sewage effluents[137] and from <2 to 125 in river water.[256] In lakes, yeast populations will fluctuate with water quality, being responsive to organic nutrients from waste discharges, stormwater runoff, seasonal vegetation decay, algal blooms, and lake turnovers.[259] The most frequently isolated yeast in this seasonal study of four lakes were *Candida* sp., *Rhodotorula glutinis, R. rubra,* and *Torulopsis* sp.

Chemical coagulation will remove 90 to 99% of yeasts from raw source waters with at least a 90% further reduction of residual yeast densities passing through sand filtration. Unfortunately, disinfection is less effective. Yeast has been found to survive chlorine residuals as high as 1.2 mg/l in residential water taps. This increased resistance of yeast to free available chlorine is primarily a result of the thickness and rigidity of the cell wall that presents a greater permeability barrier to chlorine.[260] By contrast, ozone disinfection is approximately as equally effective on yeast as it is on bacteria.[261]

Densities of yeast reported in finished drinking water may average 1.5 organisms per liter. This minimal number can be expected to increase slightly in the distribution network as the yeast cells adjust to the pipe environment and slowly colonize. Sites with higher densities of yeast may be old pipe sections, dead ends, and other stagnant water pockets in the pipe network. Yeast densities in a study of two chlorinated distribution systems in France ranged from 1 to 28 organisms per millimeter and were detected in 50% of samples examined.[251]

Specific species identified in distribution waters include *Candida parapsilosis, C. famata, C. albidus, Cryptococcus laurentii, Rhodstorula glutins,* and *R. rubra.*[244,246,251,260] Of these, the *Candida* species are of most serious opportunistic pathogen concern.[262]

REFERENCES

1. Geldreich, E.E., H.D. Nash, and D.F. Spino. 1977. Characterizing Bacterial Populations in Treated Water Supplies: A Progress Report. Proc. Amer. Water Works Assoc., Water Quality Technol. Confr., Kansas City, MO. pp. 2B-B.
2. Reasoner, D.J., J.C. Blannon, E.E. Geldreich, and J. Barnick. 1989. Non-photosynthetic Pigmented Bacteria in a Potable Water Treatment and Distribution System. *Appl. Environ. Microbiol.,* 55:912-921.
3. duMoulin, G.C., I.H. Sherman, D.C. Hoaglin, and K.D. Stottmeier. 1985. *Mycobacterium avium* Complex, an Emerging Pathogen in Massachusetts. *Jour. Clin. Microbiol.,* 22:9-12.
4. duMoulin, G.C. and K.D. Stottmeier. 1986. Waterborne Mycobacteria: an Increased Threat to Health. *ASM News,* 52:525-529.
5. LeChevallier, M.W., R.J. Seidler, and T.M. Evans. 1980. Enumeration and Characterization of Standard Plate Count Bacteria by Chlorinated and Raw Water Supplies. *Appl. Environ. Microbiol.,* 40:922-930.
6. Olson, B.H. and L. Hanami. 1980. Seasonal Variation of Bacterial Populations in Water Distribution Systems, Proc. 8th Annual American Water Works Assoc. Water Quality Technol. Confr., December 7-10, Miami Beach, FL. Amer. Water Works Assoc., Denver, CO. pp. 137-151.
7. Herson, D.S. and H. Victoreen. 1980. Identification of Coliform Antagonists. Proc. Amer. Water Works. Assoc., Water Quality Technol. Confr., Miami Beach, FL. pp. 153-160.
8. Reilly, J.K. and J. Kippin. 1981. Interrelationship of Bacterial Counts with Other Finished Water Quality Parameters Within Distribution Systems. EPA-600/52-81-035, U.S. Environmental Protection Agency, Cincinnati, OH.
9. Staley, J.T. 1983. Identification of Unknown Bacteria from Drinking Water. Project CR-807570010, U.S. Environmental Protection Agency, Cincinnati, OH.
10. Clark, J.A., C.A. Burger, and L.E. Sabatinos. 1982. Characterization of Indicator Bacteria in Municipal Raw Water, Drinking Water and New Main Water Supplies. *Can. Jour. Microbiol.,* 28:1002-1013.

11. Geldreich, E.E., J.A. Goodrich, and R.M. Clark. 1988. Characterizing Raw Surface Water Amenable to Minimal Water Supply Treatment. Proc. Amer. Water Works Annual Confr., Orlando, FL. pp. 545-570.

12. Camper, A.K., M.W. LeChevallier, S.C. Broadaway, and G.A. McFeters. 1985. Growth and Persistence on Granular Activated Carbon Filters. *Appl. Environ. Microbiol.*, 50:1378-1382.

13. Porter, K.G. 1984. Natural Bacteria as Food Resources for Zooplankton, In: *Current Perspectives in Microbial Ecology*, M. J. Klug and C.A. Reddy, Eds., Amer. Soc. Microbiol., Washington, D.C. pp. 390-395.

14. Levy, R.V., F.L. Hart, and R.D. Cheetham. 1986. Occurrence and Public Health Significance of Invertebrates in Drinking Water Systems. *Jour. Amer. Water Works Assoc.*, 78(9):105-110.

15. King, C.H., E.B. Shotts, Jr., R.E. Wooley, and K.G. Porter. 1988. Survival of Coliforms and Bacterial Pathogens within Protozoa during Chlorination. *Appl. Environ. Microbiol.*, 54:3023-3033.

16. Geldreich, E.E. 1988. Coliform Non-Compliance Nightmares in Water Supply Distribution Systems. In: *Water Quality: A Realistic Perspective*, Michigan Section Amer. Water Works Assoc. and Michigan Water Poll. Contr. Assoc., Joint Project, College of Engr., Univ. of Michigan, Ann Arbor, MI. pp. 55-74.

17. Snead, M.C., V.P. Olivieri, K. Kawata, and C.W. Kruse. 1980. The Effectiveness of Chlorine Residuals in Inactivation of Bacteria and Viruses Introduced by Post-Treatment Contamination. *Water Res.*, 14:403-408.

18. Office of Water. Cross-Connection Control Manual. 1989. EPA 570/9-89- 007. Office of Drinking Water, U.S. Environmental Protection Agency. Washington, D.C.

19. Trulear, M.G. and W.G. Characklis. 1979. Dynamics of Biofilm Processes. Presented at the 34th Annual Purdue Industrial Waste Confr., West Lafayette, IN. pp. 838-853.

20. McCabe, L.J., J.M. Symons, R.D. Lee, and G.G. Robeck. 1970. Survey of Community Water Supply Systems. *Jour. Amer. Water Works Assoc.*, 62:670-687.

21. Geldreich, E.E., M.J. Allen, and R.H. Taylor. 1978. Interferences to Coliform Detection in Potable Water Supplies. In: Evaluation of the Microbiology Standards for Drinking Water, C.W. Hendricks, Ed., EPA-570/9-78-002, Office of Drinking Water, U.S. Environmental Protection Agency, Washington, D.C. pp. 13-20.

22. Hutchinson, D., R.H. Weaver, and M. Scherago. 1943. The Incidence and Significance of Microorganisms Antagonistic to *Escherichia coli* in Water. *Jour. Bacteriol.*, 45:29 Abs. G34.

23. Armstrong, J. L., J.J. Calomiris, and R.J. Seidler, 1982. Selection of Antibiotic-Resistant Standard Plate Count Bacteria During Water Treatment. *Appl. Environ. Microbiol.*, 44:308-316.

24. Bedard, L., A.J. Drapeau, S.S. Kasatiya, and R.R. Plaute. 1982. Plasmides de Resistance aux Antibiotiques Ches les Bacteries. Isolus D'Eaux Potables. *Eau Des Quebec*, 15:59-66.

25. Grabow, W.O.K. 1978. South African Experience on Indicator Bacteria, *Pseudomonas aeruginosa*, and R+ Coliforms in Water Quality Control. Amer. Soc. Test. Materials, Special Technical Pub. 635; 168-181.

26. Rice, E.W., J.W. Mesier, C.H. Johnson, and D.J. Reasoner. 1995. Occurrence of High-Level Aminoglycoside Resistance in Environmental Isolates of Enterococci. *Appl. Environ. Microbiol.*, 61:374-376.

27. El-Zanfaly, H.T., E.A. Kassein, and S.M. Badr-Eldin. 1987. Incidence of Antibiotic Resistant Bacteria in Drinking Water in Cairo. *Water Air Soil Pollut.,* 32:123-128.

28. Armstrong, J.L., J.J. Colomiris, D.S. Shigeno, and R.J. Seidler, 1981. Drug Resistant Bacteria in Drinking Water, Proc. Amer. Water Works Assoc., Water Quality Technol. Confr., Seattle, WA. pp 263-276.

29. Geldreich, E.E., H.D. Nash, D.J. Reasoner, and R.H. Taylor. 1972. The Necessity of Controlling Bacterial Populations in Potable Waters: Community Water Supply. *Jour. Amer. Water Works Assoc.,* 64:596-602.

30. Ridgway, H.F. and Olson, B.H. 1982. Chlorine Resistance Patterns of Bacteria from Two Drinking Water Distribution Systems. *Appl. Environ. Microbiol.,* 44:972-987.

31. duMoulin, G.C. 1979. Airway Colonization by *Flavobacterium* in an Intensive Care Unit. *Clin. Microbiol.,* 10:155-160.

32. Favero, M.S., N.J. Peterson, K.M. Boyer, L.A. Carson, and W.W. Bond. 1974. Microbial Contamination of Renal Dialysis Systems and Associated Health Risks. *Trans. Amer. Soc. Artif. Intern. Organs,* 20-A:175-183.

33. Favero, M.S., N.J. Peterson, L.A. Carson, W.W. Bond, and S.H. Hindman. 1975. Gram-Negative Water Bacteria in Hemodialysis Systems. *Health Lab. Sci.,* 12:321-334.

34. Herman, L.G. 1976. Sources of the Slow-Growing Pigmented Water Bacteria. *Health Lab. Sci.,* 13:5-10.

35. Herman, L.G. and C.K. Himmelsbach. 1965. Detection and Control of Hospital Sources of Flavobacteria. *Hospitals,* 39:72-76.

36. Leifson, E. 1962. The Bacterial Flora of Distilled and Stored Water. I. General Observations, Techniques and Ecology. *Int. Bull. Bacteriol. Nomenc. Taxon.,* 12:133-153.

37. Geldreich, E.E., H.D. Nash, D.J. Reasoner, and R.H. Taylor. 1975. The Necessity of Controlling Bacterial Populations in Potable Waters — Bottled Water and Emergency Water Supplies. *Jour. Amer. Water Works Assoc.,* 67:117-124.

38. Geldreich, E.E., R.H. Taylor, J.C. Blannon, and D.J. Reasoner. 1985. Bacterial Colonization of Point-of-Use Water Treatment Devices. *Jour. Amer. Water Works Assoc.,* 77:72-80.

39. Reasoner, D.J., J.C. Blannon, and E.E. Geldreich. 1987. Microbiological Characteristics of Third-Faucet Point-of-Use Devices. *Jour. Amer. Water Works Assoc.,* 79:60-66.

40. Reasoner, D.J. and E.E. Geldreich. 1979. Significance of Pigmented Bacteria in Water Supplies. Proc. Amer. Water Works Assoc. Water Qual. Technol. Confr. December 8-12, Philadelphia, PA, Amer. Water Works Assoc., Denver, CO. pp. 187-196.

41. Olson, B.H. 1982. Assessment and Implications of Bacterial Regrowth in Water Distribution Systems. Project Report EPA R-805680010. Order No. PB 82-249368. National Technical Information Service. Springfield, VA.

42. Staley, J.T. 1985. Enumeration and Identification of Heterotrophic Bacteria from Drinking Water. EPA 1600/2-85/061. MERL-Ci-2428. U.S. Environmental Protection Agency, Cincinnati, OH.

43. Stetzenbach, L., L.M. Kelley and N.A. Sinclair. 1986. Isolation, Identification and Growth of Well-Water Bacteria. *Groundwater,* 24:6-10.

44. Symons, J.M. et al. 1981. Treatment Techniques for Controlling Trihalomethanes in Drinking Water. Chapter IX. Maintaining Bacteriological Quality. EPA-600/2-81-156. U.S. Environmental Protection Agency, Cincinnati, OH.
45. Graf. W. and L. Bauer. 1973. Red Bacterial Growth (*Corynebacterium rebrum*, n. spec.) in Tap-Water Systems. *Zentralbl. Bakteriol. Parasitenkd. Infektionskr. Hyg. Abt. 1 Orig. Reiche B*, 157:291-303.
46. Guthrie, R.K., D.S. Cherry, and R.N. Ferebee. 1974. A Comparison of Thermal Loading Effects on Bacterial Populations in Polluted and Nonpolluted Aquatic Systems. *Water Res.*, 8:143-148.
47. Cusack, F.A., H.M. Lappin-Scott, and J.W. Costerton. 1988. Plugging of a Model Rock System by Using Starved Bacteria. *Appl. Environ. Microbiol.*, 54:1365-1372.
48. Morita, R.Y. 1985. Starvation and Miniaturisation of Heterotrophs with Special Emphasis on Maintenance of the Starved Viable State. In: *Bacteria in Their Natural Environments*. M. Fletcher and G. Floodgates, Eds., Academic Press, New York. pp. 111-129.
49. Bollen, W.B. 1977. Sulfur Oxidation and Respiration in 54-Year-Old Soil Samples. *Soil Biol.*, 71:209-215.
50. Kurath, G. and R.Y. Morita. 1983. Starvation-Survival Studies of a Marine *Pseudomonas* sp. *Appl. Environ. Microbiol.*, 45:1206-1211.
51. Lappin-Scott, H.M., F. Cusack, and J.W. Costerton. 1988. Nutrient Resuscitation and Growth of Starved Cells in Sandstone Cores: A Novel Approach to Enhanced Oil Recovery. *Appl. Environ. Microbiol.*, 54:1373-1382.
52. Favero, M.S., L.A. Carson, W.W. Bond, and N.J. Petersen. 1971. *Pseudomonas aeruginosa*: Growth in Distilled Water for Hospitals. *Science*, 173:836-838.
53. Helms, C.M., R.M. Massanari, K.P. Wenzel, M.A. Pfaller, N.P. Mayer, and N. Hall. 1988. Legionnaires' Disease Associated with A Hospital Water System. *Jour. Amer. Med. Assoc.*, 259:2423-2427.
54. McFeters, G.A. 1986. Survival and Virulence of Waterborne Pathogenic Bacteria in Potable Waters. Proc. Inter. Symp. III. Biofouled Aquifers: Prevention and Restoration. pp. 79-85.
55. Henrici, A.T. 1928. *Morphological Variation and the Rate of Growth of Bacteria*. Charles C. Thomas Publ., Springfield, IL.
56. Geldreich, E.E. and E.W. Rice. 1987. Occurrence, Significance and Detection of *Klebsiella* in Water Systems. *Jour. Amer. Water Works Assoc.*, 79:74-80.
57. Dawson, M.P., B.A. Humphrey, and K.C. Marshall. 1981. Adhesion: A Tactic in the Survival Strategy of a Marine Vibrio during Starvation. *Current Microbiol.*, 6:195-199.
58. Arnold, G.E. 1936. Plankton and Insect Larvae Control in California Waters. *Jour. Amer. Water Works Assoc.*, 28:1469-1479.
59. Hart, K.M. 1957. Living Organisms in Public Water Mains. *Jour. Inst. Mun. Eng.*, 83:324-333.
60. Crabill, M.P. 1956. Biologic Infestation at Indianapolis. *Jour. Amer. Water Works Assoc.*, 48:269-274.
61. English, E. 1958. Biological Problems in Distribution Systems. Infestations of Water Mains. *Proc. Soc. Water Treat. Exam.*, 7:127- 143.
62. Tracy, H.W., V.M. Camarena, and F. Wing. 1966. Coliform Persistence in Highly Chlorinated Waters. *Jour. Amer. Water Works Assoc.*, 58:1151-1159.

63. Turner, M. E. D. 1956. *Asellus aquaticus* in a Public Water Supply Distribution System. *Proc. Soc. Water Treat. Exam.*, 5:67-80.

64. Holland, G.J. 1956. The Eradication of *Asellus aquaticus* From Water Supply Mains. *Jour. Inst. Water Engrs.*, 10:221-241.

65. Small, I.C. and G.F. Graves. 1968. A Survey of Animals in Distribution Systems. *Water Treat. Exam.*, 19:150-183.

66. Mackenthun, K.M. and L.E. Keap. 1970. Biological Problems Encountered in Water Supplies. *Jour. Amer. Water Works Assoc.*, 62:520-526.

67. Levy, R.V., R.D. Cheetham, J. Davis, et al. 1984. Novel Method for Studying the Public Health Significance of Macroinvertebrates Occurring in Potable Water. *Appl. Environ. Microbiol.*, 47:889-894.

68. O'Dell, W.D. 1979. Isolation, Enumeration and Identification of Amoeba from a Nebraska Lake. *Jour. Protozool.*, 26:265-270.

69. Chang, S.L., R.L. Woodward, and P.W. Kabler. 1960. Survey of Free-Living Nematodes and Amoebas in Municipal Supplies. *Jour. Amer. Water Works Assoc.*, 52:613-618.

70. Jacquemin, J.L., A.M. Simitzia–Le Flohic, and N. Chaneoau. 1981. Free-Living Amoeba in Fresh Water — A Study of the Water Supply of the Town of Poiters. *Bull. Soc. Pathol. Exot.*, 74:521-534.

71. Kwan, W.J., Ed. 1973. *The Biology of Amoeba*, Academic Press, Inc., New York.

72. Nelsson, J.R. 1987. Structural Aspects of Digestion of *Escherichia coli* in *Tetrahymena*. *Jour. Protozool.*, 34:1-6.

73. Drozanski, W. 1978. Activity and Distribution of Bacteriolytic *N*-Acetylmuramidase during Growth of *Acanthamoeba castellanii* in Axenic Culture. *Acta Microbiol. Pol.*, 27:243-256.

74. King, C.H., E.B. Shotte, Jr., R.E. Wooley, and K.G. Porter. 1988. Survival of Coliforms and Bacterial Pathogens Within Protozoa During Chlorination. *Appl. Environ. Microbiol.*, 54:3023-3033.

75. Anand, C., A. Skimer, A. Malic and J. Kurtz. 1984. Intracellular Replication of *Legionella pneumoniae* in *Acanthamoeba palestinensis*. In: Legionella: *Proceedings of the 2nd International Symposium*. C. Thornsberry, A. Balow, J.C. Feeley, and W. Jakubowski, Eds., Amer. Soc. Microbiol., Washington, D.C. pp. 330-332.

76. Drozanski, W. and T. Chimelewski. 1979. Electron Microscopic Studies of *Acanthamoeba castellanii* Infected with Obligate Intracellular Bacterial Parasite. *Acta. Microbiol. Pol.*, 28:123-133.

77. Fields, B.S., E.B. Shotts, Jr., J.C. Feeley, C.W. Gorman, and W.T. Martin. 1984. Proliferation of *Legionella pneumophila* as an Intercellular Parasite of the Ciliated Protozoan *Tetrahymena pyriformis*. *Appl. Environ. Microbiol.*, 47:467-471.

78. Sherr, B.F., E.B. Sherr, and F. Rassoulzadegan. 1988. Rates of Digestion of Bacteria by Marine Phagotrophic Protozoa: Temperature Dependence. *Appl. Environ. Microbiol.*, 54:1091-1095.

79. Wadowsky, R.M., L.J. Butler, M.K. Cook, et al. 1988. Growth-Supporting Activity for *Legionella pneumophila* in Tap Water Cultures and Implication of Hartmannellid Amoebae as Growth Factors. *Appl. Environ. Microbiol.*, 54:2677-2682.

80. Mitcham, R.P., M.W. Shelley, and C.M. Wheadon. 1983. Free Chlorine Versus Ammonia-Chlorine: Disinfection, Trihalomethane Formation, and Zooplankton Removal. *Jour. Amer. Water Works Assoc.*, 75:196-198.

81. Tombes, A.S. and H.R. Abernathy. 1979. Determination of Breeding Sites of Nematodes in a Municipal Drinking Water Facility. U.S. Environmental Protection Agency. EPA-600/1-79-029, Health Effects Research Laboratory, Cincinnati, OH.

82. Mott, J.B. and A.D. Harrison. 1983. Nematodes from River Drift and Surface Drinking Water Supplies in Southern Ontario. *Hydrobiologia,* 102:27-38.

83. Chaudhuri, N., R. Siddiqui, and R.S. Engelbrect. 1964. Source and Persistence of Nematodes in Surface Waters. *Jour. Amer. Water Works Assoc.,* 56:73-88.

84. Baliga, K.Y., J.H. Austin, and R.S. Engelbrecht. 1969. Occurrence of Nematodes in Benthic Deposits. *Water Res.,* 3:979-993.

85. Chang, S.L. and P.W. Kabler. 1962. Free-Living Nematodes in Aerobic Treatment Plant Effluent. *Jour. Water Poll. Contr. Fed.,* 34:1256-1261.

86. Cobb, N.A. 1918. Filter-bed Nemas-Nematodes of the Slow Sand Filter Beds of American Cities. *Contr. Sci. Nematol.,* 7:189-212.

87. Chang, S.L., J.H. Austin, H.W. Poston, and R.L. Woodward. 1959. Occurrence of a Nematode Worm in a City Water Supply. *Jour. Amer. Water Works Assoc.,* 51:671-676.

88. Chang, S.L. 1961. Viruses, Amoebas, and Nematodes and Public Water Supplies. *Jour. Amer. Water Works Assoc.,* 53:288-296.

89. Chang, S.L., G. Berg, N.A. Clarke, and P.W. Kabler. 1960. Survival and Protection Against Chlorination of Human Enteric Pathogens in Free-Living Nematodes Isolated from Water Supplies. *Amer. Jour. Trop. Med. Hyg.,* 9:136-142.

90. Smeda, S.M., H.J. Jensen, and A.W. Anderson. 1971. Escape of *Salmonella* from Chlorination during Ingestion by *Prestionchus lheritieri* (Nematoda:*Diplogasterinae*). *Jour. Nematol.,* 3:201-204.

91. Lee, D.L. 1965. *The Physiology of Nematodes.* W.H. Freeman and Co., San Francisco, p. 86.

92. Hughes, J.M., and W.R. Jarvis. 1985. Epidemiology of Nosocomial Infections. In: *Manual of Clinical Microbiology,* 4th ed., E. H. Lennette, A. Balows, W. J. Hausler, Jr., and H. J. Shadomy, Eds., Amer. Soc. Microbiol., Washington, D.C.

93. Olson, B.H. and L. Hanami. 1980. Seasonal Variation of Bacterial Populations in Water Distribution Systems. Proc. Amer. Water Works Assoc., Water Quality Technol. Confr., Miami Beach, FL. pp. 137-151.

94. Lamka, K.G., M.W. LeChevallier, and R.J. Seidler. 1980. Bacterial Contamination of Drinking Water Supplies in a Modern Rural Neighborhood. *Appl. Environ. Microbiol.,* 39:734-738.

95. Nash, H.D. and E.E. Geldreich. 1980. Effect on Storage on Coliform Detection in Potable Water Samples, Water Quality Technol. Confr. Proc. pp. 123-136.

96. Reilly, J.K. and J.S. Kippin. 1983. Relationship of Bacterial Counts with Turbidity and Free Chlorine in Two Distribution Systems. *Jour. Amer. Water Works Assoc.,* 75:309-312.

97. Bateman, J.L., R.P. Tu, M.H. Strampher, and B.A. Cunha. 1988. *Aeromonas hydrophila* cellulis and Wound Infection Caused by Waterborne Organisms. *Heart Lung,* 17:99-102.

98. Notermans, S., A. havelaar, W. Jansen, S. Kozaki, and P. Guinee. 1986. Production of "Asao Toxin" by *Aeromonas* Strains Isolated from Feces and Drinking Water. *Jour. Clin. Microbiol.,* 23:1140-1142.

99. Highsmith, A.K., T.G. Emori, S.M. Aguero, et al. 1986. Heterotrophic Bacteria Isolated from Hospital Water Systems. In: *International Symposium on Water-Related Health Issues*. C.L. Tate, Jr., Ed., Amer. Water Resources Assoc., Bethesda, MD. pp. 181-187.

100. Cross, D., A. Benchimol, and E. Mimond. 1966. Faucet Aerator — A Source of *Pseudomonas* Infection. *N. Eng. Jour. Med.,* 274:1430-1431.

101. Bassett, D.C.J., K.J. Stokes, and W.R.G. Thomas. 1970. Wound Infections with *Pseudomonas multivorans*: Waterborne Contaminant of Disinfectant Solutions. *Lancet,* 1:1188-1191.

102. Highsmith, A.K., P.N. Le, R.F. Khabbag, and V.P. Mann. 1985. Characteristics of *Pseudomonas aeruginosa* Isolated from Whirlpools and Bathers. *Infect. Control,* 6:407-412.

103. Ho, J.L., A.K. Highsmith, E.S. Wong, et al. 1981. Common-Source *Pseudomonas aeruginosa* Infection in Neurosurgery. Proc. Ann. Meet. Am. Soc. Microbiol., Dallas, TX, L10, p. 80.

104. duMoulin, G., G. Doyle, J. Mackay, and J. Hedley-Whyte. Bacterial Fouling of a Hospital Closed-Loop Cooling System by *Pseudomonas* sp. *Jour. Clin. Microbiol.,* 13:1060-1065.

105. Saepan, M.S., H.O. Bodman, and R.B. Kundsin. 1975. Microorganisms in Heated Nebulizers. *Health Lab. Sci.,* 12:316-320.

106. McGuekin, M.B., R.J. Thorpe, and E. Abrutyn. 1981. Hydrotherapy: An Outbreak of *Pseudomonas aeruginosa* Wound Infections Related to Hubbard Tank Treatments. *Arch. Phys. Med. Rehabil.,* 62:283-285.

107. Abrutyn, G.A., B.J. Collins, J.R. Babb, et al. 1974. *Pseudomonas aeruginosa* in Hospital Sinks. *Lancet,* 1:578-580.

108. Carson, L., G. Bolan, N.J. Petersen, et al. 1983. Antimicrobial and Formaldehyde Resistance Patterns of Non-Tuberculous Mycobacteria Associated with Reprocessed Hemodialyzers. Proc. Ann. Meet. Intersci. Conf. Antimicrobial Agents and Chemotherapy, Amer. Soc. Microbiol., Washington, D.C.

109. Bolan, G., A.L. Reingold, L.A. Carson, et al. 1985. Infections with *Mycobacterium chelonei* in Patients Receiving Dialysis and Using Processed Hemodialyzers. *Jour. Infect. Dis.,* 154:1013-1019.

110. Cordes, L. et al. 1981. Isolation of *Legionella pneumophila* from Hospital Showerheads. *Ann. Intern. Med.,* 94:195-197.

111. Kelly, M.T., D.J. Brenner, and J.J. Farmer, III. 1985. *Enterobacteriaceae.* In: *Manual of Clinical Microbiology,* 4th ed., E.H. Lennette, A. Ballows, W.J. Hausler, Jr., and H.J. Shadomy, Eds., Amer. Soc. Microbiol., Washington, D.C.

112. Irwin, R.S., R.R. Demers, M.R. Pratter, et al. 1980. An Outbreak of *Acinetobacter* Infection Associated with the Use of a Ventilator Spirometer. *Respir. Care,* 25:232-237.

113. Gervich, D.H. and L.S. Grout. 1985. An Outbreak of Nosocomial *Acinetobacter* Infection from Humidifiers. *Amer. Jour. Infect. Contr.,* 13:210-215.

114. Smith, P.W. and R.M. Massanari. 1977. Room Humidifiers as the Source of *Acinetobacter* Infections. *Jour. Amer. Med. Assoc.,* 237:795-797.

115. Chunha, B.A., J.J. Klimek, J. Graceuski, and J.C. McLoughlin. 1980. A Common Source Outbreak of Acinetobacter Pulmonary Infections Traced to Wright Respirometers. *Postgrad. Med. Jour.,* 56:169-172.

116. Nakashima, A.K., M.A. McCarthy, W.J. Martone, and L. Anderson. 1987. Epidemic Septic Arthritis Caused by *Serratia marcescens* and Associated with a Benzalkonium Chloride Antiseptic. *Jour. Clin. Microbiol.*, 25:1014-1018.

117. McCormack, R.C. and C.M. Kunin. 1966. Control of a Single Source Nursery Epidemic due to *Serratia marcescens*. *Pediatrics*, 37:750-755.

118. Connacher, L.A., D.C. Old, G. Phillips, et al. 1988. Recurrent Peritonitis Caused by *Serratia marcescens* in a Diabetic Patient Receiving Continuous Ambulatory Peritoneal Dialysis. *Jour. Hosp. Infect.*, 11:155-160.

119. Payment, P., L. Richardson, J. Siemiatychi, R. Dewar, M. Edwards, and E. Franco, 1991. A Randomized Trial to Evaluate the Risk of Gastrointestinal Disease due to Consumption of Drinking Water Meeting Current Microbiological Standards. *Am. J. Public Health*, 81:703-708.

120. Barnstein, N., C. Vieilly, M. Nowicki, J.C. Paucod, and J. Fleurette. 1986. Epidemiological Evidence of Legionellosis Transmission Through Domestic Hot Water Supply Systems and Possibilities of Control. *Isr. Jour. Med. Sci.*, 22:655-661.

121. Wadowsky, R.M., R.B. Yee, and L. Megmar. 1982. Hot Water Systems as Source of *Legionella pneumophila* in Hospital and Non-Hospital Plumbing Fixtures. *Appl. Environ. Microbiol.*, 43:1104-1110.

122. Plouffe, J.F., L.R. Webster, and B. Hackman. 1983. Relationship between Colonization of Hospital Buildings with *Legionella pneumophila* and Hot Water Temperatures. *Appl. Environ. Microbiol.*, 46:769-770.

123. Arnow, P.M., D. Weil, and M.F. Para. 1985. Prevalence and Significance of *Legionella pneumophila* and Hot Tap Water Systems. *Jour. Infect. Dis.*, 152:145-151.

124. Groothius, D.G., H.R. Veenendool, and H.L. Dijkstra. 1985. Influence of Temperature on the Number of *Legionella pneumophila* in Hot Water Systems. *J. Appl. Bacteriol.*, 59:529-536.

125. Botzenhart, K., W. Heizmann, S. Sedaghat, P. Huy, and T. Hahn. 1986. Bacterial Colonization and Occurrence of *Legionella pneumophila* in warm and cold water, in faucet aertators, and in drains of hospitals. *Zbl. Bakt. Hyg. B*, 183:79-85.

126. Epidemiologic Notes and Reports. 1990. Outbreaks of Diarrheal Illness Associated with Cyanobacteria (Blue-Green Algae)-Like Bodies — Chicago and Nepal, 1989 and 1990. *MMWR*, 40:325-327.

127. Geldreich, E.E. and D.J. Reasoner. 1990. Home Treatment Devices and Water Quality. In: *Drinking Water Microbiology*, G.A. McFeters, Ed., Springer-Verlag, New York.

128. Good, R.C. 1985. Opportunistic Pathogens in the Genus Mycobacterium. *Annu. Rev. Microbiol.*, 39:347-369.

129. Ganzadharam, P.R.J., J.A. Lockhart, R.J. Awe, and D.E. Jenkins. 1976. Mycobacterial Contamination Through Tap Water. *Amer. Rev. Respir. Dis.*, 113:894.

130. McSwiggan, D.A. and C.H. Collins. 1974. The Isolation of *M. Kansasii* and *M. xenopei* from Water Systems. *Tubercle*, 55:291-297.

131. Goslee, S. and E. Wolinsky. 1976. Water as a Source of Potentially Pathogenic Mycobacteria. *Amer. Rev. Respir. Dis.*, 113:287-292.

132. Bullin, C.H., E.I. Tanner, and C.H. Collins. 1970. Isolation of *Mycobacterium xenopei* from Water Taps. *Jour. Hyg. Camb.*, 68:97-100.

133. Panwalker, A.P. and E. Fuhse. 1986. Nosocomial *Mycobacterium gordonae* Pseudo Infections from Contaminated Ice Machines. *Infect. Contr.*, 7:67-70.

134. Laussueq, S. A.L. Baltch, R.P. Smith, R.W. Smithwick, B.J. Davis, et al. 1988. Nosocomial *Mycobacterium fortuitum* Colonization from a Contaminated Ice Machine. *Amer. Rev. Respir. Dis.*, 138:891-894.

135. Rosebury, T. 1962. *Microorganisms Indigenous to Man*, McGraw-Hill Book Co., New York.

136. Engelbrecht, R.S., D.F. Foster, E.O. Grenning, and S.H. Lee. 1974. New Microbial Indicators of Wastewater Chlorination Efficiency. EPA-670/2-72-082. U.S. Environmental Protection Agency, Cincinnati, OH.

137. Jones, P.W., J. Bew, M.R. Burrows, P.R. J. Matthews, and P. Collins. 1976. The Occurrence of *Salmonella, Mycobacteria* and Pathogenic Strains of *E. coli* in Pig Slurry. *Jour. Hyg.,* 77:43-50.

138. Engelbrecht, R.S. and C.N. Haas. 1977. Acid-Fast Bacteria and Yeasts as Disinfection Indicators: Enumeration Methodology. Proc. Amer. Water Works Assoc., Water Quality Technol. Confr., Kansas City, MO. pp. 2B-1–2B-20.

139. Engelbrecht, R.S., B.F. Severin, M.T. Masarek, S. Faroog, S.H. Lee, C.M. Haas, and A. Lalchandani. 1977. New Microbial Indicators of Disinfection Efficiency. EPA-600/2-77-052, U.S. Environmental Protection Agency, Cincinnati, OH.

140. Shular, J.A. 1978. The Occurrence of Indicator Organisms in the Decatur, Illinois South Water Treatment Plant and in One Branch of the Distribution System. M.S. Special Problem Report. Dept. Civil Engr., Univ. IL., Urbana-Champaign.

141. Haas, C., M.A. Meyer, and M.S. Paller. 1983. The Ecology of Acid-Fast Organisms in Water Supply, Treatment, and Distribution Systems. *Jour. Amer. Water Works Assoc.,* 75:139-144.

142. Surucu, F. and C.N. Haas. 1976. Inactivation of New Indicator Organisms of Disinfection Efficiency. Part I. Free Available Chlorine Species Kinetics. Presented at the 96th Annual Meeting, Amer. Water Works Assoc., New Orleans.

143. Pelletier, P.A., G.C. duMoulin, and K.D. Slottmeir. 1988. Mycobacteria in Public Water Supplies: Comparative Resistance to Chlorine. *Microbiol. Sci.,* 5:147-148.

144. Severin, B.F. 1976. Inactivation of New Indicator Organisms of Disinfection Efficiency. Part II. Combined Chlorine as Chloramines. Presented at the 96th Annual Meeting, Amer. Water Works Assoc., New Orleans.

145. Farooq, S. 1976. Kinetics of Inactivation of Yeasts and Acid-Fast Organisms with Ozone. Ph.D. Thesis, Univ. Illinois, Dept. Civil Engr., Urbana-Champaign.

146. Carson, L.H., L.A. Bland, L.B. Cusick, M.S. Favero, G.A. Bolan, A.L. Reingold, and R.C. Good. 1988. Prevalence of Nontuberculous Mycobacteria in Water Supplies of Hemodialysis Centers. *Appl. Environ. Microbiol.,* 54:3122-3125.

147. Bagley, S.T. 1985. Habitat Association of *Klebsiella* Species. *Infect. Control,* 6:52-58.

148. Nunez, W.J. and A.R. Colmer. 1968. Differentiation of *Aerobacter-Klebsiella* Isolated from Sugarcane. *Appl. Microbiol.,* 16:1375-1878.

149. Geldreich, E.E., H.D. Nash, and D.F. Spino. 1977. Characterizing Bacterial Populations in Treated Water Supplies: A Progress Report. Proc. Amer. Water Works Assoc. Water Qual. Technol. Confr. Kansas City, MO.

150. Ptak, O.J., W. Ginsburg, and B.F. Wiley. 1973. Identification and Incidence of *Klebsiella* in Chlorinated Water Supplies. *Jour. Amer. Water Works Assoc.,* 65:604-608.

151. Bagley, S.T., R.J. Seidler, and D.J. Brenner. 1981. *Klebsiella planticola* sp. nov.: A New Species of Enterobacteriaciae Found Primarily in Nonclinical Environments. *Curr. Microbiol.*, 6:105- 109.
152. Camper, A.K. et al. 1986. Bacteria Associated with Granular Activated Carbon Particles in Drinking Water. *Appl. Environ. Microbiol.*, 52:434- 438.
153. Edberg, S.C., V. Piscitelli, and M. Carter. 1986. Phenotype Characteristics of Coliform and Noncoliform Bacteria from a Public Water Supply Compared with Regional and National Clinical Species. *Appl. Environ. Microbiol.*, 52:474-478.
154. LeChevallier, M.W., R.J. Seidler, and T.M. Evans. 1980. Enumeration and Characterization of Standard Plate Count Bacteria in Chlorinated and Raw Water Supplies. *Appl. Environ. Microbiol.*, 40:922-930.
155. Thom, B.T. 1970. *Klebsiella* in Faeces. *Lancet*, 2:1033.
156. Davis, T.J. and J.M. Matsen. 1974. Prevalence and Characteristics of *Klebsiella* Species. Relation to Association with a Hospital Environment. *Jour. Infect. Dis.*, 130:402-405.
157. Cooke, E.M. et al. 1979. Further Studies on the Sources of *Klebsiella aerogenes* in Hospital Patients. *Jour. Hyg.*, 83:391-395.
158. Bagley, S.T. and R.J. Seidler. 1977. Significance of Fecal Coliform Positive *Klebsiella*. *Appl. Environ. Microbiol.*, 33:1141-1148.
159. Edmondson, A.E., E.M. Cooke, A.P.D. Wilcock, and R. Shinebaum. 1980. A Comparison of the Properties of *Klebsiella* Strains Isolated from Different Sources. *Jour. Med. Microbiol.*, 13:541-550.
160. Naemura, L.G. and R.J. Seidler. 1978. Significance of Low Temperature Growth Associated with the Fecal Coliform Response, Indole Production and Pectin (sic) Liquefaction in *Klebsiella*. *Appl. Environ. Microbiol.*, 35:392-396.
161. Naemura, L.G., S.T. Bagley, R.J. Siedler, J.B. Kaper, and R.R. Colwell. 1979. Numerical Taxonomy of *Klebsiella pneumoniae* Strains Isolated from Clinical and Nonclinical Sources. *Curr. Microbiol.*, 2:175-780.
162. Montgomerie, J.Z. 1979. Epidemiology of *Klebsiella* and Hospital-Associated Infections. *Rev. Infect. Dis.*, 1:736-753.
163. Smith, S.M., J.T. Digori, and R.H.K. Eng. 1982. Epidemiology of *Klebsiella* Antibiotic Resistance and Serotypes. *Jour. Clin. Microbiol.*, 16:868-873.
164. Martin, W.J., P.K.W. Yu, and T.A. Washington. 1971. Epidemiological Significance of *Klebsiella pneumoniae*: A 3 Month Study. *Mayo Clin. Proc.*, 46:785-783.
165. Jarvis, W.R. et al. 1985. The Epidemiology of Nosocomial Infections Caused by *Klebsiella pneumoniae*. *Infect. Control*, 6:68-74.
166. Huntley, R.E., A.C. Jones, and V.J. Cabelli. 1976. *Klebsiella* Densities in Waters Receiving Wood Pulp Effluents. *Jour. Water Poll. Contr. Fed.*, 48:1766-1771.
167. Brown, C. and R.J. Seidler. 1973. Potential Pathogens in the Environment: *Klebsiella pneumoniae*, a Taxonomic and Ecological Enigma. *Appl. Environ. Microbiol.*, 25:900-904.
168. Knowles, R., R. Neufeld, and S. Simpson. 1974. Acetylene Reduction (Nitrogen Fixation) by Pulp and Paper Mill Effluents and by *Klebsiella* Isolated from Effluents and Environmental Situations. *Appl. Microbiol.*, 28(4):608-613.
169. Campbell, L.M. et al. 1976. Isolation of *Klebsiella pneumoniae* from Lake Water. *Can. Jour. Microbiol.*, 22:1762-1767.
170. Dufour, A.P. and V.J. Cabelli. 1976. Characteristics of *Klebsiella* from Textile Finishing Plant Effluents. *Jour. Water Poll. Contr. Fed.*, 48:872-879.

171. Caplenas, N.R., M.S. Kanarek, and A.P. Dufour. 1981. Source and Extent of *Klebsiella pneumoniae* in the Paper Industry. *Appl. Environ. Microbiol.*, 42:779-785.

172. Hendry, G.S., S. Janhurst, and G. Horsnell. 1982. Some Effects of Pulp and Paper Waste Water on Microbiological Water Quality of a River. *Water Res.*, 16:1291-1295.

173. Niemala, S.I. and P. Vaatanen. 1982. Survival in Lake Water of *Klebsiella pneumoniae* Discharged by a Paper Mill. *Appl. Environ. Microbiol.*, 44:264-269.

174. Knittel, N.D. 1975. Occurrence of *Klebsiella pneumoniae* in Surface Waters. *Appl. Microbiol.*, 29:595-597.

175. Seidler, R.J., M.D. Knittel, and C. Brown. 1975. Potential Pathogens in the Environment: Cultural Reactions and Nucleic Acid Studies of *Klebsiella pneumoniae* from Clinical and Environmental Sources. *Appl. Microbiol.*, 29:819-825.

176. Office of Enforcement Proceedings of Seminar on the Significance of Fecal Coliform in Industrial Wastes. 1972. R.H. Bordner and B.J. Carroll, Eds., EPA Tech. Rept. 3, US EPA, Denver, CO.

177. Seidler, R.J. 1981. The Genus *Klebsiella* in the Prokaryotes. In: *A Handbook on Habitats, Isolation and Identification of Bacteria*. M.P. Starr et al., Eds., Springer-Verlag, New York.

178. Bagley, S.T. and R.J. Seidler. 1978. Comparative Pathogenicity of Environmental and Clinical *Klebsiella*. *Health Lab. Sci.*, 15:104-111.

179. Wolfe, R.L., N.R. Ward, and B.H. Olson. 1985. Inactivation of Heterotrophic Bacterial Populations in Finished Drinking Water by Chlorine and Chloramines. *Water Res.*, 19:1393-1403.

180. Franzblau, S.G., D.R. Jimenez, and N.A. Sinclair. 1985. A Selective Medium for the Isolation of Opportunistic Flavobacteria from Potable Water. *Jour. Environ. Sci. Health*, 5:583-591.

181. Dooley, J.R., L.J. Nims, V.H. Lipp, A. Beard, and L.T. Delancy. 1980. Meningitis of Infants Caused by *Flavobacterium meningosepticum*. *Jour. Trop. Ped.*, 26:24-30.

182. Hazuka, B.T., A.S., Dajani, K. Talbot, and B.M. Keen. 1977. Two Outbreaks of *Flavobacterium meningosepticum* Type E in a Neonatal Intensive Care Unit. *Jour. Clin. Microbiol.*, 6:450-455.

183. Berry, W.B., A.G. Morrow, D.C. Harrison, H.D. Hochstein, and S.K. Himmelsbach. 1963. *Flavobacterium meningosepticum* Following Intracardia Operations. Clinical Observations and Identification of the Source of Infection. *Jour. Thorac. Cardiovasc. Surg.*, 45:476-481.

184. Edmondson, E.B., J.A. Reinary, A.K. Pierce, and J.P. Sanford. 1966. Nebulization Equipment. A Potential Source of Infection in Gram-Negative Pneumonias. *Amer. Jour. Dis. Child.*, 111:357-360.

185. Werthamer, S. and M. Weiner. 1972. Subacute Bacterial Endocarditis due to *Flavobacterium meningosepticum*. *Amer. Jour. Clin. Pathol.*, 57:410-412.

186. Talley, M.W. and M.R. Alexander. 1989. Preventing Potential Health Threats of Stagnation in Potable Internal Distribution Systems. A.W.W.A. Confr. Proc. Part 2:967-970.

187. Reasoner, D.J., J.C. Blannon, E.E. Geldreich, and J. Barnick. 1989. Nonphotosynthetic Pigmented Bacteria in a Potable Water Treatment and Distribution System. *Appl. Environ. Microbiol.*, 55:912-921.

188. van der Kooij, D. and W.A.M. Hijnen. Determination of the Concentration of Maltose- and Starch-Like Compounds in Drinking Water by Growth Measurements with a Well-Defined Strain of a *Flavobacterium* Species. *Appl. Environ. Microbiol.*, 49:765-771.

189. Lennette, E.H., A. Balows, W.J. Hausler, Jr., and H.J. Shadomy. 1985. *Manual of Clinical Microbiology.* 4th ed. Amer. Soc. Microbiol., Washington, D.C.

190. Fliermans, C.B. et al. 1979. Ecological Distribution of *Legionella pneumophila* from Non-Epidemic Related Aquatic Habitats. *Appl. Environ. Microbiol.*, 37:1239-1242.

191. Morris, G.K. et al. 1979. Isolation of the Legionnaires Disease Bacterium from Environmental Samples. *Ann. Intern. Med.*, 90:664-666.

192. Oritz-Roque, C. and T.C. Hazen. 1981. Isolation of *Legionella pneumophila* from Tropical Waters. Ann. Mtg. Amer. Soc. Microbiol. (Abstract).

193. Fliermans, C.B., W.B. Cherry, L.H. Orrison, S.J. Smith, D.L. Tison, and D.H. Pope. 1981. Ecological Distribution of *Legionella pneumophila*. *Appl. Environ. Microbiol.*, 41:9-16.

194. Fliermans, C.B., G.E. Bettinger, and A.W. Fynsk. 1982. Treatment of Cooling Systems Containing High Levels of *Legionella pneumophila*. *Water Res.*, 16:903-909.

195. Spino, D.F., E.W. Rice, and E.E. Geldreich. 1983. Occurrence of *Legionella* spp. and Other Aquatic Bacteria in Chemically Contaminated Ground Water Treated by Aeration. In: Legionella: *Proceedings of the 2nd International Symposium*, C. Thornsberry, A. Balows, J.C. Feeley, and W. Jakubowski, Eds., Amer. Soc. Microbiol., Washington, D.C. pp. 318-320.

196. Tison, D.L. and R.J. Seidler. 1983. *Legionella* Incidence and Density in Potable Drinking Water Supplies. *Appl. Environ. Microbiol.*, 45:337-339.

197. Stout, J.E., V.L. Yu, and M.G. Best. 1985. Ecology of *Legionella pneumophila* Within Water Distribution Systems. *Appl. Environ. Microbiol.*, 49:221-228.

198. Colbourne, J.S. and R.M. Trew. 1986. Presence of *Legionella* in London's Water Supplies. *Isr. Jour. Med. Sci.*, 22:633-639.

199. Wadowsky, R.M. and R.B. Yee. 1985. Effect of Non-Legionellaceae Bacteria on the Multiplication of *Legionella pneumophila* in Potable Water. *Appl. Environ. Microbiol.*, 49:1206-1210.

200. Colbourne, J.S., P.J. Dennis, R.M. Trew, C. Berry, and G. Veseg. 1988. *Legionella* and Public Water Supplies. *Water Sci. Tech.*, 30:5-10.

201. Witherell, L.E., R.W. Duncan, K.M. Stone, L.J. Stratton, L. Oriciari, S. Kappel, and D.A. Jillson. 1988. Investigation of *Legionella pneumophila* in Drinking Water. *Jour. Amer. Water Works Assoc.*, 80:87-93.

202. Barrow, G.I. 1987. Legionnaire's Disease and Its Impact on Water Supply Management. *Jour. Inst. Water Environ. Manage.*, 1:117-122.

203. Wadowsky, R.M. and R.B. Yee. 1983. Satellite Growth of *Legionella pneumophila* with an Environmental Isolate of *Flavobacterium breve*. *Appl. Environ. Microbiol.*, 46:1447-1449.

204. Stout, J.E., V.L. Yu, and M.G. Best. 1985. Ecology of *Legionella pneumophila* Within Water Distribution Systems. *Appl. Environ. Microbiol.*, 49:221-228.

205. Fields, B.S., E.B. Shotts, Jr., J.C. Feeley, C.W. Gorman, and W.T. Martin. 1984. Proliferation of *Legionella pneumophila* as an Intracellular Parasite of the Ciliated Protozoan *Tetrahymena pyriformis*. *Appl. Environ. Microbiol.*, 47:467-471.

206. Geldreich, E.E. 1990. Microbiological Quality Control in Distribution Systems. In: *Water Quality and Treatment*. F.W. Pontius, Ed., McGraw-Hill, New York.

207. Gambassini, L., C. Sacco, E. Lanciotti, D. Burrini, and O. Griffini. 1990. Microbial Quality of the Water in the Distribution System of Florence. *Aqua,* 39:258-264.

208. Gavin, F. and Leclerc, H. 1975. Etude des Bacilles Gram-Pigmentes en Joune Isoles de l'Eau. *Int. Ocean. Med.,* 37:17-68.

209. Ducluzeau, R., Bochan, J.M., and Defresne, S. 1976. La Microflora autochtone de l'Eau Mineral Nature Caracteres Phyrialogiques Signification Hygienique. *Med. Nutr.,* 12:115-120.

210. Hernandez Duquino, H. and Rosenberg, F.A. 1987. Antibiotic-resistant *Pseudomonas* in Bottled Drinking Water. *Can. Jour. Microbiol.,* 33:286-289.

211. Lindberg, R.B. 1974. Culture and Identification of Commonly Encountered Gram-Negative Bacilli: *Pseudomonas, Klebsiella-Enterobacter, Serratia, Proteus* and *Providencia*. In: *Opportunistic Pathogens*. J.E. Prier and H. Friedman, Eds., University Park Press, Baltimore, MD.

212. Holder, I.A. 1977. Epidemology of *Pseudomonas aeruginosa* in a burns hospital. In: *Pseudomonas aeruginosa: Ecological Aspects and Patient Colonization*. V.M. Young, Ed., Raven Press, New York.

213. Tinne, J.E., A.M. Gordon, W.H. Bain, and W.A. Mackey. 1967. Cross-Infection by *Pseudomonas aeruginosa* as a Hazard of Intensive Surgery. *Brit. Med. Jour.,* 4:313-315.

214. Schimpff, S.C., W.H. Greene, V.M. Young, and P.H. Wiernik. 1973. *Pseudomonas septicemia*: Incidence, Epidemiology, Prevention and Therapy in Patients with Advanced Cancer. *Eur. Jour. Cancer,* 9:449-455.

215. Schimpff, S.C., R.M. Miller, S. Polkavetz, and R.B. Hornik. 1974. Infection in the Severely Traumatized Patient. *Ann. Surg.,* 179:352-357.

216. Jacobson, Jay A. 1985. Pool-Associated *Pseudomonas aeruginosa* Dermatitis and Other Bathing-Associated Infections. *Infect. Control,* 6:398-401.

217. Hunter, C.A. and P.R. Ensign. 1947. An Epidemic of Diarrhea in a Newborn Nursery Caused by *Pseudomonas aeruginosa*. *Amer. Jour. Pub. Health,* 37:1166-1169.

218. Young, V.M. 1977. *Pseudomonas aeruginosa*: *Ecological Aspects and Patient Colonization*. Raven Press, New York.

219. Stoodley, B.J. and B.T. Tom. 1970. Observations on the Intestinal Carriage of *Pseudomonas aeruginosa*. *Jour. Med. Microbiol.,* 3:367-375.

220. Buck, A.C. and E.M. Cooke. 1969. The Fate of Ingested *Pseudomonas aeruginosa* in Normal Persons. *Jour. Med. Microbiol.,* 2:521-525.

221. Ringen, L.M. and C.H. Drake. 1952. A Study of the Incidence of *Pseudomonas aeruginosa* from Various Natural Sources. *Jour. Bact.,* 64:841-845.

222. Hoadley, A.W., E. McCoy, and G.A. Rohlich. 1968. Untersuchungen uber *Pseudomonas aeruginosa* in Oberflachengewassern. I. *Quellen. Arch. Hyg. Bakteriol.,* 152:328-338.

223. Weber, G., H.P. Werner, and H. Matschnigg. 1971. *Pseudomonas aeruginosa* in Trinkwasser als Todesursache bei Neugeborenen. *Zentralbl. Bakteriol. Parasitenk. Infektionskr. Hyg. I Abt. Orig.,* 216:210-214.

224. Shinefield, H.R., J.D. Wilsey, J.C. Ribble, M. Boris, A.F. Eichanwald, and C.L. Dittmar. 1966. Interactions of Staphylococcal Colonization. *Amer. Jour. Dis. Child.,* 111:11-21.

225. Manples, R.R. and A.M. Kligman. 1976. Experimental Staphylococcal Infection of the Skin in Man. In: *Staphylococci and Staphylococcal Diseases*. J. Jeljaszewiez, Ed., Gustau Fisher Verlag, New York. pp. 755-760.

226. Centers for Disease Control. 1972. Staphylococcal Food Poisoning, Milwaukee, Wisconsin. *Morbid. Mortal. Weekly Rep.*, 21:422-423.

227. LeChevallier, M.W. and R.J. Seidler. 1980. *Staphylococcus aureus* in Rural Drinking Water. *Appl. Environ. Microbiol.*, 30:739-742.

228. Antai, S.P. 1987. Incidence of *Staphylococcus aureus*, Coliforms and Antibiotic-Resistant Strains of *Escherichia coli* in Rural Water Supplies in Port Harcourt. *Jour. Appl. Bacteriol.*, 62:371-375.

229. Seeliger, H.P.R. 1975. Increasing Spectrum of Opportunistic Fungal Infections. In: *Opportunistic Fungal Infections*, E.W. Chick, A. Balows, and M.L. Farcolow, Eds., Charles C. Thomas, Springfield, IL.

230. Gutman, A.A. 1985. Allergens and Other Factors Important in Atopic Disease. In: *Allergic Diseases: Diagnosis and Management*, 3rd ed. R. Patterson, Ed., Lippincott, New York.

231. Atterholm, I., K. Ganrot-Norlin, T. Hallberg, and O. Rengert. 1972. Unexplained Acute Fever after a Hot Bath. *Lancet*, ii:682-684.

232. Metzer, W.J., R. Patterson, R. Semerdjan, and M. Roberts. 1976. Sauna Takers Disease: Hypersensitivity Pneumonitis Due to Contaminated Water in a Home Sauna. *Jour. Amer. Med. Assoc.*, 236:2209-2211.

233. Gronoos et al. 1980. An Epidemic of Extrinsic Allergic Alvcolitis Caused by Tap Water. *Clin. Allergy*, 10:77-90.

234. Hodges, G.R., J.N. Fink, and N.P. Schlueter. 1974. Hypersensitivity Pneumonitis Caused by a Contaminated Cool-Mist Vaporizer. *Ann. Int. Med.*, 80:501-504.

235. Rosenzweig, W.D., H. Minnigh, and W.O. Pipes. 1986. Fungi in Potable Water Distribution Systems. *Jour. Amer. Water Works Assoc.*, 78:53-55.

236. Rosenzweig, W.D. and W.O. Pipes. 1986. Survival of Fungi in Potable Water Systems. Amer. Water Works Assoc., Water Quality Tech. Conf. Portland, OR. pp. 449-456.

237. West, P.R. 1986. Isolation Rates and Characterization of Fungi in Drinking Water Distribution Systems. Amer. Water Works Assoc., Water Quality Tech. Conf., Portland, OR. pp. 457-473.

238. Harvey, J.V. 1952. Relationship of Aquatic Fungi to Water Pollution. *Sew. Indust. Wastes*, 24:1159-1164.

239. Cooke, W.B. and A.F. Bartsch. 1960. Aquatic Fungi in Some Ohio Streams. *Ohio Jour. Sci.*, 60:144-148.

240. Cooke, W.B. 1967. Fungal Populations in Relation to Pollution of the Bear River, Idaho-Utah. *Utah Acad. Proc.*, 44:298-315.

241. Cooke, W.B. 1986. The Fungi of Our Mouldy Earth. In: *Beiheft 85 Zur Nova Hedwigia*. J. Cramer, Berlin.

242. Burman, N.P. 1965. Taste and Odour due to Stagnation and Local Warming in Long Lengths of Piping. *Soc. Water Treat. Exam.*, 14:125-131.

243. Bays, L.R., N.P. Burman, and W.M. Lavis. 1970. Taste and Odour in Water Supplies in Great Britain: A Study of the Present Position and Problems for the Future. *Water Treat. Exam., Jour. Soc. Water Treat. Exam.*, 19:136-160.

244. Nagy, L.A. and B.H. Olson. 1982. The Occurrence of Filamentous Fungi in Drinking Water Distribution Systems. *Can. Jour. Microbiol.*, 28:667-671.

245. Niemi, R.M., S. Knuth, and K. Lundstrom. 1982. Actinomycetes and Fungi in Surface Waters and in Potable Water. *Appl. Environ. Microbiol.,* 43:378-388.

246. Rosenzweig, W.D., H.A. Minnigh, and W.O. Pipes. 1983. Chlorine Demand and Inactivation of Fungal Propagules. *Appl. Environ. Microbiol.,* 45:182-186.

247. Nagy, L.A. and B.H. Olson. 1985. Occurrence and Significance of Bacteria, Fungi, and Yeast Associated with Distribution Pipe Surfaces. Amer. Water Qual. Technol. Conf., Houston, TX, Dec. 8-11, AWWA, Denver, CO. pp. 213-238.

248. Burman, N.P. and Colbourne, J.S. 1979. Effect of Nonmetallic Materials on Water Quality. *Jour. Inst. Water Engin. Scient.,* 33:11-18.

249. Roesch, S.C. and L.Y.C. Leong. 1983. Isolation and Identification of *Petriellidium boydii* from a Municipal Water System. Abstr. Ann. Mtg. Amer. Soc. Microbiol. p. 276.

250. O'Connor, J.T., L. Hash, and A.B. Edwards. 1975. Deterioration of Water Quality in Distribution Systems. *Jour. Amer. Water Works Assoc.,* 67:113-116.

251. Hinzelin, F. and J.C. Block. 1985. Yeast and Filamentous Fungi in Drinking Water. *Environ. Technol. Lett.,* 6:101-106.

252. Rosenzweig, W.D. and W.O. Pipes. 1988. Fungi from Potable Water: Interaction with Chlorine and Engineering Effects. *Water Sci. Technol.,* 20:153-160.

253. Gerlach, W. 1972. Fusarien aus Trinkwasserleitungen. *Ann. Agric. Fenn.,* 11:298-302.

254. Louria, D.B. 1974. Superinfection: A Partial Overview. In *Opportunistic Pathogens.* J.E. Prier and H. Friedman, Eds., University Park Press, Baltimore, MD.

255. Ahearn, D.G. 1978. Medically Important Yeast. *Ann. Rev. Microbiol.,* 32:59-68.

256. Buck, J.D. and P.M. Bubucis. 1978. Membrane Filter Procedure for Enumeration of *Candida albicans* in Natural Waters. *Appl. Environ. Microbiol.,* 35:237-242.

257. Committee on the Challenges of Modern Society (NATO/CCMS). 1987. Drinking Water Microbiology. *Jour. Environ. Pathol. Toxicol. Oncol.,* 7:1-365.

258. Cragg, J. and G.M. Clayton. 1971. Bacterial and Fungal Flora of Seagull Droppings in New Jersey. *Jour. Clin. Pathol.,* 24:317-319.

259. Niewolak, S. 1977. The Occurrence of Yeasts in Some of the Masurian Lakes. *Acta Mycolog.,* 12:241-256.

260. Engelbrecht, R.S. and C.N. Haas. 1977. Acid-Fast Bacteria and Yeasts as Disinfection Indicators: Enumeration Methodology. Proc. Amer. Water Works Assoc., Water Quality Technol. Confr., Kansas City, MO. pp. 2B-1–2B-20.

261. Farooq, S. and S. Akhlaque. 1983. Comparative Response of Mixed Cultures of Bacteria and Virus to Ozonation. *Water Res.,* 17:809-812.

262. Ahearn, D.G. 1974. Identification and Ecology of Yeasts of Medical Importance. In: *Opportunistic Pathogens.* J.E. Prier and H. Friedman, Eds. University Park Press, Baltimore.

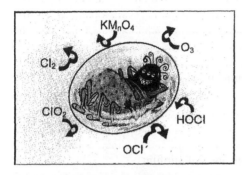

CHAPTER **4**

Biofilms in Water Distribution Systems

CONTENTS

INTRODUCTION

Seasonal coliform occurrences in the distribution system often lead to the water utility being in jeopardy of noncompliance with federal regulations and the subsequent requirement to issue a boil water notice. Intense monitoring during this event often reveals increased sporadic occurrences of coliforms, yet data from the plant suggests there are no detectable passages of coliforms beyond the treatment barriers. Furthermore, speciation of coliforms found in the distribution samples during these events are most often negative for fecal coliforms, suggesting no fecal contamination occurred as a result of a cross-contamination problem.

A series of questions immediately arise: Why the sudden proliferation of coliform occurrences in a system that has had a long history of providing acceptable water quality in the past? Has a "super bug" emerged that is resistant to disinfection or is this data merely a glitch in the laboratory analysis? If these occurrences are real, why have they been so difficult to suppress with typical disinfection practices? Finally, what is their public health significance? Answers to these questions lie in an understanding of the nature of microbial communities in a biofilm and their ability to adapt to the pipe environment and survive many adversities.

CHARACTERIZING BIOFILM

Biofilms may appear as a patchy mass in some pipe section or as a uiniform film along the inner walls of a storage tank. They may consist of a monolayer of cells in a microcolony or can be as thick as 10 to 40 mm, as in algal mats, reservoir bottom sediments, or raw water intake structures. These biofilms often provide a variety of microenvironments for growth that include aerobic and anaerobic zones due to the oxygen diffusion limitation within the biofilm.[1] Biofilm appears to be a complex structure consisting of diverse microcolonies of various organisms embedded in a matrix of exrtacellular organic polymers adhering to moist surfaces. This matrix is interlaced with water channels that have been reported to constitute as much as 40 to 60% of the total biofilm volume.[2] A conceptual model of a biofilm (Figure 4.1) has been developed from research studies at the Center for Biofilm Engineering, Montana State University. This structural model can be configurated for the coexistence of a variety of different organisms within favorable microniches and still have access to nutrients in the bulk water. What is not yet clear is the mechanism by which chloramines penetrate the biomass more effectively than free chlorine in certain situations or on different materials.[3-5] Perhaps this may be due to the fact that monochloramine reacts with DNA, tryptophane, and sulfur-containing ammino acids while having little interreaction with polysaccharides. Hypochlorous acid, however, indiscriminately attacks both polysaccharides and cell material, necessitating a greater dosage because of chlorine demand in the gelatinous matrix.

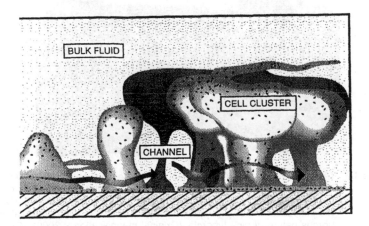

Figure 4.1   Conceptual model of a biofilm. (Courtesy of the Center for Biofilm Engineering, Montana State University.)

In the distribution system, attachment surfaces may be pipe surfaces, sediments, tubercles, water storage tank compartments, valves, gasket material in fittings, and recesses in fire plug construction.[5-7] Biofilms are initially formed when pioneering organisms enter the distribution system and become entrapped in some slow-flow area, line obstruction, or dead-end section. While some of these microorganisms are held only by random entrapment, others are more likely to succeed in their attachment because of extracellular appendages (glycocalyx) that extend from the cell membrane. These glycocalyx appendages either become enmeshed in the crevices of porous sediment or tubercle material or attach themselves to other surfaces through their gelatinous secretions. Scanning electron micrographs of the biofilm habitat (Figure 4.2) reveals the substrate to be either amorphous, porous, or crystalline with micro-organisms entrapped in sediment (Figure 4.3) or attached to fracture lines, porous areas, or encrustations (Figure 4.4) at the surface.[8-12] These sites tend to be protected from the overlying flow of water supply with the biofilm expansion generally being restricted by the shearing action of water movement.

For these organisms that can endure this harsh aquatic environment, growth proceeds slowly at first as the cells adjust to the constraints of the pipe habitat. In time, these microcolony sites attract other organisms whose more exacting nutritional needs may only be found in the by-products released in the metabolism of various organisms in the pioneering microbial community. Thus, there is a progressive diversity brought into the biofilm as the site becomes populated with a variety of bacteria, protozoans, nematodes, and worms.[8,13-20]

Figure 4.2   (Above) Pipe tubercle. (Opposite top) Cross section of pipe tubercle showing loose surface material at water interface, deposition of corrosion inhibitor in distinct layers, and compaction of deposits near the pipe wall. (Orginal magnification × 100.) (Opposite center) Electron micrograph showing porous nature of pipe tubercle. (Original magnification × 3000). (Opposite bottom) Electron micrograph showing active bacterial colonization in tubercle. (Original magnification × 3000.)

PROFILE OF HETEROTROPHIC BACTERIA

Much of the microbial flora in drinking water is composed of transient organisms that are disinfectant resistant and of no sanitary significance. They originate either from the source water and pass through the treatment barrier in varying numbers or are introduced during line repairs or new line construction or enter open storage of finished water. Of the wide spectrum of organisms that survive water treatment or enter the distribution system, some deserve

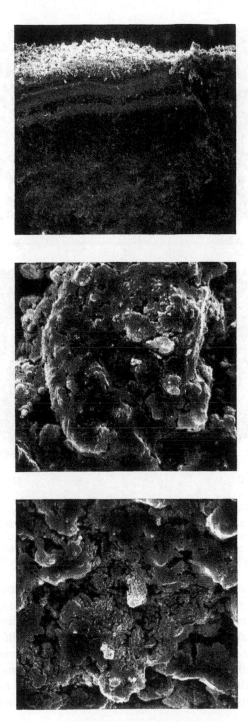

Figure 4.2 (continued)

Figure 4.3   Electron micrographs showing: (A) diatoms (Nitzschia) entrapped in surface sediment from pipe deposits in Tacoma, WA; (B) diatoms (Cyclotella)
imbedded in pipe sediment in North Bend, OR; (C) diatoms (Nitzscha)
imbedded in pipe sediment in Medford, OR; (D) diatoms (Aulacoseira)
imbedded in pipe sediment. (Original magnifications × 3000.)

special attention. Among them, *Pseudomonas, Flavobacterium, Arthrobacter,
Acinetobacter, Sarcina, Micrococcus, Proteus, Bacillus, Actinomycites,* and
some yeast have been shown to be the most troublesome once they colonize
in a biofilm because of their potential interference to coliform detection,
because they create taste and odors, or because of their acknowledged role as
opportunistic pathogens (see Chapter 3). Such organisms coexist with
coliforms in biofilms and may exert a strong antagonism to minimize populations of persisting coliform species. Any of these organisms in the established

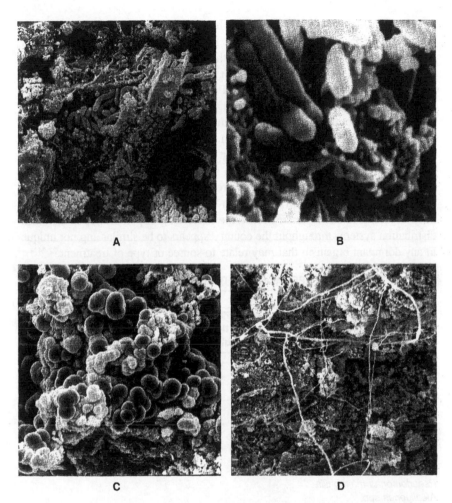

Figure 4.4 Electron micrographs showing: (A) attachment site for bacterial growth in tubercle surface material; (B) elongated bacteria in encrusted surface sediment; (C) microbial colonization in pipe sediment in Medford, OR; (D) actinomycetes colonizing pipe sediment in Portland, OR. (Original magnifications: A, × 4000; B, × 19,000; C, × 1600; and D, × 1000.)

biofilm may enter the main flow of water when expanding growth is released by water hydraulics, changes in pipe sediment adhesion, or physical action of mechanical scraping of pipe interiors.

While the profile of transient organisms may vary with seasonal factors in source water, the organisms that colonize a biofilm may be expected to be more constant in species but vary widely in density releases into the water columns. Such is the evidence demonstrated in Table 4.1. *Pseudomonas vesicularis* and *Flavobacterium* spp. were predominant in the water columns and occurred at similar frequency in zinc floc released from pipe sediments in line flushing. Heterotrophic bacterial density in the floc materia was 2,400,000 organisms per milliliter.[10] A similar predominance of these bacteria (*P. vesicularis* and *Flavobacterium* spp.) also was observed in cement-lined pipe surface scrapings taken from the same branch line. HPC density in this biofilm material was 1000 organisms per centimeter squared. *Arthrobacter* spp. were found to be predominant in both flushed sediment and iron tubercles taken from the New Jersey utility water distribution system. The HPC density for the tubercle material was 13,000,000 organisms per gram. *Arthrobacter* spp. also dominated biofilms in a pilot pipeline of iron and galvanized sections connected to a public water supply system in Delaware.[21] This genus also was isolated from 20% of the HPC population detected in biofilm taken from two water distribution systems in California.[22] In fact, the bacterial group identified in different distribution systems throughout the country appears to be similar and not unique in any dominant organism that may relate to source or type of treatment.[22-24]

Table 4.1   Predominant HPC Bacteria in Distribution System Biofilms

	Occurrence in				
Bacteria	Water column	Zinc floc sediment	Flushed sediment	Ion tubercle	Pipe surface scrapings
Pseudomonas vesicularis	++	++	+		++
Flavobacterium spp.	++	++	+		++
Pseudomonas diminuta	+			+	
Pseudomonas cepacia					+
Pseudomonas pickettii					+
Pseudomonas stutzeri			+		+
Pseudomonas fluorescens			+	+	
Pseudomonas putida				+	
Pseudomonas paucimobilis	+			+	
Pseudomonas maltophilia	+				
Alcaligenes spp.					+
Acinetobacter spp.	+				+
Moraxell spp.	+		+		+
Agrobacterium radiobacter					+
Arthrobacter spp.	+		++	++	+
Corynebacterium spp.				+	+
Bacillus spp.					+
Yeasts					+
CDC group II J	+				
Enterobacter agglomerans					+
Micrococcus spp.	+				

Note: R-2A (20 to 24°C for 7 d).
Data from LeChevallier et al.[10]

Coliform Organisms

Coliform organisms are heterotrophic bacteria that may become colonized in the distribution system biofilm. Based on coliform identification from water samples during a biofilm release period, generally one of several coliform species predominate, such as *Klebsiella pneumoniae, K. oxytoca, Enterobacter agglomerans, Enter. cloacae,* or *Citrobacter freundii.*[11,25] Recent studies on interactions between *E. coli* and strains of *Acinetobacter* and *Arthrobacter* demonstrated that while there is suppression of some coliforms, the *Enterobacteriaceae* appear able to compete successfully with other heterotrophic bacteria in a pipe sediment biofilm. Perhaps this is one reason only *Klebsiella* and *Enterobacter* strains are most often identified as the coliforms involved in public water supply noncompliance problems.

A direct search for coliforms in sediment from a New Jersey utility water distribution system revealed *Enter. agglomerans* present at a density of 0.5 organisms per gram of sediment material.[10] Mechanically scraping with a polyfoam insert (pigging) the same branch line to release tubercles provided an opportunity to test the material for coliform occurrences. Analyses disclosed fecal coliforms were present and speciation indicated *C. freundii* and *Enter. agglomerans.* It is not surprising that some of the coliforms identified had the same API 20E biochemical profile as those coliform bacteria detected in water and pipe sediments because mechanical scraping with plastic inserts moves sediment and tubercles in its path through the pipe, mixing both materials together. The coliform density from the dislodged tubercle material was reported to be greater than 160 organisms per gram. A study of pipe coupon samples from four water distribution systems in four different geographical areas revealed coliform occurrences in these pipelines were generally below detection level over a square centimeter of pipe surface (Table 4.2), suggesting coliform bacteria constitute only a very small part of the total biofilm population.

Pipe Condition

Pipe condition is often a factor in biofilm formation. Review of many case histories suggest that biofilm is generally considered to become a water quality problem in those water systems that have a predominance of pipe in service for over 50 years. Pipe composition is another consideration because corrosion areas in iron pipe are often active habitat sites. In these situations, biofilm would be expected to become more extensive and shedding films of heterotrophic bacteria more numerous. However, a study of information gathered from four water distribution networks of varying age and pipe materials (Table 4.2) indicates there are other contributing factors in pipe life and water quality deterioration.

Table 4.2 Heterotrophic Bacterial Densities for Biofilm Recovered from Pipe Coupons

Utility location	Date (1986)	Pipe composition	Pipe age (years)	Heterotrophic bacteria per cm^2	Coliform occurrence per cm^2
New Jersey	April 9	Cement lined, cast iron	57	3,000	<0.44
	April 10	Ductile iron	—	10	<0.44
	May 7	Cement lined, cast iron	55	200	<0.44
	May 29	Cement lined, ductile iron	84	172	<0.44
	May 29	Cement lined, ductile iron	8	44,000	<0.44
	June 10	Asbestos cement	37	250	<0.22
	August 14	Cement lined, ductile iron	—	740	<0.44
Indiana	April 11	Asbestos cement	17	3,500	<0.22
	April 23	Cement lined	20	1,400	<0.44
	May 7	Cement lined	31	1,700	<0.44
	June 24	Asbestos cement	9	800	<0.22
	June 28	Asbestos cement	36	60	<0.44
	July 11	Asbestos cement	13	160,000	<0.44
	July 11	Cement lined	90	2,360,000	<0.44
	July 16	Cement lined	56	19,000	<0.44
Illinois	May 1	Cement lined	65	1,060,000	1.02
	July 9	Cement lined	—	520,000	<0.62
Ohio	May 30	Cement lined, cast iron	25	240,000	<0.44

Data revised from LeChevallier et al.[10]

For instance, an 8-year-old section of cement-lined, ductile-iron pipe removed from a utility pipe network in New Jersey had 44,000 HPC per square centimeter of pipe surface while a piece of cement-lined pipe that was still in service in a water system in Indiana had 2,360,000 HPC per centimeter square of surface covered with biofilm. Yet another section of cement-lined, ductile-iron pipe collected on the same day from another part of the New Jersey utility pipe network that was on-line for 84 years had only 172 HPC per square centimeter of surface covered with biofilm. None of the pipe materials (either cast iron or ductile iron that were lined with cement or asbestos) provided any significant improvement in resisting biofilm formation. Other factors not evaluated in these examples were water flow rate in line, AOC concentration in water distribution, or frequency and approach to system-wide flushing on a scheduled basis.

NUTRIENTS: A CASE FOR BIOLOGICAL INSTABILITY OF WATER SUPPLY

Trace concentrations of nutrients are a major factor in the colonization of heterotrophic bacteria in the distribution system.[27-31] Analysis of biofilm reveals that critical nutrient requirements include carbon, nitrogen, phosphate, and manganese and the ratio for C:N:P approximates 100:10:1.[32] These substances are introduced in varying concentrations from natural waters used in water supply treatment.

Surface waters, in particular, receive an endless variety of organics discharged in municipal wastewater effluents, industrial wastes, stormwater runoff, and agricultural activities. Natural vegetation is a major contributor of humic substances in plant decay that cause problems of disinfectant demand, color, taste, and odor. Thus, it is not surprising to find total organic carbon concentrations (TOC) ranging from 1 to 20 mg/l at the water supply intake.[33] With soil barrier protection, protected groundwaters may contain TOC concentrations ranging from 0.1 to 2 mg/l.

In addition to a source of carbon, there is also a concurrent need for nitrogen in cell metabolism. Nitrogen occurs most frequently in source water as organic nitrogen, ammonia, and nitrate. Nitrite is relatively unstable in natural waters and may be a minor contributor in nutrient support. Of these nitrogen sources, ammonia from waste effluents and fertilizer loose in stormwater runoff from agricultural areas are the major contributors to raw surface water, particularly in winter and early spring when nitrification by natural self-purification processes decline. In groundwater, nitrate concentrations may range from 1.0 to over 400 mg of N per liter. Nitrate sources in these waters may be of natural occurrence in a given aquifer while in other areas the origins are soil percolations of agricultural fertilizers in stormwater runoff, feedlot manure drainage, and migration of irrigation water from farm fields.

Other inorganics needed in trace amounts for bacterial metabolism and biofilm growth in the distribution system appear to be iron, manganese, and phosphates. Appreciable amounts of iron and manganese complexes generally occur in groundwater and in water from the anaerobic hypolimnion of stratified lakes and impoundments.[34] Typical iron concentrations in groundwater are between 1 to 10 mg/l. Another source of iron may be found in pipeline corrosion, sediments, and tubercle formations.[9] Manganese concentrations appear to be lower than iron, ranging from 0.1 to 1.0 mg/l in groundwater. Like iron complexes, manganese is also a cause of biofouling of water supply lines when the concentrations are above 0.3 mg/l.[35] Phosphates are generally present in concentrations between a few tenths of a microgram and several hundred micrograms of P per liter.[32] Phosphates are not only introduced in the raw source water, but may also come from the use of certain additives used to control pipe corrosion.

While there is substantial removal of nutrients in source water through conventional treatment processes, this practice may not remove enough organic precursors to prevent elevated levels of trihalomethane formation. Applying ozonation early in the treatment train may be necessary to convert these undesirable organic residuals to complexes that either are adsorbed on GAC or consumed by organisms colonizing the filter media.

For those water utilities that use disinfection as the only treatment for a relatively clean surface-water source, a seasonal threat always exists for organic contributions from natural lignins, algal blooms, and recirculating bottom sediments during lake destratification.[36,37] These organic materials pass into distribution pipe networks and become an accumulation of nutrients for a wide range of aquatic bacteria.[38]

Most aquifers used for water supply extraction have minimal organics because of an effective soil barrier to surface contamination. Unfortunately there are other groundwater supplies that are coping with volatile organics that require air stripping to minimize the contaminant. Perhaps the reason none of these systems have coliform biofilm problems may lie with toxicity of the contaminants rather than nutrient availability.

ASSIMILABLE ORGANIC CARBON (AOC) DETECTION

The amount and type of biodegradable organic material present in drinking water or in sediment deposits is difficult to determine from data on dissolved organic carbon (DOC) or total organic carbon (TOC) measurements. These chemical tests capture only the bulk water portion of organic matter that is of concern in disinfection by-product formation, some of which will also be biodegradable. With a concern also being the suppression of biofilm development in the distribution system, there is a need for another type of test that measures only the biodegradable portion of the organic content in either treatment process water or water supply in distribution. Furthermore, this type of test should be sensitive to varying concentrations of biodegradable material that collects in site-specific locations in the pipe network that may become areas of greatest biofilm activity. Several bioassay tests have been proposed using either pure cultures of selected organisms or the mixed flora from the source water or a biomass collected from the treatment media.[32,39-46] Measurements of bacterial action in the test sample over time are determined by plate counting techniques, direct cell count, ATP, turbidity, or DOC changes. A brief description of several methods follows to illustrate the reason the databases from one test are difficult to equate to other test method results.

van der Kooij Method

The van der Kooij method for measuring AOC is probably the most frequently used pure culture approach and employs a *Pseudomonas fluorescens* (P-17 strain) and *Spirillum* (NOX strain).[47] The P-17 strain was isolated from drinking water and was selected because this organism can grow on minute concentrations of a wide variety of compounds in the aquatic environment. However, it does not utilize oxalic acid, which is frequently produced during ozonation of water supply. For this reason, a *Spirillum* strain NOX was included in the test protocol to expand the detection of more biodegradable organics.

The procedure involves pasteurization of the sample to inactivate the indigenous microbial flora, then inoculation of test portions with P-17 and NOX. The samples are then incubated at 15°C for up to 20 d without shaking, followed by plating (0, 3, 5, 7 d) to determine when peak density is achieved. Peak density results are correlated to a standard growth curve of P-17 on acetate or for NOX strain on oxalate, then reported as micrograms of acetate and oxalate equivalences per liter. The two values are combined into a "total AOC." The risk with this combining results of the two tests is in the possible overestimating of available nutrients, because both organisms may consume some organics present with equal vigor.

Kemmy Method

The Kemmy method uses a pure culture inoculum consisting of four characterized aquatic organisms (*P. fluorescens, Curtobacterium* sp., *Corynebacterium* sp., and another coryneform bacteria of unknown species).[43] The rationale for these "four specially" selected bacterial species was not given but it is known that these organisms are common to soil, vegetation, and the natural water environment. Perhaps the substitution of *Aeromonas* sp. would serve as a more sensitive indicator of assimilable nutrients and be easier to cultivate. In this procedure the water sample is dechlorinated and filter sterilized, then inoculated with the assay organisms. After incubation for 6 d at 20°C the bacterial density is determined by a spread plate technique and the results correlated to a standard mixture of peptones and yeast extract supplemented with acetate and glucose. Conversion data was reported as micrograms of AOC per liter.

Werner Method

The Werner method replaces the pure culture inoculum with the sample flora. This approach would appear to optimize regrowth potential in a specific situation, provided subsequent treatment does not seriously alter the profile of organisms arriving to the distribution network. In this procedure, the test sample is filter-sterilized, placed in a cuvette to which a sterile nutrient salt solution (carbon free) is added.[39] Thereupon a specific density of the membrane filter entrapped bacteria that was separated from the test sample after filtration is introduced to start the bioassay response. The cuvette is incubated in a specially designed turbidimeter at room temperature for 30 to 120 h, during which time turbidity measurements are made every 30 min. These data are then used to plot a logarithm of turbidity vs. incubation time. Growth rate then becomes a measure of AOC. The most serious drawback with this approach is that not all turbidity is created by cell density; some may be introduced by reformation of colloidal particles during incubation from iron or manganese in the sample.

Jago–Stanfield Method

The Jago–Stanfield method employs an adenosine triphosphate (ATP) measurement of microbial growth because of the probability that not all heterotrophic bacteria in the indigenous microflora would be detected in any given plate count procedure.[49] Again, the original microbial population was removed from the sample by filtration and a measured density of inoculum added. Ideally, this inoculum is from the raw source water or from a sample collected from the ends of the distribution system. This particular selection site would probably yield the greatest diversity in organisms colonizing the distribution pipe network. After inoculation, the test sample is analyzed each day for ATP until a maximum concentration is reached. Then the ATP maximum concentration is converted to AOC micrograms per liter by use of a standard conversion factor.

COLIFORM GROWTH RESPONSE (CGR) PROCEDURE

In another recent development, attention has been given to establishing AOC threshold levels that have the potential for coliform regrowth rather than focusing on the broad spectrum of organisms in the heterotrophic bacterial population. This coliform growth response procedure uses *Enterobacter cloacae* as the bioassay seed organism.[40] Changes in viable densities of this coliform in the test sample over a 5-d period at 20°C are used to develop an index of available nutrients to support a coliform biofilm growth. The CGR result is calculated by log transformation of the ratio between colony density achieved at the end of the incubation period compared to the initial cell concentration.

Thus:

$$CGR = \log (N_5/N_0)$$

where N_5 = number of CFU per milliliter at day 5 and N_0 = number of CFU per milliliter at day 0. Any sample that demonstrated a 1-log or greater increase is interpreted as supporting coliform growth. Calculated values of between 0.51 and 0.99 are considered to be moderately growth supportive while those less than 0.5 were regarded as not supportive of coliform growth.

It is important to note that the CGR test is only responding to those concentrations of assimilable organic materials that support growth of coliforms characteristic of regrowth in biofilms. In fact, parallel assays comparing *E. coli* response with *Enter. cloacae* indicate a significant difference in growth response among these two coliforms. *Enter. cloacae* growth can occur in nutrient concentrations far below that required by *E. coli* (Table 4.3). Note that all *E. coli* growth response values for samples taken at the water treatment plant were negative values, indicating a rapid die-off of this coliform as contrasted to parallel tests done using *Enter. cloacae*. Fluctuating raw water qualities in February and March did not appear to provide any trend in CGR assay values in this data set; however, ozonation was a factor in the increase of available assimilable organic carbon. These data are an indication of the bulk water quality prior to entering the distribution system.

Selection of a coliform isolated from the distribution system for a study of CGR responses may be more realistic than use of a similar strain kept in the laboratory cultivation. This was the approach used in an investigation of bacterial nutrients that might be supportive of coliform regrowth at the pilot water treatment plant, Rochester, NY.[40] Both the standard laboratory strain and an *Enter. cloacae* isolated from the utility distribution system were used in parallel experiments (Table 4.4). The results suggest that the use of a coliform isolate from the utility distribution system does provide a greater sensitivity to available bacterial nutrients than the laboratory strain.

BIODEGRADABLE ORGANIC CARBON (BDOC) DETECTION

Detecting any changes in the dissolved organic concentrations is another approach to obtaining a measure of assimilable organic carbon. Rather than using pure cultures, these procedures utilize the indigenous microflora of raw

Table 4.3   Coliform Growth Response (CGR) Bioassay Analysis of Water Samples
from the Hackensack Water Company, Haworth Water Treatment Plant

Sample date[b] (February 1989)	E. coli CGR[a]			Enter. cloacae CGR[a]		
	Raw	Ozonated	Finished	Raw	Ozonated	Finished
1	−0.28	−0.70	−0.50	−0.05	1.91	0.92
2	−0.20	−0.50	−0.53	−0.01	0.59	−0.08
3	−0.34	−0.60	−0.35	−0.03	1.16	1.09
4	−0.90	−0.03	−0.58	0.20	1.49	1.36
5	−1.17	−1.44	−0.99	−0.02	1.89	1.53
6	−0.05	−0.73	−0.49	0.57	1.61	1.43
7	−2.09	−0.34	−0.15	0.22	2.48	1.91
8	−0.35	−0.57	−0.91	0.07	2.08	1.51
9	−0.56	−0.01	−0.64	0.03	2.21	1.54
10	−0.19	−0.86	−0.42	−0.19	0.97	0.98
Mean value	−0.61	−0.58	−0.59	0.08	1.64	1.22

[a] CGR = $\log_{10}$ CFU (N_5/N_0).
[b] Water samples collected February to March 1989.
Data from Reasoner, Rice, and Fung.[48]

surface-water source or the biomass washed from sand used in the filter bed, provided the filter bed is not subjected to prechlorination or potassium permanganate treatment. Such microbial populations become the inoculum in the sample after the test water is passed through a small-pore membrane to remove the original flora. The advantage of using this consortium of heterotrophic microorganisms occurring in these aquatic habitats is in their acquired proficiency to degrade a diverse spectrum of dissolved organics that may be in test samples. Selection of inoculum either from raw source water or the sand filter biofilm has not been found to be critical for optimum test performance. Recent collaborative studies by seven laboratories indicate no significant difference in test values resulting from using inoculum of indigenous bacteria from different rivers or sand obtained from various drinking water treatment plants.[50] Perhaps the optimum mixed flora for the BDOC test lies with use of the HPC population in the system dead end rather than the indigenous bacteria from raw source or sand filters. The organisms in the pipe population are adjusted to survival and growth in the nutrient accumulations in pipe sediment and are survivors of treatment impact.

When interpreting results from BDOC tests, it is important to remember that DOC measurements include both nondegradable humic type materials and assimilable organic carbon substances. There is no constant factor that can be employed to sort out the proportions in DOC content. Perhaps the most appropriate use of BDOC measurements is its use to measure the effectiveness of treatment processes to reduce residual precursors that are involved in disinfection by-product formation.[51]

Table 4.4   Influence of Coliform Strain Source on Coliform Growth Response to Water Treatment Processes

Pilot-plant treatment train effluents	AOC µg/l	AOC Ratio	Lab strain Value	Lab strain Growth[a]	Utility strain Value	Utility strain Growth potential[a]
1. Raw source water — Hemlock Lake (9/27/88)	67	—	+0.20	No	+0.66	Yes
Coagulated	48	0.7	+0.09	No	+0.05	No
Coagulated + O_3 (0.5 ml/l)	61	0.9	+1.40	Yes	+2.40	Yes
Coagulated + O_3 + dual media filter	80	1.2	−0.26	No	+0.51	Yes
2. Raw source water — Hemlock Lake (10/26/88)	90	—	−0.02	No	+0.05	No
Coagulated	26	0.3	−0.21	No	−0.13	No
Coagulated + $KMnO_4$	128	1.4	+0.11	No	+0.18	No
Coagulated + $KMnO_4$ + dual media filter	84	0.9	+0.23	No	−0.13	No
3. Raw source water — Hemlock Lake (12/13/88)	156	—	+0.42	No	+0.58	Yes
Coagulated + dual media filter	50	0.3	+1.06	Yes	+1.58	Yes
Coagulated + GAC filter	41	0.3	+0.35	No	+0.67	Yes
Coagulated + chlorinated + dual media	124	0.8	+1.30	Yes	+2.78	Yes

[a] Growth defined for CGR: <0.5, no growth; 0.51 to 0.99, moderate growth; >1.0, significant growth.

Data revised from Rice et al.[40]

Hascoet–Servais–Billen Method

The Hascoet–Servais–Billen method employs a 500-ml test sample that is sterilized by filtration through a 0.2-μm cellulose acetate membrane. The first 100 to 200 ml of filtrate is discarded because of possible chemical contamination from the filter-stabilizing agents.[51] Then a 5-ml inoculum of the raw source water taken from the water plant intake is prepared by filtering it through a 2-μm membrane in order to remove most of the particulates and protozoans. This inoculum is then added to the test sample and the water incubated in the dark at 20 ± 0.5°C for at least 10 to 30 d, during which time the microbial enzymes degrade much of the inorganic constituents. DOC measurements taken initially and at every day thereafter until there is a steady minimum value established over three consecutive days are used to determine the BDOC equivalents.

Joret–Levi Method

The Joret–Levi method uses a prewashed biologically active sand inoculum taken from a plant that does not use prechlorination.[52,53] A 100-g sand sample is washed in distilled water until there is no detectable release of DOC. The test sample and sand inoculum (300 g in 300 ml of test water) are then aerated with incubation at 20°C and daily DOC measurements performed until a minimum value is obtained (generally after a few days incubation). The BDOC is then calculated by taking the difference between initial and minimum DOC values.

AOC LOADINGS IN SOURCE WATER AND TREATMENT PROCESSES

A study of field data from various investigations has revealed that total AOC values in high-quality groundwaters are low, as anticipated. Surface waters vary widely in quality and so do the associated AOC values. For example, AOC data in Table 4.5 reveals 16 to 51 μg of C equivalents per liter in untreated groundwater and over 100 μg of C equivalents in surface water. What is surprising is the observation that coliform growth responses in natural waters tested were negative. The reason for this contradiction in AOC vs. CGR test results may be related in part to differences in growth responses by *P. fluorescens* compared to *Enter. cloacae* in test waters. While *Enter. cloacae* may be more sensitive to toxic substances in polluted waters, the major factor is available nutrient concentration. *P. fluorescens* can subsist on the very minimal traces of organics in high quality distilled water, while biofilm coliforms may require at least 50 μg of C equivalents per liter or more to begin regrowth.[56] For *E. coli* regrowth to occur, the nutrient levels are several magnitudes higher, calculated to range from 14 to 30 mg/l BOD_5 or estimated

Table 4.5 Characterization of Bacterial Nutrient Conditions in Raw Source Water

Raw source water	Location	Total coliform per 100 ml	Turbidity NTU	DOC	TOC (mg/l)	AOC (mg/l)	CGR value
Groundwater							
Well HB-1	Southern California	<1	—	1.0	—	16	—
HB-2		<1	—	1.3	—	26	—
HB-6		<1	—	1.6	—	14	—
HB-8		<1	—	2.2	—	27	—
IR-12		<1	—	0.7	—	19	—
IR-13		<1	—	2.8	—	39	—
MC-5		<1	—	1.4	—	26	—
Newtown	Ohio	1	—	—	—	—	-0.1
Cheshire	Connecticut	<1	—	—	0.3	51	0.22
Lakes/reservoirs							
Hemlock	New York	2	0.4	—	2.4	140	0.14
L.A. Impoundment	California	1	2.0	—	1.7	32	0.05
L.A. Impoundment		100	1.9	—	2.3	63	-0.08
L.A. Impoundment		14	0.6	—	—	78	0.38
Hetch Hetchy		80	6.7	—	3.0	83	0.06
Hetch Hetchy		130	—	—	3.4	156	0.11
Rivers							
	New Jersey	233	3.7	—	3.8	231	0.10
Ohio	Ohio	56	—	—	2.9	163	0.55
Intercoastal	South Carolina	250	6.2	—	4.2	73	0.12

Data revised from Bradford et al.;[54] Reasoner, Rice, and Fung;[48] and Rice.[55]

to represent TOC concentrations above 100 mg/l. Such poor-quality water would be unacceptable as drinking water without adequate treatment to remove microorganisms and chemical contaminants.

Water treatment influences the status of organic residuals in source water. Using conventional treatment processes, many of the organic complexes are either removed or reduced significantly. For example, coagulation of Hemlock Lake alone or followed by filtration through dual media or GAC did reduce the raw source water AOC by 70% (Table 4.4). However, coupling these two processes to a final disinfection of surface source waters at two different California utilities resulted in only a 10% reduction in AOC (Table 4.6). Perhaps the discrepancies in removal effectiveness were due to variations in contact time for water passing through the filter biofilm where biological activity could consume some of the available AOC.

The principle of biological activated carbon (BAC) filtration applied to water treatment enhances AOC reduction by providing on-site biofilm activity to consume some AOC while also adsorbing other factions of the organic content of the process water.[57-59] While this process has been widely used in some European countries, there is reluctance to apply it in the United States because of a concern with amplification of the microbial content in the product water. Powdered activated carbon (PAC) applications, used in taste and odor control, has also been found to reduce the AOC content of process water.[57] In both carbon applications, the material is retained in service long after the adsorption capacity for many organics is exhausted. In these situations, biological activity in the carbon bed is the major driving force to reduce process water AOC.

Disinfection of water may negate much of the AOC reduction benefits just described because oxidation converts more of the nonbiodegradable organics to assimilable organic carbon.[60] Such conversions may explain the observation that there is greater AOC reduction in process water treated by coagulation and filtration than when these two treatment processes are followed by postchlorination (Table 4.4 compared to Table 4.6). Ozonation, being a more powerful oxidizing agent than chlorine, intensifies further organic conversions with the result that AOC concentrations increase.[61] As a consequence, more AOC is released into the distribution system and becomes a significant contributor to the nutrient base for biofilm growth. Thus, efforts made to reduce disinfectant by-product formation are on a collision course with the desire to discourage microbial growth in the distribution system unless higher removal efficiency is achieved for all organic residuals in process water.

AOC IN WATER DISTRIBUTION

The origin of most microbial nutrients in the distribution system is the source water, modified through various treatment processes and amended by distribution system characteristics. Some of these influences may include air

Table 4.6 Bacterial Nutrient Amplification in Water Treatment Processes

Pilot-plant treatment effluents	Total coliform per 100 ml	HPC per ml	TOC	AOC	AOC ratio (treat/raw)	CGR	Growth potential
Aqueduct source	33	535	—	105	—	-0.09	—
Coag. + filter + Cl$_2$	<1	<1	—	94	0.9	0.5	±
O$_3$ (1.3 mg/l)	<1	550	—	192	1.8	1.49	±
O$_3$ (2.2 mg/l)	<1	5	—	228	2.2	0.5	±
Coag. + O$_3$ (2.2 mg/l)	<1	56	—	233	2.2	1.49	+
Coag. + filter + O$_3$ (2.2 mg/l)	<1	9100	—	169	1.6	0.20	—
O$_3$ (1.3 mg/l) + H$_2$O$_2$	<1	530	—	186	1.8	1.31	+
Coag. + filter + O$_3$ + H$_2$O$_2$	<1	4890	—	186	1.8	1.40	+
Hetch Hetchy Reservoir	130	370	3.4	156	—	0.11	—
Cl$_2$ + filtered	<1	1	2.6	144	0.9	1.88	+
O$_3$ (3 mg/l)	<1	27	3.8	234	1.5	2.41	+
O$_3$ (4 mg/l)	<1	20	3.9	215	1.4	2.09	+
Settled + O$_3$ (3 mg/l)	<1	340	3.2	188	1.2	0.90	+
Settled + O$_3$ (4 mg/l)	<1	288	3.2	188	1.2	2.32	+
Filtered + O$_3$ (4 mg/l)	<1	570	2.6	168	1.1	0.88	+
Filtered + O$_3$ (3 mg/l)	<1	97	2.7	190	1.2	2.42	+

Data from Rice.[55]

contaminants in finished water reservoirs, soil releases in line repairs, and gray-water inputs from back-siphoning of attachment devices. As a consequence, it is not surprising to find AOC concentrations vary with travel distance in the pipe network, water demand in a given pressure zone, reversals of flow in the pipe network, and application of seasonal flushing in the system. Some examples of AOC change with distances from the first customer or change with time at the same tap may be seen in Table 4.7. Variation in AOC values is also caused by these materials becoming adsorbed to the pipe sediments and tubercles, rather than passing out of the system on water withdrawal. Without systematic flushing of the pipe network to remove sediments, many of these nutrients accumulate in slow-flow sections and dead-end areas, causing biological instability within the distribution system.

Table 4.7   Bacterial Nutrient Conditions in Water Distribution Systems

Utility	Total coliform	HPC per ml	TOC (mg/l)	AOC equivalents P-17	NOX	CGR Ratio	Growth
Rochester, NY							
Finished water	<1	6	2.3	312	—	0.79	Pos.
Rush Reservoir	<1	8	2.3	190	—	0.31	Neg.
Cobb's Hill Res.	2	105	2.1	85	—	0.08	Neg.
Distribution tap	1	1	1.9	85	—	0.17	Neg.
Utica, NY							
Hickley Reservoir							
February 4, 1992		6600		<1	210	0.02	Neg.
June 9, 1992		600		—	—	−0.14	Toxic[a]
Bareveld							
February 4, 1992		120		24	441	0.05	Neg.
June 9, 1992		3760		—	—	0.67	Pos.
Burrestone							
February 4, 1992		10		41	596	0.39	Neg.
June 9, 1992		7		—	—	0.67	Pos.
Philadelphia Suburban							
Finished water	<1	—	—	175	—	—	—
Tap A (3.4 h)	<1	—	—	125	—	—	—
Tap B (5.4 h)	<1	—	—	100	—	—	—
Tap C (7.7 h)	<1	—	—	75	—	—	—
Cincinnati, OH							
Lab Tap 8/25/88	<1	7	2.29	—	—	−0.27	Toxic
Lab Tap 3/30/89	<1	13	1.59	68	—	−0.08	Toxic
Lab Tap 4/7/89	<1	1	1.92	31	—	−0.09	Toxic
Lab Tap 4/26/89	<1	3	1.56	90	—	0.04	Neg.

[a] Copper sulfate treatment for algae control.
Data from Rice.[62]

Biological stability is an important consideration in maintaining the quality of finished water during distribution. In order for heterotrophic bacteria to grow in the pipe environment, several conditions must be met. The AOC of the water supply must contain sufficient concentrations of biodegradable organics to permit growth. The stream of nutrients must be above the minimum growth requirements for the organisms involved. Inhibitory materials such as residual disinfectants must either be absent or available at low concentrations that are insufficient to inactivate the bacterial aggregates. Finally, none of these factors will produce much change in the biofilm growth until water temperature rises above 15°C.

WATER TEMPERATURE STIMULATION

Microbial regrowth is not only keyed to nutrient accumulations in the distribution system, but also to water temperature elevations.[11,25,31,37,63-65] Field data from various water utilities that have experienced coliform regrowth in their distribution systems indicate these heterotrophic bacterial population increases are most often a late spring–summer–autumn phenomenon related to warm-water temperatures. Apparently, water temperatures above 15°C (58°F) accelerate the growth of adapted organisms persisting in the pipe environment. While these microbial inhabitants of pipe sediments, tubercles, and other corrosion sites persist precariously at low water temperatures and balance death of old cells with slow cell development, adsorption of nutrients continues because AOC degradation is minimal below 15°C. Seasonal regrowth has not occurred in areas of continual warm-water temperatures for some water systems in southern California, possibly because groundwater was supplied.[66] While groundwater temperatures are generally below 13°C, the key element involved in suppressing biofilm regrowth may be related to use of a high-quality source water that contains assimilable organic carbon (AOC) concentrations below the critical threshold for regrowth. In essence, these water supplies are biologically stable.[30]

SEASONAL ASPECTS

Warm-water months often provide the opportunity for biofilm expansion in the pipe environment provided there are sufficient bacterial nutrients (AOC) present to support accelerated growth. An example of the cyclic nature of biofilm regrowth can be seen in Table 4.8. Prior to the sudden increase in coliform occurrences many utilities have observed a period of few or no coliform occurrences. When coliforms are found, these often are less than four organisms per 100 ml. Suddenly there may be an outburst of coliform-positive samples above 5% occurrence as the water temperature rises in springtime. Most often, these coliform-positive samples are not restricted to any given area of the distribution system and densities in some instances may approach 200 coliforms per 100 ml. Repeat samples taken from the same site may be negative, suggesting the film is being released in clumps and is not uniformly dispersed into the bulk water. After a period of several weeks, the frequency of occurrence declines sharply only to return to near record levels again. Such

Table 4.8   Comparing Monthly Percent Coliform Occurrences with Quantitative Coliform Data

New Haven, CT — Lake Gaillard Source

	Percent coliform-positive	Coliform per 100 ml Mean	Coliform per 100 ml Range	Temp (°C)	Free chlorine residual	Percent coliform-positive	Coliform per 100 ml Mean	Coliform per 100 ml Range	Temp (°C)	Free chlorine residual (mg/l)
		1985					1986			
January	0.00	0	0	5.1	2.3	0.48	0.05	<1–10	5.1	2.6
February	5.35	1.21	<1–98	4.9	2.1	2.86	0.06	<1–5	4.8	2.1
March	12.56	2.43	<1–90	6.8	2.2	1.58	0.02	<1–2	6.2	2.3
April	4.62	0.29	<1–17	10.0	2.1	5.10	0.78	<1–133	9.2	1.9
May	2.63	0.04	<1–3	13.6	2.3	3.62	0.04	<1–1	11.8	2.4
June	8.33	0.84	<1–79	13.8	2.4	5.13	0.09	<1–3	12.3	2.6
July	12.68	3.62	<1–200	18.1	2.0	8.13	1.65	<1–98	17.1	1.7
August	10.91	2.35	<1–171	18.8	2.1	12.50	0.92	<1–44	21.8	1.5
September	9.41	0.96	<1–61	17.6	2.2	11.72	1.72	<1–77	20.4	1.5
October	4.19	0.52	<1–157	16.4	1.9	4.20	1.22	<1–94	18.0	1.5
November	0.00	0	0	13.1	2.0	3.25	0.17	<1–16	13.4	1.5
December	2.26	0.10	<1–7	9.7	2.5	0.00	0.16	<1–11	9.8	1.5

Rochester, NY — Lake Hemlock Source

Month	1985						1986					
Jananuary	—	—	—	—	—	—	1.30	0.03	—	<1-2	8.5	0.7
February	—	—	—	—	—	—	10.03	1.30	—	<1-80[a]	—	1.2
March	—	—	—	—	—	—	8.72	2.41	—	<1-80[a]	—	1.2
April	—	—	—	—	—	—	15.61	3.71	—	<1-80[a]	—	—
May	—	—	—	—	—	—	0.48	0.38	—	<1-80[a]	—	1.1
June	—	—	—	—	—	—	16.11	0.99	—	<1-64	—	1.1
July	—	—	—	—	—	—	3.87	0.14	—	<1-15	—	1.5
August	—	—	—	—	—	—	0.00	0.00	—	0	—	1.5
September	6.91	0.23	21.2	<1-10	—	0.7	—	—	—	—	—	—
October	0.00	0	17.9	0	—	0.7	—	—	—	—	—	—
November	4.52	0.11	13.7	<1-4	—	0.7	—	—	—	—	—	—
December	5.00	0.64	8.9	<1-48	—	0.7	—	—	—	—	—	—

[a] Maximum values were low estimate for counts above 80 coliforms per culture.

Data supplied by Darrell Smith (New Haven Water Authority) and Dale Kriewall (Rochester Water Authority).

cyclic behavior will continue until water temperature begins to decline in late autumn, whereupon coliform occurrences retreat to below the 5% occurrence level. A second-year cyclic episode of coliforms may return in the following spring, but often the peak densities are less severe, as measures taken in the action plan begin to have some effect on suppressing the biofilm.

Documenting frequency of coliform occurrences without recording densities in these samples may satisfy federal regulations but is not helpful in characterizing the magnitude of the problem at various sites, or demonstrating the effectiveness of site specific remedial measures. In these situations, a quantitative membrane filter test should be required on all repeat samples as part of the action response.

Occasionally, elevated coliform occurrences may be observed that do not relate to warm-water temperatures. These situations are more often a reflection of disturbances in the sediment, static water conditions, or repair practices in the distribution system and will be discussed in the following section.

PHYSICAL STATUS OF PIPE SEDIMENTS

Water mains, storage reservoirs, standpipes, joint connections, fire plug connections, valves, service lines, and metering devices have the potential to be sites suitable for microbial habitation. No pipe material is immune to potential microbial colonization once suitable attachment sites are established. Given sufficient time, aggressive waters will initiate corrosion of metal pipe surfaces, water characteristics may change the surface structure of asbestos–cement mains, and biological activity may create pitting on the smooth inner surface of plastic pipe materials. Not all pipe sections may show evidence of deterioration, however, even after years of active service. The reason is the nature of the water chemistry and continuous movement of water under high-velocity conditions.

ORGANISM ENTRAPMENT

Among the diverse population of heterotrophic organisms (bacteria, yeast, and fungi) capable of passage through the treatment barrier, only a small portion will adapt to the pipe environment and perhaps colonize under certain conditions. Those that succeed preferentially grow on available surfaces and not in the bulk water. There are a host of mechanisms that attract bacteria to a surface, the specific mechanism being dependent on type of bacteria and the surface properties.[67-69] Physical forces may be significant, attracting bacteria by electrostatic change of the pipe surface. Some bacteria are pulled into areas of low flow or become settled during quiescent periods of no water movement. Other bacteria may become attached to pipe walls via cellular appendages (pili) and secretions of various polysaccharides (slime material). These glycocalyx materials bind the organisms to particles, sediment coatings, tubercles, pipe joints, and wall surfaces, thereby preventing microcolonies from being completely swept away by the flowing water. Because these extracellular materials are polysaccharides, they may also serve as a protective barrier to the lethal effects of residual disinfectant contact.

HABITAT CHARACTERISTICS OF SEDIMENT AND TUBERCLES

The diverse physical structures in the distribution system and its pipe network complexities, which include locations out of the mainstream of continuously moving water, are the most opportune sites for sediment to accumulate, tuberculation to expand, and microbial colonization to develop. Sediment accumulations at these locations are created, in part, from corrosion, treatment spills of powdered carbon and unstable coagulant, and deposition of corrosion inhibitor additives. Slow-flow areas may also be the place where biological debris settles out from nonfiltered treatment systems that use surface waters.

Tubercle deposits are also attractive habitats for microbial colonization. Macroscopic examination of these deposits reveals them to have a flaky outer crust that becomes brittle upon drying. Scanning electron micrographs disclose the outer crust to be porous with the interior composed of layered depositions of materials spilled in treatment or chambered with crystals that are fractured into thin plates or structured with needle-shaped appendages. This type of deposition suggests that tubercle formation proceeds at a nonuniform rate in response to changes in water chemistry, pH, temperature, and hydraulic pressures.

Not only do sediments and tubercles provide a protective structure for microbial attachment, but they also are adsorbers of a variety of essential nutrients found in trace amounts in the bulk water. While tubercle deposits consist largely of iron oxides, there can be substantial concentrations of calcium, magnesium, aluminum, and manganese present, with lesser concentrations of other heavy metals.[9] A characterization of the heavy-metal elements in tubercles fractionated into surface and interior layers is given in Table 4.9. All of the elements in substantial concentrations on the crust of tubercles are essential in trace amounts for bacterial metabolism. Phosphates are also important to cell metabolism and can also be found in pipe corrosion products along with carbonates.[70] The other essential component to supporting cell metabolism is a source of organics. Analysis of tubercles for organic adsorption revealed that humic substances and sedimentary chlorophyll materials, characteristic of vegetation decay from surface water, are predominant.[9]

Given these conditions in the pipe environment, it is not surprising that a variety of organisms will become part of the complex microbial community in the water supply pipe network. In fact, the microbial population will include a variety of organisms involved in carbon, nitrogen, sulfur, and iron transformations that are indigenous in drinking water, particularly at corrosion sites.[71]

Table 4.9 Chemical Composition of Cast-Iron Pipe Sediment, Columbus, OH

Sample	Al	As	Ba	Cd	Cr	Cu	Fe	Pb	Mn	Se	Zn
							Heavy metals (ppm)[a]				
Surface layer	829	1.5	80.6	0.2	25.1	3.2	468,920	2.0	235.5	<0.001	65.4
	932	1.7	77.0	1.2	12.1	3.2	283,688	3.3	131.7	<0.001	91.1
Powdery interior	853	0.001	5.0	0.2	5.7	1.2	188,206	3.7	31.3	<0.001	60.2
	908	0.001	6.3	0.1	9.1	1.0	227,066	43.1	38.1	<0.001	39.9
Calcium carbonate sublayer	14,516	0.001	137.0	0.9	2.1	36.2	52,419	40.3	991.9	<0.001	806.4

[a] Silver not detected (<0.0002).

Data from Tuovinen et al.[9]

Biofilm activity in distribution system corrosion sites is most active during warm-water conditions, but will proceed even where year-round water temperatures remain near zero.[72] The highest population of bacteria and fungi were found in pipe corrosion tubercles. Among these organisms were total coliforms that were more numerous at corrosion sites than in untreated bulk water flowing past the location.[73] Specifically, *Enterobacter aerogenes* and *Klebsiella* sp. appeared to be present in greater numbers among the thiosulfate reducers. Among the most common iron reducers found in the Yellowknife, NWT (Canada) water supply distribution system were *Citrobacter* and *Klebsiella* sp. The ability of these coliforms to participate in the pipe corrosion process may benefit their survival in biofilms inhabiting chlorinated water systems.

COLIFORM RELEASES IN PIPE SEDIMENTS

The physical nature of pipe sediments is influenced by seasonal changes in raw source water chemistry and any treatment applied to suppress pipe corrosion. The adverse side effects of not forming a stable sediment leads to loose, porous deposits that are not only conducive to microbial entrapment but also to release into the bulk water by hydraulic actions.[74-82]

Such a situation was created at one utility in Illinois, serving 32,000 people.[25] Attention to the problem began when coliforms were observed in the distribution system only during the cold-water months of December to June. This winter occurrence of coliforms was unusual because most coliform biofilm releases have been observed to take place during warm-water periods. An on-site review of treatment practices and plant records indicated that there had been a pronounced shift in the source water (Lake Michigan) pH during the winter but no attempt was made to compensate for the change in water chemistry. Inspection of data on raw water characteristics revealed water pH of 7.7 in summer shifted to pH 8.2 by December, followed by a rapid decline to pH 7.4 during January to March each year. The reason for these seasonal changes in water pH was thought to be related to near shore turnover of bottom water containing partially decayed vegetation debris (humic matter). Water treatment measures used to process the lake water had little impact on stabilizing the water pH, so this characteristic was passed on into the distribution system. Implementation of recommendations to adjust the process water to pH 8.3 and add lime slowly in the treatment basin to form a more stable, firm coating on the distribution pipe walls apparently resolved the coliform occurrence problem in the following year. The successful follow-up treatment may also have been aided by the suggestion to increase the disinfection concentration during cold-water periods to compensate for the increased chlorine demand and reduced disinfection effectiveness at near-freezing water temperatures.

STATIC WATER EFFECTS

Distribution systems are often complex in their configurations, which in itself can create problems of slow flow in some areas and static water in dead ends. Static water areas are undesirable at any time because this condition provides opportunity for suspended particulates to settle into pipe sediments, biofilm development to proceed without the shearing action of hydraulic changes, and corrosion to accelerate, particularly during warm-water periods. Redesigning the distribution network to create continuous loops from numerous dead-end sections has been helpful in reducing microbial degradation and improving the efficacy of disinfection residuals in outreach areas.

Static water is also a serious problem during winter months because of the concern for water freezing, resulting in interrupted water supply to the consumer, and the increased risk of pipeline breakage. Large-volume demands for public water supply followed by months of no use does present a serious problem for maintaining the integrity of a pipeline. In temperate regions, it is advisable to bury these pipes at greater depth to provide more insulation from ambient air temperatures. Draining the water is not desirable because of the concern for ground movement that may cause more pipeline breakage when lines are dry.

PROTECTED PASSAGE IN PARTICULATES

Much has been written about the effectiveness of treatment barriers to prevent the passage of viable organisms of public health significance into public water supplies. What is known to a lesser degree are the counterforces that can defeat the ability to achieve full effectiveness for engineering processes used to reduce microbial risks. Water supplies that use disinfection as the only treatment process for surface-water sources may not be protected from occasional passages of suspended soil particles and biological matter.

SOURCE WATER PARTICULATES

For disinfection to be effective, there must be contact between the disinfectant and organisms in the water.[83-87] Some turbidity may consist of particle–microorganism complexes in which organisms are adsorbed to larger particles. In other situations, one or more smaller particles may be adsorbed to a microorganism or aggregates of cells may form, protecting organisms in the core of a clump from disinfectant contact. In any of these conditions, microorganisms may be protected from inactivation.[88,89]

While turbidity particles are often a protected transport vehicle for microbes in passage through some treatment processes, not all types of turbidity provide this mechanism. For instance, clay particles and water-flocculating agents may trap a variety of organisms but are of little value as protective shields for bacteria in water.[90-95] Thus, few viable cells are transported in these kinds of particles to potential habitat sites along the distribution system. For example, low turbidity levels of 0.3 to 2.0 NTU caused by alum floc spilling over into the distribution water from the Salem-Beverly, MA water plant did not adversely affect the efficiency of chlorination in controlling the standard plate count density, including any coliform breakthrough in treatment.[96] Similarly, a study of the North Berwick (Maine) water supply[97] indicated that a constant decrease in bacterial survival occurred with increased free chlorine residuals, in spite of plant effluent turbidity levels (1 to 5 NTU) that were consistently higher than the level set by the National Interim Primary Drinking Water Regulations.[98] Chemical characterization of this turbidity indicated it was largely inorganic with traces of tannin, lignins, and a low total organic carbon concentration. The source water had 45 fecal coliforms per 100 ml, and was a fast-flowing stream that transversed an area of low-density human population along the drainage basin.

Particle size is also an important consideration. A particle of at least 0.03-μm diameter created by adsorption of extremely small particulates may be sufficient to protect one viron. Dissolution of fecal or other material may also form similar virus-laden particles.[90,99] Those particles that can attract and protect bacteria will be larger than 10 μm, solely because these indigenous organisms are much larger than virus. Scanning electron microscopic examination suggests that less than 20% of all particles will be colonized, generally by bacteria of the same morphological configuration.[100] Why this selection process occurs is not definitely established, but may relate to the nutrients available on the particle, electrical charge differences, and surface characteristics for attachment.

Viruses and bacteria are more intimately associated with organic particles such as fecal cell debris and this material also is associated with a disinfectant demand.[94,101] As a consequence, fecal pollution always brings the risk for passage of some residual population of coliforms and pathogenic agents. While elevation of turbidity in raw water is a concern in treatment, it is the associated soluble organics that create much of the chlorine demand that interferes with microbial inactivation and provides a pathway. Evidence of this interference can be seen in a study of 160 raw source waters used by various public water supply systems throughout the nation.[102] Although the majority of these samples (48.1%) were from stream sources, 20.6% were collected from lakes and ponds and 8.8% from groundwater. The data (Table 4.10) revealed that while there was some measurable increase in chlorine demand associated with increasing turbidity, 90% of the demand was associated with the chemical nature of the water rather than with particles per se.

Table 4.10   Relationship of Turbidity to Chlorine Demand in Raw water Samples

No. of samples	Turbidity range (NTU)	Chlorine demand					
		Range	Av in raw[a]	Av in filtrate[a]	% in filtrate	Av in solids[a]	% in solids
40	0.05–1	0–1	0.68	0.53	99.1	0.005	0.9
56	>1–5	0.1–7.6	1.21	1.41	96.0	0.062	4.0
21	>5–20	0.72–2.95	1.69	1.82	93.0	0.14	7.0
12	>10–20	0.5–2.9	2.02	1.86	90.0	0.21	10.0
31	>20–170	1.08–4.9	2.69	2.09	77.0	0.60	23.0
All samples	0.05–170	0–7.6			90.0		10.0

[a] mg/l chlorine demand.

Data from Katz.[102]

BIOLOGICAL PATHWAYS

Not commonly recognized as a problem in distribution water quality is the occurrence of various larger, more complex, biological organisms including algae, crustaceans (amphipods, copepods, isopods, and ostracods), nematodes, flatworms, water mites, and insect larvae such as chironomides.[16,103-108] For example, a profile of predominant invertebrates in the Worcester, MA distribution system during 1983 and 1984 revealed copepods (48 and 55%, respectively), amphipods (19 and 15%), and fly larvae of several species (15 and 15%). Daphnia, water mites, and ostracods were found to be less common.[107] Origins of these organisms in water supply are generally raw water reservoirs and rivers, but they may also be introduced to finished water in soil particles dislodged in line repairs or in dust particles and wildlife contact in uncovered finished water reservoirs. Passage of algal cells and invertebrates into the pipe network is not limited to systems that treat source water only by disinfection. Some protozoa and invertebrates may succeed in becoming established in filter beds, releasing progeny that can successfully survive disinfection and migrate into the distribution system.[110-113]

In the process of engulfing bacteria from the sand filter biofilm, not all food-chain organisms will be digested by protozoans immediately. Some of these bacteria will remain viable for 1 to 4 h while others find the internal environment of the protozoan vacuole to be hospitable, providing nutrients for growth and a protective microenvironment in an otherwise adverse aquatic environment.[17] Coliforms disassociated with protozoa were inactivated by 0.5 mg/l chlorine (water pH 7.0 and temperature 25°C) in about 1 min; however, all coliform isolates required more than 40 min contact with 0.5 mg/l free chlorine (pH 7.0 and 25°C) when engulfed by the protozoan *Tetrahymus pyriformis* or *Acanthamoeba castellani*. To achieve a 99% inactivation of *K. pneumoniae* entrapped in *A. castellani* required a free chlorine concentration of 10 mg/l for 120 min contact time under the same water pH and temperature conditions.

Coliform persistence in San Francisco potable water, chlorinated to provide a substantial residual after hours of contact, was most severe when turbidities were between 5 and 10 units. A significant portion of this turbidity was thought to reflect the presence of crustacea that had ingested coliforms from some area of biofilm in the pipe network and thereby provided protection from disinfection contact.[115] Upon reaching the consumer's tap, many of these small crustaceans were shattered, releasing the viable coliforms entrapped inside the bacterial predator.

Laboratory experiments indicate that enteric bacteria and enteroviruses ingested by aquatic nematodes found in some public water supplies[18] may be protected from contact with high concentrations of residual chlorine even though more than 90% of the nematodes were immobilized.[114] Again, laboratory experiments have demonstrated that *Salmonella* fed to nematodes could later be recovered as viable organisms in the excreta.[116] The significance of these findings is that it establishes a possible route for bacteria and viruses to breach the disinfection barrier in water supply treatment. This route of passage would pose a threat to public health only if the nematodes were feeding on potential pathogens in the biofilm of a sand filter or microbial colonization of a pipe sediment or tubercle.[117] Carrying this biological mechanism for protection of pathogens one step further, it has been suggested that these protozoan and microinvertibrate hosts may be another survival device and virulence enhancement scheme for such fastidious pathogens as *Vibrio cholera* and *Legionella* to persist in the aquatic environment.[17,118-120]

Another important mechanism for bacterial protection from inactivation by disinfection is cell clumping or the formation of aggregates of cells. In this situation bacterial cells appear to excrete an intercellular matrix material or "Zoogloea" that promotes cell aggregation and provides a shield to the core of organisms from adverse environmental conditions.[121] Studies on *V. cholera* inactivation have shown that smooth *V. cholerae* cells are inactivated in less than 40 and 20 s by concentrations of 0.5 and 1.0 mg/l free chlorine, respectively. In contrast, cultures of a rugose variant exposed to 2.0 mg/l free chlorine for 30 min still had a viable subpopulation of cells that were virulent.[122] Particle size of the clump was an important factor in this survival.

Algal masses that form in impoundment reservoirs may also contribute to protection of entrapped bacteria as the water passes into the water treatment plant. In these situations, the bacterial cells are caught up in the gelatinous matrix of intercellular secretions that also create a substantial chlorine demand in the raw water.

TREATMENT BARRIER FAILURES

Properly operated water treatment trains are effective in providing a barrier to coliforms and pathogenic microorganisms reaching the distribution system (Table 4.11). This does not, however, preclude the passage of some aquatic organisms through the treatment scheme. Investigation of heterotrophic bacterial populations revealed that a 4-log (99.99%) or better reduction can occur through conventional treatment processes (storage, coagulation, settling, rapid sand filtration, and disinfection) for many of this diverse flora. What is most important is the achievement of a 6-log reduction in coliforms and bacterial pathogens, a 4-log reduction in any virus present in the raw water, and a 3-log reduction in pathogenic protozoans. While there are numerous treatment schemes suitable for an effective multiple barrier, passage of coliforms, opportunistic pathogens, and biofilm organisms can occur due to turbidity, carbon fines, and filter instability.

Table 4.11   Cumulative Coliform Reductions through Conventional Water Treatment[a]

Process water	Bacteria per 100 ml		
	Total coliform	*Klebsiella*	Fecal coliform
Raw surface water	100–20,000	9–3,000	1–2,000
Raw water storage (50% reduction)	50–10,000	5–1,500	0.5–1,000
Coagulation–sedimentation (62% reduction)	19–3,800	1.9–570	0.19–380
Rapid sand filtration (96% reduction)	0.8–152	0.08–23	0.008–15
Disinfection (99% reduction)	0.008–15	0.0008–0.2	0.00008–0.15

[a] Reductions based on field data.[144]

Turbidity Interference of Disinfection

Turbidity interference of disinfection appears to be associated with improper application or operation of available treatment processes.[123,124] Chlorination, application of GAC, and filter backwashing practices can provide major barrier weaknesses if not carefully applied. For instance, disinfection alone cannot provide adequate inactivation of even the most chlorine sensitive organisms (coliforms, *Salmonella*, *Shigella*, *Campylobacter*, etc.) if the raw water is subject to widely fluctuating turbidities.[36,125] This situation is particularly true when these organisms are associated with particles that are greater than 7 μm in size.[101,126] The number of bacteria either clumped, attached, or embedded to such a particle may range from five to as many as several hundred cells.[126] As a consequence, intimate contact of these organisms with a disinfectant does not readily occur, so viable passage through the treatment barrier is a possibility. For example, coliforms in water with a turbidity of 1.5 NTU that is exposed to 0.5 mg/l chlorine for 1 h resulted in more than a 99.99% inactivation while in water of a 13 NTU turbidity, there was only a 90% coliform reduction after 1 h contact with 1.5 mg of chlorine.[87] This protective effect of organic particles during disinfection has been observed for both bacteria and viruses.[95,101,127]

Carbon Fine Releases

Carbon fine releases from GAC filters and unsettled coagulants passing through the treatment train can become vehicles of transport for bacteria.[128-130] Activated carbon particles may be expected to range in size from 2 μm to greater than 40 μm in diameter. Because of the fractured nature of GAC and the organic adsorption characteristics, these particles can be a more attractive substrate than sand for bacterial colonization (Table 4.12). Note that the difference in bacterial densities for different cultivation procedures is evidence that these organisms are generally slower growing and require extended processing time. In a study on carbon fine passage into the distribution system, 41.4% of all water samples with fines contained attached heterotrophic bacteria.[132] Scanning electron microscopic analysis of particles released from GAC filters revealed bacteria had colonized 85% of the fines with densities ranging from 5 to 50 cells per particle.[133] The discrepancy in reported frequency of colonization between the two studies probably relates viable and nonviable differences in contrasting a cultivation approach with a direct microscopic scan of particles.

Profiling the heterotrophic bacterial population on particles in GAC filters has revealed the following genera: *Acinetobacter, Achromobacter, Aeromonas, Alcaligenes, Arthrobacter, Bacillus, Citrobacter, Corynebacterium, Enterobacter, Escherichia, Flavobacterium Hafnia, Klebseilla, Micrococcus, Moraxella, Proteus, Pseudomonas,* and *Serratia*.[131,132,134-138] Fungi and yeast have also been noted occasionally. Some of these bacteria have been documented to be opportunistic pathogens (Chapter 3), but the impact of such organisms amplified in a GAC bed with release to the distribution system is not known. Field evidence does indicate that the heterotrophic bacterial densities in distribution water from a treatment train using GAC (Manchester, NH) were statistically higher than for a similar full-scale treatment operation (Concord, NH) that does not employ GAC.[139]

In the search for coliform colonization on carbon fines, over 17% of the finished water samples from nine water treatment facilities were found to contain carbon particles colonized by coliform bacteria.[132] In another study, few coliform bacteria were detected in the 1000 particles examined, but on one occasion a *Klebsiella pneumoniae* strain was found attached to carbon particles released from the GAC filter.[133] The reason for this apparent checkmate on major passage of coliforms via carbon fines is the competition for nutrients and release of antagonistic substances by other heterotrophic bacteria in the biofilm community. As an example, coliform bacteria are occasionally noted to occur in virgin carbon that is introduced into a GAC contactor or filter basin.[129,131] This undesirable situation was quickly suppressed by taking the filter out of service for 10 to 12 d, during which time a biofilm of other heterotrophic bacteria in the indigenous flora of process water quickly develops and represses the coliform growth. Perhaps the same competitive mechanism also minimizes any colonization attempts by waterborne pathogens.

Table 4.12   Treated Water Bacterial Populations following Various Water Treatment Processes Using Standard Plate Medium or R-2A Medium with Extended Incubation Times (organisms per ml)

Sampling day	Lime-softened water			Sand filter effluent			GAC adsorber effluent		
	SPC, 2 d	SPC, 6 d	R-2A, 6 d	SPC, 2 d	SPC, 6 d	R-2A, 6 d	SPC, 2 d	SPC, 6 d	R-2A, 6 d
Initial	120	350	510	890	1,200	1,500	<1	140	220
7	31	202	510	820	22,000	35,000	1	24,000	95,000
14	7	7	130	<1	1,200	9,400	<1	600	4,400
21	7	18	150	2,200	2,500	33,000	<1	5,200	16,000
28	3	39	530	700	7,800	67,000	1	11,000	55,000
35	<1	490	330	100	6,000	25,000	<1	12,000	74,000
42	70	120	1,700	1,200	71,000	22,700	N.D.	56,000	52,000
49	9	1,200	23	5,000	41,000	3,000	80	4,200	100
56	<1	10	<1	<1	700	12,000	N.D.	1,900	50,000
63	29	190	170	170	2,000	3,000	N.D.	5,000	48,000

Note: All cultures incubated at 35°C; SPC, standard plate count; N.D., not done.

Data revised from Symons et al.[131]

As for the occurrence of primary pathogens in GAC filters or in released carbon fines, no extensive data has yet been developed. Preliminary findings of a *Salmonella* sp. in one GAC study and a borderline enterotoxigenic *E. coli* occurrence in another investigation on GAC particles suggests this habitat may be supportive of some waterborne pathogenic agents.[140,141]

Once the microbial community becomes established on GAC particles, there appears to be a developed protection from exposure to disinfection. In fact 2 mg/l of free chlorine and 60 min contact time failed to inactivate the heterotrophic bacterial population.[140] Much of the bacteria protection afforded by particle association against free chlorine and chlorine dioxide was thought to relate to the development of the biofilm layer over the fractured particle surface. Monochloramine, however, was more effective in penetrating the protective slimes and less reactive with the activated carbon material than other conventional disinfectants.[142,143]

It is important to mention that while all of these observations demonstrate a pathway by which bacteria can penetrate treatment barriers, none of the utilities used in these investigations had any prior history of coliform problems or experienced a documented waterborne outbreak.

Filter Instability

Filter instability is part of some scenarios for coliform entry into the distribution system and appears to be associated with improper application or operation of available treatment processes and inadequate maintenance of these barriers.[123,124] Based on field data for coliform reductions in conventional treatment (Table 4.11), engineering process barriers will provide a water quality that meets federal drinking water regulations. However, it must be noted that there may be some variations in the inactivation of coliform populations that relate to adverse changes in the processing of water. All treatment barriers must be "fine-tuned" for maximum effectiveness.

The filtration process is an important element in the physical removal of turbidity and microorganisms from water. In so doing, the process becomes a collector of particles that eventually create turbidity peaks and associated higher concentrations of cysts and bacteria at the same time. Based on a pilot-plant study, the best operating strategy would be to produce a filtered water with turbidity below 0.2 NTU and to backwash each filter when its effluent turbidity rises above this NTU level.[145,146] Wash water quality also plays an important part. Higher densities of HPC and *Klebsiella* passed through filters immediately after backwashing when an unchlorinated water supply was used to flush the unstable filters. While chlorinated backwash water results in a less apparent coliform passage, many of these organisms may be injured. This was the case at two Montana water treatment plants where injured coliforms and elevated turbidities were found in the filtered water within 30 min following each backwash.[147] Furthermore, the treatment barrier leakage continued for some time afterward. These observations point out two critical issues. The first is to optimize coliform detection in the filter effluent by using a coliform medium appropriate to stressed organisms recovery (m-T7 agar) if backwash water is chlorinated.[148] Field data (Table 4.13) suggest injured coliforms in filter effluents and finished water may account for over half of the coliforms not detected by the conventional membrane filter procedure using m-Endo LES agar. The second recommendation is to utilize a filter to waste policy to reduce the opportunity for coliform passages into the distribution system[145-147] and provide more time for the filter to stabilize at slower initial filtration rates before being returned to full service.[146]

DISTRIBUTION SYSTEM INTEGRITY

A continuous state of distribution system integrity is as essential as is the production of a safe water supply. This status is particularly critical for any groundwater system supplying untreated water to the consumers. An example in point was the occurrence of a waterborne outbreak in Cabool, MO (Chapter 8) where two major line breaks and 43 service meter replacements over a period of several weeks may have provided the opportunity for a pathogenic *E. coli* strain to enter the water supply.

Key to a protected distribution of water supply in the community is maintaining an adequate positive water pressure (20 psi or more) throughout the entire pipe network. Anything less will provide opportunities for contamination to be introduced through pipe leaks and a variety of attachment devices in the home, hospital or business. Cross-connections are always a threat to water quality in the distribution system and dictate the need for free chlorine residuals in all sections of the system to provide some measure of protection from contaminant intrusions.[150] The sudden loss of free chlorine residual is often a signal of a contamination event that may be followed quickly by elevated densities of heterotrophic bacteria and occurrences of coliforms.

Water hydraulics can contribute to the release of biofilm organisms and their migration within the pipe network. Some years ago, a city in Texas experienced a major fire that used a large volume of water in fire control. This water demand created a reduced line pressure that, upon being restored, caused a significant water hammer effect in that part of the pipe network. The sudden restoration in water pressure caused a shearing of a biofilm in the pipe section that entered the main flow of water and released 125 to 200 *Pseudomonas aeruginosa* per milliliter over a 24-h period before subsiding to nondetectable levels again. No coliform bacteria were detected during this 24-h disturbance. Distribution system managers who maintain high-pressure lines (200 to 250 psi) have also reported the dangers of rapid cutoff of these lines. This action may not only increase the dangers of line breaks in old pipe, but also cause releases of bacteria from biofilm sites. Multiple feed points to the distribution system from several water treatment plants or well fields with changes of water

Table 4.13 Injured Coliforms in Drinking Water

Location/date	Type of water	No. of samples	Mean coliforms per 100 ml m-Endo	Mean coliforms per 100 ml m-T7	Injury (%)	False negatives (%)[a]
Midwest U.S./1986	After filtration	9	1.1	3.9	71	55
Midwest U.S./1986	Distribution	13	2.4	7.9	69	23
Northeast U.S./1985–86	Distribution	86	1.9	4.8	64	54
Northeast U.S./1986	After chlorinated backwash	7	0.1	1.5	92	14
West U.S./1986–87	After unchlorinated backwash	133	2.4	8.1	70	40
West U.S./1986–87	After chlorinated backwash	37	0.03	13.4	99	97

[a] False negatives represent the percentage of coliforms that failed to produce colonies of m-Endo medium but were enumerated on m-T7.

Selected data from McFeters.[149]

demand in adjacent areas can also create reversals in flow patterns that can dislodge biofilms from pipe sediments and tubercles.

The service life of pipe, regardless of composition, is finite. Pipe deterioration is affected not only by composition, but also by treatment practices, corrosivity of the water, biological activity, soil reactivity, water hydraulics, and ground movement. As pipe sections age, corrosion accumulation or heavy scale development may become a serious problem that restricts water passage, creates taste and odor problems, produces frequent line breaks, and provides sites for bacterial colonization and biofilm development. A predominant number of the systems that have identified a problem of coliform noncompliance use surface-water sources, have no consistent history of a comprehensive flushing program on an annual basis, and have significant tuberculation in iron pipe and pipelines that have been in service over 75 years.

Water Supply Storage

Water supply storage cannot be excluded as entry points for coliform bacteria and possible pathogens. Reservoirs of treated water should be covered whenever possible to avoid recontamination of the supply from bird excrement[154,155] and surface water runoff. Birds and other wildlife may suffer intestinal infections caused by *Salmonella* and protozoans (*Giardia*, etc.) that are also pathogenic to man. Within the wildlife population in any area (as is true for a community of people) there is a continual reservoir of infected individuals that shed pathogenic organisms in fecal excretions. Sea gulls are scavengers and often are found at landfill areas and waste discharge sites searching for food, which is obviously contaminated with a variety of pathogens. At night, these birds often return inland to aquatic sites such as raw water impoundments and open finished water reservoirs, thereby introducing pathogens through their fecal excrements.[156]

An *E. coli* (serotype 02) occurrence in two areas of the Manhattan district of New York City during the summer of 1993 was thought to be associated with the colonization of 600 to 1000 sea gulls on the open finished water reservoir (Hillview) and a source water reservoir (Kensico), where stored water from two watersheds were blended. Recognizing this situation as a potential threat, the city applies more chlorine prior to the water supply entering into the huge distribution mains (city tunnels 1 and 2). After 30 to 45 min contact time, water from the tunnels enters into supply lines serving Manhattan and the other boroughs in New York City. Normally, the final disinfection application prove to be effective, but a series of events occurred that had a significant impact on distribution water quality. Blended water from the Catskill and Delaware watershed began to take on turbidities of 0.7 NTU or greater, apparent color above 10, and water pH greater than 8.0. Under these conditions, disinfection effectiveness is reduced.[157] Coincidental to the changes in water characteristics was the major construction of new transmission lines from the blending reservoir at Kensico to the Hillview balancing reservoir. The increased water movement in Hillview reservoir created turbulence and resuspension of organisms and chlorine demand substances from the bottom sediments. Finally, a near record heat wave occurred early in July and neighborhood fire hydrants were opened for kids to splash in to reduce heat-related stress. As a consequence, a peak water demand was recorded on July 8, 1993. Pulling bottom sediments through the system, possible shorter C·T values due to greater water movement through the tunnels, and elevated shearing of biofilm from pipe walls in water passage undoubtedly introduced coliforms into the bulk water movement. A week after the peak water release event, total coliforms began to appear in some distribution water samples and verification revealed most of these organisms were frequently *E. coli* serotype 02, and other *E. coli* strains were also isolated from Kensico water and the water column in Hillview reservoir. A survey of *E. coli* from gulls trapped in the Westchester County area revealed *E. coli* serotype 02 was isolated from 4% of gulls (52 samples) although none of these sea gulls were selected specifically from Hillview Reservoir. The variety of different strains isolated from Hillview Reservoir downtakes during this time period and the uniformity of strain type encountered in the distribution system remains a paradox.[150]

Covered water storage tanks should not be ignored as a potential source of new contamination. In many small western communities and national and state parks, redwood water storage tanks (Figure 4.5) are a common sight and often a source of coliform releases to the water supply. Investigations of new redwood tanks have revealed that the coliform *Klebsiella pneumoniae* could be traced to the wood tissues of redwood trees.[159] This unique association begins at embryo fertilization of tree seeds. The coliform persists throughout the life of the tree, with the organisms metabolizing wood sugars (cyclitols) passing in the wood conducting tissues.[160,161] These wood sugars continue to leach out into the stored water for several years, over which time there is a massive biofilm development of *Klebsiella* on the inner surfaces of the tank (Table 4.14). Disinfection and scraping of the wood staves in a new tank were ineffective in eliminating the bacteria because the organisms persisted deep inside the wood pores. Biofilm control requires a free chlorine residual of 0.2 to 0.4 mg/l until the nutrient supply is leached away with tank usage over a 2-year period.[161] Another solution proposed is the installation of a plastic liner in new tanks to separate water supply from the contaminating wood surfaces.[161]

Concrete and steel water supply storage tanks may also be an entry point for bacteria. Inspection of air vents may reveal that the protective screens have come loose and birds have built their nests in these locations. Bird excrement around the nesting sites may enter the stored water supply and be carried into the distribution systems before dilution dissipates the fecal debris and residual disinfectant inactivates the organisms.

Organisms may also be introduced at air vents from dust particles that slowly settle to the bottom of the compartment, forming a nutrient support

Table 4.14 Field Survey of Water Quality Emanating from Wooden Reservoir

Site	Cl_2 residual (mg/l)	Total coliforms per 100 ml	Fecal Coliforms per 100 ml	Coliform species
		Public Supplies in Public Parks		
1	<0.1	2	<1	K. pneumoniae, Enterobacter
2	<0.1	8	<1	K. pneumoniae, E. coli, Enterobacter
3	<0.1	1	1	Enterobacter
4	<0.1	5	<1	Enterobacter
5	<0.1	2	<1	Enterobacter
6	<0.1	5	1	K. pneumoniae, E. coli, Enterobacter
7	<0.1	1	<1	Enterobacter
8	<0.1	2	<1	Enterobacter
9	<0.1	6	2	K. pneumoniae, E. coli, Enterobacter
10	0.2–0.6	14	1	K. pneumoniae, E. coli, Enterobacter
11	0.2	2	<1	K. pneumoniae, Enterobacter
		Private Supplies		
		Coliforms in biofilm		
1	0.3	in biofilm	—	K. pneumoniae
2	0.3	<1	—	—
3	<0.1	2	—	Enterobacter
4	<0.1	2	—	K. pneumoniae, Enterobacter
5	<0.1	15	—	K. pneumoniae, Enterobacter
6	<0.1	150	—	K. pneumoniae
7	<0.1	14	—	K. pneumoniae, Enterobacter
8	<0.1	60	—	K. pneumoniae, Enterobacter
9	<0.1	80	—	K. pneumoniae
10	<0.1	150	—	K. pneumoniae

Data adapted from Seidler, Morrow, and Bagley.[159]

base for biofilm development on the interior walls. Materials used to coat the interior of storage tanks have also been found to support microbial growth.[162,163] For example, persistently high coliform densities in a portion of the distribution system of the East Bay Municipal Utility District (Oakland, CA) were traced to a growth-supporting organic coating on the interior of a steel tank.[164] In time, passage of these biofilms into the distribution pipe network will occur and provide opportunity for a further expansion of biofilm sites within the pipe network.

EVALUATING THE PUBLIC HEALTH SIGNIFICANCE OF BIOFILM

Historically, the total coliform group was selected for assessing the microbial quality of drinking water because these bacteria are consistently present in great numbers in feces and can be easily detected in highly diluted contaminated water. In recent years, emphasis on interpreting the significance of

Figure 4.5 Redwood tank for water supply storage.

coliform occurrences has shifted from their sanitary significance to monitoring treatment effectiveness in processing raw water to obtain a 6-log reduction in coliform concentration. Now, with the growing awareness that some nonfecal coliform strains can colonize the distribution system, there is an urgent need to carefully evaluate significance of total coliforms in a biofilm episode for public health risk.

Bacteria can be introduced into the distribution system by several mechanisms. Once organisms become established in the pipe environment, they will eventually form a biofilm community that may include *Enterobacter,* *Klebsiella,* or *Citrobacter.* Given a nutrient base, in a protected pipe sediment or tubercle and a water temperature over 15°C, growth is accelerated. With the event of the shearing force of water passage, flow reversals, and changes in the structure of pipe sediment due to water pH, biofilm fragments are torn away and coliform positive samples becomes a reality. To what extent these biofilm occurrences become a major water quality problem will often be dictated by the actions of distribution system management. Corrosion control, effective flushing programs, elimination of static water zones, and maintenance of a disinfectant residual in 95% of the pipe network are important aspects in reducing microbial colonization of the distribution network and seasonal threats of coliform biofilm releases.

While these coliform occurrences may not be of public health concern, they should not be ignored because the contamination suggests: (1) existence of a habitat that could be used by pathogens; (2) possible leaks in the treatment barrier or distribution system; and (3) the presence of excessive AOC accumulation that interferes with maintaining a disinfectant residual throughout the pipe network.

Reports of coliform biofilm in public water supply distribution systems should not be interpreted as a new problem nor the sudden emergence of a "super bug." Coliform bacteria and a chlorine residual can coexist in treated water, given certain conditions. Many older textbook discussions on the effectiveness of disinfectants to control microbial contaminants are based on published research in the past that utilized laboratory strains to determine disinfection curves. These organisms have been cultivated over time in the laboratory using ample nutrients in a favorable environment, which lowers their defensive mechanisms for survival. As a result, these laboratory strains are rapidly inactivated by a variety of disinfectants. By contrast, coliforms in the water supply pipe are constantly being subjected to a variety of stress conditions, i.e., inadequate nutrient balance, antagonistic organisms in the microflora, and varying concentrations of disinfectant residuals. Under these conditions, surviving bacteria reduce their metabolic activity and often secrete protective slime or encapsulate into agglomerates of microorganisms. These factors provide a measure of protection from disinfection exposure and the inactivation curves for these coliform strains demonstrate a greater resistance than observed with laboratory pure-culture studies. Added to this is the protective nature of corrosion sites and sediment deposition in pipe networks. It is, therefore, not unrealistic to anticipate that colonization and biofilm development may occur in water supply lines containing some detectable concentration of disinfectant residual.

In the past, utility management considered these reports of coliform occurrences in the system to be spurious; most often the result of poor sample collecting, which permitted extraneous contaminates to enter the sample bottle. If not the fault of the sample collector, then the fault was due to some laboratory error. The net result was to regard these findings as a monitoring error with little regard for the prompt resampling of the site to verify the positive finding of coliforms. The other response would be to intensify the monitoring activity for the sole purpose of driving the monthly average for coliform density to below the limit of one coliform per 100 ml.

OCCURRENCE PATTERNS

Coliform biofilms are generally most active in warm-water periods, provided there is sufficient assimilable organic carbon and a protective sediment or corrosion site for development. Characteristically, these biofilms continue to grow on pipe or reservoir walls until the shearing force of water velocity carry clumps of cells into the main flow of water. This movement results in a migration of biofilm fragments to other parts of the distribution network where new sites become centers for biofilm development.

At the start of a seasonal biofilm event, monitoring the distribution system will briefly reveal coliforms at some specific locations. Quickly the pattern of locations will change as biofilm fragments begin to move in the bulk flow of water supply. Speciation of these coliforms often reveals a predominance of one or two species, generally reported to be *Klebsiella, Enterobacter,* or *Citrobacter,* all of which are capable of slime production and encapsulation. Increasing free chlorine residual to 5 to 8 mg/l will begin to reduce both occurrences and densities per site with a cyclic pattern until water temperature conditions in autumn decline to below 10°C. If flushing is inadequate to remove sediments, AOC, and biofilm components, a renewal of the problem may occur the following spring.

REGULATION ASPECT

The specter of coliform noncompliance during the summer months emerges as a serious concern for a growing number of public water supplies. Biofilm is often the issue and the federal coliform regulation[165] exposes the problem to public awareness. While biofilm has been an issue for some utilities in the past, the problem is now more acute because of a total coliform rule that is sensitive to all coliform occurrences and triggers a noncompliance situation when these occurrences exceed 5%.

Previous use of a quantitative limit (a mean of less than one coliform per 100 ml) provided opportunities to drive the monthly average down to an acceptable level by intensifying the sampling during the monitoring period. By contrast, the new regulation focuses only on the presence or absence of coliforms in all samples examined, with the frequency of coliform occurrences being limited to a maximum of 5% for the entire month. This approach is more sensitive to coliform occurrences and is more difficult to subvert by intensifying the sampling program. Consequently, the pattern of coliform occurrences due to a breakdown in the treatment barrier or a summertime biofilm event can immediately place the water utility in a noncompliance situation.

While a true coliform biofilm event may not in itself signal a public health risk, acceptance of such occurrences must be viewed with concern because a real fecal contaminating event may be hidden in the higher densities of coliforms. Providing relief to the utility from a requirement to issue public notification of noncompliance must be dealt with carefully. The burden of proof that these noncompliance occasions are a result of biofilm release is on the utility. In developing acceptance of a "certified" coliform biofilm, attention must be directed toward providing a monitoring strategy that will characterize the predominant coliform species in the repeated occurrences.

In several cities (Muncie, IN; Springfield, IL; New Haven, CT), where the frequency of coliform occurrences resulted in a noncompliance period, careful surveillance by local medical clinics and hospitals was made on all new patients to determine if there was any indication of a waterborne disease outbreak. No increased incident of intestinal disease was noted. Also, during the period from April to September 1984, a review of nosocomial infections among intensive care patients of a large hospital in New Haven, CT revealed no increase in *K. pneumoniae* infections from a gentamicin-sensitive strain similar to the one isolated from the water supply.[166] More attempts need to be made to "fingerprint" biofilm isolates and compare these species patterns with ongoing clinical strains of opportunistic coliforms in the hospitals using water supply during a coliform biofilm event.

The most obvious observation in an incipient biofilm occurrence is that there is no previous outburst of coliforms reported from within the distribution system and no coliforms are found at first-customer locations. Sometimes these incipient biofilm events are the result of the hydraulic forces in water flow reversals that break loose pipe sediments with entrapped coliforms. In another scenario, sudden elevations in chlorine residual passing through a pipe section may dislodge part of a developing biofilm through oxidation, thereby releasing fragments of growth into the bulk water. Repeat sampling usually reveals the problem is not site specific and coliform releases vary from one to several hundred organisms per 100 ml. Verification of these transient total coliforms often discloses no fecal coliform (*E. coli*) bacteria but a predominance of *Klebsiella, Enterobacter,* or *Citrobacter.*

SEARCHING FOR FECAL CONTAMINATION

There is always a concern that biofilm occurrences could be hiding a fecal contaminating event either from inadequate treatment or distribution system contamination. For this reason, any utility experiencing coliform biofilm in the system should intensify the monitoring problem and search for evidence of fecal coliforms or *E. coli* among the positive samples. If so, verified occurrence of these coliforms should call for a boil water order until repeat samples prove their disappearance. Any occurrence of fecal coliform bacteria or *E. coli* during a biofilm episode should not be brushed aside as an aberration because these organisms are not normally able to permanently colonize biofilms.

In the fact-finding investigation, priority must be given to the search for any waterborne disease outbreak that occurred while the system was operating in the present treatment configuration. Also essential to the investigation is a review of corrective actions taken to control the coliform occurrences (biofilm) during previous episodes. Records of an active cross-connection control program also need to be studied for evidence of fecal contamination pathways into the distribution system.

Characterizing a recurring biofilm introduces the need to review historical information that may provide evidence that the biofilm situation did not evolve from a treatment barrier problem. First-customer location(s) are absolutely essential to the determination because there must never be any leakage of fecal contamination from treatment processes into the distribution system. Data from these sites must demonstrate that the water quality consistently contains less that one total coliform per 100 ml. The treatment barrier must be continually verified to be effective because a breakdown could be a critical contributor of fecal contamination or of environmental coliforms that could colonize the pipe network. Thus, the tolerance for coliforms at these sites should be zero, not 5% at the 100-ml baseline for testing. As a precaution, one-liter samples of finished water should be considered a desirable objective in the search for low-level release of coliforms into the distribution system. These special sample collections should also be timed to coincide with the return of filters to operation after routine backwashing. Treatment barrier leakage should not be tolerated if coliform biofilm occurrence is to be brought under control.

While the regulation variance rule for coliform biofilm requires a review of the historical data for fecal coliform or *E. coli* in compliance samples over the past 6 months, this time frame should be expanded to 12 months. This expanded interval would provide data from all seasonal periods to be evaluated for impact of raw source water quality fluctuations, treatment adjustments to compensate for changes in source water quality, and seasonal effect of any corrosion control program in the distribution system. All of these factors influence coliform occurrences that can reinforce biofilm growth or provide evidence of pathways for pathogen passage.

REFERENCES

1. Characklis, W.G. and K.C. Marshall. 1990. *Biofilms*. John Wiley & Sons. Inc., New York.
2. Center for Biofilm Engineering. 1994. Biofilm Heterogeneity. Center for Biofilm Engineering News. Montana State Univ. 2:(1);1-2.
3. Keevil, C.W., C. W. Mackerness, and J.S. Colbourne. 1990. Biocide Treatment of Biofilm. *Intern. Biodeterio.*, 26:169-179.
4. Rogers, J., A.B. Dowsett, P.J. Dennis, J.V. Lee, and C.W. Keevil. 1994. Influence of Temperature and Plumbing Material Selection on Biofilm Formation and Growth of *Legionella pneumophila* in a Model Potable Water System Containing Complex Microbial Flora. *Appl. Environ. Microbiol.*, 60:1585-1592.

5. LeChevallier, M.W. 1990. Coliform Regrowth in Drinking Water: A Review. *Jour. Amer. Water Works Assoc.*, 82:74-86.

6. Walker, J.T., J. Rogers, and C.W. Keevil. 1994. An Investigation of the Efficacy of a Bromine-Containing Biocide on an Aquatic Consortium of Planktonic and Biofilm Microorganisms Including *Legionella pneumophila. Biofouling,* 8:47-54.

7. Rogers, J., A.B. Dowsett, P.J. Dennis, J.V. Lee, and C.W. Keevil. 1994. Influence of Plumbing Material on Biofilm Formation and Growth of *Legionella pneumophila* in Potable Water Systems. *Appl. Environ. Microbiol.,* 60:1842-1851.

8. Allen, M.J., R.H. Taylor, and E.E. Geldreich. 1980. The Occurrence of Microorganisms in Water Main Encrustations. *Jour. Amer. Water Works Assoc.,* 72:614-625.

9. Tuovinen, O.H., K.S. Button, A. Vuorinen, et al. 1980. Bacterial, Chemical, and Mineralogical Characteristics of Tubercles in Distribution Pipelines. *Jour. Amer. Water Works Assoc.,* 72:626-635.

10. LeChevallier, M.W., T.S. Babcock, and R.G. Lee. 1987. Examination and Characterization of Distribution System Biofilms. *Appl. Environ. Microbiol.,* 53:2714-2724.

11. Geldreich, E.E. 1990. Microbial Quality Control in Distribution Systems. In: *Water Quality and Treatment*, 4th ed., F.W. Pontius, Ed. AWWA, McGraw-Hill Inc., New York.

12. Walker, J.T., D. Wagner, W. Fischer, and C.W. Keevil. 1994. Rapid Detection of Biofilm on Corroded Copper Pipes. *Biofouling,* 8:55-63.

13. Nagy, L.A. and B.H. Olson. 1982. The Occurrence of Filamentous Fungi in Drinking Water Distribution Systems. *Can. Jour. Microbiol.,* 28:667-671.

14. Rosenzweig, W.D. and W.O. Pipes. 1988. Fungi from Potable Water: Interaction with Chlorine and Engineering Effects. *Water Sci. Technol.,* 20:153-160.

15. Hinzelin, F. and J.C. Block. 1985. Yeasts and Filamentous Fungi in Drinking Water. *Environ. Technol. Lett.,* 6:101-106.

16. Levy, R.V., F.L. Hart, and R.D. Cheetham. 1986. Occurrence and Public Health Significance of Invertebrates. *Jour. Amer. Water Works Assoc.,* 77:105-110.

17. King, S.H., E.B. Shotts, R.E. Wooley, and K.G. Porter. 1988. Survival of Coliforms and Bacterial Pathogens within Protozoa during Chlorination. *Appl. Environ. Microbiol.,* 54:3023-3033.

18. Chang, S-L., R.L. Woodward, and P.W. Kabler. 1960. Survey of Free-Living Nematodes and Amoebas in Municipal Supplies. *Jour. Amer. Water Works Assoc.,* 52:613-618.

19. Smalls, I.C. and G.F. Greaves. 1968. A Survey of Animals in Distribution Systems. *Jour. Soc. Water Treat. Exam.,* 17:150-180.

20. Mackenthun, K.M. and L.E. Keup. 1970. Biological Problems Encountered in Water Supplies. *Jour. Amer. Water Works Assoc.,* 62:520-526.

21. Victoreen, H.T. 1969. Soil Bacteria and Color Problem in Distribution Systems. *Jour. Amer. Water Works Assoc.,* 61:429-431.

22. Nagy, L.A. and B.H. Olson. 1985. Occurrence and Significance of Bacteria, Fungi and Yeasts Associated with Distribution Pipe Surfaces. Proc. Water Qual. Tech. Conf., Houston, TX. American Water Works Assoc., Denver, CO.

23. LeChevallier, M.W., R.J. Seidler, and T.M. Evans. 1980. Enumeration and Characterization of Standard Plate Count Bacteria in Raw and Chlorinated Water Supplies. *Appl. Environ. Microbiol.,* 40:922-930.

24. Nagy, L.A., A.J. Kelly, M.A. Thun, and B.H. Olson. 1982. Biofilm Composition, Formation and Control in the Los Angeles Aqueduct System. Proc. Water Qual. Technol. Conf., Nashville, TN. Amer. Water Works Assoc., Denver, CO.

25. Geldreich, E.E. 1988. Coliform Non-Compliance Nightmares in Water Supply Distribution Systems. In: *Water Quality: A Realistic Perspective*. College of Engineering, Univ. Michigan, Ann Arbor, MI.

26. Herson, D.S. and H. Victoreen. 1980. Identification of Coliform Antagonists. Proc. Amer. Water Works Assoc., Water Qual. Technol. Conf., Miami Beach, FL. pp. 153–160.

27. Geldreich, E.E., H.D. Nash, and D. Spino. 1977. Characterizing Bacterial Populations in Treated Water Supplies. Proc. Amer. Water Works Assoc., Water Qual. Technol. Confr., Kansas City, MO. pp. 28-5; 1-3.

28. Allen, M.J. and E.E. Geldreich. 1977. Distribution Line Sediments and Bacterial Regrowth. Proc. Amer. Water Works Assoc., Water Qual. Technol. Confr., Kansas City, MO. pp. 3B-1; 1-6.

29. Safe Drinking Water Committee. 1982. *Drinking Water and Health*. Vol. 4. National Academy of Science, National Academy Press, Washington, D.C.

30. Rittmann, B.E. and V.L. Snoeyink. 1984. Achieving Biological Stable Drinking Water. *Jour. Amer. Water Works Assoc.*, 76:106-114.

31. LeChevallier, M.W. 1990. Coliform Regrowth in Drinking Water: A Review. *Jour. Amer. Water Works Assoc.*, 82:74-86.

32. van der Kooji, D., A. Visser, and W.A.M. Hijner. 1982. Determining the Concentration of Easily Assimilable Organic Carbon in Drinking Water. *Jour. Amer. Water Works Assoc.*, 74:540-545.

33. James M. Montgomery, Consulting Engineers Inc. 1985. *Water Treatment Principles and Design*. John Wiley & Sons, New York.

34. Knocke, W.R., R.C. Hoehn, and R.L. Sinsabaugh. 1987. Using Alternative Oxidants to Remove Dissolved Manganese from Waters Laden with Organics. *Jour. Amer. Water Works Assoc.*, 79(3):75-79.

35. Loos, E.T. 1987. Experiences with Manganese in Queensland Water Supplies. *Jour. Aust. Water Waste Assoc.*, 14:28-31.

36. Geldreich, E.E., J.A. Goodrich, and R.M. Clark. 1990. Characterizing Surface Waters That May Not Require Filtration. *Jour. Amer. Water Works Assoc.*, 82:40-50.

37. Rizet, M., F. Fiessinger, and N. Houel. 1982. Bacterial Regrowth in a Distribution System and Its Relationship with the Quality of the Feed Water: Case Studies. Annu. Conf. Proc., Amer. Water Works Assoc., Denver, CO.

38. Postgate, J.R. and J.R. Hunter. 1962. The Survival of Starved Bacteria. *Jour. Gen. Microbiol.*, 29:233-263.

39. Werner, P. 1984. Investigations of the Substrate Character of Organic Substances in Connection with Drinking Water Treatment. *Zbl. Bakt. Hyg.*, 180(1):46-61.

40. Rice, E., P.V. Scarpino, G.S. Logsdon, D.J. Reasoner, P.J. Mason, and J.C. Blannon. 1990. Bioassay Procedure for Predicting Coliform Bacterial Growth in Drinking Water. *Environ. Technol.*, 11:821-829.

41. Joret, J.C. and Y. Levi. 1986. Rapid Method for Measurement of Biologically Degradable Carbon. *Trib. Cebedeau*, 39(510):3-9.

42. Kaplan, L.A. and T.A. Bott. 1988. Measurement of Assimilable Organic Carbon in Water Distribution Systems by a Simplified Bioassay Technique. Proc. Amer. Water Works Assoc. Water Qual. Technol. Conf., St. Louis, MO. pp. 475-498.

43. Kemmy, F.A., J.C. Fry, and R.A. Breach. 1989. Development and Operational Implementation of a Modified and Simplified Method for Determination of Assimilable Organic Carbon (AOC) in Drinking Water. Water Sci. Techol., 21:155-159.

44. Servais, P., G. Billen, and M. Hascoet. 1987. Determination of the Biodegradable Fraction of Dissolved Organic Matter in Water. Water Res., 21:445-450.

45. Rice, E.W., P.V. Scarpino, D.J. Reasoner, G.S. Logsdon, and Deanna K. Wold. 1991. Correlation of Coliform Growth Response with Other Water Quality Parameters. Jour. Amer. Water Works Assoc., 83:98-102.

46. Camper, A.K., G.A. McFeters, W.G. Characklis, and W.L. Jones. 1991. Growth Kinetics of Coliform Bacteria under Conditions Relevant to Drinking Water Distribution Systems. Appl. Environ. Microbiol., 57:2233-2239.

47. Van der Kooij, D. and W.A.M. Hijnen. 1984. Substrate Utilization by an Oxalate-Consuming Spirillum Species in Relation to Its Growth in Ozonated Water. Appl. Environ. Microbiol., 47:551-559.

48. Reasoner, D.J., E.W. Rice, and L.C. Fung. 1990. Ozonation and Biological Stability of Water in an Operating Water Treatment Plant. Proc. Water Qual. Technol. Conf. San Diego, CA. pp. 1215-1228.

49. Jago, P.H. and G. Stanfield. 1985. Development and Application of a Method for Determining the Assimilable Organic Carbon Content of Drinking Water. Water Research Centre, Medmenham, United Kingdom. Internal Report 1038-H.

50. Huck, P.M. 1990. Measurement of Biodegradable Organic Matter and Bacterial Growth Potential in Drinking Water. Jour. Amer. Water Works Assoc., 82:78-86.

51. Hascoet, M.C., P. Servais, and G. Billen. 1986. Use of Biological Analytical Methods to Optimize Ozonation and GAC Filtration in Surface Water Treatment. Proc. Amer. Water Works Annu. Meet., Denver, CO.

52. Joret, J.C. and Y. Levi. 1986. Method Rapide d'Evaluation du Carbone Eliminable des Eaux par Voie Biologique. Trib. Cebedeau, 39:3-9.

53. Joret, J.C., Y. Levi, and C. Volk. 1991. Biodegradable Dissolved Organic Carbon (BDOC) Content of Drinking Water and Potential Regrowth of Bacteria. Water Sci. Technol., 24:95-101.

54. Bradford, S.M., P.A. Hacker, B.H. Olson, L. Tan, and M. Rigby. 1990. Evaluation of AOC in Surface and Groundwaters Using Two Bioassay Methods. Proc. Amer. Water Works Assoc., Water Qual. Technol. Conf., San Diego, CA.

55. Rice, E.W. 1989. Bioassay Procedures for Predicting Coliform Bacterial Growth in Drinking Water. Graduate Thesis, Department of Civil and Environ. Engr., College of Engineering, Univ. of Cincinnati.

56. LeChevallier, M.W., R.G. Lee, and R.H. Moser. 1989. Bacterial Nutrients in Drinking Water. Report Authorization 279, Formula 55-00. Amer. Water Works Service Company, Inc., Bellville, IL.

57. Eberhardt, M. 1976. Experience with the Use of Biologically Effective Activated Carbon. In: Translation of Reports on Special Problems of Water Technology. Vol. 9. Adsorption. H. Sontheimer, Ed., EPA 600/9-76-030.

58. Bourbigot, M.M., A. Dodin, and R. Lherritier. 1982. Limiting Bacterial After-growth in Distribution Systems by Removing Biodegradable Organics. Proc. Amer. Water Works Assoc. Water Qual. Technol. Conf., Miami Beach, FL.

59. Committee Report. 1981. An Assessment of Microbiol Activity on GAC. *Jour. Amer. Water Works Assoc.*, 73:447-454.

60. van der Kooij, D. 1992. Assimilable Organic Carbon as an Indicator of Bacterial Regrowth. *Jour. Amer. Water Works Assoc.*, 84:57-65.

61. Reasoner, D.J. 1990. Assimilable Organic Carbon and Distribution System Quality. Amer. Works Assoc. Annual conference. Sunday Seminar (S-11), Cincinnati, OH.

62. Rice, E.W. Personal communication.

63. Howard, M.J. 1940. Bacterial Depreciation of Water Quality in Distribution Systems. *Jour. Amer. Water Works Assoc.*, 32:1501-1506.

64. Smith, D.B., A.F. Hess, and D. Opheim. 1989. Control of Distribution System Coliform Regrowth. Proc. Amer. Water Works Water Qual. Tech. Confr., Philadelphia, PA. pp. 1009-1029.

65. Donlan, R.M. and Pipes, W.O. 1988. Selected Drinking Water Characteristics and Attached Microbial Population Density. *Jour. Amer. Water Works Assoc.*, 80:11;70-76.

66. Olson, B. 1982. Assessment and Implications of Bacterial Regrowth in Water Distribution Systems, U.S. Environmental Protection Agency, Rep. 600/52-82-072, Cincinnati, OH.

67. Costerton, J.W., G.G. Geesey, and K.J. Cheng. 1978. How Bacteria Stick. *Sci. Amer.*, 238:86-95.

68. Bitton, G. and K.C. Marshall. 1980. *Absorption of Microorganisms to Surfaces.* Wiley-Interscience, New York.

69. Costerton, J.W. and H.M. Lappin-Scott. 1989. Behavior of Bacteria in Biofilms. *ASM News,* 55:650-654.

70. Camp, T.R. and R.L. Meserve. 1974. *Water and Its Impurities*, Dowden, Hutchinson and Ross, Inc., Stroudsburg, PA.

71. Lee, S.H., J.T. O'Connor, and S.K. Banerji. 1980. Biologically Mediated Corrosion and Its Effects on Water Quality in Distribution Systems. *Jour. Amer. Water Works Assoc.*, 72:636-645.

72. Emde, K.M.E., D.W. Smith, and R. Facey. 1992. Initial Investigation of Microbially Influenced Corrosion (MIC) in a Low Temperature Water Distribution System. *Water Res.,* 26:169-175.

73. Facey, R.M., D.E. Smith, and K.M.E. Emde. 1991. Case Study: Water Distribution Corrosion, Yellowknife, N.W.T. *Proc. Microbially Influenced Corrosion.* N.J. Dowling, M.W. Mittleman, and J.C. Danko, Eds., MIC Consortium, Knoxville, TN. pp. 4-45-4-49.

74. Singley, J.E., B.A. Beaudet, and P.H. Markey. 1984. Corrosion Manual for Internal Corrosion of Water Distribution Systems, U.S. Environmental Protection Agency, Rep. 570/9-84-001, Office of Drinking Water, Washington, D.C.

75. DVGW-Forschugsstelle, AWWA Research Foundation, Internal Corrosion of Water Distribution Systems. 1985. AWWA Research Foundation, Denver, CO.

76. Langelier, W.F. 1936. The Analytical Control of Anti-Corrosion Treatment. *Jour. Amer. Water Works Assoc.*, 28:1500-1521.

77. McCauley, R. 1960. Controlled Deposition of Protection Calcite Coating in Water Mains. *Jour. Amer. Water Works Assoc.*, 52:1386-1396.

78. Stumm, W. 1960. The Corrosive Behavior of Water. *Proc. Amer. Soc. Civ. Eng.*, 86:NoSA-6.

79. Larson, T.E. and R.V. Skold. 1958. Current Research on Corrosion and Tuberculation of Cast Iron. *Jour. Amer. Water Works Assoc.*, 50:1429-1452.

80. Larson, T.E. 1966. Deterioration of Water Quality in Distribution Systems. *Jour. Amer. Water Works Assoc.*, 58:1307-1316.

81. Butler, G. and H.C.K. Ison. 1966. *Corrosion and Its Prevention in Waters*. Reinhold Pub. Corp., New York.

82. Rosenweig, W.D. 1988. Influence of Phosphate Corrosion Control Compounds on Bacterial Growth. EPA Project Summary CR-811613-01-0, U.S. Environmental Protection Agency, Cincinnati, OH.

83. Hudson, H.E. 1962. High Quality Water Production and Viral Disease. *Jour. Amer. Water Works Assoc.*, 54:1265-1274.

84. American Water Works Committee on Viruses in Water. 1969. Viruses in Water. *Jour. Amer. Water Works Assoc.*, 61:491-494.

85. Cookson, J.T. 1974. Virus and Water Supply. *Jour. Amer. Water Works Assoc.*, 66:707-711.

86. Culp, R.L. 1974. Breakpoint Chlorination for Virus Inactivation. *Jour. Amer. Water Works Assoc.*, 66:699-703.

87. LeChevallier, M.W., T.M. Evans, and R.J. Seidler. 1981. Effect of Turbidity on Chlorination Efficiency and Bacterial Persistence in Drinking Water. *Appl. Environ. Microbiol.*, 42:159-167.

88. Hoff, J.C. 1979. Disinfection Resistance of *Giardia* Cysts: Origins of Current Concepts and Research in Progress. In: Waterborne Transmission of Giardiasis. W. Jakubowski and J.C. Hoff, Eds., U.S. Environmental Protection Agency, EPA-600/9/79-001, 306 pp., Cincinnati, OH.

89. Ridgway, H.F. and B.H. Olson. 1981. Scanning Electron Microscope Evidence for Bacterial Colonization of a Drinking Water Distribution System. *Appl. Environ. Microbiol.*, 41:274-287.

90. Sanderson, W.W. and S. Kelly. 1964. Discussion of Human Enteric Viruses in Water: Source, Survival and Removability. In: *International Conference on Water Pollution Research, London, 1962*. Pergamon Press, Oxford. pp. 536-541.

91. Boardman, G.D. 1976. Protection of Waterborne Viruses by Virtue of Their Affiliation with Particulate Matter. Ph.D. Thesis. Univ. of Maine.

92. Stagg, C.H., C. Wallis, and C.H. Ward. 1977. Inactivation of Clay-Associated Bacteriophage MS-2 by Chlorine. *Appl. Environ. Microbiol.*, 33:385-391.

93. Lister, J.B. 1980. Microbiological Relationships of the 5 NTU Point Turbidity MCL. Proc. Amer. Water Works Assn., Water Quality Technol. Conf., Dec. 9-12, 1979. Philadelphia, PA. Amer. Water Works Assoc., Denver, CO. pp. 197-202.

94. Foster, D.M. 1980. Ozone Inactivation of Cell- and Fecal-Associated Viruses and Bacteria. *Jour. Water Poll. Contr. Fed.*, 52:2174-2184.

95. Hoff, J.C. 1978. The Relationships of Turbidity to Disinfection of Potable Water. In: Evaluation of the Microbiology Standards for Drinking Water. C.W. Hendricks, Ed., U.S. Environmental Protection Agency, Rep. 570/9-78-002, Washington, D.C.

96. Reilly, J.K. and J. Kippin. 1981. Interrelationship of Bacterial Counts with Other Finished Water Quality Parameters within Distribution Systems. U.S. Environmental Protection Agency, Rep. 0600/52-81-035, Cincinnati, OH.

97. Cuillo, R.H., H.E., Ferran, Jr., E.E. Whitaker, and H. Leland. 1983. Bacterial Survival in Potable Water with Low Turbidity. U.S. Environmental Protection Agency Grant R-806329, Nasson College, Springvale, ME.

98. U.S. Environmental Protection Agency. 1976. National Interim Primary Drinking Water Regulations. Rep. 570/9-76-003, Office of Water Supply. Washington, D.C.

99. Neefe, J.R., J.B. Bathy, J.G. Reinhold, and J. Stokes, Jr. 1947. Inactivation of the Virus of Infectious Hepatitis in Drinking Water. Amer. Jour. Pub. Health, 37:365-372.

100. Olson, B.H. 1982. Assessment and Implications of Bacterial Regrowth in Water Distribution Systems. EPA-600/52-82-072. U.S. Environmental Protection Agency. Cincinnati, OH.

101. Berman, D., E.W. Rice, and J.C. Hoff. 1988. Inactivation of Particle- Associated Coliforms by Chlorine and Monochloramine. Appl. Environ. Microbiol., 54:507-512.

102. Katz, E.L. 1986. The Stability of Turbidity in Raw Water and Its Relationship to Chlorine Demand. Jour. Amer. Water Works Assoc., 78(2):72-75.

103. Small, I.C. and G.F. Greaves. 1968. A Survey of Animals in Distribution Systems. Jour. Soc. Water Treat. Exam., 19:150.

104. MacKenthun, K.M. and L.E. Keup. 1970. Biological Problems Encountered in Water Supplies. Jour. Amer. Water Works Assoc., 62:520-526.

105. Gerardi, M.H. and J.K. Grimm. 1970. Aquatic Invaders, Water Eng. Manage., 10:22-23.

106. Chang, S.L., R.L. Woodward, and P.W. Kabler. 1960. Survey of Free-Living Nematodes and Amoebas in Municipal Supplies. Jour. Amer. Water Works Assoc., 52:613-618.

107. Levy, R.V., R.D. Cheethan, J. Davis, G. Winer, and F.L. Hart. 1984. Novel Method for Studying the Public Health Significance of Macroinvertebrates Occurring in Potable Water. Appl. Environ. Microbiol., 47:889-894.

108. Levy, R.V. 1990. Invertebrates and Associated Bacteria in Drinking Water Distribution Lines. In: Drinking Water Microbiology. G.A. McFeters, Ed., Springer-Verlag, New York.

109. Zrupko, G. 1988. Examination of Large Volume Samples Taken from the Municipal Water Treatment Plant. Budapesti Koregeszsezugy, 1:21-25.

110. Cobb, N.A. 1918. Filter-Bed Nemas: Nematodes of the Slow Sand Filter-Beds of American Cities. Contr. Sci. Nematol., 7:189-212.

111. George, M.G. 1966. Further Studies on the Nematode Infiltration of Surface Water Supplies. Environ. Health, 8:93-102.

112. Tombes, A.S., A.R. Abernathy, D.M. Welch, and S.A. Lewis. 1979. The Relationship between Rainfall and Nematode Density in Drinking Water. Water Res., 13:619-622.

113. Mott, J.B. and A.D. Harrison. 1983. Nematodes from River Drift and Surface Drinking Water Supplies in Southern Ontario. Hydrobiologia, 102:27-38.

114. Tracy, H.W., V.M. Camarena, and F. Wing. 1966. Coliform Persistence in Highly Chlorinated Waters. Jour. Amer. Water Works Assoc., 58:1151-1159.

115. Clarke, N.A., G. Berg, P.W. Kabler, and S.L. Chang. 1964. Human Enteric Viruses in Water: Source, Survival and Removability. In: *International Conference on Water Pollution Research, London, 1962.* Vol. 2, Pergamon Press, Oxford. pp. 523-535.

116. Smerda, S.M., H.J. Jensen, and A.W. Anderson. 1971. Escape of Salmonellae from Chlorination during Ingestion by *Pristionchus lheritieri* (Nematoda: Diplogasterinae). *Jour. Nematol.,* 3:201-204.

117. Haney, P.D. 1978. Evaluation of Microbiological Standards for Drinking Water. *Water Sewage Works,* 125:R126-134.

118. Huq, A., E.B. Small, P.A. West, M.I. Huq, et al. 1983. Ecological Relationships between *Vibrio cholerae* and Planktonic Crustacean Copepods. *Appl. Environ. Microbiol.,* 45:275-283.

119. Fisher-Hoch, S.P. J.O. Tobin, A.M. Nelson, M.G. Smith, et al. 1981. Investigation and Control of an Outbreak of Legionnaires' Disease in a District General Hospital. *Lancet,* 2:932-941.

120. Fields, B.S., E.B. Shotts Jr., J.C. Feeley, G.W. Gorman, et al. 1984. Proliferation of *Legionella pneumophila* as an Intracellular Parasite of the Ciliated Protozoan *Tetrahymena pyriformis. Appl. Environ. Microbiol.,* 47:467-471.

121. White, P.B. 1938. The Rugose Variant of Vibrios. *Jour. Path. Bact.,* 46:1-6.

122. Rice, E.W., C.H. Johnson, R.M. Clark, K.R. Fox, et al. 1993. *Vibrio cholerae* 01 Can Assume a "Rugose" Survival Form That Resists Killing by Chlorine, Yet Retains Virulence. *Int. J. Envir. Health,* 3:89-98.

123. Lippy, E.C. and S.C. Waltrip. 1984. Waterborne Disease Outbreaks — 1946- 1980: A Thirty-Five Year Perspective. *Jour. Amer. Water Works Assoc.,* 76:60-67.

124. Craun, G.F. 1984. Waterborne Outbreaks of Giardiasis Current Status. In: *Giardia and Giardiasis Biology, Pathogenics and Epidemiology.* S.L. Erlandsen and E.A. Meyer, Eds., Plenum Press, New York. pp. 243-261.

125. Hoff, J.C. and E.E. Geldreich. 1981. Comparison of the Biocidal Efficiency of Alternative Disinfectants. *Jour. Amer. Water Works Assoc.,* 73:40-44.

126. Ridgway, H.F. and B.H. Olson. 1982. Chlorine Resistance Patterns of Bacteria from Two Drinking Water Distribution Systems. *Appl. Environ. Microbiol.,* 44:972-987.

127. Hoff, J.C. and E.W. Akin. 1986. Microbial Resistance to Disinfectants: Mechanisms and Significance. *Environ. Health Perspect.,* 69:7-13.

128. Camper, A.K., M.W. LeChevallier, S.C. Broadaway, and G.A. McFeters. 1985. Growth and Persistence on Granular Activated Carbon Filters. *Appl. Environ. Microbiol.,* 50:1378-1382.

129. LeChevallier, M.W. and G.A. McFeters. 1990. Microbiology of Activated Carbon. In: *Drinking Water Microbiology.* G.A. McFeters, Ed., Springer-Verlag, New York.

130. Syrotynski, M. 1971. Microscopic Water Quality and Filtration Efficiency. *Jour. Amer. Water Works Assoc.,* 63:237-245.

131. Symons, J.M., A.A. Stevens, R.M. Clark, E.E. Geldreich, et al. 1981. Treatment Techniques for Controlling Trihalomethanes in Drinking Water. EPA-600/2-81-156, U.S. Environmental Protection Agency, Cincinnati, OH.

132. Camper, A.K., M.W. LeChevallier, S.C. Broadaway, and G.A. McFeters. 1986. Bacteria Associated with Granular Activated Carbon Particles in Drinking Water. *Appl. Environ. Microbiol.,* 52:434-438.

133. Stewart, M.H., R.L. Wolfe, and E.G. Means, 1988. An Assessment of the Bacteriological Activity on Granular Activated Carbon Particles. Proc. Amer. Soc. Microbiol. Annu. Meet., Miami Beach, FL.

134. Cairo, P.R., J. McElhaney, and I.H. Suffet. 1979. Pilot Plant Testing of Activated Carbon Adsorption Systems. *Jour. Amer. Water Works Assoc.,* 71:660-673.

135. Parson, F. Jan 10, 1980. Bacterial Populations in Granular Activated Carbon Beds and Their Effluents. U.S. Environmental Protection Agency, Cincinnati, OH.

136. Rollinger, Y. and W. Dott. 1987. Survival of Selected Bacterial Species in Sterilized Activated Carbon Filters and Biological Activated Carbon Filters. *Appl. Environ. Microbiol.,* 53:777-781.

137. Brewer, W.S. and W.W. Carmichael. 1979. Microbial Characterization of Granular Activated Carbon Filter Systems. *Jour. Amer. Water Works Assoc.,* 71:738-740.

138. Cummins, B.B. and H.D. Nash. 1978. Microbiological Implications of Alternative Treatment. Proc. Amer. Water Works Assoc. Water Qual. Technol. Conf., Louisville, KY.

139. Haas, C.N., M.A. Meyer, and M.S. Paller. 1983. Microbial Dynamics in GAC Filtration of Potable Water. *Proc. Amer. Soc. Civ. Eng. Jour. Environ. Eng. Div.,* 109:956-961.

140. LeChevallier, M.W., T.S. Hassenauer, A.K. Camper, and G.A. McFeters. 1984. Disinfection of Bacteria Attached to Granular Activated Carbon. *Appl. Environ. Microbiol.,* 48:918-928.

141. Camper, A.K., D.G. Davies, S.C. Broadway, M.W. LeChevallier, et al. 1985. Association of Coliform Bacteria and Enteric Pathogens with Granular Activated Carbon. Abstr. Annu. Meet. Amer. Soc. Microbiol. N38:p. 223.

142. LeChevallier, M.W., C.D. Cawthon, and R.G. Lee. 1988. Inactivation of Biofilm Bacteria. *Appl. Environ. Microbiol.,* 54:2492-2499.

143. Neden, D.G., R.J. Jones, J.R. Smith, G.J. Kirmeyer, et al. 1992. Comparing Chlorination and Chloramination for Controlling Bacterial Regrowth. *Jour. Amer. Water Works Assoc.,* 84:7:80-88.

144. Geldreich, E.E. 1986. Control of Microorganisms of Public Health Concern in Water. *Jour. Environ. Sci.,* 29:34-37.

145. Logsdon, G.S. and E.W. Rice. 1985. Evaluation of Sedimentation and Filtration for Microorganism Removal. Proc. Amer. Water Works Assoc. Ann. Conf., Washington, D.C. pp. 1177-1197.

146. Logsdon, G.S. 1990. Microbiology and Drinking Water Filtration. In: *Drinking Water Microbiology.* G.A. McFeters, Ed., Springer-Verlag, New York.

147. Bucklin, K.E., G.A. McFeters, and A. Amirtharajah. 1991. Penetration of Coliforms through Municipal Drinking Water Filters. *Water Res.,* 25:1013-1017.

148. McFeters, G.A., J.S. Kippin, and M.W. LeChevallier. 1986. Injured Coliforms in Drinking Water. *Appl. Environ. Microbiol.,* 51:1-5.

149. McFeters, G.A. 1990. Enumeration, Occurrence, and Significance of Injured Indicator Bacteria in Drinking Water. In: *Drinking Water Microbiology.* G.A. McFeters, Ed., Springer-Verlag, New York.

150. Snead, M.C., V.P. Olivieri, K. Kawata, and C.W. Kruse. 1980. The Effectiveness of Chlorine Residuals in Inactivation of Bacteria and Viruses Introduced by Post-Treatment Contamination. *Water Res.,* 14:403-408.

151. Buelow, R.W., R.H. Taylor, E.E. Geldreich, A. Goodenkauf, et al. 1976. Disinfection of New Water Mains. *Jour. Amer. Water Works Assoc.,* 68(6):283-288.

152. Harold, C.H.H. 1934. 29th Report Director of Water Examination, Metropolitan Water Board, London, England.

153. Hamilton, J.J. 1974. Potassium Permanganate as a Main Disinfectant. *Jour. Amer. Water Works Assoc.,* 66(12):734-735.

154. Alter, A.J. Sept. 1954. Appearance of Intestinal Wastes in Surface Water Supplies at Ketchikan, Alaska. Proc. 5th Alaska Sci. Conf. AAAS, Anchorage, AK.

155. Anonymous. 1954. Ketchikan Laboratory Studies Disclose Gulls are Implicated in Disease Spread. *Alaska's Health,* 11:1.

156. Fennel, H., D.B. James, and J. Morris. 1974. Pollution of a Storage Reservoir by Roosting Gulls. *Water Treat. Exam.,* 23:5-24.

157. Geldreich, E.E. and S. Shaw. 1993. New York City Water Supply Microbial Crisis. Public Water Supply Section, Region 2. U.S. Environmental Protection Agency, New York, NY.

158. Division of Drinking Water Quality Control. 1994. Supplemental Report to "Coliform Occurrences in the New York City Distribution System July and August 1993," New York City Department of Environmental Protection, New York, NY.

159. Seidler, R.J., J.E. Morrow, and S.T. Bagley. 1977. *Klebsiella* in Drinking Water Emanating from Redwood Tanks. *Appl. Environ. Microbiol.,* 33:893-905.

160. Knittel, M.D., R.J. Seidler, and L.M. Cabe. 1977. Colonization of the Botanical Environment by *Klebsiella* Isolates of Pathogenic Origin. *Appl. Environ. Microbiol.,* 34:557-563.

161. Talbot, H.W., Jr., J.E. Morrow, and R.J. Seidler. 1979. Control of Coliform Bacteria in Finished Drinking Water Stored in Redwood Tanks. *Jour. Amer. Water Works Assoc.,* 71:349-353.

162. Schoenen, D. 1986. Microbial Growth due to Materials Used in Drinking Water Systems. In: *Biotechnology,* Vol. 8. H.-J. Rehm and G. Reed, Eds., VCH Verlagsgesellschaft, Weinheim. pp. 628-647.

163. Schoenen, D. 1989. Influence of Materials on the Microbiological Colonization of Drinking Water. *Aqua,* 38:101-113.

164. Ellgas, W.M. and R. Lee. 1980. Reservoir Coatings Can Support Bacterial Growth. *Jour. Amer. Water Works Assoc.,* 72:693-695.

165. U.S. Environmental Protection Agency. 1989. Drinking Water: National Primary Drinking Water Regulations; Total Coliforms (including Fecal Coliforms and *E. coli*): Final Rule. *Fed. Regis.,* 54:27544-27568.

166. Ludwig, F., A. Cocco, S. Edberg, R. Jarema, et al. 1985. Detection of Elevated Levels of Coliform Bacteria in a Public Water Supply — Connecticut. *Morbid. Mortality Week. Rep.,* 34:142-144.

CHAPTER 5

Characterizing Microbial Quality of Water Supply

CONTENTS

INTRODUCTION

There is much information that needs to be discovered in the characterization of water quality trends during distribution.[1-3] Historically, all interest focused solely on the search for coliforms with the understanding that their presence was the universal indication of all unsatisfactory water quality conditions. In recent years research and field investigations have proven this interpretation needs refinement, but use of the coliform criterion continues because this bacterial group has served well as an indicator of treatment effectiveness and distribution system integrity. However, there are other situations that cannot be satisfied by information solely on coliform presence or absence.[4]

TRENDS IN WATER QUALITY

Modifying treatment processes to reduce disinfection by-product formation and lead residuals may result in undesirable trade-offs in microbial quality. For these reasons, it becomes important to expand the characterization of drinking water in the distribution system for early evidence of accelerated microbial quality deterioration. Among these concerns is the acquisition of a database on heterotrophic bacterial trends toward significant increases that would serve as an alert to flush lines to reduce sediment deposits, restore disinfection residuals, and reduce growth of nitrifying bacteria in systems using chloramination. Rapid growth of the heterotrophic bacterial population in the biofilm with subsequent early release of biofilm fragments into the bulk water is often followed in a few days with the sudden appearance of coliforms from the same habitat.

The search for stressed coliform releases past the treatment barrier may be very important during times of changing source water qualities and peak summer demand periods. Before the frequency of coliform occurrences reaches 5%, it is important to gather information on the magnitude of coliform densities occurring as well as profiling coliform species in a search for possible hidden fecal contamination in a biofilm event.

There are other organisms that appear in water supply that are of no apparent public health significance but are responsible for a variety of problems including corrosion, taste and odor, red water, and slime. These "nuisance" organisms have been reported to be the cause of a variety of operational difficulties in the transmission of groundwater or relate to many consumer complaints about the aesthetic qualities expected in a wholesome drinking water supply. Characterizing the localized problem of taste, odor, and off-color through monitoring for nuisance organisms and their metabolic by-products offers opportunities to apply preventative measures before customer complaints escalate.

Finally there is the need to formulate a strategy in the search for a pathogenic agent in a suspect waterborne outbreak. Sites near illness cases and areas of long residence time (slow-flow areas and dead ends) are prime locations for detecting evidence of the pathogenic agent. Water entering the distribution system should also be analyzed using 2-l sample portions for evidence of a treatment barrier failure. Characterizing treated water for loss of disinfectant residual, increases in turbidity, and unstable water pH are also part of the appraisal for adverse water quality trends that have impact on a safe supply.

HETEROTROPHIC BACTERIA

There are a variety of organisms to be found in a safe water supply. Many of these organisms pass through various treatment processes largely undiminished in densities because they are resistant to disinfection either as vegetative cells with impervious membranes,[2,5-9] occur in cell aggregates,[10,11] or survive in a viable spore state.[12,13] Others enter the water system in open finished water reservoirs, during line repairs, in backflows from pipelining projects, and new pipe network additions. Most often these organisms are not of immediate public health significance, but upon amplification in a protected pipe habitat become the source of taste and odor complaints or emerge as an opportunistic pathogen threat.

CHARACTERIZING POPULATION TRENDS

Characterizing density trends among the heterotrophic bacterial population may be a useful tool in regulating the frequency and design of a flushing program. Dead-end sections and static water areas are major locations for bacterial regrowth, particularly during warm-water periods. During these regrowth events, the heterotrophic plate count (HPC) may be observed to increase five- to tenfold or more over a baseline occurrence of approximately 100 organisms per milliliter. When this occurs, the most significant controlling factor is moving sediments out of the pipelines through a vigorous flushing program, thereby cutting off the bacterial nutrient supply, removing incipient biofilm formation, and restoring fresh water with a chlorine residual to the pipe environment. Another factor for bacterial regrowth may relate to type of disinfectant used. For instance, use of ozonation in water treatment may be effective in reducing organic precursors, but in so doing makes available a host of biodegradable materials that may be released to the distribution system[14] if not removed by a biological treatment process in the treatment train.[1]

Some utilities maintain a chloramine residual in their distribution system in an effort to meet the trihalomethane regulation or to achieve a detectable disinfectant residual in 95% of the pipe network.[7,15] Over time, this continuous application of chloramines may lead to increased heterotrophic bacterial densities in various dead ends and slow-flow sections of the distribution system. This change in water quality is largely due to incomplete nitrification.[16] As a consequence, nitrifying bacteria become predominant in the water containing ammonia or nitrate residuals, with pH of 7.5 to 8.0, and at water temperatures approaching 25°C.[17] This microbial response will be slow to detect because growth of nitrifying bacteria is sluggish, ranging from 8 h to several days to accomplish a mere doubling in cell count. Perhaps more important is the release of various metabolic by-products by nitrifying bacteria that enhance the growth of heterotrophic bacteria.[17,18] This latter observation explains why heterotrophic bacterial densities increase in chloraminated water supplies, reflecting the slow but deliberate biological activity associated with nitrifying bacteria. Routine monitoring for nitrifying bacteria is cumbersome.[19] For instance, detection of ammonia oxidizers requires at least 3 weeks with as much as 15 weeks incubation in an appropriate medium for nitrite utilizers.[20,21] Thus, the use of a heterotrophic plate count is still a useful indirect indicator of the negative aspects to chloramination —incomplete nitrification, excessive ammonia applied, and loss of disinfection effectiveness.

SELECTING TEST PROTOCOLS

Bacteria in drinking water supplies are frequently not physiologically vigorous. Many of these organisms have been stressed by recent passage of surface water through the disinfection process or are starved because of minimal concentrations of available nutrients in groundwater. Survivors that reach the distribution system may either die in this harsh aquatic environment, adjust to low-nutrient conditions, and slowly colonize in a biofilm or maintain a meager existence by gearing down their metabolic rate to slower generation times and increasingly smaller daughter cells. These latter organisms become the ultramicrobacteria that can be restored to full vigor and possible virulence when the environmental conditions change.

Cultivation

Cultivation of a large proportion of this diverse heterotrophic population requires careful consideration on four interrelated factors: dilute medium of diverse nutrients, incubation temperature, incubation time, and *in situ* placement (surface growth) during cultivation. The medium must provide not only a dilute nutrient base that might approach chemical constituent levels in distribution water, but also be inclusive of essential organics necessary for the fastidious organisms in the heterotrophic bacterial population. This nutrient status must also be in response to the demands of stressed organisms whose cell membranes are functioning poorly and enzyme systems damaged so that membrane osmatic gradients are erratic and metabolic pathways disrupted. Under these conditions, successful recovery in laboratory cultivation is more apt to be achieved through use of a dilute medium. For those organisms of common passage through treatment barriers and others recently introduced in soil contaminants during line repairs or in cross-connection events, utilizing any enriched medium formulation will result in rapid bacterial growth and early visualization. These conflicting purposes have resulted in the development of two different approaches to recommended media.

Standard plate count (SPC) agar (tryptone glucose yeast extract agar) and m-HPC medium (formerly known as m-SPC medium) are examples of enriched medium formulations.[22,23] Traditionally, most water plant laboratories have used SPC agar and have an extensive database on heterotrophic bacterial densities in their distribution system. In a national study on drinking water quality in 969 public water supplies, standard plate counts of ten organisms or less per milliliter were common in over 60% of those distribution systems that had a detectable chlorine residual.[24] With recognition that pour plate technology limited growth rates of obligate aerobes in agar and there was a heat shock effect from melted agar,[25] attention turned to surface cultivation of bacteria. Because the membrane filter technique provides for surface cultivation of entrapped organisms, there is better recognition of pigmented bacteria and the ability to analyze large sample volumes where densities below 20/ml are a frequent occurrence. Furthermore, the two methods produce essentially equivalent densities (Table 5.1) when compared with samples taken from process waters and distribution sites. This similarity in results would, over time, be indistinguishable and could be blended into the historical database on trends in water quality. In a more extensive evaluation of the membrane filter HPC method as a substitute for the standard plate count, data was collected biweekly for 1 year at two New England water treatment plants and associated distribution systems.[24] Again, the statistical analysis of the data from raw source water, partially treated water, finished water, and distribution water were equivalent.

While the m-HPC membrane filter method would provide a tie-in to records of treatment performance and disinfectant residual effectiveness in distribution, there is a need to capture more of the diverse heterotrophic population

Table 5.1   Comparison of the Bacterial Counts of Treated
Water Samples Using Standard Plate Count
Method and Membrane Filter m-HPC Procedure

| | | Bacterial level (count per ml) | | |
| | | | Membrane filter procedure | |
Date	Sample[a]	Standard plate count procedure	m-HPC[b] medium	Ratio SPC/m-HPC
6/22/77	Process water	1,800	3,200	0.56
	Process water	110	78	1.41
	Process water	520	480	1.08
6/29/77	Process water	2,400	3,100	0.77
	Process water	330	400	0.82
	Process water	270	1,300	0.21
	Process water	40	390	0.10
7/06/77	Process water	11,000	18,000	0.61
	Process water	710	370	1.92
	Process water	1,500	340	4.41
	Process water	650	140	4.64
	Drinking fountain (a)	35	19	1.84
7/13/77	Process water	1,300	10,000	0.13
	Process water	370	430	0.86
	Process water	760	450	1.69
	Process water	200	130	1.54
	Drinking fountain (a)	100	29	3.45
	Drinking fountain (b)	9,200	12,000	0.77
	Drinking fountain (c)	1,400	1,200	1.17
7/20/77	Process water	6,100	10,000	0.61
	Process water	6,400	7,000	0.91
	Drinking fountain (a)	1,200	310	3.87
	Drinking fountain (b)	27,000	30,000	0.82
	Drinking fountain (c)	1,900	1,800	1.06
			Geometric mean	0.99

[a] Process water samples were taken from a water treatment pilot plant after various
unit treatments including ozonation and carbon filtration.
[b] m-HPC medium was formerly known as m-SPC agar.
Table revised from Taylor and Geldreich.[23]

of bacteria in the distribution system. The reasons lie in achieving more subtle
trends in water quality deterioration in the pipe network that relate to corrosion,
taste, odors, and color and a broader range of opportunistic pathogens that are
fastidious in their nutrient requirements.

The use of R2A agar[26] is recommended as the medium of choice although
other media may be competitive when all factors are optimized. The R2A
formulation is more diverse in range of nutrients (proteose peptone no. 3,
casamino acids, glucose, soluble starch, and sodium pyruvate) with constituents being in more dilute proportions per liter. For example, component concentrations are either 0.3 or 0.5 g/l in R2A medium while constituents in SPC

agar or m-HPC medium range from 1.0 to 25.0 g/l. R2A medium also has a dedicated buffer system while the others do not, relying exclusively on high concentrations of ingredients to serve as a buffer.

Incubation Temperature and Time

Incubation temperature controls not only the spectrum of organisms that may be cultured, but also the rate of growth into colonies. The traditional incubation temperature of 35°C is unrealistic for organisms whose origins are not the intestinal tract of warm-blooded animals, but rather the aquatic environment. The optimum temperature for many heterotrophic bacteria is somewhat less than 30°C.[27] As a general rule, the detection of a greater portion of this population points to the use of 28°C as the optimum incubation temperature.[26,28]

Incubation time is interrelated with selection of temperatures for the visualization of colonies. Organisms in general grow much slower at 20°C than at 35°C (Table 5.2) so that while 2 or 3 d may be sufficient for adequate colony development at 35°C, 5 or 7 d is necessary to see developing colonies on media incubated at 20°C. For a few organisms, 7 d of incubation is not enough time to visualize these very slow-growing organisms. As a consequence, incubation for 28 to 30 d has been recommended[27] to achieve the maximum recovery of bacteria. This extensive time would make the protocol impractical for routine monitoring of many sites in a distribution system and does not substantially change data interpretation. There is also the serious problem of moisture loss in the medium and potential air contaminant occurrence during this extended holding time that will adversely impact cultivation unless strict requirements for moist, sterile air circulation in the incubator are met.

In Situ *Placement*

In situ placement of organisms on medium may also influence recovery in several ways. The pour plate technique often leads to lower bacterial colony counts because of the selective effects of heat shock from melted agar temperatures.[25,30,31] In preparation of the pour plate, the agar medium is melted in a boiling water bath or in flowing steam not exceeding atmospheric pressure. This results in a melted agar temperature of over 50°C that must be tempered to between 44 and 46°C before mixing with the measured sample volume. Unfortunately, this range of temperatures will still cause a significant loss of some organisms in the heterotrophic population.[25] Heat effect as well as shock effect on the bacterial population can be seen in Table 5.3.

The surface plate method (spread plate) circumvents these problems because the inoculum is spread over the surface of a solidified agar plate.[22] While spread plates give higher counts, pour plate replicates yield greater precision between replicates. Much of this discrepancy occurs in the spreading

Table 5.2 Effect of Incubation Temperature and Time on MPC Recoveries[a]

Temp (°C)	Medium and method	Incubation time (d)			
		2	4	6	7
20	SPC-PP	22	130	570	900
	R2A-SP	90	1100	4700	6100
	R2A-MF	75	650	3000	4900
	M-HPC-MF	48	400	1600	2000
28	SPC-PP	90	640	950	1000
	R2A-SP	360	2800	6700	7200
	R2A-MF	160	2200	3500	4000
	M-HPC-MF	140	1000	1700	1900
35	SPC-PP	22	100	110	115
	R2A-SP	200	340	500	510
	R2A-MF	41	200	270	280
	M-HPC-MF	32	140	150	150

Note: PP = pour plate; SP = spread plate; MF = membrane filter.

[a] Distribution samples.

Data revised from Reasoner.[29]

Table 5.3 Effect of Agar Temperature and Incubation Time on Recovery of Heterotrophic Microorganisms from an Aquatic Sample

Standard methods procedure (SPC agar)	Incubation time (d, at 35°C)				
	2	4	7	14	21
Spread plate	110 ± 31[a]	130 ± 38	200 ± 40	250 ± 75	300 ± 61
Pour plates					
42 C agar	33 ± 11	59 ± 13	76 ± 16	82 ± 14	84 ± 15
45 C agar	23 ± 6	50 ± 9	62 ± 15	78 ± 17	79 ± 17
50 C agar	12 ± 2	32 ± 4	43 ± 7	55 ± 6	58 ± 9

Note: All densities × 10^{-2}/ml.

[a] Standard deviation.

Data from Klein and Wu.[25]

technique. While the glass rod spreads the water sample over the agar surface to achieve dispersion, some organisms remain attached to the spreading device and are lost to the cultivation process. Because this is not a uniform loss, some variation in replicate spread plates is unavoidable.

The membrane filter procedure provides another approach to surface cultivation of organisms. The unique advantage of this technique is that it permits the analysis of larger sample volumes of high-quality water containing too few organisms to be detected by a 1-ml sample portion. The only restrictions to the size of sample analyzed are turbidity and colony growth on the filter surface. Fine sand particles as turbidity are not as serious a problem as colloidal clay or organic debris. These latter particulates plug the filter pores and inter-fere with discrete colony development. Colony size and density of organisms growing on the effective filtration surface area are also limiting restrictions. Because the diameter of a standard membrane filter is 47 mm and the effective filtration surface somewhat less, the density limit for discrete colony develop-ment will be less than 200 colony-forming units (c.f.u.).

Some indication of the impact of MF surface cultivation in comparison with pour plates and spread plates can be seen in Table 5.2. While recovery densities are significantly better than the traditional pour plate method because of temperature effect in the pour plate method,[32] spread plate results still appear to be the method of choice for optimum recovery of heterotrophic bacteria. Probably colony confluence and colony densities over 200 c.f.u. on a small MF surface is the major drawback in achieving equivalent results of the spread plate technique.

COLIFORM DETECTION

While some form of coliform standard has existed since 1912, its defini-tion, methodology, and data interpretation have changed over the years. In general, the test has provided a practical measure of treatment effectiveness, but at times a questionable indication of pathogen occurrence. Part of this weakness may be due to poor detection of stressed coliforms, growth inter-ference by other members of the heterotrophic bacterial population, and reluc-tance of the utility laboratory to analyze one-liter samples rather than limiting examination to 100-ml test portions.

REGULATION REQUIREMENTS

Despite these concerns, the federal drinking water regulations continue to recognize total coliforms as the primary microbiological health-associated criterion for public water supply quality.[33] Historically, the choice of total coliforms as a health effect criterion was related to the sanitary quality of water. More recently, the search for total coliforms has converged on demon-strating treatment effectiveness and water quality characterization in the dis-tribution system.

QUANTIFICATION VS. PRESENCE–ABSENCE

Previous use of a quantitative limit (an arithmetic average of less than one coliform per milliliter) provided opportunities to drive the monthly average down to an acceptable level by intensifying the sampling during the monitoring period. By contrast, the new regulations focus only on the presence or absence of coliforms in all samples examined during the month and limit the frequency of positive sample occurrences within that period of time to 5%. This approach is more sensitive to coliform occurrences and is more difficult to subvert by intensifying the sampling program. Consequently, the pattern of coliform occurrences attributable to chronic leakage in a treatment barrier or to a summertime biofilm growth event can immediately place the water utility in a noncompliance situation.

The presence–absence (P–A) concept involves placing equal emphasis on all positive samples regardless of density, with a limit defined by a specific percentage of positive coliform occurrences allowed during a reporting period. While the federal regulation has specified a universal limit of 5% coliform occurrence, the concept could be used by a state agency to grade utility performance differently. For example, for water systems using conventional treatment of surface waters, 5% coliform occurrence would be the maximum occurrence permissible while 3% might be a realistic maximum occurrence for surface treatments using only sand filtration and disinfection. For untreated groundwater supplies, the tolerance should be restricted to 1%, the rationale being that with less barrier defense, soil layer protection must be at a higher level of effectiveness.

The use of a simplified P–A approach to the examination of drinking water was first proposed in the late 1930s.[34] More recently, a P–A procedure has become extensively used in the Province of Ontario, Canada.[35] This newer P–A method also provides an opportunity to test for the occurrence of several indicator groups from the same culture,[35] which could be useful in further characterizing water quality in the sample.

Several approaches to achieving a P–A determination in drinking water are available. The specific P–A test described in the Standard Methods[22] is a basic modification of the multiple fermentation tube procedure and both require the same confirmation technique. Unlike the conventional multiple fermentation test that uses 5- to 10-ml test portions, the P–A procedure involves a single 100-ml sample portion. This procedural change immediately reduces the amount of processing necessary in the analysis and also provides a determination on a larger sample.

Another approach that has recently received attention is the P–A test known commercially as Colilert or Colisure, depending on the manufacturer. In this P–A test, the 100-ml sample is mixed in a bottle containing a chromogenic substrate for the detection of coliform enzymes.[36] Unlike the fermentation methods described above, growth of many other organisms is inhibited, thereby providing more selectivity and specificity for coliform metabolic activity that attacks the substrate, releasing a chromogen (color agent) in the culture after 24 h of incubation.

For many laboratories, the membrane filter technique has been used routinely over the past 25 to 30 years to examine water quality in the distribution system. Consequently, there is a substantial database available on occurrence and density of coliforms in many public water supplies. The ability to make data conversions (presence/absence or coliform density) provides the flexibility to satisfy the new coliform regulations and furnish information on the magnitude of coliform densities in a given coliform event. As another feature, the membrane filter analysis may be performed on larger samples; this is particularly appropriate for use in a search for leakage in the treatment barrier.

How equivalent these methods are with each other in detecting coliform occurrences in water supplies lies in the gathering of a sufficient database. Such comparative data is especially important for the monitoring of potable water, where the vast majority of samples contain few or no coliforms.[36,37] It is imperative, therefore, that any new procedure be shown to detect very low numbers of coliform bacteria at a level of sensitivity equal to that of other procedures acceptable by federal regulations.

Several research reports suggest that the P–A test is at least as sensitive as the multiple-tube fermentation test and the membrane filter procedure[37,38] while two other investigators suggested that the P–A test was more sensitive.[39,40] Part of the difference in these studies may relate to the size of the database that was used to insure that the procedures were statistically comparable. Geographical differences in water supply that might produce different microbial floras and subtle dissimilarities in water chemistry were investigated via comparative data gathered from four areas: New England (Vermont and New Hampshire), Pennsylvania, Oregon, and Hawaii.[37-40] The data summarized in Table 5.4 and analyzed elsewhere[41] indicates that the P–A test possesses the sensitivity required for the detection of total coliform bacteria in a variety of potable waters. Combined recoveries showed the P–A test detected a significantly higher number of positive samples than either the multiple-tube fermentation test or the membrane filter method.

Table 5.4 Comparative Results of the P–A, FT,
and MF Tests in Drinking Water
Samples Collected from Four Different
Geographic Areas

Area	Total samples	Positive results (%) by test		
		P–A	FT	MF
New England	1483	23.4	8.4	6.3
Pennsylvania	2601	23.2	—	26.7
Oregon	1560	34.5	19.6	15.1
Hawaii	200	48.3	8.8	20.5
Average		32.4	12.3	17.2

Were the positive cultures in the P–A test a reflection of coliform occurrences since the other two tests were negative? Data in Table 5.5, which identifies the organisms in the positive P–A test, confirms the nature of these results to be a true indication of coliform occurrences. What this difference in sensitivity has demonstrated is that the P–A test medium may be more

Table 5.5   Organisms Isolated from Positive P–A
Tests when the MF and MPN (FT)
Tests Were Negative

Isolate[a]	No. of times isolated
Citrobacter freundii	8
Enterobacter agglomerans	6
Klebsiella pneumoniae	5
E. cloacae	3
E. aerogenes	2
K. oxytoca	2
K. ozaenae	1
Escherichia coli	1
Serratia plymuthica	6
S. fonticola	1
S. rubideae	1
S. oderifera	1
Habnia alvei	1

[a] Identified according to the profile numbers determined by API.

Data from Jacobs et al.[39]

effective in the recovery of stressed coliforms than the lauryl tryptose broth
or m-Endo medium used in the multiple-tube fermentation test or the mem-
branè filter procedure. Key ingredients in P–A broth are more dilute (50%
less) than that found in lauryl tryptose broth. Furthermore, P–A broth does
not contain the essential selective agents (basic fuchsin and sodium deoxycholate) necessary for Endo type media performance in a membrane filter
culture. Both of these compounds may retard the recovery of any stressed
coliform bacteria in a treated water supply sample.

COLIFORM CELLULAR STRESS

Coliform bacterial stress is a result of any alteration of the normal metabolic process, specifically through loss in enzyme activity, degradation of
ribosomal ribonucleic acid (RNA), single- and double-strand breakage in
deoxyribonuclease acid (DNA), or leakage of cytoplasmic constituents. Any
of these alterations in cell structure, permeability, and biosynthetic characteristics will not necessarily kill the organisms, but necessitate extended time for
cellular repair before growth and multiplication can proceed again. Stressed
coliform detection should be of concern in water treatment operations because
they illustrate inadequacies of treatment barrier processes that might provide
passage for pathogenic agents.[42] Some evidence of this problem can be seen
in Table 5.6. These significant levels of nonfatal injury may originate in source
waters contaminated by heavy metals, phenols, water acidity, and ultraviolet
irradiation.[43-53] Other situations may be created in water treatment as a

Table 5.6 Injured Coliforms in Drinking Water

Location/date	Type of water	No. of samples	Mean coliforms per 100 ml		Injury (%)	False negatives (%)[a]
			m-Endo	m-T7		
Midwest U.S./1986	After filtration	9	1.1	3.9	71	55
Midwest U.S./1986	Distribution	13	2.4	7.9	69	23
Northeast U.S./1985–86	Distribution	86	1.9	4.8	64	54
Northeast U.S./1986–86	Raw	86	14.5	16.8	14	—
Northeast U.S./1985	During treatment	320	1.4	1.9	26	24
Caribbean/1985	Cisterns	13	15.2	20.5	26	—
East U.S./1986	After treatment	4	2.0[b]	209.0	99	100
Northwest U.S./1984–86	Small systems	552	62.0[c]	139.0[c]	55[d]	—
Northwest U.S./1984–86	Small systems	45	3.1	6.3	51	—
Northeast U.S./1986	After chlorinated backwash	7	0.1	1.5	92	14
West U.S./1986–87	After unchlorinated backwash	133	2.4	8.1	70	40
West U.S./1986–87	After chlorinated backwash	37	0.03	13.4	99	97
Southeast/1986	Raw (surface)	24	598.0	2828.0	79	—
Southeast/1986	Well (mineralized)	51	254.0	505.0	49	—
Southeast/1986	Well	122	2.0	3.0	33	—
Southeast/1986	Distribution	280	1.0	4.0	75	—

[a] False negatives represent the percentage of coliforms that failed to produce colonies of m-Endo medium but were enumerated on m-T7.
[b] MPN values.
[c] Values are percent positive for coliforms.
[d] Estimate.

Data set from McFeters.[42]

consequence of inadequate exposures to disinfectant C·T values, copper service lines, and antagonistic responses by select heterotrophic bacteria in excessive densities.[44,45,54-59] Furthermore, these bacteria subjected to sublethal inactivation become hypersensitive to the secondary stresses induced by selective cultivation in the laboratory as noted previously. As a result, lag time in normal growth rate may range from 2 to 5 h, depending on the stressful situation.

Secondary stress caused by laboratory processing of the sample should be considered the first line of defense against undetected passage of injured but viable coliform bacteria. This concern starts with sample collection. There must be sufficient dechlorinating agent in the sample bottle to quickly stop disinfection action. Additionally, the sample should be kept cool during transit to prevent accelerated growth of interfering, antagonistic organisms. Samples that are not processed promptly once received in the laboratory are subject to significant changes in the predominance of organisms in the microbial flora, possibly obscuring the metabolic response of stressed coliforms.

As noted previously, selective agents used in routine media may contribute to secondary stress that desensitizes the detection of injured coliforms unless the standard incubation time is extended for 2 to 4 h or a resuscitation step is used in preliminary cultivation[22] prior to transfer to the selective medium with surface active ingredients, such as deoxycholate.[60,61] A more promising approach is to use a medium such as m-T7 formulated specifically[62] for the improved recovery of stressed total coliforms in drinking water.[56]

Because of the occurrence of stressed coliform cells in treatment processes, water entering the distribution system or water from static areas of the distribution system is never constant. The laboratory should consider a seasonal search of plant effluent during winter (low temperature operations) and of distribution static water areas during summer when heterotrophic bacterial populations exceed 1000 organisms per milliliter. In either situation, coliform detection may be impared.

VERIFICATION

Verification of every coliform result is a good laboratory practice that eliminates all doubts that the findings are perhaps caused by some false-positive reaction in the test. In the multiple-tube fermentation test, all presumptive positive tubes are confirmed in brilliant green lactose broth for gas production. The P–A test also involves confirmation of the positive culture bottles in brilliant green lactose broth. Because the membrane filter method for coliform detection is based on aldehyde release in lactose metabolism, positive colonies are subjected to lauryl tryptose broth and brilliant green lactose broth in a demonstration of gas production. The hydrolyzable chromogenic substrate medium used in coliform tests should also be subject to the verification process. There are two concerns that justify the verification process for chemical defined media: (1) an occasional water may yield a false-positive reaction from high densities of *Aeromonas* and *Pseudomonas* species and (2) the patented nature of the formulation provides no alert to changes in medium ingredients that may alter the spectrum of detected organisms.

SPECIATION

Coliform speciation is a more in-depth verification process that has on occasion been used to demonstrate the equivalency of methods. Based on several comparative studies of coliform identities in positive cultures from three procedures (Table 5.5 and Table 5.7), it becomes apparent that there is no bias as to species detected by any of the recommended methods. These studies have also increased awareness that the total coliform population is a heterogeneous mix of bacterial species that are capable of lactose utilization.

Table 5.7   Frequency of Total Coliform Species Isolated
from the Multiple Tube Fermentation Test
and Colilert

Species	% of all isolates[a]	
	Multiple-tube fermentation test	Colilert
Klebsiella pneumoniae	31	28
Enterobacter agglomerans	19	16
Citrobacter freundii	16	12
E. cloacae	10	11
K. oxytoca	3	6
Enterobacter species	3	7
E. aerogenes	1	1
Escherichia coli	1	1
Serratia plymuthica	3	4
S. fonticola	1	1
S. rubidaea	3	4
S. odorifera	2	3
Hafnia alvei	1	1
CDC groups	3	4
Unidentified Enterobacteriaceae	3	3

[a] All isolates confirmed in BGLB.
Data from Edberg et al.[63]

Speciation as a coliform verification tool has its greatest use in the identification of a biofilm event and in further confirmation of the existence of a microbial problem in water supplies. Perhaps the occurrence is related to a momentary short-circuiting of the disinfectant C·T requirement that represents a breakdown in the treatment barrier or a reflection of a line repair contamination event. In either case, further exploration of the original sample may provide information on fecal contamination that might not be present in a repeat sample collected the next day. Most often, repeat samples are negative because the contaminating event is of short duration.

The laboratory staff should never lose sight of the basic requirement to isolate and purify coliform strains from the positive broth culture. This purification step is also important in the identity of membrane filter colonies, particularly when sheeny colonies are not discrete or are surrounded by numerous noncoliform colonies. Some noncoliform colonies may form a transparent spreading growth on the moist surface of the membrane filter culture, mixing with developing coliform colonies. As a consequence, inoculation of any of the commercial biochemical test kits with mixed bacterial species will result in erroneous metabolic responses, making identification of species impossible.

Commercial Speciation Kits

Commercial speciation kits are acceptable with few confusable identities; however, not all aquatic organisms will provide sufficient growth to generate a metabolic response. In some instances, organisms isolated may be too fastidious in their growth requirements and not respond to the incorporated nutrient base medium. Such was the case in a study of heterotrophic bacterial populations encountered in water distribution systems.[64] In this investigation, bacteria initially recovered on standard plate count agar from drinking water would not subculture unless a low-nutrient medium (R3-A) was used. Furthermore, growth and reaction on API-20E was often difficult or negative for these strains. While this problem has the greatest impact on noncoliform strains in the heterotrophic bacterial population, some species identification difficulties could be a possibility with stressed coliforms and those strains that have receded into the ultramicrobacteria phase of survival in the distribution pipe network.

Commercial test kits for speciating coliform bacteria and numerous other gram-negative organisms have become a useful tool in the water plant laboratory. This interest is largely the result of the need to identify species in a coliform occurrence that may be a fecal contamination event or a biofilm release in the water supply. Speciation of bacteria requires information on metabolic activity in a variety of biochemical tests, and as a consequence of unanticipated demand for these media, most laboratories have resorted to the use of commercial test kits that are readily available and have a substantial shelf life.

Because information obtained from coliform species identification is critical to decisions to be made on the sanitary quality of drinking waters, it is important to recognize the accuracy of commercial kit systems. Two of the most popular test systems used in water plant laboratories are the API-20E and Enterotube II. In a recent evaluation, three laboratories used a set of fresh cultures from 40 selected recent aquatic and clinical isolates to evaluate the efficacy of these two biochemical test systems.[65] Results of this comparison are given in Table 5.8. Overall, API-20E correctly identified 83% of the organisms and the Enterotube II, 70%. The majority of misidentifications were among noncoli *Escherichia* strains, specifically *E. hermanii, E. fergusonii,* and *E. vulneris.* These species were not detected as either fecal coliforms or total coliforms by traditional coliform methods (Table 5.9); however, they were detected as total coliforms only by the defined substrate system. The presence of these *Escherichia* species is not necessarily indicative of fecal contamination, but should be considered members of the total coliform group of bacteria.[66]

Table 5.8 Evaluation of Two Commercial Biochemical Test Kits by Three Laboratories

Aquatic and clinical stock cultures	No. tested	No. of strains correctly identified					
		API-20E			Enterotube II		
		Lab 1	Lab 2	Lab 3	Lab 1	Lab 2	Lab 3
Citrobacter freundii	1	1	1	0	1	1	1
Escherichia coli	2	2	2	2	2	2	2
E. coli 0157:H7	2	2	2	2	2	2	2
E. fergusonii	10	9	8	8	5	7	8
E. hermannii	10	8	9	7	9	9	10
E. vulneris	10	9	9	8	4	5	6
Enterobacter cloacae	1	1	1	1	1	1	1
Klebsiella oxytoca	1	1	1	1	1	1	1
K. pneumoniae	1	1	0	1	1	1	1
Shigella dysenteriae	1	1	1	1	1[a]	1[a]	1[a]
S. sonnei	1	1	1	1	1	1	1

[a] Correct to genus.
Data from Rice et al.[65]

Table 5.9 Results of Confirmed Total Coliform Analysis in Carbohydrate-Based and Defined-Substrate Media

Organism	No. tested	No. positive			
		MF	LTB	PA	DS
Citrobacter freundii	1	1	1	1	1
Escherichia coli	2	2	2	2	2
E. coli 0157:H7	2	2	2	2	2
E. fergusonii	10	0	0	0	10
E. hermannii	10	2	2	2	10
E. vulneris	10	0	0	0	19
Enterobacter cloacae	1	1	1	1	1
Klebsiella oxytoca	1	1	1	1	1
K. pneumoniae	1	1	1	1	1
Shigella dysenteriae	1	0	0	0	0
S. sonnei	1	0	0	0	0

Note: MF, m-Endo LES agar; LTB, lauryl tryptose broth; PA, presence–absence broth; DS, defined substrate.
Data from Rice et al.[65]

SUBSTITUTE CRITERIA: A WORD OF CAUTION

Adequate monitoring of public water supplies has always been a concern for a number of reasons. Major complaints have included the difficulty in demonstrating a disinfectant residual in 95% of the sampling sites in the distribution system and collecting a minimum number of coliform samples per month to satisfy the federal regulations. Solutions to these problems have been proposed using a concept of criteria substitutions that are innovative in concept but need to be placed in proper perspective with a word of caution.

LOW-DENSITY HPC CREDIT

Federal regulations require that a detectable disinfectant residual must be demonstrated in at least 95% of all distribution system sites where bacteriological samples are collected. For some distribution systems this may be difficult to achieve because of chlorine demand in the water and long residence time in the pipe network. In this situation, the impact of transient free chlorine residuals may still be in evidence through the presence of a minimal heterotrophic plate count population. Based on a national survey of 969 community water distribution systems in nine metropolitan areas throughout the United States, it was concluded that a density of less than 500 HPC per milliliter could be achieved in the distribution system of numerous utilities that were able to maintain a measurable chlorine residual in their pipe network.[67]

Some water systems have resorted to chloramination to maintain a more persistent disinfectant residual in the extremities of the distribution system. While these chloramines may be more persistent, they are slower-acting disinfectants than free chlorine. Furthermore, excessive application of ammonia to convert free chlorine to chloramines may encourage the survival of a higher density of heterotrophic bacteria, particularly in static water zones. These factors will eventually lead to a condition where summertime populations of heterotrophic bacteria will provide little opportunities for the substitution of low-density heterotrophic plate counts in lieu of detectable disinfectant residuals to meet the 95% requirement in the regulation.[33]

Ozone is occasionally used in water treatment because it is a powerful oxidant of organics. Unfortunately, ozone is rapidly dissipated in water so that a secondary disinfectant (free chlorine, chlorine dioxide, or chloramines) must be applied to the treated water entering the distribution system to satisfy the need for disinfectant residual protection from microbial quality deterioration in the pipe network. Pilot-plant and field studies indicate that ozonation creates more assimilable organic carbon in the oxidation process as a result of the conversion of long-chain organics such as natural humics. If these reaction products are not captured in a biological activated carbon process, the net result is the passage of a nutrient base into the pipe network where established organisms may regrow in the area of static, warm water. Here again, the ability to substitute low-density heterotrophic plate counts at sites with no detectable chlorine residual may not be possible.

RESIDUAL CHLORINE DETECTION CREDIT

An attempt to substitute a detectable disinfectant residual for 75% of the coliform monitoring requirements in water systems was proposed in the National Interim Primary Drinking Water Regulations of 1976.[68] The intent was to get better monitoring coverage for those water systems that routinely did not meet requirements for submission of specified numbers of samples each month or had a poor bacteriological quality because disinfection was not used or only superficially applied in a token dosage.

This regulation position was developed from two separate field investigations. The first study involved a "snap-shot" of bacteriological quality characteristics for 969 water supply distribution systems nationwide.[69] In essence, most of these systems were found to not meet the requirements for taking a minimum number of bacteriological samples each month. Additional support was drawn from another study that involved the Cincinnati water distribution system.[70] From this second investigation it was concluded that maintenance of a free chlorine residual of 0.2 mg/l reduced the frequency of coliform detection to about 1% in the distribution system.

Any partial substitution of a chlorine residual test for some portion of the monthly bacteriological sampling requirements must be viewed with a great amount of caution. This approach to monitoring should never be adopted automatically to cover all chlorinated water systems or accepted as an escape mechanism for those water supplies that do not meet the minimal bacteriological sampling frequency. Rather, the intent of such a precarious strategy should be to selectively shift analyses from coliform testing to disinfectant residual measurement at some sites that have a continuous history of good bacteriological record. At the same time, coliform monitoring should be redesigned to include a like number of substitute samplings into areas where disinfectant residuals are difficult to maintain and intermittent coliform releases may occur. The intent is to search out and destroy incipient biofilms and the localized loss of microbial quality of water supply in zones of static water occurrence. Under these circumstances the disinfectant residual test is highlighted as a desirable supplemental test but not a substitute for the required number of coliform tests to be done each month in the monitoring program.

One of the most significant disadvantages with disinfectant residual substitution for coliform testing is during a coliform biofilm event. In this case, coliforms may often be found in samples taken from sites with a chlorine residual. As a precaution, disinfectant residual substitution for coliform testing should be abandoned and the search for possible fecal contamination in the system intensified. Obviously, this chlorine substitution policy could not apply to protected groundwater systems that do not practice disinfection.

NUISANCE ORGANISMS

Water utilities are, at times, plagued with a variety of microbe-induced problems that have little known health significance but contribute to treatment difficulties or impart bad taste, odor, and color to the water at the consumer's tap. Such organisms are referred to as nuisance organisms. Operational difficulties associated with nuisance organisms include short filter runs due to algal bloom impediments to the filtration process or infestations by nematodes that slough into the distribution system, tuberculations in water pipe and clogging of service meter screens by various biological debris. Most consumer complaints are a reflection of the activity of nuisance organisms that may cause an off-taste, odorous, or discolored water to appear at the tap.

CHARACTERIZING NUISANCE ORGANISMS

Many organisms have been identified in this general classification, including algae, bacteria, fungi, copepoda, nematodes, sowbugs, midge larvae (bloodworms), snails, and mollusks.[71,72] These organisms frequently originate in surface waters, pass through treatment systems using disinfection as the only barrier, or infest filter beds and are released during unstable filter conditions following backwashing. While groundwaters are generally free of many of these organisms, iron and sulfur bacteria can be a serious cause for biofouling and microbial plugging at the point of extraction.[73]

As a consequence of increasing pressure from federal regulations to include filtration of surface waters, there will be less passage opportunities for algae, diatoms, and higher aquatic organisms to the distribution system where they may become a nuisance in water transmission. The import of this is that nuisance organisms will more often be confined to bacterial activities in slow-flow areas and dead ends of the pipe network. Some of the bacteria involved in the cause of stale-tasting, odorous, or colored waters include *Pseudomonas, Alcaligenes, Flavobacteria*, several strains of *Bacillus (B. cereus, B. subtilis*, and *B. firmus)*, and the iron, sulfur, and manganese bacteria. Of these organisms, perhaps the iron, sulfur, or manganese bacteria represent the nuisance groups of greatest impact because of their involvement in pipe tuberculations and highly colored waters that may appear at taps that are infrequently used. To the consumer, the by-products of iron-, sulfur-, and manganese-utilizing bacterial activity signify a water supply that is unpleasant in terms of aesthetics and affects food preparation, bathing, and washing of clothes. To the water supply purveyor, these complaints should be viewed as a signal of prolonged static water conditions in some area of the pipe network, biofilm expansion, and significant pipe corrosion.

Biofouling Iron-Precipitating Bacteria

Biofouling iron-precipitating bacteria in a biofilm consortium produces hard plaquelike formations and tubercles in iron pipe, thereby reducing water flow. In other situations, colonization of these organisms can also provide slimes released at the tap, although other nuisance bacteria, previously cited, are often the predominant slime producers. Active biofilms are not only located in pipe, but can also be found on the interior parts of valves and fire hydrants as well as the internal areas of storage tanks. Eventually, the biological activity leads to localized odors, color, increased particulates, and unpleasant tastes. Some indication of the magnitude of a nuisance biofilm community can be realized from the report of 10 million iron bacteria per milliliter at a site where the water contained a dense red sediment.[74]

These specialized bacteria are a heterogeneous collection of organisms, some of which produce sheaths of oxidized iron around the chain of cells. Examples are *Crenothrix polyspora* and *Sphaerotilus natans* (sheath without iron deposits). Others form a stalk of cells with iron deposited extracellularly. Genera in this group include *Gallionella, Hyphomicrobium*, and *Caulobacter.* The single-cellular iron bacterium *Siderocapsa treubii* should also be recognized.

Community association is also critical, for these organisms are fastidious and difficult to cultivate in pure culture. Frequent association of *Sphaerotilus* with *Gallionella* and the difficulty in separating *Aeromonas* from *Gallionella* argue for the existence of some kind of commensal relationship in a consortium of mutual benefits.[75] With this evidence of total biofilm community involvement rather than the independent aggressiveness of a single species, it becomes important to cultivate the mixed flora whenever possible.

Corrosion Induced by Sulfate-Reducing Bacteria

Corrosion is induced by sulfate-reducing bacteria not only on the soil side of buried pipe, but also as part of the bottom layer of biofouling consortia in pipe tuberculations. In a biofilm consortium, the upper layers of the biofilm contain a variety of bacteria, some of which provide the polysaccharide matrix barrier to dissolved oxygen and the inactivation potential of a disinfectant residual.[74] In this anaerobic microenvironment, sulfate-reducing bacteria thrive, utilizing short-chain organic acids that might be released by the metabolic activity of other organisms.[77] Sulfate-reducing bacteria are obligate anaerobes and their survival and mechanism for corrosion relies on interactions with other bacteria in this situation. With utilization of the available organic nutrients in the anaerobic environment, hydrogen sulfide is formed with release of associated foul odor. Oxidation of ferrous sulfide in this microenvironment releases excesses of hydrogen, which, in turn, enhances the biomediated corrosion rate of iron pipe.

Sulfur-reducing bacteria are a group of wide diversity. Characteristics include being anaerobic single-cellular, multicellular, photosynthetic green and purple pigmented, filamentous, or aerobic sulfur oxidizers. Most common sulfur-reducing bacteria in pipe corrosion include *Desulfovibrio, Thiobacillus, Beggiatoa, Thiodendran,* and *Thiothrix.* Microscopic observation of *Beggiatoa* will show the organism to be filamentous with cells filled with refractile globules of sulfur.

Black Water Caused by Manganese-Utilizing Bacteria

Black water caused by manganese-utilizing bacteria may also create aesthetic problems with water quality. This problem is usually associated with groundwater although high levels of manganese can be found in some stratified reservoirs of stored surface water.[78] Again, microbial activity in an established biofilm situated in the pipe network is the cause of manganese oxidation and a "dirty water" condition. Organisms in the biofilm that utilize manganese (>0.01 to 0.5 mg/l or above) include *Pseudomonas, Arthobacter, Hyphomicrobium,* and *Sphaerotilus discophorus.*[79] The resultant oxides of this metabolic activity coat the pipe walls to form a ripple deposit across the direction of flow, which eventually reduces water flow.[80]

DETECTING NUISANCE BACTERIAL OCCURRENCE

Monitoring zones of slow flow and areas of frequent customer complaints should be examined on a seasonal basis for significant changes. By gathering a database on sensitive locations and changes in activity level, it should be possible to anticipate areas of incipient biofilm activity and provide early indications for flushing critical areas in the pipe network. A P–A test approach might be the most practical preliminary screening approach for nuisance organism colonizations.

Recently, a simple field kit has become available for determining the presence of some of the groups of nuisance bacteria such as iron bacteria, sulfate-reducing bacteria, and slime-forming bacteria. The commercial test kits are known as BART, which is an acronym for biological activity and reaction test.[70] What is unique with this system is the innovative design that permits the development of anaerobic conditions in culture — a necessity for the cultivation of many nuisance bacteria. This condition is created by a plastic ball in the culture tube that restricts air passage to the bulk of the medium and sample mix. Incubation is done at room temperature and culture response is noted over a period of several weeks. While these individual group systems provide only a presence or absence result, the location and frequency of occurrence for nuisance bacteria in the pipe network will provide clues to problems in the pipe environment that need attention.

Another approach involves the establishment of a heterotrophic profile that delineates possible dominance of nuisance organisms in specific locations. While there is no established test for such a monitoring program, use of a comparative heterotrophic plate count procedure may provide the necessary information. Using R2A medium as the base for comparison, a second R2A medium modified by adding 112 mg/l ferric ammonium citrate would provide differentiation for iron bacteria.[81] Modified R2A pour plate cultures are then incubated at 20°C for 2 to 4 weeks (with care taken to prevent agar desiccation) because these organisms are slow to develop orange colored colonies. Cultivation in an anaerobic incubator or the application of an agar overlay to reduce available oxygen may improve the recovery of sulfur-reducing anaerobic bacteria. Parallel spread plates of unmodified R2A (28°C incubation for 7 d) would become the base to define the general heterotrophic population because ferric ammonium citrate and anaerobic culture conditions can be toxic to some subpopulations in the pipe biofilm. Ratios of these two populations would characterize the magnitude of nuisance bacteria in the biofilm area.

Cultivation of the manganese-oxidizing bacteria could be accomplished in a similar fashion. In this situation, R2A medium would be modified by the addition of 50 mg $MnSO_4 \cdot H_2O$ per liter of the original R2A medium. Manganese-oxidizing bacteria will produce brown or black colonies[79] on this medium. Further modification of suggested media formulations may be necessary to achieve optimum recovery of these slow-growing subpopulations among the heterotrophic bacteria.

Microscopic examination of presumptive cultures either in the BART kit, the colonies on a modified R2A medium, previously described, or from a suspect water sample may provide some additional information about the general type of organisms involved. While the organisms in a culture or colony will be in great abundance, the suspect water samples present a need for concentration. Centrifugation of a 10-ml sample (4000 rpm for 10 min) to create a pellet of material for microscopic examination is desirable. Minimal centrifugation is desirable to diminish distortions in cell morphology, making identification more difficult. Scans of the mixed cultures, colonies, or pellet material should reveal filamentous or stalked bacteria, often with deposits of iron or sulfur as described in the standard methods.[59]

NEW CONSTRUCTION AND LINE REPAIRS

Every distribution system is a growing, aging entity that must be carefully managed to maintain the integrity of the pipe network for delivery of a safe water supply. Many public water utilities have a long-term commitment to expand the distribution system to keep pace with continuing suburban growth. These projected plans not only include miles of new pipeline, but also water supply reservoirs in new areas of accelerated population growth to sustain supply under pressure. Urban renewal and highway construction projects may also provide opportunities for reconfiguration of portions of the distribution system. Corrosion, unstable soil, pipe faulting, land subsidence, construction/demolition vibrations, and extremely low ambient temperature often cause line breaks and necessitate water main rehabilitation.

While line repairs are generally done by utility staff, new construction is more often performed by contractors whose business is laying pipe or building water storage reservoirs. In either case, there must be a defined protocol for achieving an acceptable bacteriological quality through the new construction and water main repairs. To achieve this objective, the American Water Works Association[78] has identified six areas of concern: (1) protection of new pipe sections at the construction site; (2) restriction on the use of joint-packing materials; (3) preliminary flushing of pipe sections; (4) pipeline disinfection; (5) final flushing of concentrated disinfectant residuals; and (6) bacteriological testing for pipe acceptance.

PREVENTIVE MEASURES DURING CONSTRUCTION

Pipe sections, fittings, and valves stockpiled in yard areas or at the construction site should be protected from soil, seepages from stormwater runoff or sewage leakages into the trench, and habitation by pets and wildlife.[83,84] Each of these contamination sources may deposit significant fecal pollution in the interior of pipe sections awaiting installation. Septic tank drain fields, subsurface water in areas of poor drainage or areas of high water table, and flash flooding may also introduce significant contamination into unprotected pipe sections. Excrement transmitted by these contamination sources may become lodged in pipe fittings and valves. Thus, such sites become protected habitats from which coliforms and any associated pathogens in the impact material may be shed into the water supply. Commonsense protective measures include end covers for these pipe materials, drainage of standing water from trenches, and flushing of assembled pipe sections to remove all visible signs of debris and soil.[85-89]

Gasket seals of pipe joints can be a source of bacterial contamination in new pipes.[89] Angular spaces in joints provide a protected habitat for continued survival and possible multiplication of a variety of bacteria in the new pipe section.[90] In these instances, although the standard plate count and any coliform occurrences may be temporarily reduced by main disinfection, bacteria soon become reestablished from the residual population harbored in particulates packed in annular spaces. Research has shown that kemp, jute, and cotton yarn should not be used as packing materials because they are difficult to disinfect and provide nutritive support for bacteria.[91-95] Nonporous, nondegradable materials such as molded or tubular plastic and rubber are preferable. Lubricants used in these seals must also be nondegradable in order to avoid bacterial growth in joint connections. Efforts to develop bacteriostatic lubricants have resulted in the production and use of quaternary ammonium compounds that minimize contamination at pipe joint spaces.[88,96]

Pipe repairs and new line construction may also serve as entry locations for contamination into the distribution system. While acceptance of repaired mains and new lines often depends only on a routine laboratory report of no coliform detection in water held overnight in the pipe section, a more rigorous check on pipe cleanliness should be done. This special program should include a search for injured coliforms that may not be detected by the usual coliform methodology, particularly since disinfection is used to sanitize the pipe work. An example of this situation can be seen in data reported in Table 5.10 for a test done immediately after the overnight holding period on chlorinated water and the follow-up examination of the water in service 1 week later. While this coliform contamination may indicate the bacteria were injured, there was no inactivation of these organisms, some of which may recover to colonize the pipe sediment or an area of tuberculation in the pipe network.

The magnitude of the heterotrophic bacterial population in the pipe section can also be of value in verifying freedom from soil contaminants. Soil deposits in new pipe sections not only introduce a variety of heterotrophic bacteria, but may also provide some measure of associated protection to organisms from disinfectant contact. Some of the poor disinfection results attributed to chlorine applied in these situations may be traced to excessively dirty line sections or joints.

After the new main is assembled, it is important to thoroughly flush the line at a minimum velocity of 10 ft/s (76.2 cm/s) with treated water to remove particulates. With pipelines 16 in. (4.1 cm) or more in diameter, this velocity may not be attainable or may be ineffective. In these cases, use of cleaning devices such as poly pigs or foam swabs should be considered. There is no substitute for absolute cleanliness when installing new mains. No disinfectant can possibly be effective when lines contain sediments that provide protective habitat for bacterial growth. In some cases coliform contamination of the distribution system was traced to major line repairs and temporary losses of free chlorine residuals in those sections of the pipe network. Running water supply to waste in an effort to move fresh water with a free chlorine residual into the section of pipe was not practiced because the system did not disinfect the water, water conservation measures were in effect, or for fear of water freezing on the streets in the vicinity of the fire plugs adjacent to the line repairs.

Following flushing, disinfectant should be introduced into the new pipe sections and the water held for 24 to 48 h to optimize line sanitation. Bacteriological tests for coliforms and the heterotrophic plate count should then be performed. If the results of these tests (M-T7 agar for coliforms; R-2A agar for HPC) are satisfactory (<1 coliform per 100 ml; <500 HPC per milliliter), the line may be placed in service. If not, the line should again be flushed and

Table 5.10 Detection of Injured Coliforms in Three New England Drinking Water Treatment and Distribution Systems

Sample source	No. of samples	No. of confirmed colonies per 100 ml detected on:		Injury (%)	False negatives (%)
		m-Endo agar LES	m-T7 agar		
Filter backwash	1	18	136	86.7	82
After backwash	1	5	42	97.4	69
Water leaving treatment plants	46	0.2	5.7	96.5	79
Throughout systems	71	0.3	9.5	96.8	82
Pipe break	28	0.9	35.3	97.4	100
1 week after pipe break	11	0	67.5	100	100
After disinfection of new main	1	0	11	100	

Data revised from McFeters, Kippin, and LeChevallier.[148]

refilled with distribution water dosed with 50 mg/l available chlorine. Chlorine levels should not decrease below 25 mg/l during the 24-h holding period before repeat bacteriological testing. In pipes free of extraneous debris, free available chlorine (1 to 2 mg/l), potassium permanganate (2.5 to 4 mg/l), or copper sulfate (5.0 mg/l) have been used to meet coliform requirements.[151-153] Only free available chlorine, however, was found to eliminate large numbers of heterotrophic bacteria (Table 5.11).

PIPE DISINFECTION PRACTICES

Once the new or repaired pipe section is completed and thoroughly flushed, it must be properly disinfected to prevent contamination of the distribution network. A variety of procedures have been proposed or are in use for this purpose. The most common approach to disinfection has been the application of a heavy chlorine dosage (50 mg/l HOCl usually in the form of liquid chlorine, calcium hypochlorite granules or tablets, or sodium hypochlorite solutions) in standing water supply held in the pipe section for a 24-h period. This approach has met with mixed acceptance by the industry, because of a variety of factors. High-pH water produces chloramines, which possess slower bacterial inactivation. Inadequate flushing of new pipe sections creates a chlorine demand that lowers the effective chlorine residual, and sediment deposits impacted in pipe joints form a protective shield for associated bacterial populations. Low water temperatures also interfere with the chlorine kill rates.

Copper sulfate, which is known to be toxic to bacteria, has been used in a few situations where conventional disinfectants were inadequate. Copper concentrations used for coliform removal range from 5 to 30 mg/l with a retention time of 1 to 3 d. Because copper is toxic to humans, use of copper sulfate in this application must be followed by a rigorous flushing of the new line before it is accepted in the distribution system. A combination of chloramine and copper sulfate to form cuprichloramine has been proposed as a more effective, broad-spectrum disinfectant to sanitize new pipe.[97] Potassium permanganate has also been used occasionally as a disinfectant because of its highly oxidative nature.[98] The visible color of $KMnO_4$ is a definite aid to properly dosing new mains, but it may create disposal problems when flushing. Generally, 2.5 to 4 mg/l has been used with retention time of 1 to 4 d.

Field evaluation of free chlorine and potassium permanganate as disinfectants for new or repaired water mains indicates both of these disinfectants eliminated coliforms (Tables 5.11) from fairly clean lines; however, potassium permanganate did not reduce the heterotrophic plate count population. Most interesting was the observation that when lines were flushed to remove sediments, use of water supply containing a free chlorine residual of 1 to 4 mg/l or more was very effective and was the only disinfectant found to eliminate large numbers of heterotrophic bacteria.[88] Upon filling, this captured disinfected distribution water should be held in the line for 24 h with the free chlorine residual not declining below 0.5 mg/l.

Table 5.11   Disinfection Effectiveness of Water Mains

Application Time of sample	Sample station	Free residual chlorine (mg/l)	Coliform per 100 ml	Standard plate count per ml
1. High-dose chlorination[a]				
Before flushing and disinfection	B-1	2.4	<1	1
	B-2	1.5	<1	65
	B-3	0	>35	120,000
One day after flushing and filling	B-1	7.5	<1	1
with chlorine disinfectant	B-2	58	<1	1
(disinfectant still in main)	B-3	29	<1	3
One day after filling with	B-1	2.4	<1	4
chlorinated drinking water	B-2	2.2	<1	2
	B-3	10.0	<1	1
Two days after filling with	B-1	2.1	<1	1
chlorinated drinking water	B-2	2.1	<1	10
	B-3	8.0	<1	7
2. Potassium permanganate[b](4–5 mg/l)				
Before flushing and disinfection	A-1	0	<1	110
	A-2	0	<1	34,000
	A-3	0	<1	32,000
One day after disinfection with	A-2	(4 to 5)[c]	<1	170
potassium permanganate				
(potassium permanganate still in	A-3	(1)[c]	<1	14
main)				
One day after filling with	A-1	2.3	<1	1
chlorinated drinking water	A-2	2.2	<1	390
	A-3	1.9	<1	65
Two days after filling with	A-1	2.1	<1	2
chlorinated drinking water	A-2	1.9	<1	380
	A-3	1.6	<1	55
3. Potassium permanganate (1.5 mg/l)[d]				
Before flushing and disinfection	C-1	0	8	9,600
	C-2	0	1	3,200
One day after disinfection with	C-1	(1.5)[c]	<1	2,800
potassium permanganate	C-2	(0)[c]	<1	12,000
(potassium permanganate				
still in main)				
One day after filling with	C-1	1.25	<1	36
chlorinated drinking water	C-2	1.35	<1	5
Two days after filling with	C-1	0.95	<1	56
chlorinated drinking water	C-2	1.55	<1	86

Note: pH range 8.7 to 9.3, temp. range 12 to 13°C for applications 1 and 2; pH range 8.9 to 10.5, temp. range 18.5 to 26°C for application 3.

[a] Disinfection of 1790 m (5862 ft) of 70-cm (24-in.) prestressed-concrete cylindrical and ductile-iron water main.
[b] Disinfection of 995 m (3260 ft) of new 41-cm (16-in.) cast-iron and ductile-iron water main.
[c] Potassium permanganate concentration.
[d] Disinfection of 5980 m (19,600 ft) of 92-cm (36-in.) prestressed-concrete cylindrical water main.

Table revised from Buelow et al.[151]

Key to prevention of contamination is physical cleanliness of new pipe sections being installed so that there will be opportunity for successful disinfection. A survey of five water distribution systems in Britain indicated the failure rate of different types of materials to meet the coliform standard in new pipe installation was very similar (Table 5.12) and averaged 18.5%. Pipe diameter was not a factor in frequency of failures (Table 5.13), but more failures for all new pipelines were greater during the summer (Table 5.14). The problem of continued bacterial contamination following initial disinfection could be the result of a variety of factors including the production of slower-acting chloramines, high-pH water, inadequate flushing, sediment deposits in pipe joints that provide a protective habitat for bacteria, degradable gasket lubricants, and low temperatures that reduce chlorine kill rates during the winter months.[88] When bacteriological tests are unsatisfactory, the line should be flushed and refilled with distribution water dosed with 50 mg/l free chlorine. During the 24-h holding period, chlorine concentration should not decline to below 25 mg/l. If this additional treatment is necessary and bacteriologically successful, the new construction should be flushed until disinfectant residuals are reduced to that consistent with the pipe network disinfectant residual found in that area of the system. Flushing of this highly chlorinated water in areas where runoff will reach streams and ponds requires dechlorination to comply with federal and state regulations for aquatic wildlife protection. This may be accomplished by mixing the chlorinated water discharge with the proper feed concentration of sodium thiosulfate, sulfur dioxide, sodium bisulfate, or sodium sulfite. Circulating excessive chlorine residuals of this magnitude is undesirable because of the concern for elevated THMs in the system or release of biofilm fragments from older sections of the pipe network.

Table 5.12   New Pipeline Disinfection: Recurrence of Failures with Respect to Material

Material	No. laid	No. passed first time	Number failing chlorination					Failure (%)
			1×	2×	3×	4×	5+	
Asbestos cement	391	329	48	11	2	1	0	16
Iron	70	54	11	3	1	0	1	23
PVC	282	222	23	12	8	1	5	21
All materials	743	605	93	26	11	2	6	18.5

Data from Buelow et al.[88]

When new pipeline is to be added to an existing branch main that has little or no active withdrawals of water, it is advisable to thoroughly flush the entire branch main to remove static water and restore a disinfectant residual. If not done, there is a risk that the new pipe section will quickly become coated with

Table 5.13   New Pipeline Disinfection: Failures
with Respect to Pipe Diameter

Pipe diameter (mm)	No. passed after first chlorination	No. failing after initial chlorination	Failure (%)
50	71	10	12
75	99	25	20
100	285	61	18
150	93	29	24
200	17	1	a
225	17	6	a
300	23	6	a

a The total failure for 200, 225, and 300 mm pipe sizes is 19%.

Data from Buelow et al.[88]

Table 5.14   New Pipeline Disinfection:
Failures (for Five Water Utilities)
Expressed on Monthly Basis

Month	No. laid	No. failed initial chlorination	Failure (%)
January	78	18	23
February	80	4	5
March	63	7	11
April	70	5	7
May	81	9	11
June	52	14	27
July	49	17	35
August	48	17	35
September	82	26	32
October	42	9	21
November	53	5	9
December	45	7	16

Data from Buelow et al.[86]

sediments and colonized by bacteria brought in by water flow from the area of static water supply.

Some utilities have reported success using hypochlorite tablets securely glued to the top of the pipe as it is being installed.[90] Furthermore, a spoonful of hypochlorite powder was placed in each pipe joint at assembly time to inactivate bacteria that might be protected in crevices not reached by line flushing. Again, the new pipe was filled with water supply and held for up to 48 h to provide enough contact time. Following a satisfactory bacteriological report, the line was thoroughly flushed of the superchlorinated water, then placed in service.

BACTERIOLOGICAL TESTING PROTOCOL

Because the sanitary quality of water supply has been based solely on absence of total coliform bacteria, it has been rationalized that the acceptance of new or repaired mains depends only on a report that no coliforms are detected in water held in these new or repaired lines. The reasoning for this approach is that absence of total coliforms implies no contamination present of public health significance. A more rigorous check on new construction cleanliness would include an additional examination for heterotrophic bacteria. The heterotrophic plate count in this situation measures both the myriad of soil organisms that could have been introduced in the new construction and, if done on successive days, may provide some evidence of regrowth potential from deposited sediments.

Prior to pipeline testing for acceptance, the new line should be flushed to remove the holding water used to disinfect the line, followed by introduction of fresh water supply into the entire new section. At this time, water samples are collected from the far end of the new pipeline; thereupon, the line is valved off from further use until the bacteriological testing is completed. If the laboratory results indicate no coliforms present per 100 ml and the HPC is less than 100 organisms per milliliter, the line is considered acceptable for use. However, where possible, the in-service use should be delayed for another 24 h to provide for a repeat sampling and testing that may provide an opportunity to verify the initial bacteriological report. Holding a portion of the initial sample at room temperature for 24 to 48 h, then examining it for elevated HPC may provide an indication of regrowth potential in available pipe sediments. For example, if the HPC density within 48 h of holding at room temperature increases to 1000 or more organisms per milliliter, another line flushing is in order to dislodge protective sediments in some pipe joints.

LABORATORY CREDIBILITY THROUGH CERTIFICATION

Nothing could be more critical in the characterization of microbial quality of water supplies than laboratory credibility. Critical decisions must often be made on evidence supplied from laboratory reports. Such data is often pivotal to a proper action response, compliance concern with federal regulations, and perhaps legal recourse following extended boil water orders, illness, or death resulting from a waterborne outbreak.

CONCEPTS IN LABORATORY CERTIFICATION

Laboratory certification is thus very important. The concept of laboratory certification or evaluation of water laboratories is not new, having been initiated by the U.S. Public Health Service in 1943. The first manual on evaluation of water laboratories was published in 1966[99] and later expanded into a second edition as a handbook for evaluating water bacteriological laboratories.[100] In more recent years, the focus has been solely on a manual for the certification of laboratories analyzing drinking water.[101] The intent in these documents is to identify those laboratory procedures and personnel qualifications that have a substantial impact on data quality. The primary source for acceptable methods in water supply monitoring has traditionally been the current edition of *Standard Methods for the Examination of Water and Wastewater*;[102] however, provisions are available (through federal regulations) for acceptance of newly developed alternative procedures, if they are shown to be equivalent to standard methods.

It must be remembered that the current program of laboratory certification focuses only on a narrow range of concerns with coliform occurrences and to some extent on the limited application of heterotrophic bacterial densities in water distribution. Pathogen detection and speciation of coliforms in biofilm are not recognized as part of the laboratory services required in routine monitoring. However, recent development of the Information Collection Rule and anticipated follow-up legislation for surface-water quality enhancement will provide for expanded laboratory service in the future. Those laboratories skilled in the analysis of water samples for *Giardia, Crytosporidium,* or viruses will have the opportunity to receive a special approval, provided they can demonstrate this expertise on unknown samples and pass an on-site review of their procedures.[103] While the present laboratory certification program is limited in coverage, the opportunity to have a scheduled review of basic activities every 1 to 3 years does have a positive influence on sustaining good laboratory practices for a variety of testing services. Experience has shown that there is a continuing need for laboratory certification services in the federal, state, municipal, and private sectors to keep the number of deviations to an absolute minimum. The optimum frequency of laboratory certifications at the state level appears to be 3 years. Visits to these laboratories at more frequent intervals yield little value to either the staff or the certification program; however, longer intervals result in an increased number of deviations. On-site evaluations of local laboratories should be made on a yearly basis because these facilities often have a more frequent turnover of technicians, equipment problems, and space limitations.[104]

The state health and environmental agencies are cognizant of these problems. Some state agencies have established a vigorous program including an active laboratory certification service, a training program for the local laboratory personnel, and technical assistance directed toward improving the quality and reliability of the laboratory work performed. These progressive leaders in public health service are to be commended for their early recognition of the problems and a resolution to address the issues. Unfortunately, a few state agencies have developed little or no activity in this type of program. Much of this inactivity is the result of budgetary constraints that prohibit the necessary travel and personnel needed to operate an effective laboratory certification service.[105]

Critical to the evaluation of a laboratory for certification is defining a group of fundamental requirements absolutely essential for data development. These essentials[101] include laboratory equipment and supplies, basic laboratory practices (sterilization, glassware washing, and media preparation), sample collection and handling, analytical procedures, quality control, data collection, and response to laboratory results. Some of these critical items are also among the most frequently occurring deviations found in state, regional, and municipal laboratories (Table 5.15).

Table 5.15 Most Commonly Occurring Basic Deviations from Standard Methods in Water Laboratories

	Deviations (%)	
Items	60 state and regional laboratories	70 municipal laboratories
Sample bottles (size, air space)	56.7	30.1
Distilled water suitability	51.7	25.7
Sample transit time	45.0	8.6
Buffered dilution water	31.7	31.4
Incubator (temp, record, maintenance)	28.3	65.7
Clean glassware	18.3	25.7
Media pH (pH equipment, pH record)	15.0	70.0
Sample dechlorination	11.7	42.9
Autoclave (procedure, maintenance)	11.7	50.0
Media sterilization	8.3	20.0
Media storage	1.7	28.6
Sampling frequency	1.7	12.5

Data revised from Geldreich.[104]

Another crucial element in every effective monitoring program is the laboratory staff. They must be knowledgeable in the performance of a variety of microbiological testing procedures. Because drinking water microbiology has been rapidly expanding in the last decade, all laboratory staff members need to become involved in a continuing education program that focuses on advances in the microbiological analysis of drinking water. Appropriate courses are given at regular intervals by professional societies and occasionally by state and federal programs.[106] Of particular importance are those workshops dedicated to laboratory certification, pathogen detection, and new advances in drinking water microbiology. It is important to note that sustained skills in drinking water microbiology can only be maintained by performing frequent analyses per month, not by performing a few tests per year or by citing a certificate of past accomplishment.

Facilities used for microbiology should be clean and the temperature and humidity controlled to permit laboratory incubators to function properly. Adequate lighting and ample bench-top space are important for safety considerations and to minimize errors in processing samples. It has been recommended that the laboratory contain a space of 150 to 200 ft^2 and include 5 to 6 ft of usable linear bench space **per analyst.**[101] A significant amount of the space beyond what is needed for processing samples will be dedicated to storage space for media, glassware, positioning water baths, and portable equipment. Floor space must be designated for stationary equipment (incubators, refrigerators, file cabinets, and desks) and associated area for cleaning glassware and sterilizing materials. Without these space considerations there will be an opportunity for more laboratory accidents, poor performance of equipment, and contamination of cultures. Ultimately these situations can contribute to a compromise in data quality.

Availability of qualified technicians, approved techniques, and proper support equipment are key elements in acquiring the necessary data. However, without proper quality control procedures, the reliability of the data would be in question. Quality of commercial products used in the laboratory and stability of instrumentation and equipment must never be taken for granted. Even the most respected manufacturers may on occasion release a product that does not perform properly. Instrumentation reliability has improved greatly with the advent of solid-state electronics; however, there may be unpredicted failures in components due to voltage spikes, aging of sensing probes, and deterioration in parameter specifications that cause drift in measurement accuracies. Temperature-controlling equipment is also subject to component failure and metal fatigue in switching relays. Cyclic functions of autoclaves may become sluggish due to faulty valve operations and electric autoclaves can lose steam-generating capacity when there is partial heater coil burnout. Finally, there can be the factor of technician carelessness due to heavy workloads that create shortcut procedures and laboratory accidents. All of these situations influence data reliability.

As a safeguard against these possibilities, every laboratory must have a formal, written quality assurance plan that is followed by the analysts and technicians. This program should include adequate quality control tests on analytical procedures, commercial supplies, instrumentation, and equipment. Checks on analytical procedures should include sterility testing of materials, verification of positive results, and positive and negative controls on each series of analyses. In the daily operation of instruments and equipment, there is a need to perform pH meter standardization, check media pH, record incubator temperatures, and follow autoclaving procedures. At less frequent intervals, the laboratory should evaluate the quality of laboratory pure water, reestablish the calibration of balances, and demonstrate the freedom of toxic residues on glassware items. When quality control tests reveal an adverse condition, it is essential that any sample analyses influenced by this situation be rejected and appropriate measures be taken promptly to restore confidence in data reliability. All of these quality control activities to insure continued production of acceptable data will require about 10 to 15% of daily laboratory operations.

Laboratory certification in the EPA program for monitoring public water supplies does not extend to a variety of analyses for other organisms of significance in the distribution system. Special analytical services may be desirable for those systems that have taste and odor problems, experience nitrification in the distribution system biofilm and iron corrosion, or must satisfy an enhanced surface-water filtration rule. In these situations, there are two options: send samples for processing to a specialized laboratory or develop laboratory capability in these emerging areas of drinking water microbiology.

In developing expertise in special laboratory services, it will take time to achieve an acceptable level of competence. The program can start with participation in specialized workshops on the subject, but these offerings are not available on a frequent basis from professional societies or institutions. As an alternative, in-house training with consensus methods in the current edition of the standard methods[102] may be the most practical approach. If the analysis is for bacterial communities in biofilm, use of positive samples will provide opportunities to gain proficiency in cultivation and speciation among technicians in the laboratory. Specialized techniques specific for fecal streptococcus and *Clostridium* are other issues the laboratory might be required to consider. If the training is for the detection of *Salmonella,* pathogenic *E. coli, Legionella,* viruses, *Giardia,* and *Cryptosporidium,* then the laboratory staff must develop proficiency with concentration techniques, specific cultivation methods, fluorescent antibody procedures, and microscopic recognition of internal cyst and oocyst for protozoans. As a special condition for testing approval, performance samples containing specified pathogens should be examined every 3 months to insure the analyst is maintaining the necessary equipment and skills to analyze water samples for microbial agents during a waterborne outbreak. Finally, some form of extended laboratory certification or approval should be developed by the primary agency to cover the special equipment and procedural skills required of the analyst.

REFERENCES

1. Geldreich, E.E. 1992. Microbiological Changes in Source Water Treatment: Reflections in Distribution Water Quality. In: *Strategies and Technologies for Meeting SDWA Requirements.* R.M. Clark and R.S. Summers, Eds., Technomic Pub. Co., Lancaster, PA.
2. Gibbs, R.A. and C.R. Hayes. 1989. Characterization of Non-Enteric Bacterial Regrowth in the Water Supply Distribution Network from a Eutrophic Reservoir. *Water Sci. Technol.,* 21:48-53.
3. O'Connor, J.T., S.K. Banerji, and B.J. Brazos. 1989. A Guide to the Control of Water Quality Deterioration in Distribution Systems. *Pub. Wks.,* 104:Part I:44-46 (1988); 105:Part II:62-63, 114-115; Part III:46-48.
4. Geldreich, E.E. 1973. Is the Total Count Necessary? Proc. Water Qual. Technol. Conf., Amer. Water Works Assoc., Denver, CO. pp. VII-1–VII-8.

5. Pelletier, P.A., G.C. duMoulin, and K.D. Stottmeier. 1988. Mycobacteria in Public Water Supplies: Comparative Resistance to Chlorine. *Microbiol. Sci.,* 5:147-148.

6. Hill, J.C. 1946. Bacterial Oxidation of Ammonia in Circulating Water. *Jour. Amer. Water Works Assoc.* 38:980-982.

7. Wolfe, R.L., N.I. Lieu, G. Izaquirre, and E.G. Means. 1990. Ammonia-Oxidizing Bacteria in a Chloraminated Distribution System: Seasonal Occurrence, Distribution and Disinfection Resistance. *Appl. Environ. Microbiol.,* 56:451-462.

8. Herman, L. 1976. Sources of the Slow-Growing Pigmented Water Bacteria. *Health Lab. Sci.,* 13:5-10.

9. Reasoner, D.J. and E.E. Geldreich. 1979. Significance of Pigmented Bacteria in Water Supplies. A.W.W.A. Water Qual. Technol. Conf., December 9-12, 1979, Philadelphia, PA. pp. 187-196.

10. Katzenelson, E. et al. 1974. Inactivation Kinetics of Viruses and Bacteria in Water by Use of Ozone. *Jour. Amer. Water Works Assoc.,* 66:725-729.

11. Sharp, D.G. 1976. Viron Aggregation and Disinfection of Water Viruses by Bromine. U.S. E.P.A.

12. Bonde, G.J. 1977. Bacterial Indicators of Water Pollution. In: *Advances in Aquatic Microbiology.* Vol. 1, M.R. Droop and H.W. Jannasch, Eds., Academic Press, London.

13. Bisson, J.W. and V.J. Cabelli. 1980. *Clostridium perferingens* as a Water Pollution Indicator. *Jour. Water Poll. Contr. Fed.,* 52:241-248.

14. Hiisvirta, L.O. 1986. Problems of Disinfection of Surface Water with a High Content of Natural Organic Material. *Water Supply,* 4:53-58.

15. Wolfe, R.L., N.I. Lieu, G. Izaguirre, and E.G. Means. 1990. Ammonia-Oxidizing Bacteria in a Chloraminated Distribution System: Seasonal Occurrence, Distribution, and Disinfection Resistance. *Appl. Environ. Microbiol.,* 56:451-462.

16. Larson, T.E. 1939. Bacteria, Corrosion and Red Water. *Jour. Amer. Water Works Assoc.,* 31:1186-1196.

17. Watson, S.W., F.W. Valois, and J.B. Waterbury. 1981. The Family *Nitrobacteraceae.* In: *The Prokaryotes.* M.P. Starr et al., Eds., Springer-Verlag, Berlin.

18. Jones, R.D. and M.A. Hood. 1980. Interaction between an Ammonium Oxidizer, *Nitrosomonas* sp., and Two Heterotrophic Bacteria, *Nocardia atlanta* and *Pseudomonas* sp. *Microb. Ecol.,* 6:271-276.

19. Painter, H.A. 1970. A Review of Literature on Inorganic Nitrogen Metabolism in Microorganisms. *Water Res.,* 4:393-450.

20. Alexander, M. and F.E. Clark. 1965. Nitrifying Bacteria. In: *Methods of Soil Analysis.* C.A. Black, Ed., Amer. Soc. Agronomy, Madison, WI.

21. Matulewich, V.A., P.F. Strom, and M.S. Finstein. 1975. Length of Incubation for Enumerating Nitrifying Bacteria Present in Various Environments. *Appl. Microbiol.,* 29:265-268.

22. American Public Health Association. 1992. *Standard Methods for the Examination of Water and Wastewater.* 18th ed. American Public Health Association, Washington, D.C.

23. Taylor, R.H. and E.E. Geldreich. 1979. A New Membrane Filter Procedure for Bacterial Counts in Potable Water and Swimming Pool Samples. *Jour. Amer. Water Works. Assoc.,* 71:402-405.

24. Haas, C.N., M.A. Meyer, and M.S. Paller. 1982. Analytical Note: Evaluation of the m-SPC Method as a Substitute for the Standard Plate Count in Water Microbiology. *Jour. Amer. Water Works. Assoc.*, 74:322.
25. Klein, D.A. and S. Wu. 1974. A Factor to be Considered in Heterotrophic Microorganisms Enumeration from Aquatic Environments. *Appl. Microbiol.*, 27:429-431.
26. Reasoner, D.J. and E.E. Geldreich. 1985. A New Medium for the Enumeration and Subculturing of Bacteria from Potable Water. *Appl. Environ. Microbiol.*, 49:1-7.
27. Maki, J.S., S.J. LaCroix, B.H. Hopkins, and J.T. Staley. 1986. Recovery and Diversity of Heterotrophic Bacteria from Chlorinated Drinking Waters. *Appl. Environ. Microbiol.*, 51:1047-1055.
28. Lombardo, L.R., P.R. West, and J.L. Holbrook. 1985. A Comparison of Various Media and Incubation Temperatures Used in the Heterotrophic Plate Count Analysis. Water Qual. Technol. Conf., Amer. Water Works Assoc., Denver, CO. pp. 251-270.
29. Reasoner, D.J. 1990. Monitoring Heterotrophic Bacteria in Potable Water. In: *Drinking Water Microbiology.* G.A. McFeters, Ed., Springer-Verlag, New York.
30. Postgate, J.R. 1967. Viability Measurements and the Survival of Microbes Under Minimum Stress. *Adv. Microbiol. Phys.*, 1:1-23.
31. Van Soestberger, A.A. and C.H. Lee. 1969. Pour Plates or Streak Plates? *Appl. Microbiol.*, 18:1092-1093.
32. Stapert, E.M., W.T. Sokolski, and J.I. Northam. 1962. The Factor of Temperature in the Better Recovery of Bacteria from Water by Filtration. *Can. Jour. Microbiol.*, 8:809-810.
33. U.S. Environmental Protection Agency. 1989. Drinking Water: National Primary Drinking Water Regulations; Total Coliforms (Including Fecal Coliforms and *E. coli*): Final Rule. *Fed. Reg.*, 54:27544-27568.
34. Weiss, J.E. and C.A. Hunter. 1939. Simplified Bacteriological Examination of Water. *Jour. Amer. Water Works Assoc.*, 31:707-713.
35. Clark, J.A. 1968. A Presence–Absence (P–A) Test Providing Sensitive and Inexpensive Detection of Coliforms, Fecal Coliforms and Fecal Streptococci in Municipal Drinking Water Supplies. *Can. Jour. Microbiol.*, 14:13-18.
36. Geldreich, E.E., E.W. Rice, and E.J. Reed. 1987. Monitoring for Indicator Bacteria in Small Water Systems. *Proc. Inter. Conf. Resource Mobilization for Drinking Water and Sanitation in Developing Nations.* pp. 667-677. American Society of Chemical Engineers, New York.
37. Pipes, W.O., H.A. Minnigh, B. Moyer, and M.A. Trog. 1986. Comparison of Clark's Presence–Absence Test and the Membrane Filter Method for Coliform Detection in Potable Water Samples. *Appl. Environ. Microbiol.*, 52:439-443.
38. Fujioka, R., N. Kungskulniti, and S. Nakasone. 1986. Evaluation of the Presence–Absence Test for Coliforms and the Membrane Filtration Method for Heterotrophic Bacteria. Proc. Water Qual. Technol. Conf., Portland, OR, Amer. Water Works Assoc., Denver, CO. pp. 271-279.
39. Jacobs, N.J. et al. 1986. Comparison of Membrane Filter, Multiple Fermentation Tube, and Presence–Absence Techniques for Detecting Total Coliforms in Small Community Water Systems. *Appl. Environ. Microbiol.*, 51:1007-1012.

40. Caldwell, B.A. and R.Y. Morita. 1987. Sampling Regimes and Bacteriological Tests for Coliform Detection in Groundwater. Project Report EPA/600/287/083. Environmental Protection Agency, Cincinnati, OH.

41. Rice, E.W., E.E. Geldreich and E.J. Read. 1989. The Presence–Absence Coliform Test for Monitoring Drinking Water Quality. *Public Health Rep.*, 104:54-58.

42. McFeters, G.A. 1990. Enumeration, Occurrence, and Significance of Injured Indicator Bacteria in Drinking Water. In *Drinking Water Microbiology*. G.A. McFeters, Ed., Springer-Verlag, New York. pp. 478-492.

43. Zaske, S.K., W.S. Dockins, and G. A. McFeters. 1980. Cell Envelope Damage in *Escherichia coli* Caused By Short-Term Stress in Water. *Appl. Environ. Microbiol.*, 40:386-390.

44. Domek, M.J., M.W. LeChevallier, S.C. Cameron, and G.A. McFeters, 1984. Evidence for the Role of Copper in the Injury Process of Coliforms in Drinking Water. *Appl. Environ. Microbiol.*, 48:289-293.

45. Domek, M.J., J.E. Robbins, M.E. Anderson, and G.A. McFeters. 1987. Metabolism of *Escherichia coli* Injured by Copper. *Can. Jour. Microbiol.*, 33:57-62.

46. Chamberlin, C.E. and R. Mitchell. 1978. A Decay Model for Enteric Bacteria in Natural Waters. In: *Water Pollution Microbiology*. R. Mitchel, Ed., John Wiley and Sons, New York. pp. 325-348.

47. Kapuscinski, R.B. and R. Mitchell. 1981. Solar Radiation Induces Sublethal Injury in *Escherichia coli* in Seawater. *Appl. Environ. Microbiol.*, 41:670-674.

48. Fujioka, R.S. and O.T. Narikawa. 1982. Effect of Sunlight on Enumeration of Indicator Bacteria under Field Conditions. *Appl. Environ. Microbiol.*, 44:395-401.

49. Hackney, C.R. and G.K. Bissonnette. 1978. Recovery of Indicator Bacteria in Acid Mine Streams. *Jour. Water Poll. Contr. Fed.*, 52:1947-1952.

50. LeChevallier, M.W. and G.A. McFeters. 1985. Enumerating Injured Coliforms in Drinking Water. *Jour. Amer. Water Works Assoc.*, 77:81-87.

51. McFeters, G.A. and D.G. Stuart. 1972. Survival of Coliform Bacteria in Natural Waters: Field and Laboratory Studies with Membrane Filter Chambers. *Appl. Microbiol.*, 24:805-811.

52. Bissonnette, G.K., J.J. Jezeski, G.A. McFeters, and D.G. Stuart. 1975. Influence of Environmental Stress on Enumeration of Indicator Bacteria from Natural Waters. *Appl. Microbiol.*, 29:186-194.

53. Camper, A.K. and G.A. McFeters. 1979. Chlorine Injury and the Enumeration of Waterborne Coliform Bacteria. *Appl. Environ. Microbiol.*, 37:633- 641.

54. McFeters, G.A. and A.K. Camper, 1983. Enumeration of Indicator Bacteria Exposed to Chlorine. *Adv. Appl. Microbiol.*, 20:177-193.

55. Mudge, C.S. and F.R. Smith. 1935. Relation of Action on Chlorine to Bacterial Death. *Amer. Jour. Pub. Health*, 25:442-447.

56. McFeters, G.A., J.S. Kippin, and M.W. LeChevallier. 1986. Injured Coliforms in Drinking Water. *Appl. Environ. Microbiol.*, 51:1-5.

57. Geldreich, E.E., M.J. Allen, and R.H. Taylor. 1978. Interferences to coliform detection in potable water supplies. In: Evaluation of the Microbiology Standards for Drinking Water. U.S. Environmental Protection Agency, Washington, D.C. pp. 13-30.

58. LeChevallier, M.W. and G.A. McFeters. 1985. Interactions between Heterotrophic Plate Count Bacteria and Coliform Organisms. *Appl. Environ. Microbiol.*, 49:1338-1341.

59. Herson, D.S. 1980. Identification of Coliform Antagonists. *Proc. Amer. Water Works Assoc.*, Water Quality Techol. Confr., 153–157, Miami Beach, FL.

60. Schusner, D.L., F.F. Busta, and M.L. Speck. 1971. Inhibition of Injured *Escherichia coli* by Several Selective Agents. *Appl. Microbiol.*, 21:41-45.

61. McFeters, G.A., S.C., Cameron, and M.W. LeChevallier. 1982. Influence of Diluents, Media and Membrane Filters on the Detection of Injured Waterborne Coliform Bacteria. *Appl. Environ. Microbiol.*, 43:97-103.

62. LeChevallier, M.W., S.C. Cameron, and G.A. McFeters. 1983. New Medium for the Recovery of Coliform Bacteria from Drinking Water. *Appl. Environ. Microbiol.*, 45:484-492.

63. Edberg, S.C., M.J. Allen, and D.B. Smith. 1988. The National Collaborative Study. National Field Evaluation of a Defined Substrate Method for the Simultaneous Enumeration of Total Coliforms and *Escherichia coli* from Drinking Water: Comparison with the Standard Multiple Tube Fermentation Method. *Appl. Environ Microbiol.*, 54:1595-1601.

64. Spino, D.F. 1985. Characterization of Dysgonic, Heterotrophic Bacteria from Drinking Water. *Appl. Environ. Microbiol.*, 50:1213-1218.

65. Rice, E.W., M.J. Allen, T.C. Covert, J. Langewis, and J. Standridge. 1993. Identifying *Escherichia* Species with Biochemical Test Kits and Standard Bacteriological Tests. *Jour. Amer. Water Works Assoc.*, 85:74-76.

66. Rice, E.W. et al. 1991. Assay for β-Glucuronidase in Species of the Genus *Escherichia* and Its Application for Drinking Water Analysis. *Appl. Environ. Microbiol.*, 57:592-593.

67. Geldreich, E.E., H.D. Nash, D.J. Reasoner, and R.H. Taylor. 1972. The Necessity of Controlling Bacterial Populations in Potable Waters: Community Water Supply. *Jour. Amer. Water Works Assoc.*, 64:596-602.

68. U.S. Environmental Protection Agency. 1976. National Interim Primary Drinking Water Regulations, EPA-570/9-76-003.

69. McCabe, L.J., J.M. Symons, R.D. Lee, and G.G. Robeck. 1970. Survey of Community Water Supply Systems. *Jour. Amer. Water Works Assoc.*, 62:670-687.

70. Buelow, R.W. and G. Walton. 1971. Bacteriological Quality vs. Residual Chlorine. *Jour. Amer. Water Works Assoc.*, 63:28-35.

71. Mackenthun, K.M. and L.E. Keup. 1970. Biological Problems Encountered in Water Supplies. *Jour. Amer. Water Works Assoc.*, 62:520-526.

72. Silvey, J.K.G., D.E. Henley, and J.T. Wyatt. 1972. Planktonic Blue-Green Algae: Growth and Odor-Production Studies. *Jour. Amer. Water Works Assoc.*, 64:35-39.

73. Cullimore, D.R. 1992. *Practical Manual of Groundwater Microbiology*. Lewis Pub., Boca Raton, FL.

74. Lueschow, L.A. and K.M. Mackenthun. 1962. Detection and Enumeration of Iron Bacteria in Municipal Water Supplies. *Jour. Amer. Water Works Assoc.*, 54:751-756.

75. Christian, R. 1976. Iron Bacteria of Importance to the Water Industry. Amer. Water Works Assoc. Water Qual. Technol. Conf. (3B-3) 1-22, San Diego, CA.

76. Dowling, J.E., M.W. Mittelman, and D.C. White. 1991. The Role of Consortia in Microbially Influenced Corrosion. In: *Mixed Cultures in Biotechnology*, J.G. Zeikus, Ed. McGraw-Hill, New York.

77. Hamilton, W. 1990. Sulfate-Reducing Bacteria and Their Role in Microbially Influenced Corrosion. In: *Microbially Influenced Corrosion and Biodegradation*. J.E. Dowling, M.W. Mittelman, and J.C. Danko, Eds. University of Tennessee, Knoxville.

78. Loos, E.T. 1987. Experiences with Manganese in Queensland Water Supplies. *Water,* 13:28-31.

79. Committee on the Challenges of Modern Society (NATO/CCMS). 1987. Drinking Water Microbiology. *Jour. Environ. Pathol. Toxicol. Oncol.,* 7:1-365.

80. Schweisfurth, R. and R. Mertes. 1962. Mikrobiologische und Chemishe untersuchungen uber Bildung und Bekampfung von Manganschlamm- Ablagerungen in einer Druckleitung fur Talsperrenwasser. *Arch. Hyg. Bakteriol.,* 146:401-417.

81. Smith, Stuart. 1992. *Methods for Monitoring Iron and Manganese Biofouling in Water Wells.* Amer. Water Works Assoc. Res. Found., Denver, CO.

82. American Water Works Association. 1968. AWWA Standard for Disinfecting Water Mains. AWWA C601-68.

83. Becker, R.J. 1969. Mains Disinfection Methods and Objectives, Part I, Use of Liquid Chlorine. *Jour. Amer. Water Works Assoc.,* 61:2:79-81.

84. Russelman, H.B. 1969. Main Disinfection Methods and Objectives, Part III, Public Health Viewpoint. *Jour. Amer. Water Works Assoc.,* 61:2:82-84.

85. Suckling, E.V. 1943. *The Examination of Water and Water Supplies*, 5th ed., Blakiston & Son, Philadelphia, pp. 658-659.

86. Davis, A.R. 1951. The Distribution System. In: Manual for Water Works Operators. L.C. Billings, Ed., Texas State Dept. Health, Austin, Texas. pp. 342-363.

87. Shull, K.E. 1974. Quality Control in New Main Disinfection at Philadelphia Suburban Water Company. Water Qual. Technol. Conf., Dallas, TX. AWWA, Denver, CO. pp. VII-1–VII-4.

88. Buelow, R.W., R.H. Taylor, E.E. Geldreich, et al. 1976. Disinfection of New Water Mains. *Jour. Amer. Water Works Assoc.,* 68:283-288.

89. Hutchinson, M. 1971. The Disinfection of New Water Mains. *Chem. Indus.,* 1:139-142.

90. Rossum, J.R. 1974. Tablet Method for Main Disinfection. Water Qual. Technol. Conf., Dallas, TX. AWWA, Denver, CO. pp. VI-1–VI-3.

91. Calvert, C.K. 1939. Investigation of Main Sterilization. *Jour. Amer. Water Works Assoc.,* 31:832-836.

92. Adams, G.O. and F.H. Kingsbury. 1937. Experiences with Chlorinating New Water Mains. *Jour. New Engl. Water Works Assoc.,* 51:60-68.

93. Taylor, E.W. 1958. *The Examination of Waters and Water Supplies.* 7th ed. Little Brown & Co., Boston, MA.

94. Prescott, S.C., C.A. Winslow, and M.H. McGrady. 1946. *Water Bacteriology.* 6th ed. John Wiley & Sons, New York.

95. Schubert, R.H.N. 1967. Das Vorkmmen der Aeromonaden in Obseirdischen Gewassern. *Arch. Hyg.,* 150:688-708.

96. Hutchinson, M. 1974. WRA Medlub: An Aid to Mains Disinfection. *Soc. Water Treat. Exam.,* 23(Part 11):174-189.

97. Harold, C.H.H. 1934. Metropolitan Water Board (London) 29th Annual Report.
98. Hamilton, J.J. 1974. Potassium Permanganate as a Main Disinfectant. *Jour. Amer. Water Works Assoc.,* 66(12):734-735.
99. Geldreich, E.E. and H.F. Clark. 1966. Evaluation of Water Laboratories. PHS Pub. 999-EE-1. Department of Health, Education and Welfare, Washington, D.C.
100. Geldreich, E.E. 1975. Handbook for Evaluating Water Bacteriological Laboratories. EPA-670/9-75-006. U.S. Environmental Protection Agency, Cincinnati, OH.
101. Laboratory Certification Program Revision Committee. 1991. Manual for the Certification of Laboratories Analyzing Drinking Water. EPA 570/9- 90-008A, U.S. Environmental Protection Agency.
102. American Public Health Association. 1992. *Standard Methods for the Examination of Water and Wastewater.* 18th ed. American Public Health Association, Washington, D.C.
103. Federal Register. 1994. Monitoring Requirements for Public Drinking Water Supplies: Proposed Rule. 40 CFR Part 141. *Fed. Reg.,* 59(28):6354-6409.
104. Geldreich, E.E. 1967. Status of Bacteriological Procedures Used by State and Municipal Laboratories for Potable Water Examination. *Health Lab. Sci.,* 4:9-16.
105. Geldreich, E.E. 1971. Application of Bacteriological Data in Potable Water Surveillance. *Jour. Amer. Water Works Assoc.,* 63:225-229.
106. Bordner, R.H. 1993. Personal communication.

CHAPTER 6

Monitoring Strategies to Characterize Water Quality

CONTENTS

INTRODUCTION

The goal of a water utility is to provide a safe, reliable, aesthetically pleasing water supply to its customers. In this context, reliability implies a history of no waterborne disease outbreaks. Unfortunately, should a waterborne disease outbreak occur, remedial action takes place after the event. In order to take a more proactive rather than reactive approach, an aggressive monitoring program must be utilized. Such a program should provide a continuing database on treatment effectiveness, locate contaminant penetration, and supply early signals of microbial deterioration in the storage of finished water and its passage to the consumer's tap. Thus, the main objectives of the monitoring program are to reduce the vulnerability of a community to waterborne outbreaks by enabling early intervention, establishing quality trends, and identifying areas of infrastructure deterioration in the distribution system. Furthermore, confusion over the significance of coliform biofilm releases in distribution systems, the impact such a problem has on masking new contamination events, and the application of impending regulation changes have focused attention on considering different strategies for monitoring the bacteriological quality of water supply.

CHARACTERIZING THE DISTRIBUTION SYSTEM

There are a number of features of distribution systems that may have a bearing on water quality changes. Elements that are presumed to have an important bacteriological influence include size of distribution network for population served, predominant pipe material and age of pipelines, water pressure, number of line breaks each year, water storage capacity, and water supply retention time in the system. The information obtained from utilities (Table 6.1) suggests that while there is a general increase in network size in relation to population served, some distribution systems apparently have a greater population density per square mile (Wilmette, IL and Rochester, NY) than others because the miles of pipe are much less than in cities of similar size. Cast iron and ductile iron are the predominant pipe materials, with most pipes being in service for over 50 years. No information was available on frequency of line repairs, which would relate to several factors including pipe age, water pH (corrosion), and water pressure. Drinking water storage capacity and population served had a significant bearing on water supply retention time in the system. Other factors of importance that are not included in Table 6.1 that have an impact on bacterial quality of water are flushing frequency, distribution network turbidity, percent of dead ends, and disinfection demand near terminations of the pipe network.

MONITORING STRATEGIES

There are two levels of monitoring in public water systems that must be recognized: compliance monitoring and special-purpose monitoring. Compliance monitoring involves the development of a systematic plan to cover all major aspects and areas of the distribution network with a sampling strategy that demonstrates the continued delivery of a safe water supply in all service areas. A special monitoring program is designed to locate problem areas that need corrective actions and to better understand system characteristics.

EVALUATING THE STRATEGY

Planning a monitoring strategy to best characterize drinking water quality in the distribution system starts with site selection and projecting sampling frequencies to adequately describe the water quality over time. This need is true of raw water, water in treatment, and water supply in distribution to the public.

REVIEW OF MONITORING INTENSITY

Sample site selection in the distribution system should be carefully developed so as to characterize water supply quality in all service areas. This objective is of great importance and must take precedence over satisfying a

Table 6.1 Characterizing Distribution Systems for Selected Utilities

Water treatment system	Pop. served	Network size (mi)	Total storage (MG)	Type of pipe	Age of oldest pipe (years)	System pressure range (psi)	Retention time (d)	Routine flushing program
Brockport, NY	35,000	100	4.8	C	75	<200	1–2	Yes
Kennebunk, ME	12–50,000	181	4.0	C	90	30–110	Variable	Yes
Terre Haute, IN	66,700	320	3.05	C,D,AC	125	60–80	1–2	Yes
Muncie, IN	85,000	300	4.00	AC,C,D,S	80	50–90	2–5	Yes
Wilmette, IL	100,000	95	12.00	C,D	100	30–80	1	Yes
Springfield, IL	142,000	420	12.00	D	90	50–60	2	Yes
Lexington, KY	225,000	1,162	16.93	C,AC,P	100	50–80	2	Yes
Rochester, NY	245,000	625	245	C	>80	90	2	Yes
Monmouth, NJ	260,000	1,048	13.45	C,D	—	—	—	No
Grand Rapids, MI	263,000	864	73.75	C,D	70	<120	2	Yes
New Haven, CT	370,000	1,391	51.00	C,D	80	40–60	5–6	Yes
Fairfax, VA	820,000	2,376	25.00	C,D,CP,S	50	30–80	<1	Yes

Note: C, cast iron; D, ductile iron; AC, asbestos cement; S, steel; CP, concrete pressure; P, plastic.

Data from Geldreich, Goodrich, and Clark.[21]

minimum number of samples per month specified by federal regulations. Oftentimes, sample collections in systems serving less than 25,000 people are made at sites chosen for convenience rather than for coverage of the entire network of pipes and reservoirs. Study of sample records indicates that 75% or more of the required samples per month were taken from the same locations: the municipal building, the laboratory tap, the residence of some city official, and a favorite restaurant or taverns.[1] Only an occasional attempt was made to obtain other samples that might meaningfully measure water quality as delivered through other portions of the distribution lines. This approach could hardly be considered adequate to characterize water quality or sufficient to demonstrate continued integrity of the distribution network.

Historical Record

The historical record of site locations should be reviewed for justification to continue their inclusion in any revised sampling site plan. These sites often have a history of occasional coliform occurrences while others were chosen just because of accessibility. Site selections by utility are often centrally located, and include a site by a reservoir intake, near schools, a dead end, a factory (largest water user on system), a previous site of intermittent coliform occurrences, an area of coliform biofilm location, fire hydrants, an area of reduced flow velocity, residential taps, and a site near the pumping station.

Site Selection Rationale

The site selection rationale should be based on being representative of network structural characteristics and water quality conditions. A portion of these sites should be fixed locations that are selected with reference to first-customer location, pressure zones, interconnection with adjacent utility system, potential sources of contamination, areas of high risk (hospitals and clinics), and any sites with a past history of coliform occurrences. A recent survey of 1796 utilities on current practice in bacteriological sampling[2] reveals that approximately one third of these water supply systems use the same sampling locations from month to month (Table 6.2). About half of the utilities use some fixed locations and vary others from month to month. Additional factors that must be considered in revising the sampling plan include frequency of unsatisfactory samples, repeat-sampling response, peak water usage as related to seasonal shifts in population served, first customer locations, new line construction, and sites of frequent line breaks. Some sampling locations should be varied every month so that all sections of the distribution system are monitored systematically over time.

Having identified sensitive areas in the system to monitor, the next consideration is gaining access to these locations on the distribution network.

Table 6.2 Patterns of Sampling over Area of Distribution System

Selection of sampling location	Utilities number (%)	Utilities considering indicated factors in location selection (%)						
		Convenience	Area representation	Central points	Peripheral locations	Central and peripheral locations	Varies	No response
Fixed locations	626 (34.9)	35.5	80.2	32.3	20.6	55.9	10.2	0.3
Variable locations	295 (16.4)	25.8	84.7	27.8	29.2	44.7	54.6	2.7
Both fixed and variable locations	848 (47.2)	42.3	88.6	31.5	31.5	78.3	35.0	0.1
No response	27 (1.5)							

Data from the AWWA committee report.[2]

Sampling locations may be public sites (police and fire stations, government office buildings, and industrial facilities), private residences (single residences, apartment buildings, and townhouse complexes), and special sampling stations built into the distribution network. The AWWA survey[2] identified only six different types of locations used for collecting bacteriological samples (Table 6.3). Few of the utilities that responded to the national survey used all of the different types of locations.

Table 6.3   Types of Sampling Locations Used

Type of sampling location	Utilities reporting indicated frequency of use (%)					
	Always	Frequently	Seldom	Never	Available	Response
Fire hydrants	0.7	5.2	30.3	56.5	0.4	6.9
Storage tanks	9.1	18.5	32.4	0.4	0.7	8.8
Pumping stations	15.7	24.6	22.5	27.7	2.4	7.1
Commercial buildings	21.8	51.3	14.7	7.9	0.2	4.1
Public buildings	25.6	50.3	12.3	7.5	—	4.1
Private residences	31.8	40.1	21.1	3.8	—	3.2

Data from the AWWA committee report.[2]

Development of an effective sampling program requires not only knowledge of the hydraulics involved in the pipe network, but also inputs from the historical database and cross-connection surveys. For these reasons, the design should be a joint responsibility of local water authorities (water plant management, health officer, or municipal engineer) and the appropriate state engineering program. Additional guidance and requested input can also be found in consultations with water supply representatives in each of the regional offices of the EPA.

Sampling Frequency

The sampling frequency must be adequate to carefully characterize the water quality as being continuously safe to drink. In reality, the monitoring program will only test a minute fraction of all available water supply during some designated instance of time over a 30-d period. This is indeed a difficult quality control task, considering the many variables that can be involved.

A number of factors must be taken into consideration if frequency of monitoring is to be optimized.[3-9] Integrity of the distribution system is often an important issue with old pipe networks that have been in service for over 75 years and in the process have acquired a variety of pipe materials, numerous potentials for cross-connections, and opportunities for flow reversal as the system is expanded to meet a spreading area of population served. As a consequence, every water distribution system is unique in its strengths and weaknesses and, therefore, requires careful customizing of the monitoring frequency. While the federal regulations have redesigned the minimum sampling frequency per month using population served (Table 6.4), many utilities collect more samples per month (Table 6.5), based on a national survey,[2] than are required by law.

Table 6.4   Total Coliform Sampling Requirements Based upon Population

Population served	Minimum number of samples per month[a]	Population served	Minimum number of samples per month
25 to 1,000	1[b]	59,001 to 70,000	70
1,001 to 2,500	2	70,001 to 83,000	80
2,501 to 3,300	3	83,001 to 96,000	90
3,301 to 4,100	4	96,001 to 130,000	100
4,101 to 4,900	5	130,001 to 220,000	120
4,901 to 5,800	6	220,001 to 320,000	150
5,801 to 6,700	7	320,001 to 450,000	180
6,701 to 7,600	8	450,001 to 600,000	210
7,601 to 8,500	9	600,001 to 680,000	240
8,501 to 12,900	10	780,001 to 970,000	270
12,901 to 17,200	15	970,001 to 1,230,000	300
17,201 to 21,500	20	1,230,001 to 1,520,000	330
21,501 to 25,000	25	1,520,001 to 1,850,000	360
25,001 to 33,000	30	1,850,001 to 2,270,000	390
33,001 to 41,000	40	2,270,001 to 3,020,000	420
41,001 to 50,000	50	3,020,001 to 3,960,000	450
50,001 to 59,000	60	3,960,001 or more	480

[a] A noncommunity water system using groundwater and serving an average of 1000 persons per day or fewer must monitor in each calendar quarter during which the system provides water to the public. If the state, on the basis of a sanitary survey conducted within the last 10 years, determines that some other frequency is more appropriate, that frequency shall be the frequency required under these regulations, except that in no case must it be less than once per year. Such frequency shall be confirmed or changed on the basis of subsequent sanitary surveys. A noncommunity water system serving an average of more than 1000 persons per day during any month, or a noncommunity water system using surface water, must monitor at the same frequency as a like-sized community public water system for each month the system provides water to the public.

[b] Based on a history of no coliform contamination and on a sanitary survey every 5 years by the state showing the water system to be supplied solely by a protected groundwater source and free of sanitary defects, a community public water system serving 25 to 1000 persons may reduce this sampling frequency with the written permission of the State, except that in no case shall it be reduced to less than once per quarter.

Data from the Environmental Protection Agency.[10]

Having established some specified number of sampling locations per month, the next decision is, when should these samples be collected during the month? Is it logical to assume the monitoring program should be spread out over the entire 30-d period because of the desire to continually check on

Table 6.5   Number of Samples Collected per Month for Bacteriological Testing

Population served (thousands)	Total number of utilities	Original distribution system samples			
		Fewer than NIPDWR requirements	Equal to NIPDWR requirements	Greater than NIPDWR requirements	No response
<1	112	3[a]	33	76	
1–2.5	233	10	117	106	
2.5–4.9	226	37	67	119	3
4.9–8.5	240	49	61	129	1
8.5–12.9	194	49	23	121	1
12.9–18.1	174	51	18	103	2
18.1–24	94	19	49	25	1
24–50	218	54	33	129	2
>50	305	48	30	218	9
Total	1796	320	431	1026	9
Percentages		17.8	24.0	57.1	1.1

[a] These systems collect one sample per quarter, which meets NIPDWR requirements under certain conditions.
Data adapted from the AWWA committee report.[2]

the quality of potable waters produced? In general, this is the approach most utilities use unless they are small water systems. Because small water systems are required by regulation to submit five or fewer samples per month, these sample collections are most frequently clustered on one day, with no other monitoring for bacteriological quality during the remaining days of the month. This minimal monitoring may be acceptable if a protected groundwater supply is used, but could be less desirable if surface water is the source for minimal treatment processing.

The one aspect that is often difficult to resolve relates to appropriate time for sample collection. Theoretically, sample collection should be distributed uniformly or randomly over the month and timed to coincide with filter backwashing schedules in treatment and distribution system maintenance activities, including major repairs. This ideal approach is only occasionally achieved, more through coincidence than by intent. The AWWA survey[2] revealed some interesting facts about current monitoring practices. As noted previously, small systems cluster their sample collections on one day early in the week (Table 6.6) for shipment to a distant laboratory. Larger utilities (serving over 2500 people, but without laboratory service) do make an effort to collect samples throughout the month. These sample collections are most often made on Mondays and Tuesdays with a minimal number collected at the end of the week or on Sunday (Table 6.7). This practice is a reflection of concern for shipment by various means of parcel service (mail, bus, and other delivery services) to avoid transit delays over weekends and holidays. For those utilities that have their own laboratory, sample collections can be made more frequently and at different times during the day. What was more remarkable is that some systems collect samples in the evening or overnight (Table 6.8) for either early morning shipment or processing at the start of laboratory hours. The more common procedure is to schedule sample collections in the morning for afternoon processing or shipment to a distant laboratory. The net effect of these tendencies is that the samples are most likely to be collected at set times that do not vary and are not necessarily correlated to times when filters are returned to service or related to pollution spills into the area of raw water intakes.

Table 6.6   Distribution of Sample Collection during a Month

Population served (thousands)	Utilities reporting all samples collected in indicated period (%)				
	1 week	2 weeks	3 weeks	Throughout month	No response
<1	54.0	14.2	1.7	23.9	7.1
1–2.5	35.4	33.4	6.3	25.1	
2.5–4.9	37.4	6.4	7.5	48.1	0.5
4.9–8.5	35.8	8.8	7.1	47.1	1.3
8.5–12.9	25.4	6.4	5.5	61.9	0.8
12.9–18.1	22.9	5.8	4.2	66.7	0.5
18.1–24	15.9	5.7	6.5	71.7	0.1
24–50	12.6	5.2	4.3	78.0	
>50	6.7	1.4	1.1	90.9	

Data from the AWWA committee report.[2]

Table 6.7   Distribution of Sample Collection during a Week

Population served (thousands)	Utilities reporting that samples are always or frequently collected on indicated day (%)						
	Sunday	Monday	Tuesday	Wednesday	Thursday	Friday	Saturday
<1	0	56.6	54.0	31.8	13.3	8.8	0
1–2.5	2.1	66.1	51.0	23.2	18.5	7.3	1.7
2.5–4.9	2.0	66.0	61.6	37.9	24.6	13.8	2.4
4.9–8.5	2.2	62.4	62.4	42.9	26.1	13.7	1.3
8.5–12.9	2.8	61.8	64.1	48.6	27.6	11.6	1.7
12.9–18.1	4.8	61.4	61.4	48.8	31.3	16.3	4.2
18.1–24	2.3	56.8	63.6	42.0	29.5	21.6	2.3
24–50	5.8	62.4	63.4	58.5	38.0	24.8	5.8
>50	14.7	83.5	84.1	76.0	64.8	48.2	14.8

Data from the AWWA committee report.[2]

Several other characteristics exist that could be used to establish a realistic level of monitoring intensity. One possibility is to relate sampling frequency to the volume of water consumed by the population served.[11] This approach has the advantage that the water supply authority will know with reasonable accuracy the volume of water being supplied to particular zones. The weakness

Table 6.8   Distribution of Sample Collection during a Day

Population served (thousands)	Utilities reporting that samples are always or frequently collected at indicated time (%)			
	Morning	Afternoon	Evening	Night
<1	75.2	47.8	9.7	0
1–2.5	74.2	32.6	0	0
2.5–4.9	76.4	48.3	2.9	0.5
4.9–8.5	74.3	42.9	2.2	0
8.5–12.9	82.3	44.2	1.1	1.1
12.9–18.1	88.6	40.4	1.8	1.2
18.1–24	88	40.9	1.0	0
24–50	92.7	40.9	3.4	1.4
>50	98.2	65.8	4.9	3.9

Data from the AWWA committee report.[2]

in this concept is that the per capita consumption of water will vary seasonally as well as geographically, i.e., tropical areas contrasted to temperate regions. Furthermore, very significant portions of the water demand may relate to industrial uses of the public water supply and to expanded watering of lawns by home owners during extended dry periods.

The number of service connections, excluding fire hydrants, is known by the water utility and has received some consideration as a basis for calculating sampling frequency. The inherent weakness with using a count of service connections is that there are numerous examples to be cited where meter connections do not reflect the number of people using the water supply in multiple family residences, such as apartment buildings, residence hotels, and condominiums, or in the workplace environment of highrise office buildings and large industrial complexes.

Total length of the distribution pipe network has also been proposed as the basis for determining sampling frequency. The basis for this approach is the general assumption that increased length of pipe network will bring increased risk of contamination from multiple residential and commercial service connections and by ground disturbance. Unfortunately, total length of the distribution system can be misleading where local topography requires the use of long distribution lines to reach small clusters of homes. In these situations, a small water utility could have the same amount of total pipeline in its distribution network as that of a medium-sized city with a distribution network configuration of a compact grid of interconnecting pipelines.

Sampling frequency formulas based on population served are the accepted practice in many countries.[13,10-14] The basis for this approach is the recognition that as the population increases, so will the size and complexity of the system and the potential for distribution network contamination by cross-connections and back-siphonage.[15] The emphasis is, therefore, on a demonstration of safe water quality in distribution to the public with lesser emphasis on monitoring treatment plant effectiveness. This latter attitude will be changing in the future with more restrictive directives intended to reduce disinfection by-products and organic precursors.

Assuming the concept to utilize population served is the best alternative, the next critical consideration is in the establishment of specified numbers of samples required per given population level. While the sampling frequency established has, in general, proven to be adequate for routine monitoring of most water systems, it has not provided satisfactory monitoring of small water supplies serving a population of less than 10,000.[16-18] It is in these systems that there often is no monitoring for most days in the month. To provide for a more realistic monitoring, the bacteriological quality of the smaller water systems should be based on one coliform sample taken from the first customer's location each week, not one sample per month from anywhere in the distribution system as specified in the National Interim Primary Drinking Water Regulations.[10] A fifth sample collected near a dead end on the distribution system would serve to check on water quality in the network. While the pipe network of a small water system is less complicated, with fewer service connections there is always a danger that water quality may deteriorate from cross-connections.

RAW SOURCE WATER QUALITY

Wastewater treatment plants, urban and rural runoff, feedlots, and a host of other activities often discharge to a watercourse that may be the source for a public water supply. Even the most pristine watershed may become contaminated from wildlife activity.[19] For these reasons, monitoring intake raw water for coliforms and turbidity is required in all water systems that only use disinfection as a treatment process.[20] Sudden changes in the background density of total coliforms or fecal coliforms generally occur with elevated turbidities. These abrupt changes in water quality are caused more frequently by stormwater runoff over the watershed, but can also be a result of turnovers of bottom waters in stratified lakes and raw water impoundments. Because these events also create an increased disinfection demand in the water to be treated, water plant operators must adjust chlorine concentrations to maintain a continued level of disinfection effectiveness. If the limits for total coliform or turbidity are exceeded, decisions to stop processing raw water for a few hours may also be necessary.

Groundwaters

Groundwaters are less likely to present any microbial treatment problems because of their high quality with minimal fluctuation in characteristics. Still, it is desirable to have some database on groundwater quality to prove the supply is protected from the influences of surface-water pollution or aquifer contamination by landfill leachates or other permeating influences of waste injections.

Because groundwater is generally constant in its characteristics, less frequent monitoring of the source water is required. Sampling once per week in dry periods and at daily intervals during one storm event each season may be sufficient to demonstrate the effectiveness of a soil barrier. The constant qualities of a groundwater source can be seen in Table 6.9. Note that there is minimal chlorine demand in this quality of water, suggesting little disinfection interference from trace organics that might be present. The uniformity of data also demonstrate that storm events during this period had no impact on the quality of water withdrawn from the aquifer.

Table 6.9 Characteristic Qualities of a Water Supply Well Field

Date range	No. of samples	SPC per ml	Total coliform per 100 ml	Temp (°C)	Water pH	Turbidity (NTU)	Chlorine demand
December 1–7, 1984	7	4.7	<1	—	.7	0.19	0.1
December 8–14	7	<1	<1	11.0	6.9	0.58	0.01
December 15–21	7	14.4	<1	10.7	6.9	0.53	0.03
December 22–28	6	28.5	<1	10.0	7.0	0.27	0.1
December 29– January 4, 1985	6	155.0	<1	10.3	7.0	0.30	0.1
January 5–11	7	1.6	<1	10.4	7.0	0.18	0.1
January 12–18	7	4.1	<1	10.0	6.9	0.20	0.1
January 19–25	7	7.0	<1	11.0	6.9	0.19	0.1
January 26– February 1	7	823.0	<1	11.0	7.0	0.21	0.1
February 2–8	7	78.3	<1	11.0	7.0	0.20	0.1
February 9–15	7	0.4	<1	10.0	7.0	0.24	0.1
February 16–22	7	12.3	<1	10.0	6.9	0.27	0.1
February 23– March 1	7	8.4	<1	11.0	6.9	0.19	0.1
March 2–9	7	1.6	<1	11.0	6.9	0.41	0.1
March 9–15	5	0.4	<1	10.0	7.0	0.22	0.2

Data courtesy of the New Haven, CT water system.

Surface Waters

Surface waters receiving discharges from municipal wastewater treatment plants and industrial and agricultural activities require conventional treatment, including coagulation, flocculation, sedimentation, filtration, and disinfection, to minimize organics and fluctuating densities of microbial contaminants. Such serial treatment provides the multiple barriers needed to ensure the microbiological quality of the water. Yet, with all of these safeguards, it is still very desirable to include the raw water intake in the monitoring strategy. In these situations, the water authority needs to maintain a continued database on the status of raw water quality and look for trends that may become alarming evidence of further degradation in the raw source water. Only through a constant vigil on watershed activities can the utility be alert to undesirable commercial developments that could increase the pollution loading of their source water. There are limits in processing poor-quality source waters into a safe supply on a continuous basis.

Sampling frequency for raw surface waters that receive minimal treatment should be on a daily basis for coliform densities, turbidity levels, apparent color, and chlorine demand. To illustrate this point, coliform data from Hemlock Lake, Rochester, NY was analyzed for effect of sampling interval on characterization.[21] Hemlock Lake is monitored daily, but data in Table 6.10 was arranged by time intervals on a weekly and biweekly basis. As can be seen, the mean, standard deviation, and 90th percentile values decreased with the reduction in sampling frequency, indicating the loss of information resulting from a reduced monitoring effort. Figure 6.1 expresses the scattering of coliform values resulting from selection of three different monitoring intervals. Less frequent sample collections could also produce data at the other extreme, i.e., make the water quality look worse than it really is over the long term.

Table 6.10   Sampling Data for Hemlock Lake

Sampling frequency	Total coliforms		
	Mean (count per 100 ml)	Standard deviation	90th percentile (count per 100 ml)
Daily	64	80	256
Weekly	17	30	44
Biweekly	19	40	54

Data from the Rochester, NY water system.[21]

Chlorine demand in raw water does vary as a reflection of pollution loadings, lake turnovers, and humic materials released in seasonal decay of vegetation. Abrupt changes in chlorine demand also often follow major storm events over the watershed. Table 6.11 illustrates the varying nature of chlorine demand for a water supply using disinfection as the only treatment of a raw surface water during the month of November. Fortunately, sufficient chlorine was applied to compensate for high turbidities and elevated densities of coliform on certain days during the month. When minimal treatment is practiced, adequate monitoring is clearly a key factor in determining changing characteristics of raw water sources and the need to carefully apply adequate concentrations of disinfectant to compensate for chlorine demand and cold water temperature influences on disinfectant effectiveness.

TREATMENT EFFECTIVENESS

Conventional treatment processes in series (coagulation, sedimentation, filtration, and disinfection) are essential to the processing of river water because of widely fluctuating water quality. Where lakes and large watershed impoundments are utilized, treatment may consist only of disinfection when

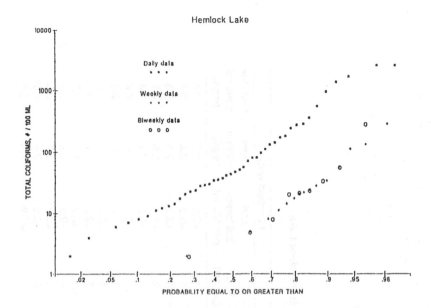

Figure 6.1   Effect of sampling on total coliform values from Hemlock Lake.

water quality meets specified quality conditions through watershed management or blending waters from different sources. More often, the concern for passage of protozoan pathogens has dictated the addition of filtration followed by disinfection as the minimal treatment scheme. Because fluctuations in raw source water quality impacts successful treatment, daily sampling should be done to provide information on stormwater runoff conditions and alert the utility to sewage treatment bypasses released upstream. In most instances this type of monitoring provides the only evidence of upstream pollution spills that could be used to demonstrate cause for intervention by a state environmental protection agency.

The most important sampling site in the treatment train is at the entry point for finished water (plant effluent) release into the distribution system. Here the purpose is to verify that treatment process barriers are effectively preventing the passage of organisms associated with fecal contamination. In some cases, this sampling site may be located at the first customer's tap, if contact time for disinfectant action must be extended beyond that available in the plant contact basin.

Composite Samples

Composite samples collected over a 24-h period are desirable to search for any pulsing releases of coliform bacteria from unstable filters brought on-line too quickly after backwashing. In process waters, bacteria may be in

Table 6.11   Impact of Raw Water Characteristics on Chlorine Demand for Hemlock Lake (Rochester, NY)

Date (November 1985)	Raw water[a]				Finished water			
	SPC (per ml)	Coliforms per 100 ml		Turbidity (NTU)	SPC (per ml)	Total coliforms per 100 ml	Free chlorine (mg/l)	Chlorine demand (mg/L)
		Total	Fecal					
1	36	4	2	1.7	<1	<2.2	2.3	0.8
2	52	8	8	1.8	<1	<2.2	2.4	0.8
3	89	<2	<2	2.1	<1	<2.2	2.5	0.8
4	14	<2	<2	2.1	2	<2.2	2.8	0.8
5	17	<2	<2	2.3	3	<2.2	2.4	0.4
6	—	11	8	2.4	<1	<2.2	2.4	0.4
7	210	2	<2	2.3	<1	<2.2	2.4	0.4
8	Spreader	22	2	2.2	1	<2.2	2.6	0.6
9	Spreader	14	9	2.3	<1	<2.2	2.7	0.9
10	Spreader	11	<2	2.2	<1	<2.2	2.6	0.7
11	14	11	2	2.1	4	<2.2	2.6	0.6
12	30	34	22	2.1	<1	<2.2	2.5	0.5
13	15	<2	<2	2.0	3	<2.2	2.6	0.7
14	26	9	5	2.5	2	<2.2	2.5	0.5

15	103	8	5	2.9	4	<2.2	2.5	0.5
16	Spreader	79	13	4.0	—	<2.2	2.6	0.6
17	161	33	4	3.2	2	<2.2	4.7	2.2
18	261	350	11	2.4	1	<2.2	3.3	1.4
19	84	180	2	2.5	1	<2.2	2.8	0.9
20	50	36	2	2.5	<1	<2.2	2.6	0.7
21	301	28	17	2.5	<1	<2.2	2.4	0.6
22	144	240	49	2.2	19	<2.2	2.4	0.7
23	36	27	2	2.1	<1	<2.2	2.4	0.6
24	170	14	2	2.0	1	<2.2	2.0	0.2
25	51	130	<2	1.9	1	<2.2	2.5	0.6
26	43	23	5	1.7	0	<2.2	2.3	0.4
27	23	240	5	1.7	2	<2.2	2.0	0.3
28	130	11	2	1.8	4	<2.2	2.1	0.2
29	60	20	<2	1.6	2	<2.2	2.1	0.3
30	14	130	<2	1.5	4	<2.2	2.1	0.5

[a] Raw water temperature average of 11.5°C for the month.

aggregate distribution rather than being randomly distributed. In this case, composite sampling should be more effective than grab sampling in detecting coliform bacteria from an equivalent volume of water tested. Composite samplers are available commercially, but may not be readily adapted to the aseptic sampling of large volumes of treated drinking water. Specifications for sample compositing must include control of collection frequency and volume, immediate dechlorination to prevent further inactivation of bacteria by residual disinfectant, and provisions for sample refrigeration to minimize subsequent changes in the microbial flora prior to examination. While compositing of plant effluents has been investigated, results have not been very promising because of technical limitations in setting up proper sampling conditions.[22] The only other option is to schedule individual large-volume sample collections throughout the 24 h interval to detect small residual densities of coliforms being released periodically.

Large-Volume Samples

Large-volume sample analysis is the other alternative. In this case a large grab sample (1 to 2 l) is collected for processing through a membrane filter procedure to concentrate the organisms, or a Balston filter is used to entrap organisms during a 24-h continuous passage of plant effluent, or sample collections (1 l) are correlated with scheduled sand filter returns to service after backwashing. Any routine collection of a sample or several samples each day may fail to detect any coliform releases because the timing sequence is wrong or the volume analyzed inadequate to detect a few coliforms per liter.

When the utility must locate the sampling site at the first customer's tap (to maximize disinfection contact time) some of these options are not practical. In this case, a permanent sampling station should be installed adjacent to the property with controlled access at all times. Because any residual coliform bacteria present in a plant effluent will be in a stressed state, it is important to process these samples within a few hours to optimize recovery and use m-T7 agar for optimum recovery of coliform survivors.

DISTRIBUTION SYSTEM INTEGRITY

Testing finished water entering the distribution system is the foundation of every compliance-monitoring program. Of equal importance is verification that a safe water is supplied to every consumer in all sections of the distribution area. To accomplish this objective requires a careful selection of sampling sites based on a study of the individual system peculiarities, and identifying the strengths and weaknesses in protecting water quality during delivery to the consumer's tap. Regardless of system size, there are some sites that need to be established as permanent locations, such as pumping stations, storage tanks, and areas of previous coliform occurrences.[2,19] Open reservoirs of finished water, particularly those that are not protected by booster chlorination need to be watched for sudden contaminating events from wild bird or other animal populations, stormwater runoff, and accidental spills of domestic and industrial wastes from truck traffic on adjacent highways.

Strategic Site Selections

Strategic site selection starts at the first-customer location. This site is a priority requirement to verify treatment effectiveness.[23] Because the treatment barrier must not be penetrated by any indicator bacteria and waterborne pathogens, daily coliform testing of water collected near the first customer is essential. Whenever any of the defined water quality characteristics (coliform, turbidity, color, and chlorine demand) of the raw water temporarily transcend recognized limits, bacteriological samples should be taken in the morning and afternoon at different sampling sites along the pipe network for a 2-d period to verify continued production of a safe water supply.[21] While this strategy may satisfy a regulation requirement, other locations should be selected to search for reactivated coliforms stressed by inadequate chlorine exposure. Stressed coliforms[24-27] that become established in the sediments of some slow-flow section or deadend may be one of the scenarios for the initiation of a biofilm in the pipe network. Associated with this kind of survival are inadequate disinfection C·T values for alternative disinfections and insufficient free chlorine residuals in the distribution system. Evidence of ineffective disinfection application can be observed in the microbial quality changes occurring in water analyzed from the ends of the system. It is here that the least amount of any disinfectant residual may be found and changes in disinfectant application will most likely be first reflected in shifts in the microbial quality. Declining free chlorine residual concentrations below 0.1 mg/l during warm-water periods may result in sudden increases in the heterotrophic bacterial population followed in a few weeks with coliform occurrences at these sites.[28] A similar effect has been observed by one utility using chloramines.[29]

Other sampling sites should be chosen carefully to reflect: water quality distributed in all subsections of the pipe network, largest number of people at a service connection (hospitals, schools, public buildings, highrise apartments, hotels, and factories), and access to taps on private property.[30] Such locations should be sampled on a rotating basis, with none being repeatedly collected for analysis in more than three consecutive periods in any month, unless a coliform-positive occurrence is detected.

For those utilities that have several water treatment plants supplying potable water to a single distribution system, the placement of sampling sites takes on another perspective. In these situations, there are multiple entries for finished water that must be monitored near each first-customer location. Furthermore, blending waters of different characteristics and reversals of flow in the pipe network must be evaluated as contributing factors to difficulties in locating a contaminating event. Knowledge of transit time in the system, the extent of water demand, and pressure variations are important aspects in detecting and tracing a contamination event.

SAMPLE COLLECTION AND STABILITY DURING TRANSIT

Water samples must be collected with care, otherwise the information reported on water quality will not be valid and could have serious repercussions for the utility in devising an action response or in complying with regulations. Properly trained sample collectors are cognizant of procedures for correct collection of samples so there is no chance of mislabeling or contamination at time of sampling or preparation for shipment. This attention to detail is important to avoid the management response that the unsatisfactory laboratory report was false, being caused by sample contamination during collection. For those individuals who must review sample collector protocols periodically, there are a number of aspects that should be evaluated before judgment can be made on the adequacy of the procedures and legitimacy of samples submitted to the laboratory.

SAMPLE COLLECTION RESPONSIBILITY

Bacteriological samples are collected by designated employees of most utilities. In some states, part of the samples per month are collected by district sanitarians or county health department employees who forward these samples to the state laboratory for processing. In the case of small water systems, samples are collected by the operator and mailed to the state laboratory. Other less common sample collections are performed by a university laboratory or a private laboratory under contract to monitor the system and process the samples. A study of the responses from 1796 utilities in the AWWA national survey[2] indicates most water systems use their own personnel to collect bacteriological samples (Table 6.12). This would suggest the utilities can directly supervise the training and any periodic refresher course on the proper techniques of sample collection.

Sample Collection Procedures

Sample collection procedures are critical to obtaining a representation of water quality within a given segment of the distribution network; therefore, taps selected for sample collection should be supplied with water directly from a main rather than from a storage tank in the building. Other concerns in the selection of a tap involve location of the faucet in proximity to environmental contamination, leakage from the stem valve, and the inclusion of attachment devices. The sampling tap must be protected from exterior contamination

Table 6.12 Personnel Responsible for Collecting Bacteriological Samples

Population served (thousands)	Utilities reporting samples collected by indicated personnel (%)						
	Utility employees	State agency employees	Commercial laboratory employees	Both utility and state employees	Both state and commercial laboratory employees	No response
<1	77.6	3.6	6.3	8.9	0	3.6
1–2.5	87.1	0	3.0	4.7	3.0	2.1
2.5–4.9	81.3	3.4	3.9	8.9	1.5	1.0
4.9–8.5	84.5	3.9	3.5	6.2	1.3	0.5
8.5–12.9	85.1	3.9	1.7	8.1	0.6	0.6
12.9–18.1	81.8	4.5	4.5	5.7	1.1	2.3
18.1–24	89.1	2.9	1.7	2.9	1.7	1.7
24–50	85.4	2.9	4.4	3.4	1.5	2.4
>50	93.0	1.4	1.4	3.9	0.4	0

Data from the AWWA committee report.[2]

associated with being too close to the sink bottom or to the ground. Contaminated water or soil from the faucet exterior may enter the bottle during the collection procedure because it is difficult to place a bottle underneath a low tap without grazing the neck interior against the outside faucet surface. Leaking taps that allow water to flow out from around the stem of the valve handle and down the outside of the faucet or taps in which water tends to run up on the outside of the lip are to be avoided as sampling outlets. Aerator, strainer, and hose attachments on the tap must be removed before sampling. These devices can harbor a significant bacterial population if they are not cleaned routinely or replaced when worn or cracked. Whenever an even stream of water cannot be obtained from taps after such devices are removed, a more suitable tap must be sought. Taps whose water flow is not steady should be avoided because temporary fluctuation in line pressure may cause sheets of microbial growth, lodged in some pipe section or faucet connection, to break loose. The chosen cold-water tap should be opened for 2 or 3 min or for a sufficient time (using a smooth-flowing water stream at moderate pressure without splashing) to permit clearing the service line. Then, without changing the water flow, which could dislodge some particles in the faucet, sample collection can proceed. Because length of service lines vary in each situation, extended flushing done to draw fresh water from a street lateral main cannot be generalized. There are just too many variations in building-pipe network configurations to establish a specific flushing time.

A check of tap-flushing practices done by sample collectors[2] reveals that from 4 to 13% of these individuals run water to waste for less than 1 min before taking a sample (Table 6.13). This practice will not discharge sufficient static water from most service lines to bring in fresh water from the street mains. As a result, the bacteriological quality of water tested is not representative of the distribution system. Depending on how long the water has been in the pipe and whether ambient air temperatures have caused a water temperature rise, heterotrophic bacteria may begin to multiply to perhaps ten times the original density in fresh distribution water.

Treating water taps before collecting potable water samples is not necessary if reasonable care is exercised in the choice of sampling tap (clean, free of attachments, and in good repair) and if the water is allowed to flow adequately at a uniform rate before sampling. Superficially passing a flame from a match or an alcohol-soaked cotton applicator over the tap a few times may have a psychological effect on observers, but it will not have a lethal effect on attached bacteria. The application of intense heat with a blow torch may damage the valve-washer seating or create a fire hazard to combustible materials adjacent to the tap. If successive samples from the same tap continue to contain coliforms, however, the tap should be disinfected with a hypochlorite solution to eliminate external contamination as the source of these organisms.[31]

This negative position on a protocol for flaming taps before sample collection is supported by several independent studies. After a study of 253 samples from farm water supplies, Thomas et al.[32] reported that flaming taps before sampling

Table 6.13 Flushing of Tap before Sample Collection

Population served (thousands)	Utilities reporting tap flushing for indicated time (%)			
	<1	Between 1 and 3 min	>3 min	No response
<1	9.8	33.0	51.7	5.4
1–2.5	7.0	43.2	45.4	4.5
2.5–4.9	5.9	40.8	49.3	3.9
4.9–8.5	4.9	40.3	49.6	5.3
8.5–12.9	6.1	46.0	41.2	6.6
12.9–18.1	13.6	37.5	40.9	8.0
18.1–24	4.8	50	38.6	6.7
24–50	7.8	37.6	46.3	8.3
>50	4.5	41.9	46.8	6.7

Data from the AWWA committee report.[2]

resulted in no significant differences in the multiple-tube test (five-tube MPN) for both total coliforms and fecal coliforms, or in the standard plate counts incubated at 37 or 22°C. There was, however, a tendency for the bacterial densities to be lower, but the trend was not significant and could have occurred by chance. In a second study involving 527 distribution samples collected without tap flaming from the Chicago public water supply, only two samples (or 0.4%) contained coliform bacteria.[33] For a third study, water was flushed from taps located in 76 gasoline service stations in Dayton, OH, but again, the taps were not flamed or otherwise disinfected.[34,35] The results showed no coliform-positive samples from 40 of the 76 stations, and MF coliform counts in excess of 4 per 100 ml occurred in only 4 of the 10,916 samples tested. In spite of this evidence, more than 50% of the 1796 utilities surveyed[2] reported they flame the sample tap before taking a sample (Table 6.14). Only the larger utilities, serving populations of more than 24,000, were less convinced that this practice was necessary.

When sample glass bottles fitted with ground-glass stoppers are used, a string or paper wedge must be inserted between the bottle and closure before sterilization to facilitate easy opening during sample collection. Upon opening

Table 6.14 Flaming of Tap before Sample Collection

Population served (thousands)	Utilities reporting indicated practice (%)		
	Tap flamed	Tap not flamed	No response
<1	50.9	46.4	2.7
1–2.5	57.9	40.7	1.3
2.5–4.9	60.6	38.4	1.0
4.9–8.5	62.8	35.8	1.3
8.5–12.9	56.4	41.4	2.2
12.9–18.1	57.9	40.0	2.2
18.1–24	68.1	30.7	1.2
24–50	46.3	53.2	0.4
>50	38.7	59.5	1.8

Data from the AWWA committee report.[2]

the bottle, the string or paper wedge should be discarded without being allowed to touch the inner portion of either the bottle or stopper. Reinserting this item into the sample bottle after sample collection will increase the risk of water sample contamination. Use of a screw cap, wide mouth, plastic (autoclavable) bottle may be the better option for this reason.

Regardless of the type of sample bottle closure used, the bottle should not be laid cap down or placed in a pocket. Rather, the bottle should be held in one hand and the cap in the other, keeping the bottle cap right-side up (threads down) and using care not to touch the inside of the cap. Likewise, contaminating the sterile bottle with fingers or permitting the faucet to touch the inside of the bottle should be avoided. The bottle should not be rinsed or wiped out or blown out by the sample collector's breath before use. Such practices may not only contaminate the bottle, but also may remove the thiosulfate-dechlorinating agent. During the filling operation, care should be taken that splashing drops of water from the ground or sink do not enter into either the bottle or cap. The stream flow should not be adjusted while sampling in order to avoid dislodging some particles in the faucet. The bottle should be filled to within 1 in. of the bottle top or to the shoulder of the container and capped immediately. The tap should then be turned off. An excellent illustrated booklet on distribution system bacteriological sampling is available from the California–Nevada Section, American Water Works Association.[36]

SAMPLE STABILITY

Maintaining the microbial stability of a water sample once it has been collected is difficult, but important for an accurate assessment of water quality. Soon after water is collected, environmental conditions begin to change in the sample. Ambient air temperatures alter the original water temperature, thiosulfate neutralizes the inactivating effects of residual chlorine on bacteria, and some organisms in the flora begin to successfully compete for minimal nutrients in the water while other organisms become stressed or die from adversities (inadequate nutrients, antagonism, and predation). Coliform bacteria are more likely to be poor survivors in this competitive situation, particularly when water temperatures exceed 10°C and transit time is extended beyond 24 h.[38-42] By preference to insure the least amount of unpredictable change, water samples should be iced and analyzed within 6 h of collection to prevent changes in coliform densities.[42]

Sample Transit Time

Sample transit time logistics should not be a problem for those utilities that have their own laboratories. This situation is not the case for other water systems that must send their samples to a state laboratory for analysis. In this latter situation, every effort must be made by sample collectors to time mail shipments of drinking water samples with existing mail, truck, bus, or air schedules. Because of the probability that samples mailed toward the end of the week or before a holiday will be held in shipment for 1 or 2 d, sample collectors are forced to confine sample collections to the first three days of the week. If the postal service is unacceptable, shipment by truck, bus, bank clearing house service, private courier service, or other alternative means of transportation should be investigated. For those water supplies located within 2-h driving time of the laboratory, every effort should be made by the sample collector to bring sample collections directly to the laboratory rather than resort to mail service. When samples are to be transported by car, delivery should be done promptly and not postponed to some more convenient time during the next few days. Transporting samples for several hours in the high temperature of a car trunk or on the back seat of the non-air-conditioned automobile during the summer can drastically alter the bacterial population. Long exposure to sunlight should be avoided because of the lethal effects of sunlight on bacteria.[43]

The limitations of sample stability present a serious problem for monitoring programs in some of the larger western states (Table 6.15). A part of this problem is due to adverse winter weather conditions that slow movement of all forms of transportation. From the nationwide survey,[2] the problem of excessive sample transit periods over 24 h appears to be a serious problem for all utilities that do not have a certified laboratory (Table 6.16); the worst record is for those samples that must travel a greater distance to reach a state central laboratory. To a large extent, this problem is related to the lack of coordination between sample collection and timing of mailing to meet the diminishing frequency of mail movement. The problem can be further aggravated in remote areas by the movement of mail to regional centers, sometimes outside a specific state for sorting and consolidation, prior to shipment back into the originating state and on to the laboratory.

In subtropical and tropical areas, special effort should be made by sample collectors to refrigerate or insulate all water samples during transit to the laboratory because of the warm water and air temperatures. Keeping water samples cool (not frozen) will retard changes in the bacterial density and, thereby, yield laboratory results that are a more meaningful measurement of water quality at the time of collection.

Samples shipped by commercial carrier must be adequately protected in suitable shipping cases to avoid breakage or spilling. Where sample collections are made within a reasonable driving distance of the laboratory, sample collectors may use large picnic coolers as sample cases with a 4 to 10°C temperature maintained through use of ice, dry ice, or prefrozen chemical cold packs. If only ice is available as a refrigerant, it may be necessary to modify the sample case by constructing a watertight center compartment to contain the ice and, thus, avoid any contamination of water samples with melted ice. The laboratory staff should also be authorized to reject any samples submerged in reservoirs of melted ice. This requirement eliminates any doubt concerning the integrity of the collected sample during transit.

Table 6.15 Municipal Water Sample Transit Time Observed in Five Western States

Time (d)	Occurrence (%)				
	Idaho	Utah	Oklahoma	Colorado	Missouri
1	88.6	88.0	82.8	68.6	56.6
2[a]	7.2	6.5	9.2	18.1	30.1
3	3.0	1.8	5.2	11.3	6.5
4	0.7	1.1	1.8	2.0	6.8
>4	0.5	2.6	1.0	None	None
Total samples	915	384	499	204	7185
Time period	(January to October 1968)	(6 d in September 1968)	(11 d in February 1969)	(January to October 1968)	(July to October 1968)

[a] Maximum transit time limit for potable water samples.

Data from Geldreich.[44]

Table 6.16   Length of Time between Sample Collection and Analysis

Parameter	Utilities reporting indicated time between sample collection and analysis (%)			
	<24 h	>24 h	Variable	No response
Population served (thousands)				
<1	42.9	50.4	1.7	5.0
1–2.5	34.1	60.0	4.5	1.4
2.5–4.9	42.9	50.4	3.1	3.5
4.9–8.5	47.1	43.9	5.7	3.3
8.5–12.9	51.0	43.5	1.0	4.5
12.9–18.1	47.2	43.8	3.9	5.1
18.1–24	55.3	37.2	5.3	2.1
24–50	71.1	21.6	5.0	2.3
>50	83.0	10.5	5.6	0.9
Type of laboratory				
Utility	82.5	11.5	5.0	1.2
Regional	76.4	12.5	5.6	5.6
State	27.1	65.9	3.9	3.1
Commercial	58.6	34.9	2.4	4.1
No response	49.0	35.5	6.5	9.0

Data from the AWWA committee report.[2]

Laboratory Processing

Laboratory processing of samples must be initiated within 2 h of the arrival time. Where laboratories receive samples throughout the day, the staff should plan to initiate the processing of all morning samples by 11:00 a.m. and samples received during the afternoon after 3:00 p.m. Late sample collection arrivals may necessitate using a flexible work schedule for some staff members. When samples are delivered too late to be examined during the regular work day, overnight refrigeration of these late arrivals is a permissible alternative, provided processing is done promptly the next morning. Under no circumstances, should samples be stored in the refrigerator during the weekend for processing on the following workday.

SPECIAL CONSIDERATIONS

The development of a monitoring strategy to fully characterize a distribution system will not easily fit some generic pattern because each system is unique unto itself. Very small water systems serving less than 100 people generally have an uncomplicated distribution network, but as the population served increases and service area in miles grows, monitoring distribution networks becomes very complex for a variety of reasons. In some instances the utility may continue to expand the service area coverage to a point where another source and treatment plant become necessary to meet demand. The other alternative in this situation is to purchase water from an adjacent utility and deliver that water to the system during peak seasonal periods through an interconnection of the two distribution systems. As an outgrowth of the continued growing need for more water and better quality, consolidation of adjacent systems may occur. The result of this consolidation is to bring together a system of multiple sources and multiple treatments, forming a merged distribution network. These additions of source, treatment, and service coverage require a redesign of the monitoring strategy to achieve representative coverage. Some idea of the magnitude of the problem can be seen in the size of the pipe network involved for different systems (Table 6.1). As a general rule, size of the pipe network in miles of pipe may be estimated by determining the square root of the population served.

SMALL-SYSTEM NETWORKS

This category of distribution networks comprise the very small public water supplies serving less than 500 people, small subdivisions in a rural housing development, mobile home parks, and a variety of noncommunity systems that provide potable water for some industrial requirement. The distribution pipe network may be described as being less than 70 mi in length, containing a single finished water reservoir that often delivers water by gravity from a groundwater supply that may or may not require disinfection.

Monitoring Strategies

Monitoring strategies for small water systems must focus on water quality at the well or, if disinfected, bacteriological quality near the first customer. If coliforms are found sporadically in an untreated groundwater supply, monitoring should focus on the well because untreated source water quality affects all users of the distribution system. Keep in mind that when an occasional coliform is found in an untreated groundwater, it is imperative to establish that the few organisms (less than ten coliforms per 100 ml) are environmental strains, not coliforms of fecal origin.

Untreated well supplies should be monitored, preferably, once per week to verify that water quality released to the distribution system does satisfy the maximum contaminant level (MCL) for coliform bacteria (no detectable coliform bacteria per 100 ml in 95% of all samples analyzed over a running 12-month period). To sample the untreated water supply less frequently during the month will require an on-site sanitary survey every 5 years by the state water authority or its designated representative. Where a sanitary survey indicates the available water supply must be chlorinated, evidence of treatment effectiveness is critical and thus the major sample collection emphasis during the month should be placed on water quality at the first water supply tap, with only one sample suggested per quarter being collected from different areas of the pipe network to insure no contaminating defects in water supply distribution.

Monitoring water quality in distribution is necessary to verify the integrity of the pipe network. Collection and analysis of one sample from an untreated groundwater supply at a randomly selected site on the pipe network each month should suffice. Placing the distribution system monitoring into a subordinate role for determining water quality delivered to the consumer assumes there is minimal risk of contamination because line repairs are infrequently required, there is a limited building of new structures in the community, and urban housing renovation is a rarity.

What has been described here is a comprehensive monitoring program for the smallest water systems with uncomplicated distribution networks of less than 70 mi of pipeline. The federal regulation has mandated a sampling program that is even less intense for these systems only because of monitoring cost considerations and the logistics of sample collection, mailing costs, and significant increase in tests that would be required to be analyzed by the state laboratory. However, the concern for waterborne outbreaks remains and the need to get an early warning of unsatisfactory water quality persists.

Alternatives

Alternatives for some of the required coliform monitoring have been proposed for small systems without sacrificing public health protection. For example, a triannual sanitary survey of small water system operations (serving <3000 persons) could be substituted for some of the increased monitoring requirements, provided the utility utilizes a disinfected groundwater or filter their raw surface water prior to disinfection. This type of survey (Appendix) involves an on-site inspection of the watershed, raw source quality, treatment processes, and distribution network and review of the monitoring database for the utility by qualified personnel. The intent is to determine effectiveness of the water authority in providing safe water to the community.

Another approach suggested has been to apply a chlorine measurement substitution rule for part of the coliform monitoring requirement.[11] Under this rule, the water authority could substitute up to 75% of bacteriological samples required, provided the chlorine residual measurements are taken at points representative of conditions within the distribution system and at a frequency of at least four chlorine measurements for each substitute coliform analysis. None of these chlorine substitution sites should replace locations where coliforms have been reported in previous sample examination. Furthermore, the water authority must maintain no less than 0.2 mg/l free chlorine throughout the distribution system and daily determinations of chlorine residual are mandatory. While this option would successfully demonstrate the presence of a disinfectant residual throughout the distribution network, there is no certainty that coliforms might not be present at the same time because of ineffective contact time, particulate protection, or biofilm development.

LARGE-SYSTEM CHARACTERISTICS

Water utilities in metropolitan areas have more complex distribution systems and must respond to population growth as well as react to a variety of water line breaks incurred from accidents in urban housing rehabilitation, major demolitions for new construction, and highway repairs. The total length of pipe in the service area may range from 100 to more than 1000 mi of branching lines, interconnections, closed loops, and dead-end sections. The number of finished water reservoirs and standpipes increases to provide a water supply reserve to meet demand and improve water pressure throughout the far reaches of the service area. Some customer service areas have higher water demands than others, so the general belief that water flows out through the system in one direction is not true. There can be water flow reversals as a result of uneven demand, temporary bypass of reservoirs for maintenance service, or from excessive demand created in fighting a major fire. Considerations should also be given to the effects of hourly periods of peak demand as well as higher water usage during certain days of the week. As a result of these complexities the retention time for water passing through the distribution system may vary from hours to several weeks in different parts of the distribution network. In fact, it is not unexpected to note that any change in disinfection application at the treatment plant may not be observed in the far reaches of the distribution system until several days or weeks later.

Perhaps the first step in designing a monitoring plan for extensive distribution systems is to determine flow patterns for major zones in the pipe network and retention time for water passage to the end of the system. Changes in water quality in the Cincinnati distribution system following a modification in the point of disinfection application at the treatment plant were not observed immediately.[45] There was an approximate 15-d delay before some decrease in free chlorine residual concentration and pH occurred. In other water systems, the ripple effect of treatment modifications ranged from 3 to 7 d.[45,46] Unfortunately, not all utilities have an accurate understanding of flow patterns and retention time for water within their pipe network. Availability of this type of information can be useful in identifying zones of slow-moving water as likely areas of bacterial regrowth and flow patterns provide guidance on locating direction of contamination passage.

The next step in developing the monitoring plan is to identify characteristics of the system that will impact bacteriological quality. Each distribution system is unique and, therefore, it is difficult to utilize a generic model approach without requiring some significant changes or addition of factors to the equation. With respect to microbiological quality, attention must focus on sites that monitor the following: first-customer location to test the treatment barrier, areas of greatest population density, locations of previous coliform occurrences during the year, major slow-flow or dead-end sections where bacterial regrowth may occur, open finished water reservoirs (risk from wildlife contamination), and regions of lost chlorine residual.

Permanent Site Selections

Permanent site selections must be sensitive to sudden changes in drinking water quality, system-wide. Examples of such sites are the first-customer location, open finished water reservoirs, interconnection with a neighboring water utility, and areas of continual low line pressure. There is no specific percent of the total samples collected each month recommended to be from fixed sites; rather, the emphasis should be to select sufficient key locations that will provide a constant watch over potential places for contamination intrusion. This flexibility leaves the bulk of the monitoring effort directed toward random selection of accessible sites within all districts that comprise the distribution service area. The intent is to continually reappraise water quality throughout the distribution system in a logical and repetitive manner.

CONSOLIDATED SYSTEMS

For those utilities that have either multiple treatment plants or were created by consolidation of two or more neighboring public water systems through interconnection, the approach to a monitoring strategy becomes more complex. In these situations, there are multiple entries of finished water that are being processed from different sources and perhaps using variations on treatment approach. Combining distribution systems may also introduce additional concerns about pipe age, structural status of storage reservoirs, water flow restrictions, and increased frequency of coliform occurrences.

Sorting out these potential problem areas and redesigning the monitoring plan to fully characterize bacterial quality of water in distribution will require additional fixed sampling sites. Finished water released into the distribution system from each treatment plant or well field must be monitored near the sites for each first customer down-line. This will provide an alert of which water treatment barrier needs immediate review for process correction. Water quality in storage at each reservoir should not be overlooked as a potential contributor to coliform persistence in accumulated sediments and side wall slimes. A permanent sampling site should also be established at the points of pipe network interconnections from different systems and related to flow direction through the junctures. Study of flow patterns may also reveal new areas of slow-flow or static water that must be monitored for bacterial growth, particularly when water temperature exceeds 15°C.

Combined systems present a complex pattern of water flow in the pipe network with the potential for numerous flow reversals during any 24-h period. Plotting the zone of water supplied by any particular water treatment plant or well field can be accomplished by use of some water quality characteristic unique to the supply. For instance, hardness in groundwater supplying a system that also receives treated surface water is a contrasting marker. Total trihalomethanes or fluoride in treated surface water are other approaches to tracing the movement through the mixed water system. Knowledge of water transit time, flow pattern, and extent of water demand are important factors that can be very helpful in tracing a contamination event in combined systems.

COLIFORM OCCURRENCES IN COMPLIANCE MONITORING

Federal regulation specifies that total coliform detection is the primary bacteriological measure of quality for public water supplies.[23] When regulation revisions on the total coliform rule became effective (December 31, 1990), emphasis shifted from a quantitative measurement, with a monthly limit of one coliform per 100 ml in all samples, to a qualitative test for coliform presence or absence, with a limit of 5% positive occurrences over each month for most systems.[23] Small water systems must analyze data from a minimum of 60 consecutive samples on a continual basis before any meaningful coliform percent occurrence can be established.

FREQUENCY OF OCCURRENCES VS. QUANTITATION

Much of the information (Table 6.17) on the relationship of coliform frequency of occurrences to the monthly mean coliform limit of one organism per 100 ml was gathered from data supplied by one utility (New Haven, CT) with multiple treatment systems feeding the same distribution network. Careful investigation of water movements in the common distribution network was done to select the sample locations for treated water origin and characteristic of each treatment plant. Data in Table 6.17 also include monitoring information from another utility (Rochester, NY) that applies disinfection as the only treatment process to surface water originating in a large lake.

These data were arranged in several ranges of coliform percent occurrences to compare with calculations of the coliform mean values within each grouping. In general, increasing percent of coliform occurrence paralleled higher coliform monthly mean values; however, there were exceptions. This increase in coliform percent embraced higher coliform densities from an occasional sample in the month and was sensitive to large clusters of samples containing one to three coliforms per 100 ml. A hypothetical example can be seen in Table 6.18 (Case 1 vs. Case 2). Interesting was the fact that maximum individual coliform densities in the different ranges of coliform occurrence level varied independently. Turbidities and chlorine residual concentration appeared to have no noticeable influence on the presence of coliforms in any frequency range, nor was there any significant impact from source water or how it was treated. The crossover point where percent coliform occurrence equates to an average of one coliform per 100 ml for all samples examined during a 30-d period appeared to approximate 8%.

Seasonal changes in distribution water quality were also investigated (Table 6.19) using the percent coliform occurrence concept (presence and absence). Warm-water months often produce more total coliform occurrences related to biofilm growth in the distribution system. During these months, frequency of coliform occurrence may exceed the 5% limit as well as the previous quantitative standard of one total coliform per 100 ml, but could not provide any indication of the magnitude of the problem other than that the water quality was out of compliance. Applying increased chlorine alone to achieve higher chlorine residuals in the distribution system had little impact on suppressing coliform occurrence in these instances.

REPEAT-SAMPLING PROGRAM

The intent of a repeat-sampling program is to confirm the existence of a coliform occurrence in the public water supply and simultaneously establish whether these coliforms are of fecal origin. Because the repeated occurrence of fecal contamination is serious, there should be a prompt report by the laboratory on every unsatisfactory sample result to the water utility plant operations or engineering division. Slow communication of these results and lack of immediate action to resample the same site defeats efforts directed toward maintaining a rapid response to an early alert of possible contamination of the public water supply.

Upon initial detection of coliforms in a sample, there should be a request for immediate repeat collection and analysis of water from the site and several additional samples collected above and below this location. This tactic will not only provide evidence that the coliform occurrence is persisting, but also information as to the extent of the contamination event in the pipe section. Intensity of the resampling effort should be continued at least on a daily basis until the problem is corrected and there are two consecutive negative samples.

These special sample collections should be sent to the laboratory for total coliform and fecal coliform analyses. Information acquired on the origin of these coliforms (fecal or nonfecal), field data collected on chlorine residuals, and site inspection of the line section and associated service connections is needed to formulate the proper response that could also eventually involve a decision on public notification.

Not all repeat-sampling efforts will confirm the initial report of coliforms found in the water system. This inconsistency is particularly a common occurrence if repeat sampling is postponed until weeks later. Rarely does the difference

Table 6.17 Range of Percent Coliform Occurrence vs. Monthly Coliform Mean Values

Treatment system[a]	Source	No. of samples	Coliform positive (%)	Coliforms per 100 ml		Distribution	
				Monthly mean	Monthly range	Turbidity (NTU)	Free chlorine residual (mg/l)
Cheshire (New Haven, CT)	Wells	843	0–4.9	0.07	<1–31	0.21	0.93
		321	5–9.9	0.93	<1–132	0.17	1.00
		177	10–14.9	2.85	<1–169	0.17	0.67
		112	15–19.9	0.96	<1–36	0.26	1.32
		190	20+	3.31	<1–153	0.25	1.15
Saltonstall (New Haven, CT)	Lake	858	0–4.9	0.24	<1–193	0.27	1.25
		515	5–9.9	0.47	<1–41	0.22	1.10
		335	10–14.9	1.23	<1–127	0.20	1.00
		403	15–19.9	2.42	<1–108	0.30	1.27
		50	20+	1.48	<1–27	0.22	1.33
Hemlock (Rochester, NY)	Lake	1333	0–4.9	0.12	<1–>80	1.42	1.18
		498	5–9.9	1.20	<1–>80	1.35	0.92
		269	10–14.9	1.30	<1–>80	0.95	1.20
		423	15–19.9	2.30	<1–>80	1.71	1.08
Sleeping Giant/Mt. Carmel (New Haven, CT)	Wells	675	0–4.9	0.16	<1–67	0.22	1.23
		342	5–9.9	0.93	<1–148	0.21	1.07
		329	10–14.9	1.60	<1–94	0.25	1.17
		235	15–19.9	3.80	<1–160	0.22	1.21
		314	20+	3.54	<1–165	0.29	1.26

Source	Type						
Gaillard (New Haven, CT)	Lake	2615	0–4.9	0.21	<1–157	0.58	1.95
		1626	5–9.9	0.70	<1–133	0.40	1.82
		1147	10–14.9	2.10	<1–200	0.46	1.81
West River (New Haven, CT)	River	586	0–4.9	0.16	<1–25	0.40	1.25
		224	5–9.9	0.44	<1–26	0.28	1.21
		263	10–14.9	1.60	<1–149	0.35	1.28
		70	15–19.9	3.80	<1–128	0.36	1.20
		34	20+	1.12	<1–12	0.61	1.16
Gaillard/West River (New Haven, CT)	Blend	1964	0–4.9	0.23	<1–159	0.49	1.64
		1217	5–9.9	0.84	<1–117	0.34	1.41
		319	10–14.9	0.82	<1–55	0.35	1.41
Gaillard/Whitney (New Haven, CT)	Blend	1053	0–4.9	0.06	<1–16	0.46	1.47
		721	5–9.9	0.68	<1–113	0.37	1.44
		204	10–14.9	0.67	<1–35	0.31	1.27
		282	15–19.9	2.20	<1–174	0.37	1.36
		113	20+	1.33	<1–93	0.20	1.26
Gaillard/Whitney/Wells (New Haven, CT)	Blend	1178	0–4.9	0.08	<1–12	0.80	1.36
		297	5–9.9	0.56	<1–40	0.30	1.41
		327	10–14.9	1.47	<1–125	0.33	1.37
		214	15–19.9	1.50	<1–100	0.34	1.34
		142	20+	4.40	<1–174	0.42	1.33

[a] Samples for each treatment system collected over a 3-year period, except for Hemlock Lake database of 12 months.

Data from Geldreich, Goodrich, and Clark.[47]

Table 6.18   Hypothetical Presence–Absence Test Violation

Individual samples in a month	Results (density per 100 ml)
Case 1	
1	0
2	0
—	—
—	—
—	—
94	—
95	1
96	1
97	1
98	1
99	1
100	1
Total coliforms per month	6
Density av per 100 ml (old regulation)	0.06 = No violation
Presence–absence (new regulation)	6% = Violation
Case 2	
1	0
2	0
—	—
—	—
—	—
—	—
96	0
97	100
98	75
99	50
100	0
Total coliforms per month	225
Density av per 100 ml (old regulation)	2.25 = Violation
Presence–absence (new regulation)	3% = No violation

in results reflect a sample contamination during the initial collection and even less likely is the initial positive result due to laboratory error in processing. Irregular contamination events that cannot be confirmed are generally related to repairs on the distribution network, sudden changes in water pressure that break loose coliform attached to particles in pipe sediment, or a brief cross-connection event on an associated service line. Line flushing to resolve consumer complaints of taste and odors produce water turbulence that may instigate a temporary resuspension of viable coliforms entrapped in corrosion by-products. Other brief contamination histories have been found to correlate with inadequate minimal treatment of surface water during major storm events and to contamination of open finished water reservoirs from wildlife or stormwater runoff.

Follow-Up Action

Follow-up action should be initiated as soon as the laboratory reports presumptive evidence that total coliforms are detected in a sample. A set of repeat samples is collected within 24 h of notification and sent to the laboratory for analysis.[23] At least one sample must be from the same location as the original positive sample while two other samples should be from adjacent service connections. If these repeat samples indicate the presence of coliforms, a duplicate set is again collected from the same sites, analyzed, and data submitted to the state. Intensity of the resampling effort should be continued at least on a daily basis until the problem is corrected and there are two consecutive negative samples. All repeat samples results become part of the compliance record and are factored into the coliform percent occurrences.

When total coliform bacteria are detected in either a routine or repeat sample, the laboratory must determine by further examination of the culture, whether fecal coliform or *E. coli* are present. The detection of fecal coliform, the predominant organism being *E. coli*, indicates a high probability that fecal contamination has mixed with the water supply. Information acquired on the suspect origins of these coliforms, field data collected on chlorine residuals, and site inspection of the pipeline section and associated service connections is needed to formulate the proper response that also involves a decision on public notification. While not absolute, there is sufficient risk of pathogen occurrence to warrant a notification to the public that fecal contamination has been detected in a portion of the distribution system and a recommendation made to boil water until new laboratory data have verified the contamination no longer exists in the pipe network. The chance that a fecal contamination event is part of a biofilm occurrence is remote because *E. coli* requires more nutrient support than biofilm organisms, does not effectively encapsulate, and is sensitive to free chlorine residuals (0.2 mg/l) often found in distribution water.

Trends in Repeat Sampling

Trends in repeat sampling were investigated in treatment systems using groundwaters, lakes, rivers, and a blend of these treated source waters (Table 6.20). Data for the year 1985 was taken at the peak of a biofilm coliform period so there was a significant increase in repeat sampling of all coliform occurrences. Even under this adverse situation, 76 to 82% of these repeat samples were negative. The following two years showed a dramatic reduction in coliform-positive samples. As a consequence, the mandatory repeat-sampling

Table 6.19 Comparing Monthly Percent Coliform Occurrences with Quantitative Coliform Data[a]

| | 1985 | | | | | 1986 | | | | |
	Percent coliform-positive	Coliforms per 100 ml Mean	Coliforms per 100 ml Range	Temp (°C)	Free chlorine residual (mg/l)	Percent coliform-positive	Coliforms per 100 ml Mean	Coliforms per 100 ml Range	Temp (°C)	Free chlorine residual (mg/l)
Cheshire (Well Source)										
January	0.00	0	0	6.1	1.1	0.00	0	0	5.8	1.1
February	3.92	0.22	<1–6	5.9	1.1	0.00	0	0	5.5	0.9
March	0.00	0	0	7.8	1.2	2.04	0.29	<1–14	6.6	1.0
April	0.00	0	0	10.4	1.1	5.56	0.50	<1–24	8.1	0.8
May	0.00	0	0	13.3	1.1	2.27	0.05	<1–2	10.0	0.9
June	25.37	1.81	<1–11	15.7	1.0	2.56	0.03	<1–1	10.5	0.8
July	30.77	5.55	<1–153	17.6	1.5	14.29	5.55	<1–169	14.6	0.7
August	15.71	1.27	<1–32	18.3	1.7	16.67	0.43	<1–9	19.2	0.8
September	5.45	4.11	<1–132	16.6	1.5	22.22	1.73	<1–25	15.7	0.8
October	6.67	0.18	<1–5	14.4	1.5	4.88	0.80	<1–31	14.5	0.7
November	0.00	0	0	11.9	1.4	2.78	0.03	<1–1	12.6	0.7
December	0.00	0	0	8.6	1.2	5.41	0.38	<1–12	11.2	0.7
Gaillard (Lake Source)										
January	0.00	0	0	5.1	2.3	0.48	0.05	<1–10	5.1	2.6
February	5.35	1.21	<1–98	4.9	2.1	2.86	0.06	<1–5	4.8	2.1
March	12.56	2.43	<1–90	6.8	2.2	1.58	0.02	<1–2	6.2	2.3
April	4.62	0.29	<1–17	10.0	2.1	5.10	0.78	<1–133	9.2	1.9

Month	1985					1986				
May	2.63	0.04	<1–3	13.6	2.3	3.62	0.04	<1–1	11.8	2.4
June	8.33	0.84	<1–79	13.8	2.4	5.13	0.09	<1–3	12.3	2.6
July	12.68	3.62	<1–200	18.1	2.0	8.13	1.65	<1–98	17.1	1.7
August	10.91	2.35	<1–171	18.8	2.1	12.50	0.92	<1–44	21.8	1.5
September	9.41	0.96	<1–61	17.6	2.2	11.72	1.72	<1–77	20.4	1.5
October	4.19	0.52	<1–157	16.4	1.9	4.20	1.22	<1–94	18.0	1.5
November	0.00	0	0	13.1	2.0	3.25	0.17	<1–16	13.4	1.5
December	2.26	0.10	<1–7	9.7	2.5	0.00	0.16	<1–11	9.8	1.5

West River (River Source)

Month	1985					1986				
January	0.00	0	0	5.7	2.1	0.00	0	0	4.4	1.1
February	3.70	0.26	<1–7	4.6	1.3	14.71	0.41	<1–6	4.0	1.3
March	10.00	2.80	<1–67	6.5	1.3	2.63	0.03	<1–1	5.6	1.5
April	3.70	0.44	<1–12	10.6	1.6	10.00	0.20	<1–4	8.5	1.4
May	3.85	0.08	<1–2	14.5	1.2	12.77	1.72	<1–28	12.0	1.3
June	2.65	2.25	<1–53	16.2	1.1	7.32	0.12	<1–3	13.8	1.2
July	26.47	1.12	<1–12	20.3	1.2	0.00	0	0	18.6	1.3
August	12.12	5.06	<1–149	22.6	1.2	7.50	0.95	<1–22	23.7	1.2
September	4.17	0.13	<1–3	20.4	1.1	16.67	4.19	<1–128	19.5	1.3
October	0.00	0.00	0	16.1	1.2	6.98	0.70	<1–21	16.5	1.3
November	4.17	0.08	<1–2	10.8	1.2	2.86	0.03	<1–1	12.0	1.3
December	0.00	0.00	0	8.7	1.3	0.00	0	0	8.5	1.3

[a] Except for ranges, all data were derived from mean values.

Table 6.20   Trends in Repeat Sample Results for 3 Years

Treatment system	Source	Year	Number of repeat samples	Positive repeat samples	
				%	No.
Cheshire	Wells	1985	88	20.4	18
		1986	10	10.0	1
		1987	6	0.0	0
Saltonstall	Lake	1985	111	23.4	26
		1986	25	0.0	0
		1987	10	10.0	1
Sleeping Giant/Mt. Carmel	Wells	1985	100	19.0	19
		1986	11	0.0	0
		1987	26	27.9	7
Gaillard	Lake	1985	213	17.8	38
		1986	21	9.5	2
		1987	22	4.4	1
West River	River	1985	36	19.4	7
		1986	16	6.2	1
		1987	2	0.0	0
Gaillard/West River	Blend	1985	86	10.4	9
		1986	12	13.5	2
		1987	6	25.0	2
Gaillard/Whitney	Blend	1985	96	14.4	14
		1986	8	12.5	1
		1987	4	25.0	1
Gaillard/Whitney/Wells	Blend	1985	143	24.2	35
		1986	12	9.3	1
		1987	3	0.0	0

Data from the New Haven, CT Water Authority.

requirement was required so infrequently that it is difficult to detect any statistical differences in coliform occurrence between treated source waters.

One of four patterns emerge from repeat sampling: (1) the repeat samples are negative; (2) there is only an initial repeat positive and succeeding follow-up samples are negative; (3) a linear decline in coliform densities occurs over several days until reaching a nondetectable limit; or (4) coliform densities demonstrate a declining cyclic tendency over several weeks. This quantitative information can be useful in attempting to understand the reason for a coliform occurrence.

Most repeat-sampling analyses will yield negative results even when samples are examined from the same site on the following day. This situation suggests that the coliform occurrence was caused by a minor instability of pipe sediment in the service line or building plumbing during sample collection, because other samples taken from adjacent sites were negative. When coliforms are found in repeat samples but their densities promptly decline to nondetectable limits (Figure 6.2), along with similar observations on any adjacent sampling sites, the inference is that there was either a momentary break in the treatment barrier, a contaminating event associated with line repair, or a temporary cross-connection nearby. A cyclic decline in repeat coliform occurrences (Figure 6.2) is associated with biofilm release in warm-water periods and may also be associated with similar cyclic coliform density patterns at other sites in a pipe network section. In severe cases of biofilm development, similar cyclic patterns may be noted throughout the system during warm-water periods.

EARLY WARNING OF COLIFORM OCCURRENCES

Heterotrophic bacterial densities in public water supplies may also become an issue in the total coliform monitoring program. There is evidence that heterotrophic densities of over 1000 organisms per milliliter may interfere with the detection of a few coliforms per 100 ml in either the multiple-tube fermentation test or the membrane filter procedure.[48] In the multiple-tube test, excessive, rapid growth by some heterotrophic bacteria (*Pseudomonas, Flavobacterium,* and others) can suppress the growth of total coliforms, preventing detection of gas production. Interference to the membrane filter test occurs when some strains of noncoliform bacteria are not restricted by the medium suppressive agents and develop either spreading growth over the membrane filter surface or numerous colonies that interfere with coliform growth and differentiation as sheeny colonies. These uncommon coliform interference problems may be avoided with the use of the Colilert coliform test,[49] although high densities of *Aeromonas* and *Flavobacterium* may give occasional false readings in the total coliform and *E. coli* tests, respectively.[50]

For those systems using disinfection as the only controlling treatment process, monitoring the distribution network for a disinfectant residual is especially important. When a measurable residual is not detected by conventional tests for disinfectant concentration in the distribution system, minimal heterotrophic plate count densities may provide secondary evidence of its existence in trace amounts. In this application, any heterotrophic bacterial density below 500 organisms per milliliter would be interpreted as evidence that there was a "detectable" residual for compliance purposes. This concept is based on observations from a survey of distribution water quality and monitoring practices, nationwide (Table 6.21), in which the density of heterotrophic bacteria could be held to less than 500 organisms per milliliter in 88% of 923 distribution systems supplying a disinfected water supply.[30]

While the national survey provided data on a wide variety of systems and water qualities in distribution, the information was limited to sample collections done over only 1 d for each system. Furthermore, no attempt was made to sort out free chlorine residual from the total chlorine measurements. An opportunity to review the influence free chlorine residuals has on heterotrophic plate count (HPC) was provided in a review of 4 months of data from Monmouth, NJ and 3 years of data from New Haven, CT. Data was reduced to

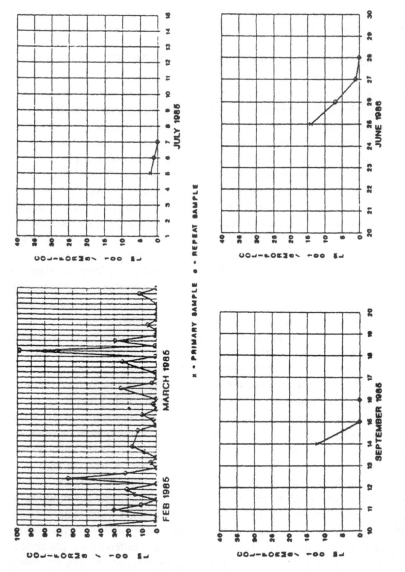

Figure 6.2   Repeat coliform-positive record in New Haven, CT.

Table 6.21   Effect of Varying Levels of Residual Chlorine on the Total Plate Count
in Potable Water Distribution Systems

Standard plate count[a]	Standard plate count (%)							
	Residual chlorine (mg/l)							
	0.0	0.01	0.1	0.2	0.3	0.4	0.5	0.6
<1	8.1[b]	14.6	19.7	12.8	16.4	17.9	4.5	17.9
1–10	20.4	29.2	38.2	48.9	45.5	51.3	59.1	42.9
11–100	37.3	33.7	28.9	26.6	23.6	23.1	31.8	28.6
101–500	18.6	11.2	7.9	9.6	12.7	5.1	4.5	10.7
501–1000	5.6	6.7	1.3	2.1	1.8	0	0	0
>1000	10.0	4.5	3.9	0	0	2.6	0	0
Number of samples	520	89	76	94	55	39	22	28

[a] Standard plate count (48 h incubation, 35°C)
[b] All values are percent of samples that had the indicated standard plate count.

Data in percent from a survey of community water systems in nine metropolitan areas;
Geldreich et al.[30]

three ranges of HPC for varying concentrations of free chlorine residual and plotted in Figures 6.3 and 6.4. As anticipated, HPC frequencies in density ranges of 1 to 10; 101 to 500, and over 1000 organisms per milliliter in Monmouth data declined with increasing free chlorine residuals. At New Haven, the pattern was somewhat different, in that free chlorine residuals of approximately 1 mg/l were associated with increased numbers of samples with 1 to 10 organisms per ml. Increasing free chlorine residual above 1 mg/l resulted in a rapid decline for any HPC density. The cause of this pattern distortion is not definitely known, but may have been due to change in microbial flora profile, with chlorine-resistant organisms gaining a temporary dominance before higher chlorine concentrations finally caused a general inactivation of the entire microbial population.

SPECIAL-PURPOSE MONITORING

Special-purpose monitoring may take several different directions: search for origins of a contamination event, circumscribe the affected area of the distribution system, and verify the effectiveness of responses to correct the problem. Such strategies go beyond a routine monitoring program by expanding the spectrum of laboratory analyses, increasing the frequency and changing the location of sampling sites, and by reviewing information obtained from a sanitary survey of all components in the production and delivery of a safe water supply to the consumer's tap.

To pursue these objectives will require an expansion of monitoring activities to include new sampling sites, more samples collected at an increased frequency, exploratory laboratory analyses, and daily data reviews for impact of treatment adjustment and distribution system operations.

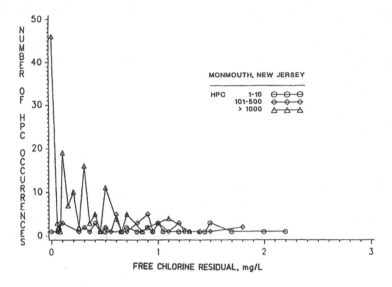

Figure 6.3 Heterotrophic plate count occurrences vs. free chlorine residual, Monmouth, NJ.

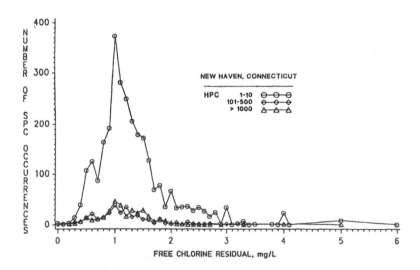

Figure 6.4 Heterotrophic plate count occurrences vs. free chlorine residual, New Haven, CT.

SEARCHING FOR ORIGINS OF A CONTAMINATING EVENT

Repeated occurrences of low densities of coliforms bacteria (1 to 10 coliforms per 100 ml) indicate chronic contamination that needs urgent attention. Searching for origins of a contaminating incident poses the question, Was there a break in the treatment barrier or a loss of protection in the distribution system? In either case, these positive findings should be viewed as an early warning of a break in the protective barrier against pathogen entrance into the potable water supply. This concern is supported by a study of causes of waterborne outbreaks in the United States and the suggestion that there is the potential for an outbreak somewhere nationwide every month.[52]

Treatment Barrier Instability

Treatment barrier instability is always a threat.[53] Normally, treatment barrier redundancy provides adequate protection from a major passage of contamination. The problem is more often one of barrier leakage associated with sudden, unexpected chlorine demand in the raw water, high-turbidity passages, filter instability, and improper disinfectant CT values. As a consequence, monitoring the quality of processed water as it enters the distribution system or just before it arrives at the first customer is essential. This is accomplished either through frequent sample collections and analyses every 4 h, gathering 24-h composites of water, or large (1 l or more) sample volume analyses correlated to process adjustments.

Rapidly fluctuating quality of raw source water beyond expected limits is often the cause of treatment difficulties. For this reason there should be a more frequent update on source water characteristics following major storm events, organic spills, and sewage bypasses upstream of the intake. Springtime raw water quality may also have a high chlorine demand as a result of agricultural fertilizer runoff, resuspended materials in seasonal lake destratification, and increased humics from algal blooms and vegetation decay.

Filter backwashing operations create temporary instability of the media. If these rehabilitated filters are brought back on-line too quickly, they may not be effective in preventing the passage of turbidity and organisms. Scheduling plant effluent sampling to coincide with filter changes would be an excellent quality control test to insure inactivation of any coliform passages from filter start-up operations. Eliminating this factor as a cause of repeated low-density coliform occurrences in the distribution system is very important.

Intervals of inadequate disinfectant CT values will most often release stressed coliforms that may not be detected in finished water from the plant using conventional MF or MT testing procedures. In this situation, the laboratory should select either the m-T7 stressed coliform MF method or the presence–absence fermentation test. Both techniques are described in Standard Methods.[54]

Distribution Vulnerabilities

Distribution vulnerabilities are always a matter of concern. Contamination of a satisfactory supply may occur in some portion of the pipe network due to repercussions of line failures, poor line repairs, or cross-connections in low-pressure zones. Reservoirs of finished water can be another source of contamination if there is a long period of stratified water conditions. Attention should be given to finished water reservoir protection and possible areas for cross-connections. Open reservoirs of finished water are prime suspects for microbial contamination; therefore, the effectiveness of associated booster chlorinators on the water released into the pipe network should be monitored frequently. Even covered reservoirs and standpipes are suspect locations of contamination from animal activity in air vents, seepage of rainfall drainage, and passage of atmospheric dust.

Locating the extent of a coliform contaminating event in the pipe network is part of the repeat-sampling program. The first reaction is to verify that a coliform contamination exists at the site and expand the search by including sampling sites adjacent to this location to define the extent of the contamination event. Upon confirmation that the contamination is not system-wide, a special monitoring program should be initiated to locate the cause, define the area affected, and apply appropriate corrective measures. Line pressure conditions, cross-connection potential, and recency of line breaks and new tap-ins on the service mains should be investigated as possible causes. Limiting the testing to coliform analyses is not enough. These carefully selected sites need to have a water pressure test, chlorine residual determination, turbidity measurement, and supplemental analyses done for fecal coliforms and heterotrophic bacteria. Loss of disinfection residual,[51] abrupt changes in pipeline turbidity, and elevated heterotrophic bacterial densities may be indications of a blending of contaminated water with water supply in distribution.[19] If fecal coliforms or *E. coli* are also found, the contamination is of recent fecal origin and the customers along that section of the pipe network must be promptly alerted to boil water.

While disinfection protocols are generally effective in removing bacterial contamination encountered from general construction,[55-56] there may be later repercussions if parts of the wood form remain exposed to water contact.[57] As in the case of wood used in storage tanks, the exposed wood in new pipelines can support coliform growth for months, even in the presence of a high chlorine residual.

Leaking pipelines also need to be reviewed as possible suspects in the coliform search. Small pinhole openings in corroded pipe walls and poor joint connections may be a pathway for bacterial migration from drainage around the pipe exterior during periods of water pressure decline. Because the coliform incursion is quickly swept downstream of the site and sampling is rarely at the precise site of entry, leak detection must be involved in the search along the zone of contamination defined by intensive bacteriological sampling.

Cross-connections from service lines is another source of coliforms in a localized zone of the pipe network. Any sudden loss of disinfectant residual in one zone of the pipe network should be interpreted as an indication of a contamination event, which may be followed quickly by elevated densities of heterotrophic bacteria and the occurrence of coliforms. Field investigation should include a search for cross-connections in attachment devices on service lines in the area. Particular candidates are commercial establishments using large volumes of water such as car-wash operations and laundromats. Coliforms would be common to the dirt removed from automobiles and soil removed in washing clothing.

DEMONSTRATING THE EXISTENCE OF A COLIFORM BIOFILM

Innovative monitoring strategies are required if the locations of coliform biofilm occurrences in the pipe network are to be identified. In general, there are several common characteristics associated with all coliform biofilms in water supply lines. Coliform colonization in a microbial biofilm occurs over time in the stable pipe environment. Periodic releases of coliforms into the water generally occurs only when the water temperature rises above 15°C. Furthermore, most of these coliform problems occur in surface-water systems located in geographical areas of pronounced seasonal weather changes.[58]

Speciation of coliforms can be an important contribution to verifying biofilm existence in these special monitoring programs.[59] The predominant coliforms are species of *Klebsiella, Enterobacter,* and *Citrobacter* and the profile is often dominated by one species for several months before the dominance changes to another coliform. *Escherichia coli* has not been found to compete successfully in water supply biofilms and, when found, is only a transient organism that quickly disappears. Coliform biofilm releases are generally characterized as being unpredictable, making it difficult to localize the focal point of biofilm development in a complex network of pipes.

Locating Biofilms

Locating biofilms is difficult and requires innovative monitoring strategies. Where a coliform biofilm is suspected, special sampling sites should be selected, within the defined pressure zone, in an effort to verify the initial data and isolate the problem area in the pipe network. Isolating the specific area can be challenging because of flow reversals in the distribution system and the transient nature of those coliform releases.

Developing a database on heterotrophic bacterial densities each quarter of the year will provide some reference point to potential locations of seasonal regrowth.[60] These organisms often are survivors of water treatment that have become established in the pipe sediments, utilizing available assimilable organic carbon in their adjustment to a hostile water environment. During warm water periods, persisting heterotrophic bacteria begin to regrow, often followed in several weeks by a similar regrowth of coliforms that slough from the expanding biofilm due to the shearing action of water passage. Because the general bacterial population comprising the heterotrophs is much larger than that of the coliform portion, more consistency of increased heterotrophic occurrence may be anticipated at biofilm sites.

Prime suspect locations for biofilm development are areas where the total coliform occurrence repeatedly exceeds 5%. To explore this aspect, coliform data, gathered over three consecutive years, from the New Haven distribution system was sorted into a range of months in which the coliform occurrence rate exceeded 5% for each location. These ranges were then plotted on a map of sampling locations (Figure 6.5). Inspection of the map reveals nine locations had coliforms over 16 to 20 months and seven others had coliforms in more than 20 of 36 months, strongly suggesting prime areas from which the biofilm coliforms might be originating. The plot also reveals many of these sites to be in areas of slow flow or near dead ends where static water conditions could encourage colonization in pipe sediments.

A search for stressed coliforms during a biofilm occurrence event revealed a significant increase in the percent of coliforms detected on m-T7 medium as compared with the results of parallel testing with m-Endo agar. Data in Table 6.22 were grouped into two categories; samples collected from main lines and other samples taken at customer taps. Two observations emerged: (1) increasing free chlorine to reduce the biofilm may lead to more stressed coliform occurrences and (2) once a biofilm becomes active during warm-water periods, coliforms may be detected in any size water line.

Table 6.22   Comparison of Coliform Results Obtained by m-Endo and m-T7 Media

Month	Positive samples	m-Endo medium Coliforms		m-T7 medium Coliforms	
		Percent positive	Average per 100 ml	Percent positive	Average per 100 ml
Main Lines (Fire Hydrants)					
May	41	7.3	0.1	19.5	0.4
June	86	20.9	2.6	32.6	2.0
July	115	27.0	1.6	39.1	1.6
August	58	20.7	0.4	29.3	0.4
Service Lines (Customer Taps)					
May	16	0.0	0.0	0.0	0.0
June	44	22.7	1.0	34.1	2.2
July	43	25.6	1.0	39.5	1.3
August	31	25.8	1.1	29.0	1.0

Data from the Monmouth, NJ distribution system.

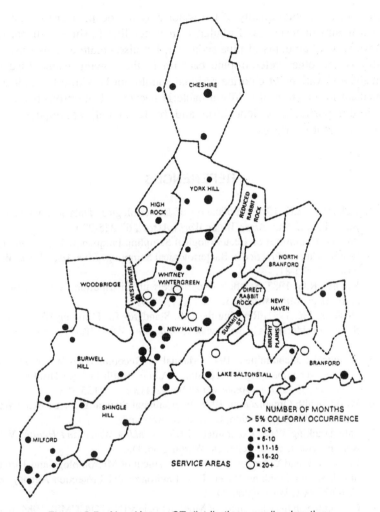

Figure 6.5   New Haven, CT distribution sampling locations.

Plotting the locations for high trihalomethanes (THM) in the distribution system may also be a useful predictor for locating probable sites of biofilm development.[61] Biofilm is a deposit of organic material and when exposed to high concentrations of chlorine, generally used to control coliform occurrences, will react to create THM in the pipe network. Areas of accumulated assimilable organic carbon (AOC) should also be suspected as potential sites for heterotrophic bacterial regrowth, coliform establishment, and diminished residual chlorine.

Demonstrating effectiveness of the action response will require a follow-up monitoring for water quality changes based on coliform occurrence, turbidity, chlorine residual, and heterotrophic plate count. Immediate improvement in the microbial quality of the water may not occur over the first 48 h after a treatment response. Turbulence in water line flushing will not only remove varying amounts of pipe sediment, but also create a resuspension of coliforms and other heterotrophic bacteria in the flowing water. Likewise, immediate elevations of chlorine residual should not be viewed as indicating immediate inactivation of coliform bacteria because of the protective effects of biofilm particulates deposition and bacterial cell aggregates in the water–sediment interfaces.

REFERENCES

1. Geldreich, E.E. 1971. Application of Bacteriological Data in Potable Water Surveillance. *Jour. Amer. Water Works Assoc.*, 63:225-229.
2. AWWA Committee on Bacteriological Sampling Frequency, 1985. Committee Report: Current Practice in Bacteriological Sampling. *Jour. Amer. Water Works Assoc.*, 77:75-81.
3. Donner, R.G. 1987. Seattle's Experience with Distribution System Sampling. *Jour. Amer. Water Works Assoc.*, 79(11):38-41.
4. Hoskins, J.K. 1941. Revising the U.S. Standards for Drinking Water Quality: Some Considerations in the Revision. *Jour. Amer. Water Works Assoc.*, 33:1804-1831.
5. Technical Subcommittee. 1943. Manual of Recommended Water Sanitation Practice Accompanying United States Public Health Service Drinking Water Standards, 1942. *Jour. Amer. Water Works Assoc.*, 35:135-188.
6. World Health Organization. 1971. International Standards for Drinking Water. World Health Organization, Geneva, Switzerland.
7. Safe Drinking Water Committee. 1977. *Drinking Water and Health*, Vol. 1, National Academy of Sciences, Washington, D.C.
8. Berger, P.S. and Y. Argaman. 1983. Assessment of Microbiology and Turbidity Standards for Drinking Water, U.S. Environmental Protection Agency, EPA 570-9-83-001, Washington, D.C.
9. Committee on the Challenges of Modern Society (NATO/CCMS). 1984. Drinking Water Microbiology. NATO/CCMS Drinking Water Pilot Project Series CCMS128, U.S. Environmental Protection Agency, EPA 570/9-84- 006, Washington, D.C.
10. Environmental Protection Agency. 1976. National Interim Primary Drinking Water Regulations. EPA-570/9-76-003. Office of Water Supply, Washington, D.C.
11. World Health Organization. 1984. Guidelines for Drinking Water Quality, Vol. 1, World Health Organization, Geneva, Switzerland.
12. Department of Health and Social Security. 1969. The Bacteriological Examination of Water Supplies. Reports on Public Health and Medical Subjects, Vol. 71, London, U.K.
13. Council of the European Communities. 1975. Proposal for a Council Directive Relating to the Quality of Water for Human Consumption, Official Jour. European Communities, 18 No. C214/2.

14. Ministry of National Health and Welfare. 1977. Microbiological Quality of Drinking Water. Health and Welfare, Ottawa, Canada.
15. World Health Organization. 1970. European Standards for Drinking Water, World Health Organization, Geneva, Switzerland.
16. Craun, G.F. 1978. Impact of the Coliform Standard on the Transmission of Disease. In: Evaluation of the Microbiology Standards for Drinking Water. C.W. Hendricks, Ed., U.S. Environmental Protection Agency, EPA-570/9-78-002, Office of Drinking Water, Washington, D.C.
17. Pipes, W.O. 1983. Monitoring of Microbial Water Quality. In: Assessment of Microbiology and Turbidity Standards for Drinking Water. P.S. Berger and Y. Argaman, Eds., U.S. Environmental Protection Agency, EPA-570/9-83-001, Office of Drinking Water, Washington, D.C.
18. Jacobs, N.J., W.L. Ziegler, F.C. Reed, T.A. Stukel, and E.W. Rice. 1986. Comparison of Membrane Filter, Multiple-Fermentation-Tube, and Presence–Absence Techniques for Detecting Total Coliforms in Small Community Water Systems. *Appl. Environ. Microbiol.,* 51:1007-1012.
19. AWWA Organisms in Water Committee. 1987. Committee Report: Microbiological Considerations for Drinking Water Regulation Revisions. *Jour. Amer. Water Works Assoc.,* 79:81-84, 88.
20. U.S. Environmental Protection Agency. 1989. Drinking Water: National Primary Drinking Water Regulations; Filtration, Disinfection, Turbidity, *Giardia lambia,* Viruses, *Legionella,* and Heterotrophic Bacteria. *Federal Register,* 54: 27486–27568, July 29, 1989.
21. Geldreich, E.E., J.A. Goodrich, and R.M. Clark. 1988. Characterizing Raw Surface Water Amenable to Minimal Water Supply Treatment Proc. AWWA Annual Conference, Orlando, FL. Part 1, 545-570.
22. Pipes, W.O. 1988. Composite Sample Study of Plant Effluent. WQTC, St. Louis.
23. Office of Drinking Water. 1989. Drinking Water; National Primary Drinking Water Regulations; Total Coliforms (Including Fecal Coliforms and *E. coli*); Final Rule. 40CFR Parts 141 and 142. NPDR. *Federal Register,* Thursday, June 29, 1989.
24. LeChevallier, M.W., S.C. Cameron, and G.A. McFeters. 1983. New Medium for Improved Recovery of Coliform Bacteria from Drinking Water. *Appl. Environ. Microbiol.,* 45:484-492.
25. LeChevallier, M.W. and G.A. McFeters. 1985. Enumerating Injured Coliforms in Drinking Water. *Jour. Amer. Water Works Assoc.,* 77:81-87.
26. McFeters, G.A., J.S. Kippin, and M.W. LeChevallier. 1986. Injured Coliforms in Drinking Water. *Appl. Environ. Microbiol.,* 51:1-5.
27. McFeters, G.A., M.W. LeChevallier, A. Singh, and J.S. Kippin. 1986. Health Significance and Occurrence of Injured Bacteria in Drinking Water. *Water Sci. Technol.,* 18:227-231.
28. Geldreich, E.E. 1981. Maintaining Bacteriological Quality. In: Treatment Techniques for Controlling Trihalomethanes in Drinking Water. J.M. Symons et al., Eds., EPA-600/2-81-156, Municipal Environmental Research Laboratory, U.S. Environmental Protection Agency, Cincinnati, OH. pp. 194-227.

29. Hudson, L.D., J.W. Hankins, and M. Battaglia. 1983. Coliforms in a Water Distribution System: A Remedial Approach. *Jour. Amer. Water Works Assoc.,* 75:564-568.

30. Geldreich, E.E., H.D. Nash, D.J. Reasoner, and R.H. Taylor. 1972. The Necessity of Controlling Bacterial Populations in Potable Waters: Community Water Supply. *Jour. Amer. Water Works Assoc.,* 64:596-602.

31. Caldwell, B.A. and R.V. Morita. 1987. Sampling Regimes and Bacteriological Tests for Coliform Detection in Groundwater. Project Report EPA/600/187/683. U.S. Environmental Protection Agency, Cincinnati, OH.

32. Buelow, R.W. and G. Walton. 1971. Bacteriological Quality vs. Residual Chlorine. *Jour. Amer. Water Works Assoc.,* 63:28-35.

33. Thomas, S.B., C.A. Scarlett, W.A. Cuthbert, et al. 1975. The Effect of Flaming of Taps before Sampling of the Bacteriological Examination of Farm Water Supplies. *Jour. Appl. Bacteriol.,* 17:175-181.

34. McCabe, L.J. 1969. Trace Metals Content of Drinking Water from a Large System. Presented at Symposium on Water Quality in Distribution Systems, Amer. Chem. Soc. National Meeting, Minneapolis, MN.

35. Geldreich, E.E. 1975. Handbook for Evaluating Water Bacteriological Laboratories. U.S. Environmental Protection Agency, EPA-670/9-75-006, Cincinnati, OH.

36. System Water Quality Committee, California–Nevada Section, Amer. Water Works Assoc. 1978. Distribution System Bacteriological Sampling and Control Guidelines.

37. Caldwell, E.L. and L.W. Parr. 1933. Present Status of Handling Water Samples. *Amer. Jour. Pub. Health,* 23:467-472.

38. The Public Health Laboratory Service Water Subcommittee of Great Britain. The Effect of Storage on the Coliform and *Bacterium coli* Counts of Water Samples. 1953. Storage for Six Hours at Room and Refrigerator Temperatures. *Jour. Hyg.,* 51:559-571.

39. Geldreich, E.E., P.W. Kabler, H.L. Jeter, and H.F. Clark. 1955. A Delayed Incubation Membrane Filter Test for Coliform Bacteria in Water. *Amer. Jour. Pub. Health,* 45:1462-1474.

40. Lonsane, B.K., N.M. Parhad, and N.V. Rao. 1967. Effect of Storage Temperature and Time on the Coliforms in Water Samples. *Water Res.,* 1:309-316.

41. McDaniels, A.E. and R.H. Bordner. 1983. Effects of Holding Time and Temperature on Coliform Numbers in Drinking Water. *Jour. Amer. Water Works Assoc.,* 75:458-463.

42. McDaniels, A.E. et al. 1985. Holding Effects on Coliform Enumeration in Drinking Water Samples. *Appl. Environ. Microbiol.,* 50:755-762.

43. Fujioka, R.S. and O.T. Narikawa. 1982. Effect of Sunlight on the Enumeration of Indicator Bacteria under Field Conditions. *Appl. Environ. Microbiol.,* 44:395-401.

44. Geldreich, E.E. 1971. Application of Bacteriological Data in Potable Water Surveillance. *Jour. Amer. Water Works Assoc.,* 63:225-229.

45. Symons, J.M., A.A. Stevens, R.M. Clark, E.E. Geldreich, O.T. Love, Jr., and J. DeMarco. 1981. Treatment Techniques for Controlling Trihalomethanes in Drinking Water. EPA-600/2-81-156, U.S. Environmental Protection Agency, Cincinnati, OH.

46. Geldreich, E.E. 1993. Microbiological Changes in Source Water Treatment: Reflections in Distribution Water Quality. In: *Strategies and Technologies for Meeting SDWA Requirements*. R.M. Clark and R. Scott Summers, Eds., Technomic Pub. Co., Lancaster, PA.

47. Geldreich, E.E., J.A. Goodrich, and R.M. Clark. 1989. Strategies for Monitoring the Bacteriological Quality of Water Supply in Distribution Systems. Proc. Amer. Water Works Confr., Los Angeles, CA.

48. Herson, D.S. 1980. Identification of Coliform Antagonists. Proc. Amer. Water Works Assoc., Water Quality Technol. Conf., Miami Beach, FL. pp. 153-160.

49. Edberg, S., M.J. Allen, and D.B. Smith. 1988. National Field Evaluation of a Defined Substrate Method for the Simultaneous Enumeration of Total Coliforms and *Escherichia coli* from Drinking Water: Comparison with the Standard Multiple Tube Fermentation Method. *Appl. Environ. Microbiol.,* 54:1595-1601.

50. Covert, T.C., L.C. Shadix, E.W. Rice, J.R. Haines, and R.W. Freyberg. 1989. Evaluation of the Autoanalysis Colilert Test for the Detection and Enumeration of Total Coliforms. *Appl. Environ. Microbiol.,* 55:2443-2447.

51. Snead, M.C., V.P. Olivieri, K. Kawata, and C.W. Druse. 1980. The Effectiveness of Chlorine Residuals in Inactivation of Bacteria and Viruses Introduced by Post-Treatment Contamination. *Water Res.,* 14:403-408.

52. Craun, G.F. and L.J. McCabe. 1973. Review of the Causes of Waterborne-Disease Outbreaks. *Jour. Amer. Water Works Assoc.,* 65:74-84.

53. Geldreich, E. E. 1994. Microbial Water Quality Concerns in Distribution Systems: Problems and Solutions. *Rivista Italiana D'Igiene,* 54:3-24.

54. American Public Health Association. 1985. *Standard Methods for the Examination of Water and Wastewater.* American Public Health Association, Washington, D.C.

55. AWWA Standard for Disinfecting Water Mains. 1986. AWWA C651-86, American Water Works Association, Denver, CO.

56. Buelow, R.W., R.H. Taylor, E.E. Geldreich, et al. 1976. Disinfection of New Water Mains. *Jour. Amer. Water Works Assoc.,* 68:283-288.

57. Talbot, Jr., H.W., J.E. Morrow, and R.J. Seidler. 1979. Control of Coliform Bacteria in Finished Drinking Water Stored in Redwood Tanks. *Jour. Amer. Water Works Assoc.,* 71:349-353.

58. Geldreich, E.E. 1988. Coliform Non-Compliance Nightmares in Water Supply Distribution Systems. In: Water Quality: A Realistic Perspective. University of Michigan, College of Engineering Seminar, Ann Arbor, MI. pp. 55-74.

59. Geldreich, E.E. and E.W. Rice. 1987. Occurrence, Significance, and Detection of *Klebsiella* in Water Systems. *Jour. Amer. Water Works Assoc.,* 79:74-80.

60. Geldreich, E.E. 1990. Microbiological Quality in Distribution Systems. In: *Water Quality and Treatment*, 4th Ed., American Water Works Association, Denver, CO. pp. 1113-1158.

61. Rittmann, B.E. and V.L. Snoeyink. 1984. Achieving Biologically Stable Drinking Water. *Jour. Amer. Water Works Assoc.,* 76:106-114.

CHAPTER 7

Microbial Breakthroughs in Changing Source, Treatment, or Supply Parameters

CONTENTS

INTRODUCTION

To fully understand the proliferation of coliform occurrences in some public water supplies, there must be a better appreciation for varying characteristics of source waters, in particular, the microbial component, particulates, soluble organics, and ammonia content. It must be realized that the microbial flora in drinking water has its origins in source water, although modified in density and diversity by treatment processes, with persistence of survivors being conditioned by the distribution system environment.

TREATMENT BARRIERS

Treatment barrier effectiveness can change as a consequence of not being responsive to fluctuating raw water conditions or as a result of operational modifications at the plant. Many water utilities are seriously reviewing the need to modify treatment operations in an attempt to reduce the formation of disinfectant by-products either by reducing organic precursors or changing the type of disinfectant for less reaction products. These moves must be considered carefully because of possible adverse repercussions on treatment barrier effectiveness in terms of coliform compliance and, ultimately, the occurrence of a waterborne outbreak.

FACTORING SOURCE WATER CHARACTERISTICS

It is obvious that without adequate treatment, there will be noncompliance problems with distribution water. Treatment needs to be designed to adequately process a given source water into a safe potable water supply. Excluding those situations where enhanced treatment has been found to be essential for *Giardia* and *Cryptosporidium* control, there are other more subtle changes in water characteristics that may degrade treatment barriers and over time release coliforms into the distribution system.

CHLORINE DEMAND

Abrupt changes in raw source water chlorine demand may result in ineffective disinfection for a brief period, thereby providing opportunities for coliforms to pass into the distribution system. Springtime snowmelt, major storms over the watershed, seasonal turnovers in impoundments, algal blooms, drought conditions, and application of agricultural fertilizers to fields in the watershed can introduce a variety of substances that exert a chlorine demand and reduce chlorine availability to treat water. While immediate attention to increased disinfectant dosage is desirable, system operations may be slow to respond to the changing water quality. Two case histories illustrate possible scenarios that lead to coliform biofilm problems in distribution systems.

Case History: Rochester, NY

The Rochester water supply is derived from surface water that was, in 1986, disinfected without the benefits of filtration. Major sources of this surface water are Hemlock Lake and Canadice Lake in a watershed covering 66.4 mi². The city owns 20% of the watershed; public use of city-owned property is restricted to hunting and fishing, with no camping or swimming permitted.[8] A cement-lined brick tunnel, 2 mi long and 6 ft in diameter, brought water from Lake Hemlock to the treatment plant. Water entering the tunnel was prechlorinated to discourage bacterial colonization on the tunnel surfaces; when water temperature reached 50°F. Periodically, 0.07 to 0.1 mg/l of copper was added to combat biofilm. The goal was to achieve 1.2 to 2.0 mg/l free chlorine in the raw water, but chlorine demand often limited this to 1.0 mg/l.

At the treatment plant water was again chlorinated, then fluoridated and gravity-fed into the storage and distribution system. Finished water was held in open storage reservoirs resulting in the immediate reduction of chlorine residual because of sunlight exposure and introducing the potential for contamination from runoff, algal blooms, and aquatic bird inhabitation. Although water leaving these reservoirs was rechlorinated, residuals in the distribution system rarely exceeded 0.5 mg/l and were often found to be in the range from trace to 0.1 mg/l. Chlorine demand in the cement-lined mains between the system reservoirs was often high enough that influent water to the next reservoir was undetectable. These facts suggested the distribution pipe network had significant accumulation of sediments from the unfiltered supply and needed frequent flushing.

First evidence of a biofilm problem could be traced in laboratory records to 1981 when bacteriological tests revealed the presence of atypical colonies that were verified to be *Enterobacter cloacae*. For several succeeding years, these coliform occurrences were restricted to periods during May to September. In a surprise event, coliform bacteria were suddenly detected in the Rochester distribution system during February 1986. Densities ranged from 1 to 161 coliforms per 100 ml and the predominant coliform species was *Enterobacter cloacae*. Frequency of these coliform occurrences varied, as did the location of positive samples in the pipe network. The problem appeared to subside until severe rainstorms in mid-June 1986 caused turbidity levels to increase and bacterial problems to recur. Seven fecal coliforms (*E. coli*) were confirmed in samples collected on June 18 and 30. A boil water order was issued to the public on July 7. Residual chlorine concentrations were increased to 2 to 3 mg/l. During the following weeks, there were no more reports of *E. coli* although fecal *Klebsiella* were reported in samples on July 24 and July 26. By July 31, the boil water order was lifted.

A task force of utility, local health department, state, and federal water supply representatives was created to review the problem and devise an action strategy. Major task force conclusion was that the problem was caused by heavy rainfall that washed material into Hemlock Lake and resulted in a rise in turbidity, bacteria, and chlorine demand. It was recommended that if turbidities exceeded action levels of 3 and 5 NTU, respectively, chlorination should be increased and raw water pumpage decreased accordingly. It was also recommended that the utility consider removing accumulated sludge from the bottom of Highland and Cobbs Hill reservoirs because algae and aquatic invertebrates colonized these reservoirs. Furthermore, it was agreed to gradually reduce residual chlorine in the distribution system back to 1 mg/l. The reasoning was that elevated chlorine levels, while providing greater disinfection, were also making water more aggressive and possibly causing sloughing of attached bacteria. It was also noted that while the water bureau had a regular distribution system flushing program, it took 6 years before all the lines could be flushed. A systematic flushing program should be done annually with localized flushing of areas done whenever a coliform occurrence is detected. Long-term recommendations included construction of a filtration plant to improve treatment barrier protection and covering the two finished water reservoirs to avoid recontamination of the water supply by birds as well as to improve retention of disinfectant residuals in the finished water supply.

Case History: Muncie, IN

The source water for the Muncie Water Works Company is a small stream, the White River. Upstream of the intake there are a few municipal and industrial discharges to the river. The major activity on the watershed is intensive agriculture. Storms over the drainage basin bring runoff from this agricultural area, which in springtime involves elevated turbidities, high bacterial densities, and ammonia nitrogen concentrations from fertilizers that reach 7 mg/l.

Pretreatment of this water involved alum for coagulation, chlorine for oxidation of iron and manganese, and disinfection. Lime was added on occasion to optimize coagulation. Product water then flowed through both a rapid-mix coagulation basin and a slow-mix basin. Thereupon half of the water is clarified before filtration through conventional rapid sand filters. The other portion was passed through granular activated carbon filters for taste and odor control. All process effluents from the filter were then combined and chlorine applied to achieve a free residual of 2 mg/l in the plant effluent.[2]

In the spring of 1980, when the ammonia concentration rose in the raw water, chlorine dioxide (0.5 mg/l) was applied for several days in February and March as a partial substitute for the normal 2 mg/l combined chlorine used in the previous disinfection strategy. The intent was to avert taste and odor formation. Shortly after the application of chlorine dioxide–combined chlorine ended, coliforms suddenly appeared in the distribution system routine sampling program. This situation suggested that the application of 0.5 mg/l chlorine dioxide as a partial substitute for the 2 mg/l combined chlorine had not been fully effective during the cold-water period and heavy spring runoff.

By the end of March, 14% of the 240 samples analyzed had more than 4 coliforms per 100 ml. Subsequently, the distribution system became colonized with an active biofilm that contained *Enterobacter cloacae* as the predominant coliform species. During that summer, coliform episodes reappeared for brief periods on several occasions, ranging from 1 to 51 organisms per 100 ml.

Laboratory studies on this strain of *Enterobacter cloacae* indicated the organism was inactivated by 0.5 mg/l free chlorine residual when held for a 2-h contact period at room temperature. In the slime condition, this strain was reported to be resistant to chlorine dosages up to 10 mg/l for a 2-h contact time.

Action response to this coliform occurrence included raising free chlorine residual from 2 mg/l in the plant effluent to 4.5 mg/l, then later to 15 mg/l for a 2-week period. All 300 mi of the distribution lines were flushed over a 4-d period and the distribution storage tank drained, cleaned of sediment with high-pressure sprays, disinfected, and returned to service. As a result of these actions, coliform occurrences and their densities declined over the remainder of the summer.

While plant data indicated the treatment system was supplying a satisfactory water quality to the distribution system, there is concern that coliform "bleed across" the disinfection treatment barrier at levels below a detection level set at 100 ml. As a matter of precaution, sand media in the conventional filters were replaced with a more effective mixed media. The GAC filter was also suspect because it had been in service for several years and could have released coliforms that were not inactivated during the chlorine dioxide–combined chlorine exposure period.

RAW WATER pH SHIFTS

Seasonal changes in raw water pH may contribute to the instability of sediments coating the pipe walls. This change in sediment stability can lead to sporadic releases of coliform bacteria and other viable organisms in the attached biofilm or entrapped in the accumulation of particulate deposits.

Case History: Wilmette, IL

Such a situation was created at the utility in Wilmette, serving 32,000 people and using Lake Michigan as source water. The problem began when coliforms were repeatedly observed in the distribution system during the cold-water months of December to June. Winter occurrence of coliforms in the pipe network is unusual; most coliform biofilm release problems have been observed to take place only during warm-water periods. An on-site review of treatment practices and plant records indicated that there had been a pronounced shift in the Lake Michigan source water pH during the winter. Inspection of data on raw water characteristics revealed water pH of 7.7 in summer leveled off at pH 8.2 by December, followed by a rapid decline to pH 7.4 during January to March each year. By the end of March, 14% of the 240 samples analyzed had coliform counts that exceeded four organisms per 100 ml. The reason for these seasonal changes in water pH was thought to be related to near-shore turnover of bottom water containing partially decayed vegetation debris (humic matter). Water treatment measures used to process the lake water had little impact on stabilizing the water pH, so this characteristic was passed on into the distribution system.

Implementation of recommendations to adjust the water to pH 8.3 prior to release from the plant and to add lime slowly in the process basin to form a more stable, firm coating on the pipe walls apparently resolved the coliform occurrence problem the following winter. Successful follow-up treatment may also have been aided by the suggestion to increase the disinfectant concentration during cold-water periods to compensate for the increased chlorine demand and reduced disinfectant effectiveness at near-freezing water temperatures.

COLD SOURCE WATER TEMPERATURES

Cold water temperatures have an influence on disinfectant effectiveness. At temperatures of 5°C and below, inactivation of organisms by disinfectants requires a longer contact time or an increase in concentration to achieve the same kill rate as at 20°C. The problem is of particular concern for surface-water supplies in northern latitudes. Because contact time is generally fixed, the only alternative available to achieve adequate disinfection (C·T) values is to increase the disinfectant concentration applied in the contact basin.

Case History: Fairbanks, AK

In this instance, the utility is part of a military base that supplies water to two separate military communities. Surface water is generally processed by conventional treatment most of the year. However, in late autumn surface source water is passed through a rapid sand filter without the benefit of coagulation, then disinfected prior to entry into the two distribution systems.

With raw water temperatures in January and February stabilized at approximately 1 to 5°C at the intake, the applied chlorine dosage in the contact basin was not increased. As a consequence, a species of *Klebsiella* began to be detected in the distribution system nearest to the utility and the water supply was in noncompliance for 2 to 3 months. The other distribution system, serving the second military base, was not affected. The long travel time created in water supply transmission to this military base provided an extended contact time beneficial for coliform inactivation. Under wintertime conditions in northern latitudes, the applied chlorine dosage to surface source waters should be increased to achieve effective inactivation of bacteria and viruses.

PLANT OPERATIONAL PRACTICES

Operational practices at the public utility can be a source of coliform contamination in water supply. Many of these problems have been traced to poor maintenance, construction defects that lead to short-circuiting in the process basin, and financial constrains that impede improved treatment or result in reduced operation time in an effort to cut back production costs. Some systems have been expanded for future growth in the community and are currently oversized for customer demand or must adjust for peak customer demand in cyclic seasonal population growth and decline (recreational area or college town). These situations require operator skills in marshalling water through process basins so that treatment barriers remain effective and water supply is kept moving in the distribution system to avoid adverse water quality changes. Two case histories are presented to illustrate some of the dangers facing all operators as they cope with the day-to-day production of a safe water supply. One case history illustrates the importance of keeping the plant free of microbial colonization and the other shows the potential dangers of surface-water contamination at the clear-well.

MICROBIAL COLONIZATION IN THE PLANT

Systems that use rivers or small impoundments require more frequent removal of scum deposits at all water–air interfaces and sludge removal from process basins. Infrequent and poor filter backwashing practices can lead to filter instability. Media deposition may become uneven, channelization may occur, and air binding may inhibit uniform passage of process water. Other important considerations include the removal of scums that form at air–water interfaces in treatment basins, connecting flumes, baffles, and agitator paddles and sludges that accumulate in mixing and settling basins.

Case History: Iowa City, IA

The Iowa City public water system experienced a low-frequency coliform event that recurred at various times from January 1983 to July 1985.[3] Prior to that time, laboratory records indicated only two to seven individual positive samples over the years 1979 to 1982 (Table 7.1). Throughout the 1983 to 1985 episode, the predominant coliform from most positive samples was *Klebsiella oxytoca*. This observation alone suggested a colonization was prevalent somewhere in the water system.

Table 7.1   Number of Water Samples Positive for
Total Coliforms by Calendar Year

Year	Number of positive coliform samples MPN					Total
	2.2	5.1	9.2	16	16+	
1979	2	0	0	0	0	2
1980	3	0	2	0	0	5
1981	3	2	0	1	0	6
1982	3	1	0	2	1	7
1983	19	10	4	3	2	38
1984	27	11	6	9	11	64
1985	10	4	3	5	5	27
1986	0	0	0	0	0	0

Data from Moyer and Hall.[3]

The municipal water utility operates two treatment plants that are interconnected at the clear-well. Treated water supply is transmitted through 58 mi of pipes and the distribution system has 1541 hydrants, 12,100 water meters, and three storage tanks located at strategic places around the city.

Iowa City draws 95% of its raw water from an impoundment of the Iowa River that also supports multiple recreational uses in this largely agricultural area. A slurry of powdered carbon was injected into the raw water for color, taste, and odor control before being sent to either or both water treatment facilities for conventional processing. At the old plant, water passed through a mixing chamber with addition of alum, then on to a sedimentation basin where chlorine and lime were added. Clarified water was then sent through rapid sand filters before reaching the clear-well for final addition of free chlorine. At the new plant, raw water is clarified, chlorinated, and filtered prior to reaching the clear-well for final application of free chlorine and a dose of fluoride. Groundwater supplies 5% of the source water and is pumped without treatment directly to the shared clear-well where it blends with effluent from both treatment plants.

With the increase in positive coliform samples during July and August 1984, a series of samples were collected throughout the system and analyzed for total coliform and *Klebsiella* content. *K. oxytoca* was isolated on the same day from the mixing chamber in the old plant, plant effluent, two routine distribution samples, and two of the storage tanks. *K. pneumoniae* was isolated from the raw source water, clarifier sludge, and clarifier effluent of the new plant (Table 7.2). *Klebsiella* could not be detected in samples from the treatment stages after chlorination, so it was evident that the plant processes before disinfection had become colonized.

Table 7.2   In-Plant Sample Results

	Sample	Total coliform	*Klebsiella* species
New plant	Raw water	≥2400	*K. pneumoniae*
	Clarifier effluent	540	*K. pneumoniae*
	Clarifier sludge	≥2400	*K. pneumoniae*
	Filter influent	<1	
	Filter effluent	<1	
	Filter backwash	<1	
Old plant	Mixing chamber	1600	*K. oxytoca*
	Filter influent	<1	
	Filter effluent	<1	
	Filter backwash	<1	
	Finished water	<1	
	Filter sand	<1	
	Filter gravel	<1	
	Filter rock	<1	

Data from Moyer and Hall.[3]

Inspection of plant records indicated a history of intermittent plant operation for short periods, extended periods of plant disuse, failure to maintain filter effluent turbidity below 1 NTU, failure to prevent low sand depth in filter beds, and failure to backwash the filters on a regular basis. These conditions led to turbidities exceeding 5 NTU between filter backwashings on several occasions during February and June 1983. Upon recommendation, the plant mixing chamber was drained to remove 4 ft of accumulating sludge. This operation alone reduced the positive samples by 50% but *K. oxytoca* continued to occur from October 1984 to March 1985. By this time, the distribution system had become colonized by microbial breakthroughs in the releases of excessive turbidities (periodically up to 5 NTU).

In January 1986, samples from the plant processes in the new plant were again collected and examined. *Klebsiella* was again detected (Table 7.3) at various sites. These results verified that the sedimentation basin and mixing chamber were colonized with *Klebsiella*, but the coliform releases in plant effluent had stopped. This was no doubt due to improvements in regular maintenance and scrupulous attention to plant operation that involved providing a filter effluent turbidity of less than 1 NTU and a free chlorine residual of 2 mg/l.

Table 7.3   Plant Follow-Up Study, 1986

Sample	Result
Raw water	*K. oxytoca*
Mixing chamber water	*K. pneumoniae*
Mixing chamber slime	*K. oxytoca, K. pneumoniae*
Mixing chanber scrapings	*K. oxytoca*
Mixing chamber sludge	*K. pneumoniae*
Baffle slime	*K. pneumoniae*
Baffle wood scrapings	*K. oxytoca*
Baffle wood core	No growth
Filter effluent	No growth
Filter core (no. 10)	*Enterobacter cloacae*

Data from Moyer and Hall.[3]

As a result of the *Klebsiella* colonization of the treatment processes and the release of plant effluent with turbidities up to 4 NTU, it was certain the distribution system would become colonized unless system flushing of lines was done. This possibility was successfully avoided following the hydrant-flushing campaign in March 1985. Some indication of this successful response can be seen in Table 7.4. A total of 29 sample pairs were collected from the periphery of the system. Each pair consisted of a sample collected immediately after the hydrant was opened and a follow-up sample collected 10 min later from the same hydrant. Apparently, pipe sediment removal and the restoration of free chlorine residuals in flushing were effective because coliform occurrence faded away a few months later.

Table 7.4   Hydrant Flush Study, 1985

Sample no.	Location	Free chlorine residual	MPN Before flush	MPN After flush	Culture results
5	N. Dodge St.	2.2	16+	0	*Klebsiella oxytoca* *Citrobacter* sp. *Proteus* sp.
7	Oaks Dr.	2.1	16	0	*Enterobacter* sp.
15	Industrial Rd.	2.2	16+	0	*Citrobacter* sp.
23	Miller	3.0	2.2	0	Negative
26	Cambria Ct.	2.7	16+	0	*Citrobacter* sp. *Enterobacter* sp.

Data from Moyer and Hall.[3]

SURFACE-WATER PENETRATION OF CLEAR-WELL

Short-circuiting of process treatment is always a concern. This may occur in poor settling of coagulants, channel formation in filter media, and development of compartment cracks in common wall barriers between basins. Any of these situations create pathways for short-circuiting of process water treatment and lead to irregularities in water quality. Another source of uncontrolled contamination may be stormwater runoff that is not adequately diverted away from process basins. Such was a possible cause of a coliform episode reported in the next case history.

Case History: Escondido, CA

During the week of January 9, 1995, a sudden total coliform contamination appeared in the Escondido public water supply. Total coliforms were first detected on Monday, January 9 in a distribution sample collected from a site near the water treatment plant. After the initial finding of coliforms in that location, contamination progressed further into the system within the next few days. By Thursday of that week, coliform densities of over 100 organisms per 100 ml were detected in three samples and several sites had repeat coliform-positive occurrences. Based on that information, the utility immediately issued a boil water order (January 12 at 4:00 p.m.), which remained in effect for the entire community until 6:00 p.m. on Sunday. A smaller portion of the distribution area continued on the boil advisory until 6:00 p.m. on Monday. No fecal coliforms or *E. coli* were detected, but speciation of the organisms revealed the predominant coliform to be *Klebsiella*.

Raw water sources for treatment originate from two local impoundment (Lake Wohford and Dixon Reservoir) and imported water from the Metropolitan Water District of Southern California. The 7000-acre-ft Lake Wohford impounds water from a largely rural watershed. Dixon Reservoir (2600 acre-ft) serves as a multipurpose recreational resource and water storage for imported raw water. This imported water supply consists of a blend of Colorado River and northern California water delivered through approximately 50 mi of large-diameter pipeline.

These raw waters are blended and treated at the Escondido-Vista water treatment plant (90-mgd capacity). Treatment consists of prechlorination, followed by coagulation and flocculation using alum and polymers, dual media filtration (anthracite coal and sand with a gravel support), and postdisinfection. During the winter, filter runs average about 70 h before backwashing. Backwash water is returned to the headworks for processing. A minimum of 2 mg/l free chlorine residual is maintained in the 5.4 million-gal plant clear-well. The clear-well was drained and cleaned a year ago.

The distribution system (serving approximately 115,000 people) consists of approximately 350 mi of pipe (mortar-lined steel, asbestos cement, and

PVC) and eight concrete or steel constructed water storage reservoirs. There are seven major pressure zones recognized in the system with 90% of the system served by gravity from the high-elevation treatment plant.

At the time of the coliform episode, water supply was being processed at a rate of 24 mgd (27% of plant capacity) from water stored in Dixon Reservoir. Water production is routinely reduced during the rainy winter months because of reduced demand from irrigating gardens, lawns, and landscaping areas. Rainfall at the start of January far exceeded "normal" amounts and reached a state of emergency in the community. During this period, plant effluent turbidity increased from 0.03 to 0.05 NTU. Plant operators increased the free chlorine residual leaving the plant to 2.5 mg/l as a safety measure.

The pattern of coliform occurrences in the distribution system did not suggest a release of pipe biofilm for several reasons. The first coliform occurrence was near the plant and over the next several days had progressed toward the ends of the system. Based on the analysis of 533 water samples examined over the next 7 d, there was no evidence of an explosion of coliforms throughout the distribution system in a random fashion. Water temperatures remained at 13°C and there was no evidence of a sudden change in water pH, reported to be 8.0.

A cross-connection near the site of the first coliform occurrence seemed unlikely because no fecal coliform or *E. coli* was detected. No reports of low water pressure or reversals of flow that might have resulted from a major fire or broken water mains. Vacuum breakers on the system could not clearly be ruled out as a source of cross-connection based on available data.

Inspection of the treatment plant revealed a break in the integrity of the clear-well that could have been the cause of the brief coliform episode. Location of the Escondido-Vista water treatment plant is on high ground slightly below and adjacent to Dixon Reservoir with the clear-well being partially recessed into the earthen retention wall that forms part of the dam. This buried concrete structure has a cement cover that was added in recent years. To prevent stormwater runoff from penetrating the clear-well, the abutting cover slab and side walls were caulked and a small, shallow drainage culvert constructed to divert surface water away from the site. Unfortunately, expansion of the concrete slab under the hot summer sun had broken the caulking seal in many places. With intense, sustained rainfall that occurred prior to the coliform event, surface drainage over the cement cover and drainage from the hillside overflowed the shallow culvert and could have entered the clear-well through the defective barrier seal.

Normally, this area of southern California receives on 6 in. of rainfall per year, of which 4 in. occurs during the winter period. In 1995, however, 6 in. of rain was recorded during January and February, filling Dixon Reservoir to overflowing, a rare occurrence over the years. The runoff from the concrete cover over the clear-well and the release of raw water from small, transitory springs in the soggy earthen dam combined to create a water volume greater than the carrying capacity of the culvert. As a consequence, the drainage culvert needs to be enlarged to increase its carrying capacity for high-volume stormwater runoffs and the caulking seal around the clear-well should be replaced periodically to maintain an effective surface-water barrier. Flushing the entire system was recommended to remove all vestiges of the contamination in the pipe system and minimize sediments where *Klebsiella* might attempt colonization.

TREATMENT MODIFICATIONS

Modifications in water treatment unit processes or in their sequential placement to optimize reductions in disinfection by-product formation must be cautiously evaluated and monitored for impact on microbial barriers and on distribution water quality. Four major treatment concepts, observed either in pilot-plant or full-scale operations, may cause changes in microbial quality: (1) changing the point of free chlorine application; (2) applying granular activated carbon (GAC) adsorption for organic removal; (3) using biological activated carbon (BAC) for further reduction of dissolved organics through microbial activity; and (4) employing of alternative disinfectants (chloramines, chlorine dioxide, and ozone) to reduce trihalomethane (THM) formation.

DISINFECTION: POINT OF APPLICATION

In the trade-off to minimize disinfection by-product formation by moving the point of disinfection application, some migration of organisms deeper into the treatment train may occur, or changes in the microbial flora of process waters will evolve. In the worst-case scenario, microbial colonization of the process media materials may result in periodic releases of biofilm aggregates containing coliform bacteria into the finished water.

Case History: Cincinnati, OH

Changing the site for chlorine application was investigated at the Cincinnati Water Works (Table 7.5) in a series of 2-week study periods.[4] During routine treatment plant operations, chlorine was applied to the source water after 48 h of open reservoir storage. Adequate retention time of raw source waters is a beneficial first step in microbial population reductions through self-purification processes and can be a buffer against temporary impairment of water quality from an upstream accidental spill of industrial chemicals. In the Cincinnati water treatment operation, coagulant was added to the water as it exited the open reservoir, and chlorine was routinely applied ahead of in-plant treatment processes. In the modified treatment operation, chlorination was delayed until after an additional 4-h clarification process consisting of coagulation and settling.

The results of both the routine and modified treatment schemes showed that 48-h source water storage with alum treatment reduced the total coliform densities by approximately 97% and turbidities by approximately 90%.

The coagulation and settling process, however, had little effect on further turbidity reductions, and the further decrease in the coliform population was only approximately 50% when chlorination was delayed (Table 7.5). Moving the point of chlorination to after coagulation and settling resulted in an intrusion of coliforms into early stages of water treatment. This change placed increased importance on providing a high-quality process water at this point, so that final disinfection would be effective in the inactivation of residual densities of various organisms of public health concern. Neither a measurable change in the bacterial quality of the finished water nor any apparent in-plant problems developed.

GRANULAR ACTIVATED CARBON (GAC) ADSORPTION

Carbon filtration, either in the GAC or BAC treatment mode, may provide opportunities for specialized microbial populations to become predominant, some of which may be resistant to conventional disinfection practices. This consideration, coupled with the fact that there are health risk limitations on chlorine dioxide concentration and slower inactivation rates for chloramines, makes it apparent that disinfection concentration and contact time values are of critical importance.

GAC has been in use for many years to remove a variety of synthetic organics and naturally occurring taste and odor compounds. Optimizing removal of organics in process water to lessen disinfection product formation would suggest that the GAC process be placed early in the progressive treatment of polluted surface waters. In fact, powdered carbon is often applied in some source water impoundments to control taste and odor problems, but is not adequate for removal of many other organic compounds.[5] Obviously, the GAC filtration process cannot be applied directly to raw surface waters with significant turbidities (above 1 NTU) because silt in these raw waters quickly coats the carbon particles and rapidly reduces organic adsorption capacity. Thus, settling of raw water and chemical treatment with clarification generally precedes the GAC process. These treatment processes also remove much of the turbidity-associated microbial flora, which include a wide range of environmental organisms, some of which are capable of aggressive colonization of GAC particles.

In the adsorption of organic substances, including those that may be trihalomethane precursors, granular activated carbon particles become focal points for bacterial nutrients and also provide suitable attachment sites for habitation. Although the portion of organic removal in the GAC process

Table 7.5 Chlorine Application Point Study — Cincinnati Water Works

Parameter	Mean values before treatment modification					Mean values after treatment modification				
	Sample point					Sample Point				
	Source	Stored source	Coagulated and settled	Filtered	Finished	Source	Stored source	Coagulated and settled	Filtered	Finished
Flow time (h)	0	48	52	52.5	55.5	0	48	52	52.5	55.5
Turbidity (NTU)	32	1.0	1.2	0.1	0.1	14	0.80	1.1	0.07	0.06
Total coliform per 100 ml	9,600	200	<1	<1	<1	84,000	2,400	1,400	<1	<1
SPC per ml	NR	NR	500	<1	5	NR	NR	5,500	15	<1
pH	7.3	7.1	8.5	8.3	8.7	7.6	7.2	8.1	8.1	8.2
Free Cl_2 residual (mg/l)	NR	NR	1.8	1.6	1.5	NR	NR	0	1.8	1.4
Total Cl_2 (ClO_2) residual (mg/l)	NR	NR	2.0	1.8	1.6	NR	NR	0	2.0	1.5

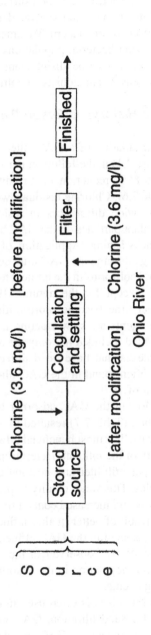

possible attributed to biodegradation is small (compared with physical adsorption to the activated carbon surface), there is a substantial microbial population present at the water–activated carbon surface interface. This process can, therefore, be of concern. Treatment barriers must remain effective against these colonizing bacterial populations that can include indicator organisms, organisms known to be disinfectant resistant, and others that are opportunistic pathogens or antagonists to coliform detection.

Case History: Jefferson Parish, LA

Installation of a GAC filter adsorber on-line after chemical clarification (Figure 7.1) in the Jefferson Parish full-scale operation did result in an approximate 85% reduction in turbidity to values ranging from 0.3 to 0.9 NTU (Table 7.6). Chlorine residuals were also reduced to virtually zero.[6] The occasional wide differences in residual turbidities reflected the entrapment of coagulant particles on the GAC bed and their migration through the filter prior to backwashing. Application of chloramines to the clarified water, before passage through the GAC filter adsorber, did not result in a complete reduction of total coliforms in the influent to below the one organism per 100 ml detection level, except for the autumn 1979 period. Disinfectant concentration and contact time become more critical when chloramines are applied, because these agents are slower acting than free chlorine. The total chlorine residual data cited included not only the active disinfectant components but also some complexes that have no disinfection power. Consequently, a few coliforms were often found in the GAC filter adsorber effluent except during the winter period of 1979.

Moving the GAC adsorber treatment process to a point following sand filtration (Table 7.7) resulted in an improvement (0.1 to 0.3 NTU) of effluent turbidity. The most beneficial effect was an improvement in the bacteriological quality of the influent. Heterotrophic bacterial densities were below 75 organisms per milliliter and no total coliforms were found in any of the effluent samples. This water quality improvement was a result of sand filtration effectiveness and increased contact time with chloramines added after clarification. As a result of better quality influent water to the GAC adsorber, no coliforms were detected in the GAC effluent over a 3-year study period. However, little difference was observed in the cyclic rise and decline of the heterotrophic bacterial population in the adsorber effluents associated with either treatment arrangements.

While GAC is often used in conventional treatment beds designed originally for sand filtration, GAC may also be used in pressure contactors. In pressure contactors, GAC bed depth is usually more than 36 in. (0.9 m) to provide the contact time necessary to remove certain classes of organics as the water is pumped through each unit. Factors to be considered are the extent of bacterial colonization in fast-flowing water through a deeper GAC bed and containment in a closed cylinder.

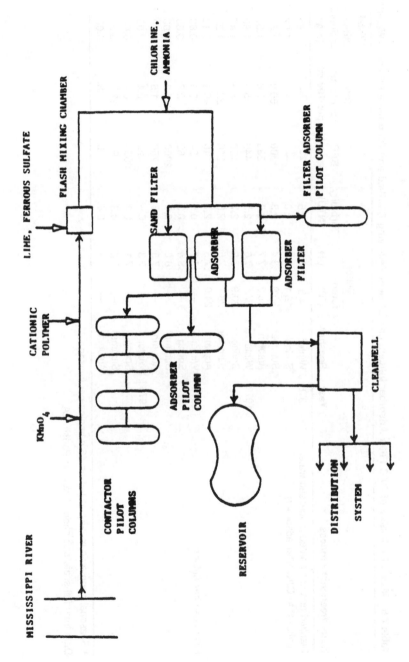

Figure 7.1 Full-scale filter and pilot-column flowchart for Jefferson Parish, LA.

Table 7.6 Microbial Quality of GAC Filter Adsorber Influent and Effluent Receiving Clarified and Chlorinated Process Water

Water treatment process	Year	Season[a]	Water		Turbidity (NTU)	SPC per ml	Total coliforms per 100 ml	Total chlorine residual (mg/l)
			Temp (°C)	pH				
Clarified and chlorinated processed water (influent to GAC filter adsorber)	1978	Winter	4.8	8.7	2.5	—	—	1.64
		Spring	16.1	8.5	2.0	—	—	1.62
		Summer	30.0	8.4	2.8	1300	60.0	1.57
		Autumn	18.3	7.7	3.9	530	9.6	1.57
	1979	Winter	7.1	7.6	2.9	440	15.3	1.90
		Spring	20.9	7.3	2.7	120	8.3	1.45
		Summer	28.7	7.5	2.8	210	22.1	1.29
		Autumn	24.0	7.5	2.7	140	<0.1	2.13
Filter adsorber effluent	1978	Winter	—	8.7	0.5	24	0.4	0.83
		Spring	—	8.4	0.3	62	2.7	0.63
		Summer	—	8.0	0.3	2800	5.8	0.00
		Autumn	—	7.5	0.8	750	1.5	0.00
	1979	Winter	—	7.3	0.9	425	<0.1	0.00
		Spring	—	7.1	0.4	2850	6.0	0.00
		Summer	—	7.2	0.3	98	12.7	0.00
		Autumn	—	7.1	0.4	110	11.8	0.00

[a] Seasonal geometric means based on 46 samples per season.

Data from full-scale operation, Jefferson Parish, LA.

Two aspects of microbial response in GAC contactors were explored: amplification potential for heterotrophic bacteria, including coliform persistence, and species profile for various *Pseudomonas* and *Flavobacterium* strains that could be opportunistic pathogens. Inspection of data in Figure 7.2 reveals that chloraminated influent source water (used as feed water through the GAC contactors in series) had standard plate counts (SPC) ranging from 4 to 170 organisms per milliliter. Lower maximum densities occurred during late autumn and winter cold water temperature conditions. After the influent water passed through the first GAC contactor, densities of heterotrophic bacteria (SPC) increased by 3 to 4 logs and remained at these higher levels with subsequent serial passage through the following contactors.

Residual coliform populations surviving the impact of source water chloramination recovered sufficiently to pass from one contactor to the next, resulting in colonization and occasional release in the effluents from each contactor. Apparently, assimilable nutrients in the GAC contactors were the limiting factor that prevented a more aggressive growth of coliforms and higher densities of heterotrophic bacteria in a stepwise fashion from one contactor to another. From these data, it quickly becomes obvious there were a large number of organisms being released in this process water that we know little about in terms of public health significance.

Does the GAC contactor environment become a habitat for various *Pseudomonas* species? Periodic speciation of isolated bacteria from GAC contactor effluent water indicate that a variety of *Pseudomonas* species (Table 7.8) colonized the GAC contactor with recurrent or continuous releases to the process water. Occasional quantitation of these organisms over a 3-year period (1985 to 1987) suggested that *Pseudomonas* species represented only 1 to 2% of the entire population of heterotrophic bacteria detected in effluent waters from GAC contactors. The extent of regrowth and density amplification of *Pseudomonas* species were not measurable because of numerous indeterminate high densities in the contactor effluents. However, these events appeared to increase as the water passed through contactors in series, suggesting that a progression of colonizing organisms migrated to each of the contactors in series.

One of the areas of greatest confusion in studying the microbial ecology of GAC adsorbers lies in the selection of a cultural protocol (medium, incubation time, and temperature) to optimize recovery of these organisms. In many of the pilot- and full-scale studies the standard plate count procedure (plate count agar [PCA], 35°C incubation for 48 h) has been used. The question arises, is this procedure optimizing recovery of the heterotrophic bacterial population? A comparative analysis of process waters by two different culture media and extending the incubation time illustrates the problem (Table 7.9). The traditional SPC procedure does not adequately detect either the magnitude of bacterial growth in adsorber beds or in other process waters. These organisms need a medium with a diversity of nutrients in low concentration, such

Table 7.7 Microbial Quality of GAC Filter Adsorber Influent and Effluent Receiving Sand Filter Process Water

Water treatment process	Year	Season[a]	Water Temp (°C)	pH	Turbidity (NTU)	SPC per ml	Total coliforms per 100 ml	Total chlorine residual (mg/l)
Sand filter process water (influent to adsorber)	1977	Winter	10.0	10.0	0.4	10	<0.1	1.60
		Spring	23.5	9.9	0.2	1.7	<0.1	1.63
		Summer	30.5	9.9	0.2	10	<0.1	1.57
		Autumn	13.0	8.7	0.9	34	<0.1	1.70
	1978	Winter	4.8	8.6	0.4	10	<0.1	1.67
		Spring	16.1	8.5	0.3	10	<0.1	1.62
		Summer	30.0	8.4	0.4	17	<0.1	1.53
		Autumn	18.3	7.7	0.9	33	<0.1	1.83
	1979	Winter	7.1	7.5	1.1	73	<0.1	1.90
		Spring	20.9	7.3	0.5	27	<0.1	1.40
		Summer	28.7	7.5	0.3	26	<0.1	1.27
		Autumn	24.0	7.4	0.5	23	<0.1	2.17
Adsorber effluent	1977	Winter	—	9.9	0.3	16	<0.1	0.00
		Spring	—	9.6	0.2	96	<0.1	0.00
		Summer	—	9.6	0.2	685	<0.1	0.00
		Autumn	—	8.5	0.8	138	<0.1	0.01
	1978	Winter	—	8.5	0.4	31	<0.1	0.20
		Spring	—	8.1	0.3	40	<0.1	0.00
		Summer	—	7.9	0.2	647	<0.1	0.00
		Autumn	—	7.4	0.7	305	<0.1	0.00
	1979	Winter	—	7.3	0.8	415	<0.1	0.00
		Spring	—	7.1	0.4	2,500	<0.1	0.00
		Summer	—	7.2	0.3	130	<0.1	0.00
		Autumn	—	7.1	0.4	63	<0.1	0.00

[a] Seasonal geometric means based on 46 samples per season.

Data from full-scale operation, Jefferson Parish, LA.

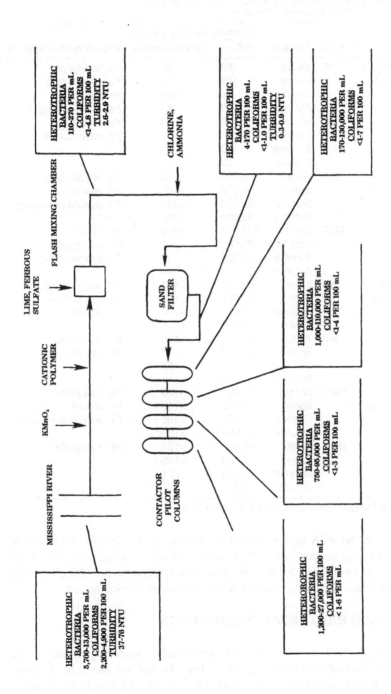

Figure 7.2 Microbial quality of effluents from contacts in series.

Table 7.8 Characterizing the *Pseudomonas* Population in GAC Contactor Effluents

Process	Year	Season	Maximum density per ml SPC	*Pseudomonas*	*Pseudomonas* species
Plant influent	1986	Winter	52	22	*Ps. alcaligenes*
chlorinated		Spring	100	—	*Ps. pseudoflava*
		Summer	44	>100	*Ps. pickittii*
		Autumn	53	44	*Ps. pseudoalcaligenes*
	1987	Winter	38	33	*Ps. maltophila*
		Spring	170	57	*Ps. paucimobilis*
		Summer	110	23	
		Autumn	23	—	
Contactor	1986	Winter	2,300	86	*Ps. pseudoflava*
no. 1 effluent		Spring	80,000	>100	*Ps. pickittii*
		Summer	53,000	>100	*Ps. pseudoalcaligenes*
		Autumn	55,000	>100	
	1987	Winter	130,000	>100	*Ps. mallei*
		Spring	85,000	—	*Ps. maltophila*
		Summer	19,000	70	*Ps. paucimobilis*
		Autumn	—	—	
Contactor	1986	Winter	62,000	>100	*Ps. paucimobilis*
no. 2 effluent		Spring	27,000	>100	*Ps. pseudoflava*
		Summer	77,000	>100	*Ps. pickittii*
		Autumn	110,000	>100	
	1987	Winter	43,000	>100	*Ps. maltoshilia*
		Spring	22,000	73	*Ps. paucimobilis*
		Summer	—	90	
		Autumn	—	—	
Contactor	1986	Winter	64,000	>100	*Ps. maltophila*
no. 3 effluent		Spring	38,000	>100	*Ps. pickittii*
		Summer	11,000	>100	*Ps. pseudoalcaligenes*
		Autumn	95,000	>100	
	1987	Winter	85,000	>100	*Ps. maltophila*
		Spring	55,000	>100	
		Summer	61,000	71	
		Autumn	4,400		

Data from full-scale operation, Jefferson Parish, LA.

as found in the R2A agar formulation.[7] Increasing the length of incubation time at a lower temperature (28°C) further enhances the recovery of a wide spectrum of organisms that may be present in GAC adsorber effluents, other stages of water treatment, finished water, and water in distribution.

ENHANCED BIOLOGICAL DEGRADATION

Many of the industrial chemical compounds found in polluted source waters and naturally occurring organics released to surface waters by decaying vegetation and algal blooms are nonbiodegradable and poorly adsorbed in GAC filtration. Because these organics may also react with disinfectants to

Table 7.9   Bacterial Populations in Water Treatment Processes Using Standard
             Plate Count Medium or R-2A Medium with Extended Incubation Times
             (Organisms per ml)

Sampling day	Lime-softened water			Sand filter effluent			GAC adsorber effluent		
	SPC, 2 d	SPC, 6 d	R2A, 6 d	SPC, 2 d	SPC, 6 d	R2A, 6 d	SPC, 2 d	SPC, 6 d	R2A, 6 d
Initial	120	350	510	890	1,200	1,500	<1	140	220
7	31	202	510	820	22,000	35,000	1	24,000	95,000
14	7	7	130	<1	1,200	9,400	<1	600	4,400
21	7	18	150	2,200	2,500	33,000	<1	5,200	16,000
28	3	39	530	700	7,800	67,000	1	11,000	55,000
35	<1	490	330	100	6,000	25,000	<1	12,000	74,000
42	70	120	1,700	1,200	71,000	22,700	N.D.	56,000	52,000
49	9	1,200	23	5,000	41,000	3,000	80	4,200	100
56	<1	10	<1	<1	700	12,000	N.D.	1,900	50,000
63	29	190	170	170	2,000	3,000	N.D.	5,000	48,000

Note: All cultures incubated at 35°C; SPC, standard plate count; N.D., not done.
Data revised from Symons et al.[7]

form undesirable by-products, attention has been directed toward the conversion of these complex, refractory compounds into more readily biodegradable
substances that can be adsorbed by GAC or consumed by microorganisms
established in the filter.[9-12] This combined process is sometimes described as
biological activated carbon (BAC) treatment, which frequently involves the
use of preozonation to enhance the BAC process.[13-15]

Case History: Shreveport, LA

A 10-gpm pilot plant (Figure 7.3) was constructed at the Shreveport treatment facility to evaluate THM precursor removal through a conventional
treatment train using a "Waterboy" package plant (without disinfection) and
GAC adsorption, or with ozonation prior to GAC adsorption in a BAC mode.[16]
The purpose of investigating BAC in this pilot study was to enhance the
biodegradation of high levels of THM precursors in Cross Lake water, the
principal water supply for Shreveport, LA. The microbiological concern was
the possible loss of effective barriers to coliform penetration further into the
system, because final disinfection was the only barrier to coliform migration
and elevated heterotrophic bacterial densities reaching the distribution system.

While preozonation of the process water was used primarily to convert
recalcitrant organics to shorter-chain carbon compounds, the process, as
applied, did have some impact on bacterial densities and profiles of organisms
entering the GAC filter bed (Table 7.10). During the cold-water periods of
autumn and winter, there was a 2- to 3-log reduction in the source water
bacterial densities applied to BAC as compared to nonozonated raw water.
Smaller reductions were noted during warm-water months and on one occasion

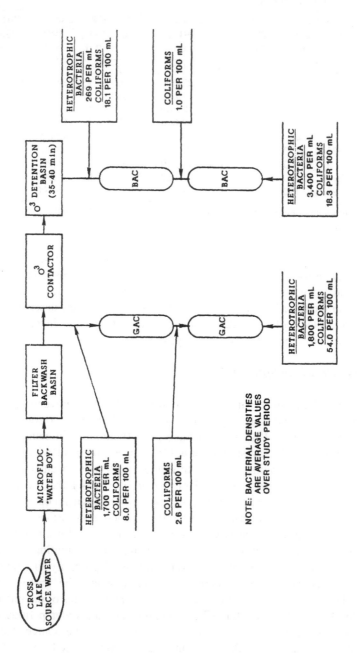

Figure 7.3 Microbial quality of effluents — GAC or O_3 + GAC — at the Shreveport, LA pilot treatment facility.

Table 7.10 Heterotrophic Plate Counts and Total Coliform Densities in Pilot Water Treatment Facility, Shreveport, LA.

Date	Temp (°C)	Nonozonated water Influent–GAC–Effluent HPC	T.C.	HPC	T.C.	Ozonated water Influent–GAC–Effluent HPC	T.C.	HPC	T.C.
1980									
July 8	32	3,300	<1	4,100	4	290	<1	13,000	1
August 12	31	2,500	1	550	8	30	<1	1,100	<1
September 15	29	400	9	1,600	7	78	1	3,400	2
October 6	21	130	4	750	2	9	<1	550	1
November 12	14	1,700	9	300	1	240	5	1,500	<1
December 8	14	1,900	N.D.	7,600	N.D.	2	<1	1,300	<1
1981									
January 13	10	82	2	67	2	3	<1	3,600	<1
February 10	9	110	4	70	5	5	<1	1,700	15
March 24	19	660	1	150	1	55	5	2,900	1
April 15	23	170	<1	240	1	230	7	2,200	N.D.
May 14	24	300	8	130	7	200	1	4,700	28
June 24	34	550	33	140	455	870	135	2,900	80
July 16	31	700	25	160	215	2,900	81	2,400	91

Note: Raw lake water quality, 10^3 to 10^4 total coliforms per 100 ml; HPC, heterotrophic plate count per milliliter using soil extract agar; T.C., total coliform density per 100 ml; N.D., no data.

(July 16, 1981) the ozonated influent contained four times the density of heterotrophic bacteria found in nonozonated influent water. Ozonation exposure may have caused the breakup of bacterial aggregates or algal masses in the source water, thereby increasing the dispersion of viable bacteria.

The BAC mode intentionally encourages greater microbial activity in the BAC bed for the purpose of assimilating much of the recalcitrant organic conversions by ozonation. While ozonation exposure initially suppressed heterotrophic bacterial densities, there was a tenfold increase in these organisms in the effluent from the second BAC contactor. The increase in biodegradable organics created by ozonation stimulated growth. Coliform growth did not follow this pattern in the BAC contactors, probably because the general population of other heterotrophic organisms rapidly became dominant and suppressed coliform development and detection. In both GAC and BAC treatment modes, the pilot study revealed that no treatment barrier protection was provided at these latter stages of water processing. In such situations, final disinfection must be 100% effective at all times to achieve the necessary 6-log reduction of bacteria for a safe water supply.

Examination of process water for coliform bacteria was also done and provided evidence that total coliforms may persist in both BAC and GAC columns (Table 7.10).[16] The raw source water contained 100 to 10,000 coliforms per 100 ml that were not completely inactivated in the pretreatment of influent waters going to the pilot plant, nor by preozonation in BAC treatment. Coliforms were also occasionally isolated from BAC treated in a similar pilot-plant study conducted in Philadelphia, PA.[15] In the Philadelphia study, the river source water contained 48,000 total coliform organisms per 100 ml and pretreatment was not very effective in the inactivation of coliforms. It is important to note that the ozone concentration used was selected to obtain maximum removal of dissolved organic carbon and was not necessarily optimum for disinfection of the raw source water.

Identifying the coliform strains isolated from the GAC filter effluents revealed that *Klebsiella, Enterobacter,* and *Citrobacter* were the genera involved. These organisms are the same coliforms that have been reported to predominate in biofilm growth within some water distribution systems.[17] How much of a case can be made for relating coliform occurrences in distribution systems to treatment barrier penetration from a process water in unknown, but should not be overlooked.

The observation that profiles of bacterial groups and species in BAC effluents show a remarkable similarity to those present in GAC effluents is not surprising. The similarity in bacterial profiles is a reflection of the way carbon treatment technology is utilized in this country. Rather than extending the service life of the filter to encourage development of specialized bacterial populations that are more efficient in assimilation of dissolved organics, greater reliance is placed on the carbon adsorption aspect, with more frequent reactivation of the carbon media. The Shreveport pilot study demonstrated that after 52 weeks of BAC operation, only microbial metabolism was responsible for removal of organics, and the rate of trihalomethane formation potential was not sufficiently reduced to meet a 0.1 mg/l maximum contaminant level (MCL) for trihalomethanes.[18]

Amplification and acclimatization of a diverse and specialized microbial population through extended service life of a BAC filter presents another concern. Under these conditions, a biofilm of specialized organisms develops through successional changes in dominant species and is similar to biologically active floc development in sewage treatment processes. While final disinfection can be effective in inactivating many of these diverse organisms, others will be resistant to applied disinfection and pass into the distribution system. Neither the health effect significance of this diversified population of organisms entering the potable water supply nor the contribution these organisms make to the development of biofilm in the distribution pipe network and associated reservoirs has been clearly demonstrated.

DISINFECTANT ALTERNATIVES

Another approach to minimize trihalomethane production in water treatment is the use of a disinfectant alternative to free chlorine. Preformed chloramines (chloramination), chlorine dioxide, and ozone have been proposed as practical disinfectant alternatives. In addition, potassium permanganate has been suggested as a preoxidant for some raw source waters. Because of the desire to maintain a disinfectant residual in distribution water, chloramines and chlorine dioxide have received considerable attention. Some surface-water systems are seriously considering ozonation because of its more favorable disinfection C·T values for inactivating *Giardia* cysts. While ozone is a powerful inactivating agent for waterborne pathogens, it does not have a lasting residual to provide protection in distribution water. Furthermore, ozone is known to create additional assimilable organics that stimulate the growth of heterotrophic bacteria in the distribution system. Each alternative disinfectant candidate has specific advantages over free chlorine application, but also some significant disadvantages that must be understood in the trade-off.

CHLORAMINES

For many utilities, conversion to chloramines rather than applying free chlorine provides the simplest solution to the trihalomethane problem. Despite the poor biocidal efficiency of chloramines, there is evidence that the chloramination process has been used successfully for years by a number of utilities. Chloramines have also proven beneficial in suppressing biofilm development, particularly in systems situated in Sunbelt locations, where the climate has extended months of temperatures exceeding 15°C. One negative aspect, however, has been the development of high heterotrophic plate counts accompanied with taste and odor problems in slow-flow sections and dead ends. For this reason, frequent system flushing and return to a free chlorine residual during the few weeks of seasonally cooler weather is desirable to suppress the growing population of nuisance organisms and their associated taste and odor releases.

In conventional chloramination, ammonia is added to the water first. Chlorine addition then follows, usually in the form of chlorine gas. The rate of conversion of free chlorine to chloramines is dependent on pH, temperature, and the ratio of chlorine to ammonia present. While the reaction occurs in hundredths of a second at warm-water temperatures and optimum pH (8.3), it can occur at much slower rates at winter temperatures and lower or higher pHs. Therefore, the presence of free chlorine for several minutes could result in rapid inactivation of microorganisms, particularly at lower pH levels, because free residual chlorine is present in the form of HOCl.

The use of chloramines also provides for a longer-lasting, measurable disinfectant in the distribution system. This is a point not lost on those systems that have difficulty in maintaining a free chlorine residual in 95% of all sites monitored in the pipe network. Monochloramine is definitely a less effective disinfectant than free chlorine, when compared at comparable low-dose concentrations and short contact periods. Thus, any proposed use of chloramines requires provisions for the application of higher disinfectant concentrations or longer contact time to achieve an effective treatment barrier.

Case History: Jefferson Parish, LA

The Jefferson Parish Water Department has relied on monochloramine as the sole water disinfectant for over 30 years. In a study of data collected over an 18-month period from this water treatment plant, Brodtmann et al.[19] reported only two total coliform occurrences in 6720 samples of finished water. This treatment scheme provided a 30-min contact time before filtration, with 1.1 to 1.0 mg/l combined chlorine residual measured in the gravity sand filter effluent. Initial processing of the river source water with potassium permanganate and polyelectrolyte addition lowered the SPC by an average 84% during water clarification (Figure 7.4). Clarification together with 8 to 10 min of monochloramine contact resulted in an average 96.1% reduction of the source water population of SPC organisms. Continued processing with sand filtration in combination with a total combined residual contact time of 30 min lowered the initial level of measured organisms 99.7%. The average monthly SPC, reported to be below 50 organisms per milliliter in the distribution system, may be misleading because the problem of regrowth is generally associated with warm-water temperature conditions, areas of slow flow, and dead-end sections of the distribution system. The data used in calculating the monthly average were not inclusive of samples collected at sites under these adverse conditions.

Case History: Louisville, KY

The Louisville Water Company, Louisville, KY investigated trihalomethane concentration generated by three different disinfectant treatments.[20] Normal plant operations used free chlorine applied to gravity-settled source water before the clear-well. During modified treatment, ammoniation of the free chlorine residual was practiced at the clear-well during several weeks of data gathering. Later in the year, application of chloramines occurred following coagulation. When ammonia was added at the softening basin, it was in some excess so that further chlorination at the clear-well would restore the chloramine residual. The net result was that a combined chlorine residual was maintained throughout the latter stages of treatment and into the distribution system.

A study of process water quality indicated that the application of chlorine to the gravity-settled source water effected a complete reduction in both total coliforms and SPC densities (Table 7.11). Densities remained low in all subsequent in-plant samples. Injecting ammonia into the clear-well at the end of the treatment train or adding ammonia in the softening basin followed by filtration and clear-well chlorination resulted in no further bacterial penetration of the treatment train. In all cases, the data demonstrated finished water of acceptable quality.

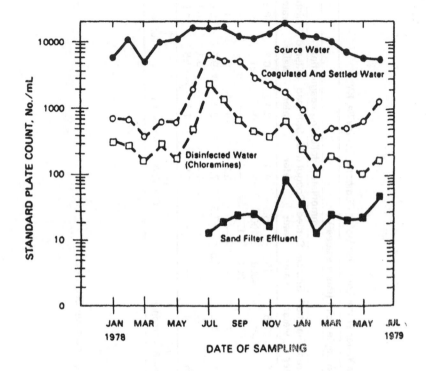

Figure 7.4   Standard plate count at various stages of water treatment at the Jefferson Parish Water Department, Jefferson Parish, LA. (From Epton, M. A. and Becnel, J. F., EPA 600/52-81-027, U.S. Environmental Protection Agency, Cincinnati, OH, 1981.)

CHLORINE DIOXIDE

Chlorine dioxide as a disinfectant alternative to free chlorine is worthy of consideration for minimizing trihalomethane production. Furthermore, chlorine dioxide can oxidize organic complexes of iron and manganese and suppress a variety of taste and odors in water supply.[21,22] Bactericidal efficiency toward *E. coli*, *Salmonella typhosa*, and *Salmonella paratyphi* was either equal or greater than free chlorine[23] and is superior to combined chlorine as a viricide. Unfortunately, chlorine dioxide must be generated on-site and there is a restriction on the concentration of chlorite and chlorate tolerated in the product water because these breakdown products are capable of oxidizing hemoglobin.[24] As a consequence of health effects, the amount of chlorine dioxide applied must be carefully managed.

Table 7.11 Chloramine Application Point Study Using Ohio River Source Water at the Louisville Water Company, Louisville, KY

| | Sample point (mean values)[a] | | | | | | | | | | | |
| | Ammoniation at clear-well[b] | | | | | | Ammoniation following coagulation[c] | | | | | |
Parameter	Source water	Settled water	Coagulated water	Softened water	Filtered water	Finished water	Source water	Settled water	Coagulated water	Softened water	Filtered water	Finished water
Turbidity (NTU)	19	23	4.7	3.8	0.4	0.5	15	18	4.6	2.4	0.3	0.2
Total coliforms per 100 ml	3200	4900	<1	<1	<1	<1	4000	1100	<1	<1	<1	<1
Fecal coliforms per 100 ml	62	104	—	—	—	—	204	177	—	—	—	—
SPC per ml	—	—	<1	51	9	6	—	—	2.2	4	2	1
pH	7.3	7.5	7.0	9.2	9.1	8.2	7.5	7.6	7.2	9.3	9.0	8.6
Free Cl₂ residual (mg/l)	—	—	2.6	0.6	0.4	0.2	—	—	1.7	—	0.1	<0.1
Total Cl₂ residual (mg/l)	—	—	2.8	0.7	0.5	1.4	—	—	1.9	1.8	1.5	1.9
			↑ Chlorine		↑ Chlorine & ammonia				↑ Chlorine	↑ Ammonia		↑ Chlorine

Free Cl₂ residual uses Cl_2; Total Cl₂ residual uses Cl_2.

Note: —, no analysis run.

a Based on five samples over a 9-d period.
b Source water temperature, 29°C (84°F).
c Source water temperature, 16°C (61°F).

From Glaze, W. H., EPA 600/2-82-046, U.S. Environmental Protection Agency, Cincinnati, OH, 1982.

Case History: Western Pennsylvania

The Western Pennsylvania Water Company Hays Mine plant presented an opportunity to study the alternative use of chlorine dioxide as the primary disinfectant.[7] For this investigation, the routine practice (Table 7.12) was source water chlorination, potassium permanganate treatment, coagulation, settling, activated carbon filtration/adsorption, and free chlorine application in the clear-well. Later, the treatment train was modified (Table 7.12) to inject chlorine dioxide and potassium permanganate into the source water entering the coagulation basin. Free chlorine was applied as a secondary disinfectant in the clear-well prior to distribution. Chlorine dioxide dosage to the source water was 1.5 mg/l and contained less than 0.1 mg/l chlorine.

During source water chlorination, mean total coliform and SPC densities in the activated carbon/filter adsorber influent were one per 100 ml and 50 per milliliter, respectively. When chlorine dioxide was the applied disinfectant prior to coagulation and settling, a disinfectant residual could not be maintained. As a result, mean bacterial densities reaching the activated carbon filter/adsorber were 43 total coliforms per 100 ml and 7100 SPC organisms per milliliter. In-plant survivors of the total coliform population passed through the 2.5-year-old granular activated carbon filter/adsorber essentially unchanged in density. In both treatment trains, the secondary application of chlorine in the clear-well was, however, an effective barrier to total coliform penetration into the distribution system.

From these data, 1.5 mg/l of chlorine dioxide was not equal to the disinfection effectiveness of 2.6 mg/l free chlorine during source water disinfection. Increasing the dose of chlorine dioxide was not economically feasible and might exceed the limit of 0.5 mg/l residual chlorine dioxide, chlorite, and chlorate recommended by the U.S. Environmental Protection Agency.[17] New information on disinfection by-product formation may further reduce this total oxidant limit to 0.3 mg/l in the future.

A further modification of treatment that utilized source water disinfection with a low concentration of both disinfectants was effective in reducing the bacterial densities in the GAC filter/adsorber influent at the Hays Mine plant. Some regrowth of total coliforms and the heterotrophic bacterial population did occur in the filter/adsorber and appeared in the effluent. With the application of chlorine at the clear-well, however, the finished water met the bacteriological standard for total coliforms and a low mean SPC of eight organisms per milliliter was present.

POTASSIUM PERMANGANATE

Potassium permanganate is often used in the water supply treatment for taste and odor control or for removal of iron and manganese.[25,26] Because it

Table 7.12 Chlorine Application Point Study—Western Pennsylvania Water Company

Parameter	Mean values before treatment modification						Mean values after treatment modification					
			Sample point						Sample point			
	Source	Plant influent	Coagulation	Settled	GAC filtered	Finished	Source	Plant influent	Coagulation	Settled	GAC filtered	Finished
Flow time (h)	0	0.5	3.75	12.5	13.5	14.8	0	0.5	3.8	12.5	13.5	14.7
Turbidity (NTU)	51	38	5.7	8.5	0.6	0.2	6.8	5.2	6.3	2.3	0.3	0.2
Total coliform per 100 ml	21,000	4	1	1	8	<1	14,000	4,200	100	43	44	<1
SPC per ml	NR	490	200	50	150	3	NR	29,000	4,790	7,100	850	1
pH	7.2	7.1	7.3	7.1	7.2	7.1	7.1	7.1	7.5	7.4	6.9	6.8
Free Cl_2 residual (mg/l)	NR	0.4	<0.1	<0.1	<0.2	0.6	NR	<0.1	<0.1	<0.1	0.1	<0.4
Total Cl_2 (ClO_2) residual (mg/l)	NR	0.8	0.4	0.3	0.2	0.8	NR	<0.1	<0.1	<0.1	<0.1	<0.1

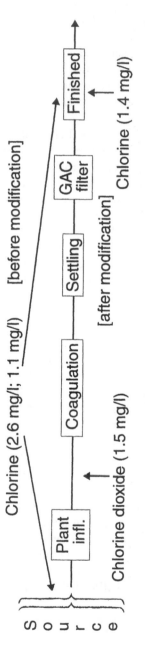

Chlorine (2.6 mg/l; 1.1 mg/l) [before modification]

Chlorine dioxide (1.5 mg/l)

[after modification]

Chlorine (1.4 mg/l)

Monongahela River

Source — Plant infl. → Coagulation → Settling → GAC filter → Finished

is an oxidant, other applications suggested have included the disinfection of process basins (concrete, cement mortar lining, and asbestos cement surfaces) and water lines after repairs.[27,28] Because potassium permanganate has a limited disinfection efficacy, application in the disinfection of water lines is not as effective as use of chlorinated water.[29,30] Nevertheless, there may be some measurable benefit achieved in using potassium permanganate as a preoxidant in early stages of the treatment train. In this situation, the preoxidant may reduce growth of algae and slime bacteria in the treatment basins plus provide some abatement in the bacterial population.

Case History: St. Joseph, MO

The St. Joseph Water Company uses clarification, sedimentation, filtration, and chlorine disinfection in the processing of Missouri River water.[31] In August 1982, 1.1 mg/l potassium permanganate was applied for 4 weeks (Table 7.13) at the discharge of the clarifiers (Figure 7.5), prior to entering settling basin no. 1. On the fifth week, no potassium permanganate was applied so as to provide percent reduction data for settling alone. Data in this preliminary study suggest that settling produced 59% of the bacterial reduction. Application of potassium permanganate apparently accounted for an additional 40.8%. While potassium permanganate would not be satisfactory for application in final disinfection, use as a preoxidant early in the treatment train may provide significant reduction of total coliforms early in water processing. What impact this predisinfectant has on shaping the resultant microbial flora entering the distribution system is unknown.

Case History: Davenport, IA

The Davenport Water Company processes raw water from the Mississippi River (Figure 7.6) using clarification, sedimentation, filtration through GAC, and disinfection. In 1983, 0.61 mg/l potassium permanganate was added to control odor in the flocculator basins and keep the sedimentation basin sludge from turning septic.[30]

Data collected over 10 months (Table 7.14) indicate that the combination of preoxidant application and settling for 35.9 h could provide a significant reduction in both total coliforms and the SPC. How much of this reduction was due to settling vs. preoxidant contact time of 35.9 h was not determined. The treatment approach did eliminate odor in the flocculator buildings that appeared after the discontinuance of prechlorination. The effects this preoxidant might have had on selective survival of bacteria and conversion of various chemical complexes to assimilable organic compounds released to the distribution system is not known.

Table 7.13   Potassium Permanganate as a Preoxidant — St. Joseph, MO Water Treatment Application[a]

	KMnO$_4$ (1.1 mg/l) applied			KMnO$_4$ not applied		
	Source[b]	Settling basin no. 1	Reduction (%)	Source[b]	Settling basin no. 1	Reduction (%)
Retention time (h)	—	8.5	—	—	10.8	—
Turbidity (NTU)	755	30	96.0	2,520	62	97.5
Total coliform per 100 ml	52,000	88	99.8	87,000	35,700	59.0

[a] Average of four runs (August 5 to 10, 1982) applied; one run (August 15) as control.
[b] Missouri River.

Data from Blanck.[31]

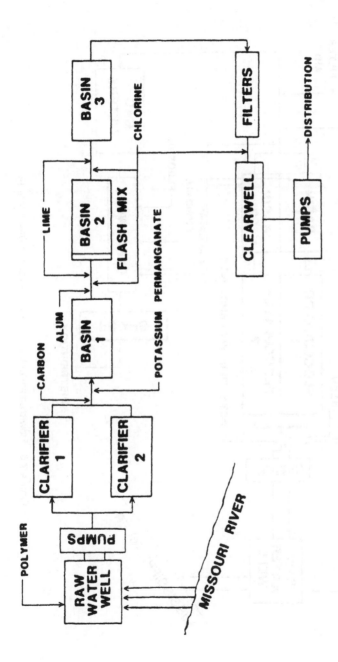

Figure 7.5 Flow diagram of the St. Joseph Water Company purification plant in St. Joseph, MO.

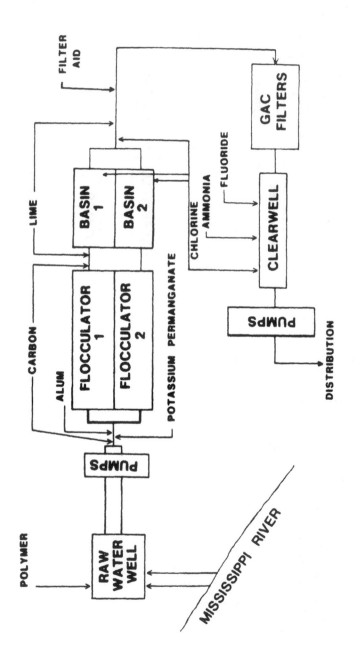

Figure 7.6   Flow diagram of the East River station in Davenport, IA.

OZONATION

Ozone has frequently been used in water supply treatment to remove taste, odor, and color because many of the compounds responsible for these characteristics are unsaturated organics.[32] Other uses include the removal of iron and manganese, or use as a coagulant aid to reduce coagulant requirements and increase filtration rates.[33,34]

There is a growing interest in the use of ozone because it is a powerful disinfectant. For instance, ozone is far more effective in the inactivation of *Giardia* cysts than chlorine. Unfortunately ozone residuals are quickly dissipated, with a lifetime of less than an hour in most drinking water systems.[34] As a result, secondary application of chlorine is necessary to provide disinfectant residual protection in the distribution system.

Treatment train application of ozone generally includes GAC filter-sorbers. Ozone exposure maximizes the breakdown of complex organics to shorter-chain compounds that are then either absorbed in the GAC filter bed or degraded by the bacterial flora in a BAC filter. The net result will be less THM precursors to react with chlorine in final disinfection. However, there is a trade-off to consider: increased bacterial densities released from the GAC contactor and increased levels of assimilable organic carbon in the finished water support seasonal regrowth of the heterotrophic bacteria in the pipe network.

Table 7.14   Potassium Permanganate (0.61 mg/l) as a Preoxidant — Davenport, IA Water Treatment Application[a]

	Source[b]	Flocculator no. 2 effluent	Reduction (%)
Retention time (h)	—	35.9	—
Turbidity (NTU)	11.8	1.9	83.9
Total coliforms per 100 ml	8660	17	99.9
SPC per ml	7960	1030	87.1

[a] Average values for 10 months (1983).
[b] Mississippi River.

Data from Blanck.[31]

IMPACTS ON DISTRIBUTION SYSTEM WATER QUALITY

Major changes in source water quality, treatment modifications, and operational practices are reflected in distribution water quality. The beneficial aspects of some treatment modifications may, in the long term, lead to reduced assimilable organic carbon and less biofilm development in the pipe environment. However, adverse effects as a result of reduction in treatment barrier redundancies may eventually lead to biofilm colonization of pipe sections, taste and odor complaints, and increased coliform occurrences. Therefore, it is essential to carefully monitor the microbial quality of water in distribution, particularly at the end of the system and in areas of slow flow where disinfectant residuals are marginal or nonexistent. Several case histories will help illustrate this concern.

TREATMENT MODIFICATIONS

Modifications in treatment train processes to reduce trihalomethane production may ultimately change the character of the bacterial populations passing through the distribution system. These quality changes may be of immediate concern if the last barrier to bacterial passage into the finished water is interrupted, if changes occur seasonally with increased water temperature or slowly with time as habitats develop.

Case History: Cincinnati, OH

The Cincinnati Water Works stopped chlorination of the Ohio River source water and began chlorination at the influent to the treatment plant (Table 7.5) on July 14, 1975 as an initial step in changing the in-plant water treatment process to control trihalomethane concentrations. Chlorination at the clearwell was used to inactivate any residual coliform population that might have penetrated other processes in the treatment chain. With careful control of chlorine dose, point of application, and water pH, a significant decrease in trihalomethane concentration was realized. The impact that this treatment modification might have on the bacteriological quality of drinking water at the distribution system dead ends and other slow-flow sections in the distribution network was determined from an intensive 2-year study.

With the cooperation of the Cincinnati Water Works Water Distribution Maintenance Section, samples from 32 dead-end water mains were examined on a rotating basis of eight sites per week. These sites are among a number of troublesome dead-end water mains that are flushed out each week to clear accumulated sediments and bring fresher water with free chlorine residuals into these distribution lines. Samples from these flushes were iced immediately and processed within 5 h of collection. Analyses of 613 water samples over the 2-year period included a ten-tube, three-dilution total coliform MPN test and an SPC incubated at 35°C for 48 h. Physical/chemical parameters measured were free chlorine residual, turbidity, water temperature, and pH.

Changes in distribution system water quality were not observed immediately on the day of the treatment change. Approximately 15 d passed before some decrease in free chlorine residual concentrations, turbidity, and pH occurred. Before the change in the point of disinfection application, increased chlorine residuals were inconsistent in limiting some coliform occurrences, probably because of sediment accumulations that resulted in an average turbidity of 20.7 NTU in these deadend sections. The most extreme example occurred during one week in December 1974, when the total coliform density averaged 138 organisms per 100 ml in the eight samples collected from selected dead-end flushings. Once the turbidity decreased to an average of 10.1 NTU, this interference with disinfection was not apparent. Why the turbidity in the dead ends was reduced following the treatment change is not known; the protocol and frequency of main flushing remained unchanged. Perhaps this reduction in turbidity was a result of more water flow created by new service tap-ins from residential developments or it may have been a result of more stable scale formation on the pipe walls (pH shifted from 8.0 to 7.8) following treatment modifications.

After the point of chlorination was moved, a free chlorine residual concentration of at least 0.2 mg/l was effective in controlling coliform occurrences in the dead-end sections of the distribution network. When free chlorine residual concentrations declined to 0.1 mg/l or less during warm-water periods, however, viable coliforms in these protected pipe habitats were detected in densities as great as 30 organisms per 100 ml. Water temperatures during these periods of low free chlorine residual concentrations fluctuated from 20 to 25°C (68 to 77°F). Sudden increases in SPC densities often occurred a few days to a week in advance of the appearance of coliforms in these waters. Thus, increased SPCs could serve as an early signal of a loss of disinfection effectiveness or other undesirable quality changes occurring in water distribution systems.

Case History: Manchester, NH

The effects that GAC or BAC treatment have on distribution water quality are largely undocumented. Several coliform species (*Klebsiella, Enterobacter,* and *Citrobacter*) have been found to colonize GAC filters, regrow during warm-water periods, and discharge into the process effluent. Carbon particles have also been detected in finished water from several water plants using powdered carbon or GAC treatment. Over 17% of finished water samples examined from nine such water treatment facilities contained carbon particles colonized with coliform bacteria.[35] These findings confirm that carbon fines provide a mechanism by which microorganisms penetrate treatment barriers and reach the distribution system. Other mechanisms that could be involved in protected transport of bacteria include aggregates or clumps of organisms from colonization sites in GAC or sand filtration and the protected nature of particulates in water.

Another important finding was that full-scale GAC treatment (Manchester, NH) resulted in a statistically significant increase in heterotrophic bacterial densities in distribution water as compared to a similar water treatment operation (Concord, NH) that does not employ GAC.[36] Furthermore, water temperature, pH, and turbidity had a positive influence on heterotrophic bacterial densities.[37] These physical/chemical conditions of water are key factors that also impact disinfection effectiveness.

DISINFECTION MODE

Stability of disinfectant residuals during water distribution is important for a number of purposes, particularly to prevent colonization of surviving organisms and to disinfect contaminants that intrude into the pipe network. Microbial colonization may lead to corrosive effects in the distribution system and aesthetic changes in taste, odor, and appearance. Regrowth of potential health-related opportunistic organisms and their impact on coliform detection should not be dismissed as a trivial problem. Further, the maintenance of a disinfectant residual to the consumer's tap keeps the system clean and protects against some cross-connection contamination.

Analysis of data taken from the national community water supply study of 969 public water systems,[38] revealed that SPCs of ten organisms or less were obtained in over 60% of these distribution systems that had a measured chlorine residual of approximately 0.1 to 0.3 mg/l. In an extensive study involving 986 samples taken from the Baltimore and Frederick, MD distribution systems, the maintenance of a free chlorine residual was found to be the single most effective measure for maintaining a low SPC.[39]

The sudden disappearance of disinfectant residuals is a sensitive indication of distribution system problems. Although the maintenance of a disinfectant residual in the distribution system will not combat massive levels of external gross contamination that are detectable through odors, color, and milky turbidity changes, the residual may quickly inactivate pathogens in situations that are involved with contaminants seeping into large volumes of high-quality potable water.[39]

Recent studies conducted in a small abandoned distribution system on a military base indicate that at tap water pH 8, with an initial free chlorine residual of 0.7 mg/l, and wastewater added to levels of up to 1% by volume, 3 logs or greater bacterial inactivation were obtained within 60 min. Viral inactivation under these conditions was less than 2 logs. In laboratory reservoir experiments, where the residual chlorine is replenished by inflow of fresh uncontaminated chlorinated tap water, greater inactivation was observed at the higher wastewater concentrations tested. Furthermore, a free chlorine residual was more effective than a combined chlorine residual in the rapid inactivation of microorganisms contained in the contaminant.[39]

The drive toward lowering the formation of trihalomethanes in drinking water and the difficulty in maintaining a free chlorine residual in the distribution system have caused some utilities to switch from traditional use of free chlorine to combined chlorine in water treatment.[40-42] This change has not come without some coliform biofilm occurrences that lead into a noncompliance issue. A change to combined chlorine is characterized by an initial satisfactory phase in which chloramine residuals are easily maintained throughout the system and bacterial counts are very low. Over a period of years, however, problems may develop including increased densities of heterotrophic bacteria, loss of chloramine residuals in the pipe network extremities, and increased taste and odor complaints.

Case History: Boca Raton, FL

The Boca Raton water utility experienced such a problem during January and February 1991.[43] For many years, disinfection of the water supply was achieved through the application of free chlorine. No problems resulted immediately after the switch to chloramines in 1983. However, in 1989–1990, there was a severe drought, so the regular flushing program was suspended. While an occasional sample might be found to have 1 or 2 coliforms per 100 ml, repeat samples were negative and percent of coliform occurrences never approached 5% in any month. On January 31, 1991, seven of 14 samples were found to have over 20 coliforms per 100 ml and positive repeat samples confirmed to be *Enterobacter* and *Citrobacter* type coliforms. What followed was a typical pattern for a coliform biofilm event (Table 7.15): coliform occurrences in disinfected water samples (chloramine residuals ranging from 3.5 to 5.0 mg/l). A boil water order for the entire system was announced on February 3. As the monitoring program expanded to include over 200 samples per week, positive samples continued to appear and disappear on repeat sampling all over the service area. Intense system-wide flushing and the switch to free chlorine (7.5 mg/l) for several weeks was effective in bringing the problem under control. By February 10, the number of positives had begun to decline and on February 15 the boil water order was lifted.

Table 7.15 Boca Raton Coliform Biofilm Event, February 1991

Date	Range of coliform densities per 100 ml[a]				
	<1	1–5	6–10	11–80	Confluent
February 1	10	11	5	5	4
February 3	57	24	4	7	—
February 4	118	29	4	3	—
February 5	55	4	—	—	—
February 6	119	17	4	4	3
February 7	53	11	—	3	1
February 9	45	18	8	6	2
February 11	59	11	1	1	1
February 12	38	2	—	—	—

[a] Coliform speciation identified these organisms to be strains of *Enterobacter* and *Citrobacter*. No *E. coli* were detected.

Data from Boca Raton Public Utilities Department.

Review of information suggested this outburst of biofilm was due to a series of events that culminated in the release of biofilm fragments into the bulk flow of water supply:

1. The system-wide flushing program was rescinded during the severe drought of 1989–1990. This provided more opportunities for excess ammonia from chloramination to adsorb to pipe sediments and pockets of biofilm to expand.

2. A new line was accepted after pressure testing, but it was not disinfected or tested for coliforms prior to being valved into the distribution network. What, if any, coliform contribution occurred in this event that could have introduced coliforms into the pipe network remains unknown.
3. System-wide power failures occurred in the afternoon on January 30 and January 31. While the power outage was said to have lasted for approximately 1 min before standby generators restarted the pumps, water pressures did drop to less than 20 psi. Records indicate the water pressure loss in some parts of the system declined to 5, 8, and 15 psi as a consequence. Interesting was the observation that these remote monitors were not connected to the operations control room at the water plant. Thus, the information could only be found by a site visit to each location in the field. The loss of pressure below 20 psi triggered a boil water order specified in the local ordinance. Loss of adequate pressure may have provided opportunities for infiltration of contaminants while the sudden return to normal pressure caused water hydraulics to shear off biofilm fragments.

The Boca Raton coliform problem appeared to be the result of transient instability of pipe sediments that revealed an incipient biofilm that needed a prompt response. In these types of situation, thorough system flushing is essential even at the risk of actually aggravating the problem by releasing more coliform bacteria in neighboring sampling sites for 24 to 48 h. Alternating disinfectant application on an annual or semiannual basis may be beneficial in preventing the entrenchment of coliforms in the pipe network.

STATIC WATER STORAGE

Long-term storage of water supply in reservoirs is not desirable. During these quiescent water periods, persisting waterborne heterotrophic bacteria begin to grow in sediments, biofilm expands in size and thickness over the inner wall surfaces, and the water stratifies, creating opportunities for anaerobic nuisance organisms to expand their activity.

Case History: Florissant, MO

The city of Florissant buys water from the St. Louis County Water Company. This private company uses conventional treatment in processing raw water from the Missouri River. While free chlorine was used in prechlorination, free chlorine in final disinfection was converted to chloramines by the addition of ammonia. This approach significantly reduced the formation of trihalomethanes.

Purchased water from the St. Louis County Water Company enters the distribution system, owned and operated by the city of Florissant, at four locations. There are 20 dead-end areas in the system which also has 923 fire hydrants. Prior to the coliform occurrence, there had never been a line-flushing program or a corrosion control strategy. A storage tank in the center of the distribution system stores water for use during peak periods in the summer.

At other times the water was held for 2 weeks then released to the distribution system and refilled during nighttime hours to take advantage of the off-peak service charge.

Unfortunately, the water in storage was bypassed during October 1983 through February 1984, thereby becoming static. In January 1994, new lines were added to the system. Apparently, contractors had no requirements to cap pipe ends lying near the trench or to disinfect lines before acceptance by the water system management. This policy provided opportunities for soil organisms (heterotrophic bacteria) to enter the distribution system. No cross-connections were observed in the business district; however, a cross-connection program was not in place to cover residential areas. Only one major line break occurred that winter.

During the first part of February, a new water system operator released some of the static water from the storage tank into the distribution system. On February 6, the laboratory noticed the sudden occurrence of noncoliform colonies (heterotrophic bacteria) on m-Endo medium from all samples in the system. No coliforms were detected in those samples. On February 8, routine examination of water from the same 15 sites revealed coliforms to be present in six samples. No repeat samples were collected that week. On February 14, the six sites were again tested and coliforms were again detected. None of the other sites were monitored during the week. As the month progressed, coliforms began to appear at other sites and densities ranged from 2 to 87 coliforms per 100 ml. With the realization that the Florissant public water supply was not going to meet the federal coliform limit for February, booster chlorination was begun on February 24. Total chlorine residual of 0.8 to 2.0 mg/l could then be detected in the pipe network, but the slower-acting chloramines were not effective in inactivating the coliform contamination. By February 29, coliforms were found in samples from 14 routine monitoring sites. The first boil water order was issued on March 3.

Toward the latter part of March, a task force was called in to study the problem and make recommendations on an action plan. This task force (composed of water authorities from the utility, state, and federal agencies) recommended a plan to systematically flush the distribution pipe network to remove biofilm sites and accumulating sediments in the pipe network. The stored water supply was also to be disinfected by free chlorine rather than by adding more choramines. Static water containing free chlorine residual was then to be pumped from the tank into the distribution pipe network only during the daytime — the period of highest water demand. Refilling the tank would be done in the nighttime hours to minimize inflow of water with a combined chlorine residual from the pipe network.

SEASONAL SERVICE LINE

Static water areas are undesirable at any time. This condition provides opportunities for suspended particulates to settle into pipe sediments, biofilm development to proceed without the shearing action of hydraulic changes, and corrosion to accelerate, particularly during warm-water periods. Redesigning the distribution network to create continuous loops from numerous dead-end sections has been helpful in reducing microbial degradation and improving the efficacy of disinfectant residuals in outlying areas.

Cold weather can be a major factor in line breaks, particularly in areas of slow flow. As water temperature drops to near freezing, water expansion will increase the risk of pipeline breakage. A solution to this problem is to bury pipelines at greater depth to provide more insulation from extended periods of subfreezing air temperatures. Draining the water from shallow lines that are used only in the summer is not desirable because of possible ground movement that may cause pipeline breakage in dry pipe sections.

Case History: Summerfest, Minneapolis, MN

Rocky subsoil and a high water table were reasons a water service line to a summer festival site in Minneapolis was placed in a shallow trench and covered. Because static water left in the pipeline over winter would be subject to freezing, festival officials introduced a food-class antifreeze into the line to keep it stable.

Even though the line was "flushed" to remove visible evidence of the dye-colored additives the next spring, sufficient nutrients remained to support coliform regrowth the following summer. This situation quickly led to the issuance of a notice to boil the water supply for all uses on the festival grounds. Elevated levels of free chlorine were not successful in eliminating coliforms in this water supply line. The problem was solved by continuous flushing of the line in the spring until the laboratory data revealed a low heterotrophic bacterial count per milliliter (similar to the city water supply at the point of entry into this private line) and no coliforms were detected in 100 ml. As a safeguard, booster chlorination was also applied during the summer months to maintain a constant low concentration of chlorine residual. This additional disinfection application also provided protection from any potential cross-contamination introduced through temporary water line service connections to numerous food and beverage vendor operations on the festival grounds.

REFERENCES

1. Kriewall, D.F. and L.G. Schanz. 1985. The Laboratory Role in Watershed Management. Proc. Amer.Water Works Assoc. Water Qual. Technol. Conf. Dallas, TX. pp. 319-338.
2. Earnhardt, Jr., K.B. 1980. Chlorine Resistant Coliform — The Muncie, Indiana Experience. Proc. Amer. Water Works Assoc. Water Qual. Technol. Conf., Miami Beach, FL. pp. 371-376.

3. Moyer, N.P. and N.H. Hall. 1986. *Klebsiella oxytoca* in a Public Water Supply. Proc. Amer. Water. Technol. Conf. Portland, OR. pp. 317-335.

4. Ohio River Valley Sanitation Commission. 1980. Water Treatment Process Modifications for Trihalomethane Control and Organic Substances in the Ohio River. Cincinnati, Ohio: U.S. Environmental Protection Agency, EPA- 600/2-80-028, (NTIS: PB301-222).

5. Epton, M.A. and J.F. Becnel. 1981. Evaluation of Powdered Activated Carbon for Removal of Trace Organics at New Orleans, Louisiana, Cincinnati, Ohio: U.S. Environmental Protection Agency, EPA 600/S2-81- 027.

6. Lykins, B., Jr., E.E. Geldreich, J.Q. Adams, J.C. Ireland, and R.M. Clark. 1984. Granular Activated Carbon for Removing Nontrihalomethane Organics from Drinking Water, Cincinnati, Ohio: U.S. Environmental Protection Agency, EPA 600/S2-84-165.

7. Symons, J.M., A.A. Stevens, R.M. Clark, E.E. Geldreich, O.T. Love, Jr., and J. DeMarco. 1981. Treatment Techniques for Controlling Trihalomethanes in Drinking Water, Cincinnati, Ohio: U.S. Environmental Protection Agency, EPA-600/2-81-156.

8. Reasoner, D.J. and E.E. Geldreich. 1985. A New Medium for the Enumeration and Subculturing of Bacteria from Potable Water. *Appl. Environ. Microbiol.*,49:1-7.

9. Benedek, A. 1979. The Effect of O_3 on Activated Carbon Adsorption: A Mechanistic Analysis of Water Treatment Data. *Ozonews*, 6(1)Part 2:1-6.

10. Benedek, A., J.J. Bancsi, M. Malaiyandi, and E.A. Lancaster. 1980. The Effect of Ozone on the Biological Degradation and Activated Carbon Adsorption of Natural and Synthetic Organics in Water. Part II. Adsorption. *Ozone Sci. Engr.*, 1:347-356.

11. Rice, R.G., G.W. Miller, C.M. Robson, and W. Kuhn. 1978. A Review of the Status of Pre-Ozonation of Granular Activated Carbon for Removal of Dissolved Organics and Ammonia from Water and Wastewater. In: *Carbon Adsorption Handbook*. P.M. Cheremisinoff and F. Ellerbusch, Eds., Ann Arbor, MI: Ann Arbor Science, pp. 485-537.

12. Stephenson, P., A. Benedek, M. Malaiyandi, and E.A. Lancaster. 1979. The Effect of Ozone on the Biological Degradation and Activated Carbon Adsorption of Natural and Synthetic Organics in Water. Part 1. Ozonation and Biodegradation. *Ozone Sci. Engr.*, 1:263-279.

13. Miller, G.W., R.G. Rice, and C.M. Robson. 1980. Large Scale Applications of Granular Activated Carbon with Ozone Pretreatment. In: *Activated Carbon Adsorption of Organics from the Aqueous Phase, Vol. 2*. M.M. McGuire and I.H. Suffet, Eds., Ann Arbor, MI: Ann Arbor Science, pp. 323-347.

14. Rich, R.G., C.M. Robson, C.W. Miller, J.C. Clark, and W. Kuhn. 1982. Biological Processes in the Treatment of Municipal Water Supplies, Cincinnati, Ohio: U.S. Environmental Protection Agency, EPA 600/2-82-020.

15. Neukrug, H.M., M.G. Smith, J.T. Coyle, J.P. Santo, and J. McElkaney. 1983. Removing Organics from Philadelphia Drinking Water by Combined Ozonation and Adsorption, Cincinnati, Ohio: U.S. Environmental Protection Agency, EPA 600/-83-048.

16. Glaze, W.H. 1982. Pilot Scale Evaluation of Biological Activated Carbon for the Removal of THM Precursors. U.S. Environmental Protection Agency, EPA 600/2-82-046.

17. Geldreich, E.E. 1990. Microbiological Quality Control in Distribution Systems. In: *Water Quality and Treatment.* F.W. Pontius and J.M. Symons, Eds., New York, NY: McGraw Hill Publishers.

18. Carswell, J.K., R.G. Eilers, and D.J. Reasoner. 1984. Pilot Scale Extramural Research on Biological Activated Carbon: A Summary Report to the Office of Drinking Water, Cincinnati, Ohio: Drinking Water Research Division, U.S. Environmental Protection Agency (Sept.).

19. Brodtmann, Jr., N.V., W.E. Koffskey, and J. DeMarco. 1980. Studies of the Use of Combined Chlorine (Monochloramine) as a Primary Disinfectant of Drinking Water. In: *Water Chlorination: Environmental Impact and Health Effects,* Vol. III, Jolley, R.L. et al., Eds., Ann Arbor, MI: Ann Arbor Science, pp. 771-788.

20. Hubbs, S.A., D. Amundsen and P. Olthius. 1981. Use of Chlorine Dioxide, Chloramines, and Short-Term Free Chlorination as Alternative Disinfectants. *Jour. Amer. Water Works Assoc.,* 73(2):97-101.

21. White, G.C. 1986. *Handbook of Chlorination.* New York, NY: Van Nostrand Reinhold Co.

22. J.M. Montgomery, Consulting Engineers, Inc. 1985. *Water Treatment Principles and Design.* New York, NY: John Wiley & Sons.

23. Malpas, J.F. 1973. Disinfection of Water Using Chlorine Dioxide. *Water Treat. Exam.,* 22:209-221.

24. Bull, R.J. 1982. Health Effects of Drinking Water Disinfectants and Disinfectant By-Products. *Environ. Sci. Technol.,* 16, 554A.

25. Cherry, A.K. 1962. Use of Potassium Permanganate in Water Treatment. *Jour. Amer. Water Works Assoc.,* 54:417-424.

26. Shull, K.E. 1962. Operating Experiences at Philadelphia Suburban Treatment Plants. *Jour. Amer. Water Works Assoc.,* 54:1232-1240.

27. Sonneborn, M. and B. Bohn. 1978. Formation and Occurrence of Haloforms in Drinking Water in the Federal Republic of Germany. In: *Water Chlorination — Environmental Impact and Health.* R. Jolley, H. Gorchev, and D. Hamilton, Jr., Eds., Ann Arbor, MI: Ann Arbor Science, pp. 537-542.

28. Hamilton, J.J. 1974. Potassium Permanganate as a Main Disinfectant. *Jour. Amer. Water Works Assoc.,* 66:734-735.

29. Cleasby, J.L., E.R. Bauman, and C.D. Black. 1964. Effectiveness of Potassium Permanganate for Disinfection. *Jour. Amer. Water Works Assoc.,* 56:466-474.

30. Buelow, R.W., R.H. Taylor, E.E. Geldreich, A. Goodenkauf, L. Wilwerding, F. Holdren, M. Hutchinson, and I.H. Nelson. 1976. Disinfection of New Water Mains. *Jour. Amer. Water Works Assoc.,* 68:283-288.

31. Blanck, C.A. 1983. Total Coliform Reduction during Treatment of Mississippi River Water Using Potassium Permanganate. In: *Proceeding — Water Quality Technology Conference, Norfolk, VA, Dec. 4-7,* Denver, CO: American Water Works Association, pp. 309-318.

32. Anselme, C., I.H. Suffet, and J. Mallevialle. 1988. Effects of Ozonation on Tastes & Odors. *Jour. Amer. Water Works Assoc.,* 80:45-51.

33. Prendivilla, P.W. 1986. Ozonation at the 900 cfs Los Angeles Water Purification Plant. *Ozone Sci. Eng.,* 8:77-93.

34. Glaze, W.H. 1987. Drinking Water Treatment with Ozone. *Environ. Sci. Technol.,* 21:224-230.

35. Camper, K., M.W. LeChevallier, S.C. Broadaway, and G.A. McFeters. 1986. Bacteria Associated with Granular Activated Carbon Particles in Drinking Water. *Appl. Environ. Microbiol.*, 52:434-438.
36. Haas, C.N., M.A. Meyer, and M.S. Paller. 1983. Microbial Dynamics in GAC Filtration of Potable Water. *Proc. Amer. Soc. Civil Eng., Jour. Environ. Eng. Div.*, 109:956-961.
37. Haas, C.N., M.A. Meyer, and M.S. Paller. 1983. Microbial Alterations in Water Distribution Systems in Their Relationships to Physical-Chemical Characteristics. *Jour. Amer. Water Works Assoc.*, 75:475-481.
38. Geldreich, E.E., H.D. Nash, D.J. Reasoner, and R.H. Taylor. 1973. The Necsssity of Controlling Bacterial Populations in Potable Waters: Community Water Supply. *Jour. Amer. Water Works Assoc.*, 64(9):596-602.
39. Snead, M.C., V.O. Olivieri, K. Kawata, and C.W. Kruse. 1980. The Effectiveness of Chlorine Residuals in Inactivation of Bacteria and Viruses Introduced by Post-Treatment Contamination. *Water Res.*, 14:403-408.
40. Vendryes, J.H. 1962. Experiences with the Use of Free Residual Chlorination in the Water Supply of the City of Kingston, Jamaica. In: Proceedings AIDIS Congress of Washington, D.C.
41. Buelow, R.W. and G. Walton. 1971. Bacteriological Quality vs. Residual Chlorine. *Jour. Amer. Water Works Assoc.*, 63:28-35.
42. Brodeur, T.P., J.E. Singley, and J.C. Thurrott. 1976. Effects of a Change to Free Chlorine Residual at Daytona Beach. In: *Proceeding — Fourth Water Quality Technology Conference, San Diego, California, Dec. 6-7, 1976*, paper 34-5. Denver, CO: Amer. Water Works Assoc.
43. Kree, D. and J. Chansler. 1993. Practical Evolution of Biofilm Theory. *Florida Water Resources Jour.*, 45:26-28.

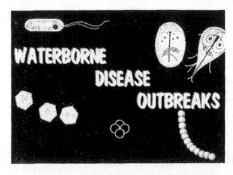

Invasions of waterborne pathogens.

CHAPTER **8**

Waterborne Pathogen Invasions: A Case for Water Quality Protection in Distribution

CONTENTS

INTRODUCTION

Pristine water resources, both surface and groundwaters, are becoming scarcer because of global increases in population and man's active intervention in the environment. In the process, an increasingly adverse impact on the earth's water resources has developed in terms of available reserves, quality, and natural self-purification capacity. Once it has been determined that a given water resource can satisfy the maximum daily demand for water supply in the community, the next important consideration is water quality and the degree of treatment necessary. Treatment requirement is determined by the level of chemical contaminants and microbial agents in the source water and establishes the level of health risk to consumers in the community. The entry of pathogenic agents into the water resource intended for water supply can become a serious concern when treatment barrier protection is inadequate.

Recently, there began an increasing public health concern about the occurrence of waterborne outbreaks, particularly as a consequence of the cholera spread in South and Central America.[1] In North America and in Europe, outbreak occurrences are self-contained in a given community and receive more immediate corrective action than in Third World nations. Perhaps the reason for this apparent indifference to outbreaks in Third World areas is the acceptance of a water quality with greater microbial risk. These recognized risks may be a consequence of economic problems that impact source water quality protection, water supply treatment, and distribution system operations, monitoring voids, and lack of effective remedial action plans.

In this country, public water supplies are meeting current regulation requirements but may be facing a growing deterioration in source water quality, treatment barrier limitations, and distribution system infrastructure deteriorations. Any of these issues may lead to pathogen penetration, as is evident in a variety of waterborne outbreak occurrences (Table 8.1). Three case histories of waterborne disease transmission caused by a loss of distribution water quality protection are examined following an overview of the pathogen threat in source waters.

Table 8.1 Waterborne Outbreaks of Disease in the United States of America

Etiologic agents	Outbreaks	Cases	Deaths
Bacterial			
Shigella	52	7,462	6
Salmonella	38	25,286	8
Campylobacter	5	4,773	0
Toxigenic *E. coli*	6	1,431	8
Cholera	1	17	0
Vibrio	1	17	0
Yersinia	1	16	0
Viral			
Hepatitis A	51	1,626	14
Norwalk	16	3,973	0
Rotavirus	1	1,761	0
Protozoan			
Giardia	84	22,897	0
Entamoeba	3	39	2
Cryptosporidium	3	418,000	2
Chemical			
Inorganic (metals, nitrate)	29	891	0
Organic (pesticides, herbicides)	21	2,725	7
Unidentified agents	266	86,740	0
Total	578	613,417	34

Adapted from Craun.[64] Additional data from Lippy and Waltrip,[87] Hayes et al.,[88] and Geldreich et al.[67,75]

CHARACTERIZING FRESHWATER QUALITY

SURFACE WATER

In remote areas there is little evidence of fecal contamination in surface-water resources. The low residual of fecal pollution in these situations are the consequence of wildlife. Wildlife populations are variable and could include beaver and muskrats along the river banks, and significant concentrations of deer, elk, and other game animals in forest reserves. Eventually, many of these fecal deposits entrapped in soil will be flushed into the drainage basin by storm events. As surface waters travel further down the watershed, human activity introduces agricultural and industrial wastes and the river receiving streams become laden with a variety of farm, domestic, and industrial wastes.

Lake water quality is also influenced by the surrounding environment. High-elevation lakes in remote portions of the watershed are of high quality (until inhabited by flocks of migratory aquatic birds). Lakes and impounds in the lower part of the watershed are more variable in quality because they are fed with surface waters containing sewage effluents and stormwater runoff from intense agricultural activities. The Great Lakes provide long holding time and vast volumes of water to dilute contaminants introduced from the impact of stormwater runoff. Unfortunately, pollution plumes around wastewater discharges are not immediately dispersed into receiving lakes. In fact, the resultant waste plume may spread to areas of the water supply intake as a consequence of wind-driven currents. At greatest risk of water quality deterioration are the numerous small lakes. These water resources become very sensitive to sporadic drainage from residential septic systems and the seasonal runoff of fertilizers from farm fields, neighborhood lawns and gardens.

GROUNDWATER

The microbial quality of groundwater is often superior because of an effective barrier of soil on top of impervious rock strata that caps the aquifer. As a consequence, quality is uniformly excellent with little influence from climatic changes and stormwater migrations through shallow depths of soil. Unfortunately, the geology of karst areas reveals the rock strata to be limestone. This material is very porous and often results in sink holes and numerous caverns through which surface-water passage occurs without effective entrapment of microorganisms. In other situations, excessive land application of minimally treated wastewaters may inundate the natural soil barrier.[2] Sanitary landfills of treated animal wastes from feedlot operations, water plant sludges, or garbage wastes that are improperly located or have a poor containment barrier can also contribute significant pathogen releases in leachates. Once the aquifer becomes contaminated, restoration of water purity is very slow, even with pumping of the underground water to a treatment site and then returning it to the aquifer.

Much of the groundwater contamination problem is found in shallow wells, less than 100 ft deep. In these wells, the source water is influenced by surface-water runoff that percolates through the soil. Because there is no protective bedrock perched on top to seal off surface contaminants, water quality is erratic and fluctuates as a consequence of storm events.

WATERBORNE PATHOGENIC AGENTS

Numerous pathogenic agents have been isolated from ambient waters used for water supply treatment.[3-5] The serious nature of these pathogen occurrences and the consequences to water supply source protection and treatment failure can be seen in Table 8.2. The list of waterborne agents both in temperate and tropical regions of the world will continue to increase as new methodology is found to detect the more elusive organisms that cause gastroenteritis or other human illnesses. Perhaps *Helicobacter pylori*, which is associated with stomach ulcers and water ingestion, should be included in this expanding list of recognized waterborne pathogenic agents.[6,7]

All of these microorganisms can be classified within four broad groups: bacteria, viruses, protozoans, and helminths (Table 8.2). Beyond this grouping there are additional pathogens that are, at the moment, "nonculturable", largely because of their fastidious requirements for unique nutrients. In nature, these substances are found in the metabolic wastes of other organisms in the aquatic flora. *Legionella* can be cited as an example of an organism that is ubiquitous in the environment yet could not be cultured in the laboratory until a few years ago. *Giardia* and *Cryptosporidium* are examples of organisms that can be identified under the microscope but cannot be successfully cultivated to prove their viability.

Table 8.2   Major Infectious Agents Found in Contaminated Drinking Waters Worldwide

Bacteria	Viruses	Protozoa
Campylobacter jejuni	Adenovirus (31 types)	*Balantidium coli*
Enteropathogenlc *E. coli*	Enteroviruses (71 types)	*Entamoeba histolytica*
Salmonella (1700 spp.)	Hepatitis A	*Giardia lamblia*
Shigella (4 spp.)	Norwalk agent	*Cryptosporidium*
Vibrio cholerae	Reovirus	
Yersinia enterocolitica	Rotavirus	
	Coxsackie virus	
Helminths		
Ancylostoma duodenale		
Ascaris lumbricoides		
Echinococcus granulosis		
Necator americanus		
Strongyloides stercoralis		
Taenia (spp.)		
Trichuris trichiura		

Data from Geldreich.[89]

ORIGINS OF WATERBORNE PATHOGENS

Varying inputs of fecal wastes from all warm-blooded animals (humans, domestic pets, farm animals, and wildlife) reach the water and soil environments either directly or indirectly through seepage from sanitary landfills, poorly processed sewage effluents from municipal treatment facilities, and malfunctioning septic tanks. Estimates of the number of infected individuals, domestic animals, and wildlife present in the environment range from less than 1 to 25% of the total population for a given warm-blooded animal type (Table 8.3). Although pathogenic bacteria, protozoans, and parasitic worms may be found in a wide range of warm-blooded animal hosts, human viral pathogens are shed only by infected people. Several other significant pathogens such as *Legionella* and *Naegleria* do not have an animal host reservoir, but appear to be ubiquitous in the environment (free-living organisms) and comprise both nonpathogenic and pathogenic strains.[8,9] The reason some strains in the environment are predisposed to be nonpathogenic, while others of the same genus or even of the same species (but different in serotype reaction) are pathogenic, is unclear. These latter strains can quickly become opportunistic, invading the human body under conditions of stress, weakened immune system, or through general physical degeneration of the body associated with advancing age.

Table 8.3 Percentage of Warm-Blooded Animals Excreting Enteric Pathogens in the Population

Pathogen	Animal	Percentage of individual excreters				
		North America	Europe	Asia	Australia	Africa
Salmonella	Human	1	1	3.9	3	—
	Cattle	13	14	—	—	—
	Sheep	3.7	15	—	—	—
	Pig	—	7–22.0	—	—	—
Shigella	Human	0.46	0.33	2.4		
Leptospira	Human	<1.0–3.0	<1.0–3.0	<1.0–3.0	—	—
	Cattle	2	—	—	—	—
	Pig	—	—	—	2.5	—
	Mice	33	—	—	—	—
	Dog	26.6	—	—	—	—
Enteropath. E. coli	Human	1.2–15.5	2.4–3.3	2	—	—
	Pig	9	—	—	—	—
Vibrio cholera	Human	—	—	1.9–9.0	—	—
V. cholera biotype El Tor	Human	—	—	9.5–25.0	—	—
M. tuberculosis	Human	—	3.1–5.6	—	—	—
Enteroviruses	Children	0.88[a]	—	—	—	—
Entamoeba histolytica	Human	10	10	16	—	17
Giardia lamblia	Human	7.4	—	—	—	—
	Domestic cat	2.5	—	—	—	—
	Dog	13	—	—	—	—
	Coyote	6	—	—	—	—
	Cattle	10	—	—	—	—
	Beaver	24	—	—	—	—

[a] Seasonal occurrence in young children.

Data from Geldreich.[3] Additional data on *Giardia lamblia* from Burke,[90] Davies and Hibler.[91]

The degree and frequency of pathogen exposure is exacerbated by expanding human populations worldwide. Crowding in both old cities (Mexico City) and modern cities (Sao Paulo) and the unregulated development of satellite communities (colonias) puts undue pressures on sanitation infrastructural barriers (water supply and distribution, sewage collection and treatment, and solid waste disposal). Add to these problems the mobility of people on an international scale and it becomes apparent that disease can quickly reach epidemic proportions.

For example, the recent cholera outbreak in South and Central America illustrates this public health problem and the role poor water quality plays in a community at risk. The pathogenic agent, *Vibrio cholera* 01, biotype El Tor was first detected in Indonesia during 1934. Initially, this pathogen remained a regional pandemic occurrence in southeast Asia until the 1960s, whereupon it appeared in several countries in the Near East. By 1970 it was detected in outbreaks in Russia and South Korea. Until January 1991 the Americas had not been exposed to this pathogenic strain. Suddenly, the first cases of El Tor cholera were identified in the port city of Lima, Peru and, within a few days, the disease had spread north to coastal cities in Peru and then quickly moved into neighboring countries in South America. Within weeks, cholera was reported in almost all countries of Central America.

While initial exposures were thought to be due to eating contaminated raw fish, the pathogenic agent rapidly spread due to poor sanitation practices and appeared in contaminated groundwater and surface waters used for water supply. Inadequate treatment of public water supplies and compromised integrity of safe water supply in distribution then hastened the escalation of illness cases. The epidemic quickly spread to other communities and neighboring countries in South America as a direct result of infected people travelling by bus or airplane to previously unaffected areas. No quarantine was in effect at the borders to screen out ill people nor to inspect for the presence of raw fish products in personal luggage. As a consequence, isolated cases of cholera in the U.S. were traced to consumption of raw shellfish brought back by travelers from outbreak areas of South America.

The rapid spread of disease beyond a community is not as great a probability in developed nations because of an established public health policy to protect water resources and utilize multiple treatment barriers. The multiple barrier concept includes: the collection and treatment of municipal sewage and food-processing wastes to reduce pathogen content; enhancement of natural self-purification in rivers and lakes by management of point source effluents and nonpoint source releases of stormwater runoff; application of appropriate treatment processes for raw waters used to produce water supply; and continuous protection of drinking water in distribution systems.

Agricultural activities that involve cattle feedlots and poultry operations result in a concentration of farm animals and their fecal wastes in a confined space. In feedlot operations, the density of beef cattle per square mile may approach 10,000 animals. Under such restrictions in space, removal of fecal wastes is a major disposal operation. The closeness of these animals in a confined feeding operation invites the spread of disease in a healthy herd or poultry flock. Some farm animal pathogens such as *Salmonella, E. coli*

O157:H7, *Cryptosporidium* and *Leptospira interrogans* (from cattle), and *Balantidium coli* (from swine) are human pathogens also. Unless proper landfilling disposal of the treated animal wastes is practiced, fecal material in stormwater runoff from feedlots and poultry farms becomes a major source of fecal contamination in rural watersheds, polluting streams and lakes in its drainage path.

Wildlife refuges are also a significant source of fecal contamination, often on a seasonal basis, because many of these animals migrate seasonally in their search for food. The largest threat of pathogens released from wildlife is due to those warm-blooded animals such as beavers, deer, coyotes, and sea gulls that are permanent residents of a watershed. These animals serve as reservoirs for *Giardia, Cryptosporidium, Salmonella, Campylobacter,* and *Yersinia* among the waterborne bacterial and protozoan pathogens.

Wildlife is also attracted to protected watershed areas because human activities are more restricted in these locations.[10] In these situations, near-shore water environments are often the location of large beaver colonies including individual animals infected with *Giardia*. Infected coyotes, muskrats, and voles are other wild animals living in remote areas that may be involved in the shedding of *Giardia* cysts and other pathogens into the aquatic environment. Terrestrial birds and waterfowl can be sources of bacterial pathogens. The songbird population includes individuals that may be infected with *Salmonella*. Sea gulls are scavengers that frequent open garbage dumps, eat contaminated food wastes, and contribute *Salmonella* in their fecal droppings to lakes in coastal areas.[11,12] In one instance, sea gulls were the contributors of *Salmonella* to an untreated surface supply in an Alaskan community that resulted in several cases of salmonellosis.[13] Wildlife is also believed to be the source of *Campylobacter* that contaminated untreated or poorly treated surface water (streams and reservoirs) of low turbidity in small communities in Vermont[14] and British Columbia.[15] Vacationers to national parks in Wyoming frequently became ill after drinking water from mountain streams, resulting in a 25% increase statewide in campylobacter enteritis.[16] Therefore, watershed management policies should include an annual program to assess the potential pathogen threat and identify carriers in the wildlife population. This information can be invaluable to water supply treatment operations and to the laboratory strategy for monitoring water quality in raw source and open reservoirs.

PATHOGEN SURVIVAL IN THE AQUATIC ENVIRONMENT

Upon discharge into a receiving stream or through migration in surface runoff into an unprotected groundwater, pathogens will have a variable survival in the aquatic environment. This condition is predicated on many factors, including elements in the natural self-purification process of the aquatic environment and the concentrating effects of pollutional discharges to the water body. Some examples of enteropathogenic *E. coli, Salmonella,* and *Vibrio cholerae* survivals will illustrate the importance of pollution abatement, the need for treatment barriers, and the continual management of clean water environments, particularly that of the water supply distribution system.

Enteropathogenic E. coli

Enteropathogenic *E. coli* strains are present in streams and lakes polluted with warm-blooded animal feces, the occurrence being probably less than 1% of the fecal coliform population.[17] The Fyris river in Sweden was found to contain enteropathogenic *E. coli* on 12 different occasions involving ten different "O" serogroups.[18] In fact, *E. coli* serogroup occurrences were used in an investigation of a lake and stream drainage basin in northeastern Pennsylvania to trace the probable sources of microbial pollution.[19] A total of ten serotypes were found in 224 samples from five waste stabilization ponds in North Dakota and the number of serotypes recovered per sample was apparently related to the population served for a given pond.[20]

E. coli enters the aquatic environment from the discharge of fecal contamination introduced by some warm-blooded animal source.[21] Once separated from the intestinal tract, survival of any *E. coli*, pathogenic or nonpathogenic, is influenced by a host of environmental factors including water pH,[22] metal-ion toxicity,[23] additions of bacterial nutrients,[24] water temperature,[25] sunlight exposure,[26] intermittent stream riffles,[24] bacterial adsorption with sedimentation,[27] and predation.[28] Multiplication of *E. coli* is rarely observed, with growth being confined to warm-water discharges of large volumes of bacterial nutrients from untreated cannery wastes, poultry processing wastes, and raw domestic sewage.

Salmonella typhimurium

Salmonella strains are a frequent occurrence in polluted waters and can persist for extended periods in highly nutrient waters. For example, *Salmonella* was detected in surface water up to 250 m downstream of a wastewater treatment plant with regularity, but never at sample sites 1.5 to 4 km upstream.[29] *Salmonella* transported by stormwater through a wastewater drain at the University of Wisconsin experimental farm were isolated regularly at a swimming beach 800 m downstream.[30] Excessive biochemical oxygen demand (BOD) or total organic carbon (TOC), low stream temperatures, and a source of *Salmonella* discharge in wastewater effluent can produce an impact that will depress stream self-purification processes. For example, *Salmonella* was isolated in the Red River of the North (between the states of North Dakota and Minnesota), 22 mi downstream of sewage discharges from Fargo, ND and Moorhead, MN during September and prior to the sugar beet-processing season.[31] By November, *Salmonella* strains were found 62 mi downstream of these two sites. With increased sugar beet processing, wastes reaching the stream in

January, under cover of ice, brought high levels of bacterial nutrients and *Salmonella* were then detected 73 mi downstream of, or 4 d flow time from, the nearest point source discharges of warm-blooded animal pollution.

Sedimentation is one of the natural self-purification factors that can rapidly remove pathogens from the overlying water into bottom deposits. This action takes place in quiet segments of a stream or during long retention periods in lakes and impoundments. Over a 1-year period, approximately 90% of *Salmonella* species obtained from the North Oconee River (Georgia) were recovered in bottom sediments. Adsorption of the organisms to sand, clay, and sediment particles resulted in their concentration in the bottom deposits.[32] A maximum density of 11 *Salmonella* organisms per 100 ml was detected in pond water while pond sediments yielded *Salmonella* densities ranging from 3 to 150 organisms per 100 g.[33]

Vibrio cholerae

V. cholerae viability in surface waters has been observed to vary from 1 h to 13 d.[34] Natural water alkalinity between a pH of 8.2 to 8.7 is more favorable to *V. cholerae* persistence than acid waters of pH 5.6 or less.[35] In other studies using synthetic water, both chlorides and organic matter were essential for persistence. When *V. cholerae* was introduced into either a synthetic water containing common saprophytes or into the high-density bacterial flora of sewage or activated sludge, there can be a significant suppression of the pathogen often resulting in 99% kill in 6 h.[36]

Although cholera vibrios may persist for only a short time in the grossly polluted aquatic environment, fecal contamination from victims of epidemics, carriers in the population, and amplification in nutrient-rich wastes may continue to reinforce the cholera population in wastewater. This fact is verified by isolation of *V. cholerae* from the Hoogly river and associated canals in Calcutta, India during epidemic and nonepidemic periods.[37] Mutation of bacteria does occur over time when exposed to adverse environments to create a more resistant population. Such appears to have happened with the appearance of a rugose variant subpopulation of the Peruvian strain of *V. cholerae*, 01, Inaba, biotype El Tor. This strain has the characteristic of clumping to form a protective mechanism from exposure to chlorine and other toxic substances in the aquatic environment.[38] This attribute becomes a powerful mechanism for survival as much as recycling the organism through the intestinal tract–wastewater–water supply route.

PATHOGEN PATHWAYS

Although sewage collection systems have decreased public health risk in urban centers, this practice only serves to transport the collected wastes to some selected destination, hopefully, where treatment is applied prior to release into a watercourse. Raw sewage discharges to receiving waters have often been shown to contain a variety of pathogens. The density and variety of human pathogens released is related to the population served by the sewage collection system, seasonal patterns for certain diseases, and the extent of community infections at a given time. Some indication of the relative occurrences of various pathogens in raw sewage is given in Table 8.4 for two cities in South Africa.[39] Because methodology differs for the wide spectrum of pathogens that might be present, these findings in sewage represent only a portion of the health threat that was identified. If socioeconomic factors and epidemic eruptions are overlaid on these expected occurrences of pathogenic agents in sewage, the risk to people, farm animals, and wildlife exposed will be seen to be unacceptable.

Table 8.4   Microbial Densities in Municipal Raw Sewage
from Two Cities in South Africa

| | Average count per 100 ml | |
Organism or microbial group	Pietermaritzburg sewage	Worcester sewage
Aerobic plate count (37°C; 48 h)	1,110,000,000	1,370,000,000
Total coliforms	10,000,000	
E. coli, type 1	930,000	1,470,000
Fecal streptococci	2,080,000	
C. perfringens	89,000	
Staphylococci (coagulase positive)	41,400	28,100
Ps. aeruginosa	800,000	400,000
Salmonella	31	32
Acid-fast bacteria	410	530
Ascaris ova	16	12
Taenia ova	2	9
Trichuris ova	2	1
Enteroviruses and reoviruses (TCID$_{50}$)	2,890	9,500

Note: TCID, tissue culture infective dose.
From Grabow, W.O.K. and Nupen, E.M., Water Research, 1556, 6, 1992. With permission.

It has been hypothesized that a sewage collection network of 50 to 100 homes is the minimum size required before there is a reasonable chance of detecting Salmonella.[40] Salmonella strains were regularly found in the sewage system of a residential area of 4000 persons.[41] In another study, 32 Salmonella serotypes were found in sewage effluent samples and downstream sites on the Oker River in Germany.[42]

In major river systems receiving discharges of meat-processing wastes, raw sewage, and effluents from ineffective sewage treatment plants, the densities of Salmonella species may be substantial. It has been calculated that the Rhine and Meuse Rivers carried approximately 50 million and 7 million Salmonella bacilli per second, respectively.[43] The Missouri River represents another example of a pollution conduit, transporting a fecal pollution load from raw sewage, effluents from wastewater treatment plants of differing efficiencies, runoff from cattle feedlots, and waste discharges from meat and poultry-processing plants. As a consequence, it is not surprising that various *Salmonella* serotypes and viruses have been detected at the public water supply treatment plant intakes at Omaha, St. Joseph, and Kansas City (Table 8.5).

Documentation of gross pollution of surface-water quality in developing countries is somewhat sparse, but, nevertheless, represents an even greater threat because treatment of domestic and food-processing wastes is often marginal or nonexistent. In India, for example, the high population densities along river valleys has led to considerable pollution problems. Of India's 3119 cities, only 209 have partial treatment of sewage and just 8 utilize secondary treatment schemes.[44] In this situation, it is not surprising to note that for a 48-km stretch of the Yamuna River (above New Delhi) there may be at least 7500 coliforms per 100 ml. After receiving an estimated discharge of 200 million liters of raw sewage from New Delhi, the coliform density in the Yamuna River suddenly escalates to 24 million coliforms per 100 ml. Along with such a dramatic rise in fecal waste discharges there is also the certainty of continuous releases of a variety of pathogens shed by infected individuals living in this densely populated area.

Stormwater and regional floods are often the major cause of transient water quality deterioration in water resources. The impact of stormwater runoff on water quality relates to all land uses over the drainage basin. Rural runoff of stormwater can be a very significant contributor of *Giardia, Cryptosporidium, Campylobacter,* and *Yersinia* from the wildlife and farm animals living on the watershed. Heavy loads of fecal pollution are common to stormwater runoff from cattle and poultry feedlots and was equated in one study of feedlot operations (near Hereford, TX) to be similar to the discharge of raw sewage for a city of approximately 10,000 people.[45] Prevalent pathogens shed by infected cattle are *Salmonella, E. coli* O157:H7, and *Cryptosporidium,* all of which are also pathogenic to man.

Urban stormwater runoff can be a major factor in fluctuating quality of surface waters. For those cities that collect both stormwater and domestic wastes in the same pipe system, the problem of treatment capacity is critical. Sudden large inflows of stormwater from major storms may overwhelm the capacity to treat, resulting in the need to bypass some portion of the mixed waste directly to receiving waters. In recent years, Chicago and several other large cities have constructed or are designing huge underground storage tunnels to hold untreated combined wastes for a few days until treatment of the excess can be accomplished. Separate collection systems for stormwater and domestic wastes are also a common alternative (Table 8.6), but often the storm drainage is discharged to receiving waters untreated. The most concentrated opportunistic pathogens in urban stormwater around the Baltimore area were found to be *Pseudomonas aeruginosa* and *Staphylococcus aureus* at levels of 1000 to 100,000 per milliliter and from 10 to 1000 per milliliter, respectively.[46]

Table 8.5 Fecal Coliform Densities and Pathogen Occurrence at the Missouri River Public Water Supply Intakes

Raw water intake	River mile	Date	Fecal coliforms per 100 ml[a]	Pathogen occurrence
Omaha, NB	626.2	October 7–18, 1968	8300	N.T.
		January 20–February 2, 1969	4900	N.T.
		September 8–12, 1969	2000	Salmonella enteritides
		October 9–14, 1969	3500	Salmonella anatum
		November 3–7, 1969	1950	N.T.
St. Joseph, MO	452.3	October 7–18, 1969	6500	N.T.
		January 20–February 2, 1969	2800	N.T.
		September 18–22, 1969	4300	N.T.
		October 9–14, 1969	N.T.	Salmonella montevideo
		January 22, 1970	N.T.	19 virus PFU[b]
				Polio types 2,3
				Echo types 7,33
Kansas City, MO	370.5	April 23, 1970	N.T.	3 virus PFU,[b] not typed
		October 28–November 8, 1968	6500	N.T.
		January 20–February 2, 1969	8300	N.T.
		September 18–22, 1969	3800	Salmonella newport
				Salmonella give
				Salmonella infantis
				Salmonella poona

Note: N.T., No test for pathogens done.

[a] Geometric mean.
[b] Plaque forming units.

Data from the U.S. Environmental Protection Agency.[93]

Table 8.6   Summary of Microbiological Data from Detroit and Ann Arbor Overflows

Month	Analysis	Separate system — Ann Arbor			Combined system — Detroit		
		Density per 100 ml[a]	Fecal coliform (%)	FC/FS ratio	Density per 100 ml[a]	Fecal coliform (%)	FC/FS ratio
April	Total coliform	340,000	—	—	2,400,000	—	—
	Fecal coliform	10,000	2.9	—	890,000	37.1	—
	Fecal streptococci	20,000	—	0.5	—	—	—
May	Total coliform	510,000	—	—	4,400,000	—	—
	Fecal coliform	51,000	10.0	—	1,500,000	34.1	—
	Fecal streptococci	200,000	—	0.26	320,000	—	4.7
June	Total coliform	4,000,000	—	—	12,000,000	—	—
	Fecal coliform	78,000	2.0	—	2,700,000	22.5	—
	Fecal streptococci	120,000	—	0.65	740,000	—	3.7
July	Total coliform	4,000,000	—	—	37,000,000	—	—
	Fecal coliform	120,000	3.0	—	7,600,000	20.5	—
	Fecal sstreptococci	390,000	—	0.31	350,000	—	21.7
August	Total coliform	1,700,000	—	—	26,000,000	—	—
	Fecal coliform	350,000	20.6	—	4,400,000	16.9	—
	Fecal streptococci	310,000	—	1.1	530,000	—	8.3

[a] Geometric means per 100 ml.

Data of 1964 adapted from Burns and Vaughan.[94]

Salmonella and enteroviruses were frequently isolated but at much lower densities ranging from 10 to 10,000 per 10 l of urban runoff. These pathogens originated in sewage that was mixed with rainwater in storm drains and through the combined sewer overflows.

The Midwest flood of 1993 is an example of a natural disaster that had a tremendous impact on water quality. Among the public health concerns cited were manure and dead animals from cattle and poultry feedlots, fertilizers, herbicides, pesticides, and other chemical industrial products. Among the more bizarre concerns in this flood was the movement of coffins and body parts out of a cemetery. A total of 707 coffins were washed out of an old rural cemetery near Hardin, MO and carried as far as 14 mi. During the peak of the flood, 18 wastewater plants in the Missouri River basin were subject to flooding. In Janesville, WI, the city sewage treatment plant became inundated and spilled more than 1 million gal of raw sewage into the Rock River. Floodwaters overflowed the water works in Des Moines, IA, St. Joseph, MO, and several smaller water utilities along the Des Moines and Missouri Rivers.[47] Throughout the flooded areas of Missouri, Kansas, Illinois, and Iowa, there were 250 contaminated water supplies.[48] Sand filters became clogged with a thick layer of fine silt and the distribution pipe networks had to be flushed of sediments and high dosages of chlorine applied to decontaminate the distribution system. After reaching a flood crest, dilution effect of the floodwaters and the relatively quiescent state of the slowly receding waters led to sedimentation of fine silt and associated microorganisms. Only the heroic efforts of public health officials in emergency operations were able to prevent waterborne outbreaks by providing a safe water supply and curbside portable toilets to a sanitation-conscious public.

Thousands of private water supplies were inundated by flooding and had to be vigorously pumped to flush out contaminants, then chlorinated to restore water quality. In a nine-state water quality survey (Missouri, Minnesota, Nebraska, North Dakota, South Dakota, Illinois, Iowa, Kansas, and Wisconsin), done 6 months after the flood period, several facts emerged. Private wells that were less than 100 ft deep were slow to recover from contamination because of soil saturation. These are wells that are most often under the influence of surface-water penetration because of poor soil barrier conditions. Deep wells that were properly protected by casing and wellhead protection were quick to be restored to excellent quality with little permanent damage to the aquifer. Apparently, hydrostatic pressure from the aquifer was an effective barrier to floodwater penetration. The Iowa Hygienic Laboratory surveillance for waterborne outbreaks observed a decrease in *Salmonella, Shigella, Campylobacter,* and *Cryptosporidium* episodes during the flood period, probably because recreational activities were shut down. Three cases of leptospirosis did occur in the state and may have been the result of wading in floodwaters. Missouri health officials noted that self-reported diarrheal illness was two to three times higher within households that used contaminated wells compared with other households that had noncontaminated private wells.

Flooding, sewage and stormwater contaminate both surface-water and groundwater systems and many private wells may become unsafe drinking water supplies because the soil barrier is breached. Over the years, a variety of waterborne disease outbreaks have been attributed to contaminated aquifers or poorly protected well sites that contained pathogenic bacteria, virus, protozoans, and worms. Specific bacterial pathogens that have been isolated from well waters include enteropathogenic *E. coli, Vibrio cholerae, Shigella flexneri, S. sonnei, Salmonella typhimurium, Yersinia enterocolitica,* and *Campylobacter.*[49-60] Poliovirus and enterovirus were isolated from a well used to supply water to restaurant patrons.[61]

Alternative methods of waste disposal may also contribute concentrated reservoirs of pathogens that contaminate the aquatic environment. Land applications of minimally treated wastewaters can contaminate groundwaters in areas where there is rapid infiltration of soils (high percolation rates) and surface waters may also become polluted through excessive overland spraying in the drainage basin.[2] Under these conditions, the natural self-purification processes (desiccation, acid soil contact, sunlight exposure, soil organism competition, antagonism, and predation) become either inoperative or ineffective. Improperly located or poorly engineered sanitary landfills for animal wastes from feedlot operations, water plant sludges, or refuse wastes can also contribute significant pathogen released to ground- and surface waters.[62] Urban solid wastes (refuse) contain not only food discards, trash, plastics, cloth, paper products, soil, rock, and ash residues, but also fecal material. This fecal material is derived from disposable baby diapers and pet animal wastes in litter materials and the droppings of rodents and birds (sea gulls etc.) foraging for food in exposed waste dumps. Dredging of river channels has also been noted to recirculate viable pathogens from the sludge banks that accumulate around sewage outfalls and boat harbors.[63]

UTILIZING MULTIPLE BARRIERS TO PATHOGEN PASSAGE

Unfortunately, the growth of human activities along the rivers and lakes of the world has become so intense that often there is not a sufficient recovery zone downstream for natural barriers (self-purification) to be effective. Furthermore, the magnitude and complexity of waste discharged to receiving waters overcome the capacity for self-purification or negate some of the critical natural factors.[45] Because of the unpredictability of this natural self-purification barrier, a controlled treatment approach is essential.

Controlling pollution must start with the collection of all wastes for treatment at specified sites. Diversion of all residential wastes to a pipe network for collective treatment and disposal has been found to be an effective means of reducing fecal contamination from migrating into the public water supply. In many developing nations, this action is one of the first steps toward improved community public health.

Upon collecting raw sewages, the next objective is application of a variety of carefully controlled treatment processes to improve the effluent quality. Engineering treatment is the controlled utilization of various physical, chemical, or biological processes to recondition contaminated water by holding it in a compartment for a specified time to maximize removal, sequestering, or conversion of undesirable impurities. Many of these treatment processes such as settling, aeration, and biological activity have their origins in the natural self-purification phenomenon. The difference lies in controlling the event through engineering to optimize the rapid removal of contaminants and permit treatment of greater waste loadings to produce higher-quality effluents.

Several factors justify a policy of mandatory disinfection of wastewater effluents discharged above a raw water intake for public water supply. First is the preventive philosophy of maximizing public health protection through a multiple treatment barrier concept. Removal of any part of this barrier simply transfers an additional burden to the public water purveyor. The second concern is that wastewater treatment often is neither precise nor reliable, with some municipal plants being unable to meet permit limitations. Third is the trend toward a shrinking natural self-purification capacity of receiving waters brought on by population explosions within a drainage basin.

Effective water supply treatment has had a major impact on the reduction of waterborne disease. Where waterborne disease outbreaks have occurred, inadequacies in source water quality, treatment deficiencies, and distribution barrier protection have been at fault (Table 8.5). While the majority of illness cases (224,973) were related to inadequate or interrupted treatment problems, 83,577 illness cases were traced to a breakdown in the protection of a safe water supply through 233 different distribution systems in the United States.[64]

DISTRIBUTION WATER QUALITY PROTECTION

Distribution water quality protection is an essential aspect of public water supply management. The two basic factors that have been very helpful in providing a safe water to the consumer's tap have been maintaining a positive line pressure system-wide at all times and providing a continuous disinfectant residual in the distribution system. Research on the protective effects of disinfectant residuals in a model distribution system indicate that for an initial 0.2 mg/l free chlorine residual and a sewage concentration of 0.1%, approximate bacterial inactivations of 99.9% and viral inactivations of 90% could be expected after 2 h of contact at pH 8.0.[65] Other important aspects include covering of all finished water reservoirs and prohibition of unauthorized tap-ins. Uncovered finished water reservoirs can be a source of posttreatment contamination whenever there is contact with bird populations.[12,13] Contamination through cross-connections, construction failures, line repairs, and inadequate separation of water supply lines and sewers are major contributors to waterborne illness in outbreak situations (Table 8.7). In fact, 15% of all giardiasis cases in the U.S. are related to cross-connections, pipeline damage, or repair of mains.[66]

Table 8.7 Outbreaks of Waterborne Illness Caused by Cross-Connections, Back-Siphonage, or Contamination of Water Mains, 1920 to 1984

Disease	No. of outbreaks associated with:	
	Cross-connections/ back-siphonage	Contamination of water mains
Gastroenteritis, etiology unknown	74	19
Typhoid fever	37	6
Chemical poisoning	14	0
Hepatitis A	8	4
Shigellosis	6	1
Giardiasis	4	2
Salmonellosis	3	1
Viral gastroenteritis	2	1
Amoebiasis	0	1
Other	2	0
Total	150	35

Data from Craun.[66]

Fortunately, in systems that have good monitoring programs most contaminating events in the pipe network are quickly recognized by coliform occurrences detected at representative sites during the frequent monitoring of the distribution network, followed by a prompt corrective response strategy. In the worst-case scenarios, the contaminating events may go undetected because of infrequent monitoring, poor pipe sanitation practices in line repairs or meter replacements, or lack of attention to laboratory reports of unsatisfactory water quality.

WATERBORNE PATHOGEN INVASIONS: CASE HISTORY EXAMPLES

When waterborne outbreaks occur, illness rates in the community increase above the norm and some people may die from the exposure. Perhaps the most important lesson to be learned from outbreaks is that waterborne pathogens are a worldwide problem that needs urgent control through environmental protection to avoid further escalation of their occurrences.

Because everyone drinks water, most reported outbreaks have involved contamination of water supply. Many of these outbreaks go unreported in developing countries; thus, much of the documentation describes case histories in North America and Western Europe. How serious are the exposures to pathogenic agents in the aquatic environment? Three case histories are noted to illustrate the consequences of the loss of the protective integrity of the distribution network.

HEMORRHAGIC *E. COLI* OUTBREAK

An outbreak of hemorrhagic *E. coli* serotype O157:H7 occurred in the small farm community of Cabool, MO (population 2090) during the period from December 15, 1989 to January 20, 1990[67] and resulted in four deaths, 32 hospitalizations, and a total of 243 known cases of diarrhea. This pathogen organism was found in the feces of some infected individuals and the initial investigation sought to locate a common source of contaminated raw beef or tainted milk. These foods had been implicated as the source of this agent in two published studies.[68,69] When this investigation proved negative, attention focused on the water supply. Based on a household survey conducted by the Centers for Disease Control, it was concluded that persons living inside the city (using municipal water) were 18.2 times more likely to develop bloody diarrhea than those persons living outside the city and using private well water supplies. At this point a boil water order was issued and the number of new illness cases rapidly declined.

This observation provided impetus to redirect the investigation to a study of the quality of the public water supply even though 4 weeks had passed after the main impact of the outbreak had subsided. The strategy was immediately revamped to focus on a review of field data, conduct a sanitary survey, search for any residual contaminated water by collecting water samples from ends of the distribution system, and attempt to model the hydraulics of the distribution system for movement of "contaminated" water through the system. Based on limited monitoring data from the water system, there were no coliforms detected in the source wells over several years and coliform occurrences in the distribution system were very infrequent. Unfortunately, this water utility is a small system and regulations require only two bacteriological samples for analysis be taken per month.

The source water for Cabool was untreated groundwater supplied by four municipal wells drilled between 1000 and 1300 ft into an aquifer capped by limestone formation and sink holes. While this type of fractured soil barrier is always a prime suspect, monitoring data revealed no coliforms in any samples tested from the two primary wells (well no. 5 and well no. 6) during the period from November 9, 1989 to January 11, 1990. Perhaps the 440 ft of well casing and protective housing around the well sites (Figure 8.1) were effective barriers to surface-water contamination. The other two wells were used only in the high-demand period and had been valved off (electrical power disconnected) since the late summer of 1989. As a further check on water quality in the aquifer, records at the local dairy on their private wells drilled 1000 ft into the same aquifer revealed the water supply met the total coliform standard. These findings suggested that the source water quality was adequately protected by 440-ft casings driven through the borehole of each well. It was concluded that groundwater was not a factor in the waterborne outbreak.

The reason source water was given first consideration in this investigation is that untreated groundwater supplies are always at risk from surface water contamination, particularly during major rainfall events and seasonal flooding. During these occasions, soil becomes saturated and horizontal migration of surface runoff near the point of extraction can introduce contamination at the well head. Site inspection at the well site and records of well construction should reveal: ground level is higher than the nearby contaminants (septic tanks, pasture land, etc); open space outside the casing is filled with a watertight cement grout applied as deep as necessary to seal off surface water migration; and a sanitary seal plus concrete cover slab around the well casing are in place.

Figure 8.1   Well house in Cabool, MO.

Attention then focused on the distribution system components (Figure 8.2), including the reservoirs and pipe network, and on utility maintenance of the system. Water storage was provided by two storage tanks, a 500,000-gal tank (T500) shown in Figure 8.3 and a 60,000-gal tank (T60). A third elevated tank (YT) designed to store 100,000 gal of water was not being utilized because the water level would only reach the base of the elevated storage compartment when the overflow elevations of the other two storage tanks was attained. Thus, this tank provided storage only in the column rising up to the compartment (bulb) and could be viewed as a substantial dead end.

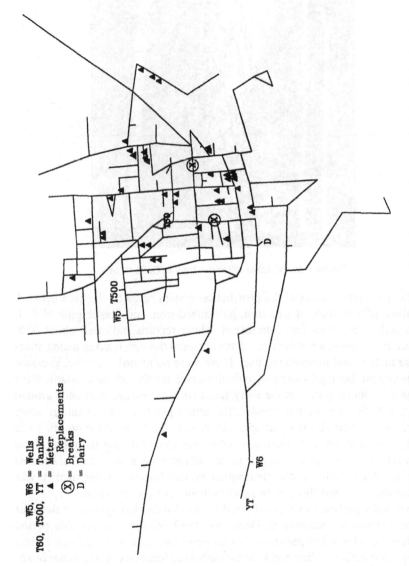

W5, W6 = Wells
T80, T500, YT = Tanks
▲ = Meter
      Replacements
Ⓧ = Breaks
D = Dairy

Figure 8.2 Drinking water distribution system including main breaks and meter replacements in Cabool, MO.

Figure 8.3   500,000-gal water tank in Cabool, MO.

On-site investigation of the distribution system revealed the pipe network consisted of a mixture of cast-iron, galvanized-iron, and plastic pipe of 2, 4, 6, 8, and 10-in. diameters intermixed. Meter records indicated about 35% unaccounted-for-water, which may have occurred through leaking mains, inaccurate meters, and nonmetered use. There were no records of water pressure in the system, but a pressure gauge on the water supply side of an alarm valve to the fire control system at the dairy revealed that pressure fluctuated around 110 psi. Following all line repairs, the utility practice was to return water supply into the break area and open the nearest fire hydrant beyond this site for 15 min in an effort to flush any debris introduced during the repair. Line disinfection was not practiced and there had never been an annual scheduled flushing of the entire water distribution system because water quality was considered good and there were no complaints of taste and odor.

Several significant events occurred in the distribution system at the start of the waterborne outbreak in December 1989. Weather during this period reached record low temperatures that caused blockages in 43 service meters and major distribution line breaks at two suburban locations. Some meter boxes were reported to be partially submerged in surface drainage during replacement, which could have introduced contamination. Based on customer recollections, the two line breaks did not reduce water pressure system-wide although localized low pressure created opportunities for back-siphonage through the breaks and a pathway for contaminant infiltration during several hours that elapsed before breaks were repaired.

A possible source of pathogen contamination was thought to be the deteriorating sewage collection system that provided opportunities for stormwater infiltration during periods of heavy rainfall. On-site inspections of sewer manholes revealed evidence of overflow problems. Various paper products associated with sewage littered the area around a few manhole covers and several other entry structures had erosion gullies around their peripheries (Figure 8.4), indicating overflow conditions were repetitive. Clearly, there were infrastructural problems with the sewage collection system, which provided entry of stormwater in broken pipes. This pipe network was in need of repair or replacement perferably with pipe of a larger carrying capacity. The sewage collection box at the head of the waste treatment lagoons also showed evidence of routine overflows. In fact, there were periodic overflows during the time it took to walk around the lagoons. This overflow wastewater ran overland to the nearby stream (Big Piney River) and a water main ran directly underneath the overflow of waste releases. As a consequence of the deteriorating sewage collection system, there remained a continual health threat to residential homes and commercial establishments from overflowing sewage onto the streets. This situation provided a means for fecal contamination passage via surface-soil drainage into the untreated water supply distribution pipe network whenever there were main repairs or service meter replacements.

Figure 8.4 Sewage overflow at distribution box in Cabool, MO.

Verifying the existence of this surface contamination pathway took two directions: a search for residual pockets of contaminated water at the ends of the system and the use of a distribution system hydraulic model to show patterns of contaminant flow from the line breaks toward the locations of the illness cases. Because this investigation of the water system was conducted 4 weeks after the main impact of the outbreak had subsided, an effort was made to capture a segment of the water that might be representative of water quality from the period of the outbreak. One potential source of suspect water might have been found in the tanker truck used by the fire-fighting department. This tanker supply could have stored city water that might had been collected during the outbreak period. Unfortunately, the stored water supply had been used and replenished after the outbreak event.

Another possibility for recovering some water from the outbreak period might have been found at the extremities of the distribution system where water usage was low. For this reason, three samples were collected at selected sites in the southwestern part of the pipe network and from the static water adjacent to the unused yellow storage tank. These samples had total coliform densities of 55, 68, and 95 organisms per 100 ml (Table 8.8). Verification and

Table 8.8   Bacteriological Results (Organisms per 100 ml) for Water Samples from Cabool, MO

Site	Date (1990)	Membrane filter tests		Multiple-tube tests	
		m-Endo LES	m-T7	LTB-BGB	Colilert
Well no. 5	February 14	<1	<1	<1.1	<1.1
Well no. 6	February 14	<1	<1	<1.1	<1.1
Grandview Terrace	February 14	<1	<1	<1.1	<1.1
Rt. 60 at M Hgwy	February 14	<1	<1	<1.1	<1.1
Kalco Manufacturing	February 15	55[a]	N.D.	N.D.	N.D.
Cedar Bluff	February 15	68[a]	N.D.	N.D.	N.D.
Yellow Tower	February 15	95[a]	N.D.	N.D.	N.D.

Note: N.D., not determined.

[a] Verified counts.

Data from Geldreich et al.[67]

identification of the coliform species isolated from these samples revealed that all three samples had a *Enterobacter* sp. and *Escherichia hermanii*, the latter being a possible fecal organism.[70] Although *E. hermanii* is not known to cause gastroenteritis, its presence is significant because this organism closely resembles *E. coli* O157:H7 in its biochemical profile and serological reactions[71] and has been found in raw milk, ground beef, and feces.[72] Some of the coliform isolates were tetracycline resistant, a characteristic shared with the outbreak strain *E. coli* O157:H7, suggesting these organisms originated from a common source of contamination. In a secondary study of these same distribution samples, *Klebsiella pneumoniae* was also isolated and grew at 44.5°C, fitting the definition of a fecal coliform. The pathogenic *E. coli* O157:H7 was not detected in any samples. No coliforms were detected in samples collected from either source wells (no. 5 or 6).

Not having detected the pathogenic *E. coli* in any samples, but demonstrating coliforms present in the distribution water and not in the groundwater supply, attention was then directed to the use of a distribution hydraulic model.[73,74] The objective was to plot water movement in the system and simulate a contamination entry at each of the two line break sites. Perhaps the contamination followed a pattern in the system where illness cases occurred in progression.

Initially, a steady-state model was used to determine where the water from wells no. 5 and 6 would be found under "normal" or average cold-weather demand conditions. To use this model, the distribution system was represented as a network of links and nodes. Pipes were considered as links and the nodes as points at which major changes in water movement occurred such as at wells, tanks, intersections, and areas of pipe diameter reductions. Utilizing information obtained from monthly meter readings on water demand, the flow directions and velocities of water in links were determined. Based on the computer-generated patterns of the water flow directions in the pipe network, it became evident that the public water sources (wells no. 5 or 6) or a possible dairy plant interconnection were not in the contamination pathway. A more likely scenario was a disturbance in the system in close proximity to most of the outbreak cases, such as the cluster of meter replacements and two line breaks.

Dynamic modeling to simulate varying the system conditions over a given period was constructed to track contaminant propagation following line repair. A conservative contamination level of 100,000 organisms per milliliter in a 10-gpm flow for a period of 4 h of continuous flow to match the normal hydraulic demand in the area was assumed at each of the break sites. No die-off of organisms was assumed in this simulation. The plot of contaminated water movement in the pipe network that resulted from these simulation assumptions provided a distribution pattern that overlaid most of the outbreak case locations with at least 10 to 200 organisms (representing a 4-log reduction) still present (Figure 8.5 and 8.6). Combining the main-break pattern for both breaks provided a simulated distribution that overlaid 85% of all illness case locations. Applying the simulation model to areas of meter replacements revealed that only three homes that had meters replaced also had illnesses. Meter replacement, therefore, did not appear to be a major cause of the outbreak but could have accounted for several cases prior to the line breaks.

Because several hours elapsed before the main breaks were repaired, the tanks had been drawn down quite extensively. Thus, it required nearly 36 h of continuous operation of both wells to refill the tanks. This scenario would have enabled contaminated water from both break locations to be distributed to an extensive area, exposing much of the service area to contaminated drinking water.

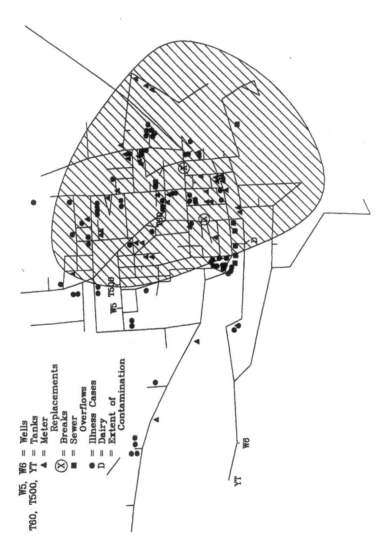

Figure 8.5 Dynamic simulation of contamination introduced at first break.

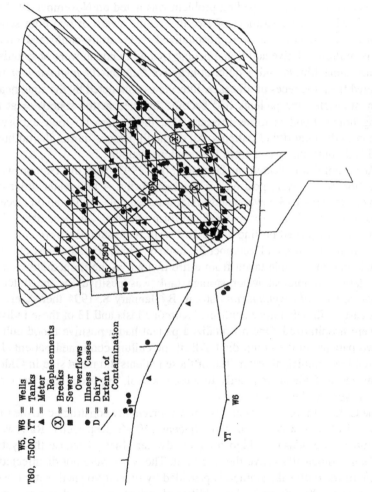

W5, W6 = Wells
T60, T500, YT = Tanks
▲ = Meter
Replacements
Ⓧ = Breaks
■ = Sewer
Overflows
● = Illness Cases
D = Dairy
/ = Extent of
Contamination

Figure 8.6 Dynamic simulation of contamination introduced at second break.

SALMONELLA TYPHIMURIUM OUTBREAK

Salmonella typhimurium was the cause of a community outbreak in Gideon, MO during November and December 1993.[75] This city of 1104 inhabitants is located in the southern corner of the state, where growing cotton is the major agricultural activity. It was estimated that 44% of the community became ill during the outbreak.[76]

An early alert to the impeding problem was noted on November 29, 1993 when the Missouri Department of Health became aware of two high school students from Gideon who were hospitalized with culture-confirmed salmonellosis. Within 2 d, five additional people living in Gideon were hospitalized with the same illness and a dulcitol-negative *Salmonella typhimurium* was recovered from the feces of all cases. Interviews conducted by the state health agency suggested the patients had become infected at the school or at the nursing home. Food histories revealed that there were no food exposures common to the majority of patients. All of the ill persons during the outbreak period had consumed water from the municipal supply.

Water samples collected on December 16, 1993 from the Gideon water distribution system were positive for fecal coliforms. A boil water order was issued on December 18. Several additional water samples collected on December 20 were also found to be fecal coliform-positive. On December 23, a chlorinator was placed on-line at the city well. Nine samples were collected from various sites also on December 23, none of which contained any chlorine residual, and one sample taken from a fire hydrant in the immediate vicinity of the largest municipal water storage tank was positive for the dulcitol-negative *Salmonella typhimurium* strain. By January 8, 1994 there were 27 illness cases with laboratory confirmed salmonellosis and 13 of these individuals were hospitalized. One hospitalized patient had a positive blood culture and two patients in the group died. All of the culture-confirmed patients had drunk Gideon municipal water. Ten of these patients did not reside in Gideon but traveled to Gideon frequently to attend school, use a day-care center in town, or work at the nursing home.

The Gideon municipal water system was originally constructed in the mid-1930s and obtains water from two adjacent 1300-ft deep wells. Raw water temperatures are unusually high for groundwater (14°C) because the system overlies a geologically active thermal fault. The wells were not disinfected at the time of the outbreak. Storage is provided by two city-owned water towers (50,000 and 100,000 gal). An additional privately owned water tower (100,000 gal) is connected to the system for filling purposes, but is separated by a backflow prevention valve. The pressure gradient between the Gideon system and the Cotton Compress water supply is such that the private water storage tank will overflow when the municipal tanks are filling. To prevent this from occurring, a special valve was installed in the influent line to the Cotton Compress tank. Water demand in the Cotton Compress operation is used for equipment washing and to provide a reservoir for fire suppression around a series of warehouses storing cotton bales.

Field inspection of the water storage tanks revealed the impact of corrosion deterioration and minimal maintenance. Shortly after the outbreak, a tank inspector found holes at the top of the Cotton Compress tank, rust on the tank and rust, sediment, and bird feathers floating in the water. According to the inspector, the water in the tank looked black and was so turbid he could not see the bottom. The municipal water storage tanks were inspected after the outbreak and found to have similar deficiencies. The 100,000-gal municipal tank was in such a state of disrepair that bird droppings and feathers could be seen in the stored water. During the field investigation, a large number of pigeons were observed roosting on both the largest municipal water tank and the Cotton Compress water storage tank. Not surprising was the observation of bird feathers in the vicinity of both storage structures and the detection of *Salmonella typhimurium* in the sediment collected from the riser pipe (Figure 8.7) of the Cotton Compress water storage tank. From this evidence it appears that the source of *Salmonella* and coliforms was from feces of pigeons observed to be roosting in the water towers. Among every bird population there are always a few individuals who are shedding pathogenic organisms in their feces.[11-13,77-81]

The Gideon distribution pipe network consists primarily of small-diameter (2, 4, and 6 in.) unlined steel and cast-iron pipe. Tuberculation and corrosion are a major problem in the distribution pipes. Under low-flow or static conditions, the water pressure is about 50 psi. However, under high-flow or flushing conditions, the pressure drops dramatically and was speculated to be involved in the movement of contaminants from the water storage tanks into the pipe network on November 10, 1993.

In early November there was a sharp climatic change in weather temperature that caused a destratification of water in the storage tanks. The resultant mixing of stratified water introduced a variety of taste and odor complaints from customers. As a consequence, a sequential flushing of the water supply was conducted over a 12-h period and involved all 50 hydrants on the system. Each hydrant was flushed for 15 min at an approximate rate of 750 gal/min. During this occurrence, it was noted that the pump at well no. 5 was operating at full capacity, which would indicate that the municipal water storage tanks were discharging. If stagnant or contaminated water was floating on top of the water column in one of the tanks, the thermal inversion could have caused this water to be mixed throughout the tank and the intensive flushing activity would then pull this poor-quality water into the pipe network. At this point, a hydraulic and contaminant propagation model[82] was constructed to test the hypothesis.

The system's pipe network configuration (Figure 8.8), water demand information, characteristic pumping curves, tank geometry, flushing program, and other information required for the analysis was obtained from maps and

Figure 8.7 Collecting residual water from riser pipe of a private water storage tank in Gideon, MO.

demographic information and numerous discussions with consulting engineers and city and state water supply officials. After initiating the hydraulic model for 48 h in the simulation, it was observed that the well pump operated at over 800 gal/min during the flushing program and then reverted to cyclic operation thereafter. The water elevation for both municipal tanks fluctuated and both tanks discharged during the flushing program. At the end of the flushing period nearly 25% of the water from the 100,000-gal municipal tank was found to be in the smaller 50,000-gal tank. Pressure drops during a field flushing program at selected hydrants were used for calibration.

The hydraulic model simulation indicated that water movement from the 50,000-gal tank tends to dominate the immediate area during the first few hours of the flushing period until it is drawn down. Water from the larger tank normally supplies most of the northern and western portions of the system. Almost all of the water in the southern and eastern portions of the system is supplied directly by the wells.

Applying the case location data collected from the CDC survey to the distribution system map, it was noted that the earlier cases and the *Salmonella*

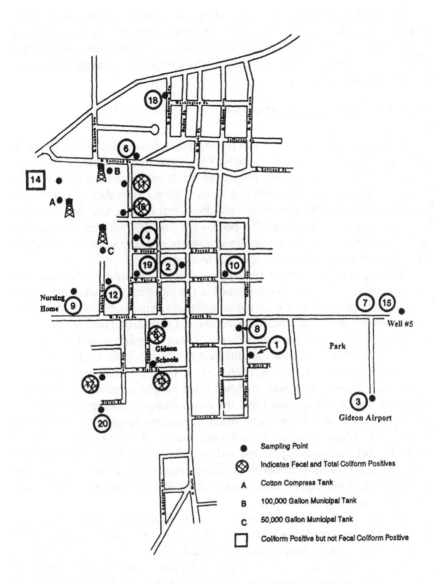

Figure 8.8 Street map and sampling points in Gideon, MO.

hydrant sample were located in the area that was primarily served by the 100,000-gal tank (Figure 8.9). One can conclude that during the first 6 h of the flushing period, the contaminated water reached residences and the Gideon school where illness cases were first located. After the flushing period, contaminated water began to spread throughout the system as noted from the apparent random progress of the disease within the community (Figure 8.10).

These findings suggest that the pathogenic agent was in the public water supply tanks and was pulled through the system during a vigorous flushing program that started near the tanks. The lack of funding for infrastructure repair and maintenance of storage tanks was a major contributing factor associated with this outbreak.

VIBRIO CHOLERAE OUTBREAK

V. cholerae, like other intestinal pathogens, is transmitted via the ingestion of contaminated food or water supply. The initial source of the El Tor strain of cholera was thought to have been from an ill merchant seaman on board a Chinese cargo ship in the harbor at Callao, Peru. From there the pathogen quickly spread to different population centers, first in Peru, then through much of Latin America.[83] Thousands of people became ill from either consuming raw seafood processed in contaminated water or drinking water from a supply that was not adequately protected from fecal wastes.

Barriers to fecal pollution are always essential. Because V. cholerae, like other intestinal pathogens, is transmitted in the ingestion of contaminated food or water, the prevention of waterborne outbreaks depends on adequate sanitation practices, including sewage collection and treatment, protection of water resource quality, and effective processing of water supplies to minimize health risks.[83] All of these conditions are the essence of a multiple treatment barrier concept. Wastewater treatment is the initial barrier to the transmission of waterborne disease. Removal of this barrier simply transfers an additional burden to the receiving waters and the downstream water supply purveyors. As is often the case in many Third World nations, wastewater is collected, but discharged without treatment to receiving streams. As a consequence, rivers and lakes become grossly polluted and are unable to provide any significant self-purification to downstream users. This situation creates a raw water quality at the public water supply intake that is often difficult to process into a safe drinking water supply.

The distribution system barrier plays an equally important part in the delivery of a protected water supply to the consumer's tap.[85] In many Third World countries, water supply treatment and distribution are subject to reduced operation schedules because of funding limitations. Compounding factors are the problems of infrastructure deterioration of the distribution system and unaccounted water losses that are often between 40 to 60%. This latter issue is often a result of unmetered service to customers who are charged a fixed price. What is more alarming is the occurrence of frequent illegal tap-ins on water supply lines and perpetually leaking water pipes that need repair or replacement. With sewage collection systems that are also in need of infrastructure repair and capacity expansion, it is not surprising to find sewage overflows that may lead to contamination of the defective water supply pipe-

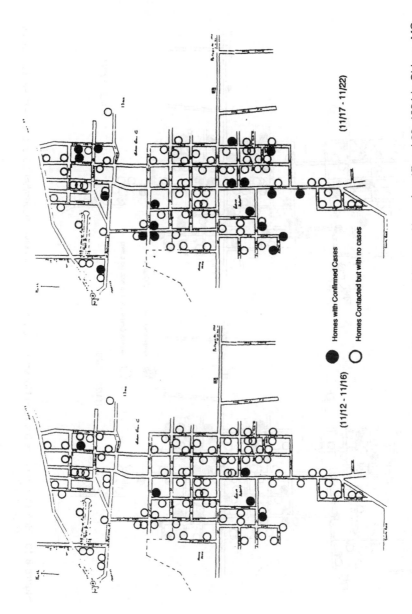

Figure 8.9 Homes with cases between November 12 and 16 and between November 17 and 22, 1994 in Gideon, MO.

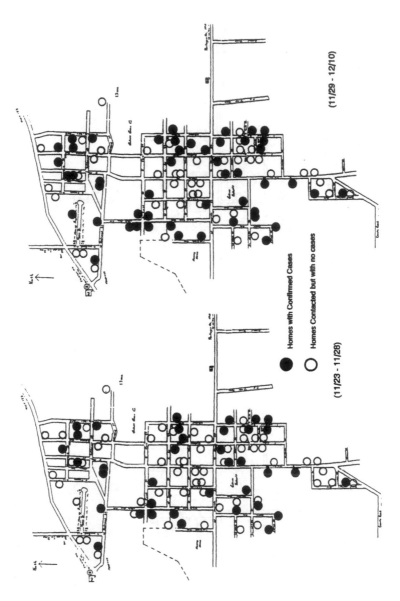

Figure 8.10 Homes with cases between November 23 and 28 and between November 29 and December 10, 1994 in Gideon, MO.

line. In the Peru outbreak, the absence of effective pollution barriers took their toll on public health protection in water supply.

The public water supply in Lima, Peru is obtained from a surface-water source and a series of wells in various locations throughout the city. The quality of the river water at the water plant intake (Figure 8.11) is poor because of the discharge of untreated sewage from various upstream farm communities and excessive soil erosions in the distant mountain watershed during major storms. While there are 100 point source discharges within 20 mi of the intake, 30 of these discharges are responsible for 90% of the river pollution. During March 1991, raw water turbidities at the plant intake ranged from 400 to 60,000 NTU, with 20% of the turbidity values being over 10,000 NTU. The plant intake is closed whenever turbidity exceeds 5000 NTU. Microbial quality was also poor, with fecal coliform densities ranging from 50,000 to 130,000 organisms per 100 ml (Table 8.9). Not surprising was the detection of the cholera agent in water collected at the intake. Processing such grossly polluted water requires extensive pretreatment even before conventional treatment can be applied. At the Lima water plant, pretreatments include raw water storage (Figure 8.12) for 6 h, polymer addition, sedimentation, prechlorination, and more settling, followed by additional polymer treatment before conventional treatment and final disinfection are applied. Even so, when summer rains in the upland watershed create periods of excessively high turbidities in the river,processing at the plant must be reduced by over 85% to avoid microbial and turbidity breakthroughs.

Considering the gross nature of this raw source water, it is remarkable that the water plant operation was able to produce a water quality that met international drinking water standards of less than one coliform per 100 ml during the flash-flooding period of March 1991. Not only were the redundancies in treatment barriers effective for coliform reduction, but the water turbidity was reduced to approximately 0.5 NTU during this same period. *Vibrio cholerae* was not detected in plant effluent entering the distribution system.

Unfortunately, the quality of water supply produced at the plant does deteriorate in distribution. In a study on distribution water quality, performed by the Pan American Health Organization (PAHO), 183 water samples were collected and promptly tested in February 1991. Fecal coliforms were found in 18.7% of all samples (Table 8.10). One sample had a maximum fecal coliform density of 240 organisms per 100 ml. Cholera was not detected in any of the special samples collected from the distribution system although two wells also supplying water to the pipe network had detectable *V. cholerae*.

There are a variety of factors that could have contributed to this loss of quality. Failure in surface water treatment barriers and loss of soil barrier protection in well fields are constant threats, but there are also serious concerns with loss of distribution system integrity. Because water supply production in Lima barely meets the demand, there is no scheduled flushing program for the distribution network. As a result of the daily occurrence of a loss of line

Figure 8.11 Clearing debris from river water intake at the water treatment plant in Lima, Peru.

Table 8.9   Raw Surface-Water Treatment,[a] Lima, Peru

| | Coliforms per 100 ml | | Heterotrophic plate count |
Source	Total	Fecal	(2 d at 35°C)
Raw	110,000–350,000	50,000–130,000	70,000–900,000
Prechlorination	500–16,000	500–3,000	1,800–40,000
After coagulation	1,900–2,000	—	400–5,800
Filtration	30–1,400	<1–17	10–2,450
Postchlorination	<1	<1	<1–11

[a] Range of values for March 4–9, 1991.

pressure, infiltration of contamination by back-siphonage from pipe breaks and increased chlorine demand are significant. An attempt was made to apply 0.5 mg/l free chlorine at different areas in the system, but free chlorine residuals were infrequently detected in many areas because of contaminant incursions during the periods of no water pressure. The loss of water pressure

Figure 8.12 Sediment deposits and vegetation debris dredged from settling basins at the water treatmnet plant in Lima, Peru.

Table 8.10 Lima Water Supply Distribution Results of 183 Distribution Samples[a]

Fecal coliforms per 100 ml	Occurrence (%)
<1	81.3
1–2	11.9
>3[b]	6.8

[a] Samples examined by PAHO Lab during February 1991.
[b] 240 FC/100 ml maximum density.

is a daily scheduled event due to pump stoppage to reduce the high cost of electrical power in the water treatment budget. This situation creates many opportunities for reversals of flow and suction that bring contaminated surface water into the water supply pipe network.

For the 15% of the population that lives beyond the public water supply distribution network service area, there are additional threats to the drinking water supply. Government and private enterprise use tanker trucks to haul water to homes in the new satellite communities that may have electrical service, but no municipal water supply or sewage system. Bacteriological analysis of tanker water has revealed no introduced contamination from these water carriers; however, such purchased water supply is often contaminated in home storage. Each home has an outside water storage tank of 250 gal that is constructed of brick material and with some makeshift cover (Figure 8.13). Inspection of these home water storage tanks frequently revealed a sediment accumulation from the prevailing dust in this arid region. Some tanks also contained various insect larvae, and, at one site, children were playing in the stored water supply. Because tank delivery of water is expensive and there is no household plumbing, water supply usage is greatly reduced and limited sanitation measures introduce additional pathways for the spread of disease.

Figure 8.13 Home storage of water supply in the "colonia" district of Lima, Peru.

Trujillo is a Peruvian city of 325,000 people and their public water supply is obtained from 61 wells throughout the city. Well depths range from 20 to 140 m. Chlorination of these groundwaters has not been consistent nor of proven effectiveness. Drip chlorinators were in place at two wells during the field investigation, but not in use because liquid chlorine was not available on a continuous basis. Because many of the wells are of low yield, water supply was restricted to 16 h in the downtown area and only 4 h every 2 d in some of the districts. Shallow wells of low yield often foretell of surface-water contamination and passage of fecal contaminants into the distribution system because soil barrier protection is not adequate and disinfectant application is not consistent.

The field investigation of water quality conditions in Trujillo by CDC field epidemiologists provided an opportunity to characterize the contamination impact from source, through distribution, to home storage of drinking water supplies. A section of the city was selected where there was a high incidence of cholera cases and the source water supplying the district could be identified.[83] Data from this study (Table 8.11) indicates that the quality of source water was unsatisfactory. Of 52 samples from the well source, 19% contained total coliforms and 13% were positive for fecal coliforms. Water supply transport through the intermittently pressurized distribution system was subject to opportunities for the microbial contamination, and it was not suprising that cholera was found at three home water taps within the district service area.

Table 8.11 Drinking Water Quality Changes from Source to Consumer Storage, in Trujillo, Peru

Source[a]	No. of samples	Total coliforms per 100 ml		Fecal coliforms per 100 ml	
		Density[b]	Range	Density[b]	Range
Municipal well	9	1.2	<1–3	1.1	<1–1
Distribution sites	21	9	<1–1700	4	<1–75
Home storage	21	118	5–1800	8	<1–150

[a] Victor Larco District, March 25 to April 29, 1991.
[b] Geometric mean values for M.F. test results.
Data revised from Swerdlow et al.[83]

At other locations, water was obtained from a neighborhood community water tap using a variety of containers (Figure 8.14), often not specifically designated for that purpose. A study in Trujillo, Peru reported fecal coliform counts were lowest in source wells, higher in distribution water, and highest in water from home storage containers.[83] The use of narrow-necked water storage containers in the home was recommended to provide a better stored water quality that would be less susceptible to contamination in handling.[86]

In Piura, Peru, *Vibrio cholerae* was found at three water supply taps and in a bucket used by one family to carry water from the community tap to their home. The city water supply consists of 18 wells located throughout the service area and interconnected by the distribution pipe network. All of the wells (120 to 140 m in depth) were subject to periodic contamination because of poor soil barrier protection. Thus, the distribution system became contaminated

Figure 8.14 Community water tap in Piura, Peru.

from inputs of poor source water quality in addition to contamination introduced in pipe leaks, daily losses of water pressure, and unauthorized tap-ins.
An estimated 70% of the early cholera cases occurred in the service area of
two wells that were subject to river water infiltration. Fecal coliform concentrations during this time ranged from 6 to 48 organisms per 100 ml.

Chlorination was unsuccessfully applied at several well sites. At one site,
gaseous chlorine was added in the wellhead pipeline, not into the well water.
As a consequence, pipe corrosion caused the submersible pump to fall to the
bottom of the well. At another site, chlorine applied at the well was undetectable at the first-customer location (community tap) only 200 m distant. Much
of the observed loss of chlorine residual was either caused by error in chlorine
application at the well or from a backflow of surface drainage around the
submerged shut-off valve at the community tap.

PATHOGEN SEARCH IN WATERBORNE OUTBREAKS

For various reasons, water supply has never been monitored continuously for changes in microbial quality nor for instantaneous recognition of any pathogen occurrence, so the search for a pathogenic agent is a special event. Continuous monitoring of turbidity after filtration and on-line recordings of applied disinfectant and system residuals in storage tanks are helpful in characterizing treatment operations. However, these characterizations do not completely insure that an effective barrier to pathogens is maintained in distribution pipe networks. For example, there is the need for instantaneous monitoring of water pressure changes in the pipe network, along with alerts to sudden losses of chlorine residual in various pressure zones, and the strategic placement of booster chlorination stations to react to the situation. As a consequence, any sudden short-duration burst of pathogen contamination is not always going to be inactivated in time to prevent nearby consumer exposure.

STRATEGIES FOR FIELD INVESTIGATION

First indications of a community health problem that might be associated with contaminated water supply may be detected in the increased caseloads appearing at hospitals, clinics, and in private medical practice. Often there is also a concurrent rise in the purchase of specific pharmaceuticals used for diarrhea control, respiratory illness, and flulike symptoms. Patient medical records should be carefully studied to determine possible route of exposure (food, person-to-person contact, and water) pattern of illness, and suspect agent among a wide group of bacteria, viruses, and protozoans (more complex parasites should also be considered if the water supply is in tropical regions). Depending on the symptoms, specimens of feces, blood, sputum, or scrapings of skin lesions must be analyzed in the laboratory for a common causative agent. The local health and water authority should not ignore the possibility of exposure to multiple pathogens because water treatment barriers may have been breached or a cross-connection may have provided passage of sewage into the water supply.

Compounding the problem of pathogen detection in the distribution system is the fact that illness from water supply exposure may take 24 h to several weeks to infect the exposed person. By this time, the pathogenic agent is often no longer a threat in water supply because of the transient nature of a treatment barrier breakthrough. Such penetrations may be caused by an unstable sand filter brought back on-line or the interrupted processing of water in an oversized water treatment system. When the intrusion has been in the distribution system, contamination may suddenly cease when line pressure is restored or the line break is repaired.

In the search for a water supply-contaminating event, the laboratory will be processing samples for both fecal contamination indications and for the suspected pathogen(s). The intent is to isolate the agent, better characterize the extent of system contamination, and demonstrate the effectiveness of remedial actions. Perhaps the repeat-sample sites or the mapping of locations where illness cases reside may provide a clue to the zone of contamination. Detection of the pathogenic agent in the distribution system may be an elusive goal because waterborne pathways are generally of short duration, pathogen survival is precarious after a few hours of passage in this aquatic environment, and laboratory methodology is so limited in sensitivity that negative results prove inconclusive. Searching for a water supply connection in the Cabool, MO outbreak[67] of E. coli O157:H7 illustrated some of the handicaps that may be encountered. As a consequence of delays in initiating a field investigation, the search for a pathogen may result in the gathering of circumstantial evidence from a variety of approaches. Because contamination input to the distribution system has probably ceased at this point, best opportunities to determine water quality during the outbreak may lie in areas of dead ends where there is little water movement, storage tanks with minimal draw-down, standpipes in low-population areas, water in tanker storage for fire control, cisterns recently filled with city water supply, and refrigerated tap water stored in the home of illness cases. Ice cubes may provide some opportunities to isolate viruses or protozoan cysts, but the freezing process is lethal to many bacterial pathogens within hours of storage.

Profiling coliform species and establishing antibiotic resistance patterns for organisms associated with the suspect pathogen in the contaminant may provide linkage. In the case of the waterborne outbreak involving the pathogenic E.coli,[67] all three coliform-positive samples collected from the ends of the distribution system contained E. hermanii, a possible fecal organism.[70] Although E. hermanii is not known to cause gastroenteritis, its presence is significant because this organism closely resembles E. coli O157:H7 in its biochemical profile and has been found in raw milk, ground beef, and feces.[72] Further study using an enrichment process revealed that some of these coliform isolates were tetracycline resistant, a characteristic shared with the outbreak strain of E. coli O157:H7.

Of the numerous opportunistic pathogens that may occur in water supply, Legionella, Mycobacterium, and Pseudomonas aeruginosa are a serious concern because of the increased exposure risk to senior citizens, AIDS patients, and infants. Interesting is the fact that the route of exposure in each instant is different: Legionella aerosol inhalation in shower water, Mycobacterium exposure in body contact (bathing), and Ps. aeruginosa through water ingestion. Generally, these organisms are found to be present in very small densities passing through a clean distribution system. However, amplification may occur in static water areas and in sediment-laden water pipes during warm-water periods. The greatest amplifiers of these and other heterotrophic bacteria will be in water-utilizing devices (shower heads, humidifiers, air coolers, and medical/dental equipment) attached to the plumbing system in homes, hospitals, and public buildings. Legionella is heat tolerant so may be found growing in the sediment of hot-water tanks, particularly in those set at lower temperature to conserve energy.

Unlike the search for primary pathogens, the detection of opportunistic pathogens in areas of poor water quality will not be difficult. The organisms will be shedding continuously from attachment sites in large densities and there are selective media available that have acceptable sensitivity. What is different is the occurrence site — the building plumbing network rather than in a public water distribution system, unless there is a serious problem of static water in storage or a filthy pipe network that has been neglected in maintenance and flushing.

IN RETROSPECT

The occurrence of waterborne outbreaks has demonstrated the need to be vigilant because pathogenic agents (bacterial, viral, and protozoan) are frequently found in the aquatic environment. Pathogens are constantly being released at variable concentrations from infected humans, pets, farm animals, and wildlife, in essence, the warm-blooded animal population. Domestic sewage and stormwater runoff became the conduits for the passage of pathogens into surface waters and through inadequate soil barriers into groundwater aquifers. Because the release of pathogens in water is a constant threat that will never be eliminated, management of watershed activities, treatment of sewage, control of surface runoff, adequate water supply treatment, and assured protection of the distribution system is essential for a safe water supply.

Access to a high-quality groundwater or safe water produced through treatment does not in itself insure that the water will remain free of microbial risk while in the distribution system. The waterborne outbreaks reported here and many others will testify to the need to include the careful management of the distribution system.

MUNICIPAL SEWAGE COLLECTION SYSTEMS

Municipal sewage collection systems are buried in the city streets along with water supply pipe networks and other utility services. Infrastructure deterioration of sewage collection systems and water supply pipe networks offer opportunities for pathogen migration given the right set of circumstances. Separating sewage collection from stormwater runoff has proven to be of significant cost advantage in that peak loading of pipelines is greatly reduced and treatment works are better able to process a uniform volume of waste at all times. Unfortunately, the occurrence of broken sewer pipes in Cabool, MO certainly contributed to a public health risk whenever there were major storm events. The collection system became overloaded periodically from stormwater infiltration during heavy rainfall, and sewage overflowed around manhole covers, contaminating soil surfaces in the adjacent area. As a consequence, if water service meters were to be repaired or if line breaks occurred, contaminated surface soil could enter the nonchlorinated water system.

STORMWATER

Stormwater can be a major contributor of pathogens in drainage over the watershed and needs to be diverted and preferably treated before release to receiving lakes and streams. Stormwater runoff was a serious problem in Gideon, MO because the ground is flat, the water table is approximately 3 m (10 ft) below the surface, and runoff drainage in the city from storms is slow. Prior to the outbreak period, heavy rains on November 15, 1993 caused flooding of one residential area (South Anderson Street). A project was underway at that time to increase the grade of the storm ditches along the roads and clear obstructions. Unfortunately, this activity did cause several water service lines to be torn away from the street main by a backhoe. Because this was an untreated groundwater system, maintaining distribution system integrity was very critical. The poor drainage in the area and the proximity of the water table to the buried pipe network probably was a factor contributing to some of the coliform occurrences in the past.

NONMETERED SERVICE

Nonmetered service encourages not only waste of the water supply resource, but also encourages a careless attitude toward water spillage from outside faucet attachments. During the field investigation at Gideon, several locations were observed where garden hoses were lying in puddles of water. Water was flowing from the hoses, probably to prevent the outside faucets from freezing. Because the residents are charged a flat fee for their water, there is no motivation to conserve water and a hose left running is of no concern. During a low-pressure period, however, these running hoses could also be possible sources of contamination.

UNACCOUNTED WATER LOSSES

Unaccounted water losses are another source of potential contamination. Lack of attention to reducing water loss in itself defeats water conservation objectives, increases cost of water production, and provides opportunities for water supply contamination. Poor pipeline construction practices, lack of attention to repairing water line leaks, illegal tap-ins on the pipe network, and growing infrastructure deterioration of the distribution system in Lima, Peru illustrate the impact of no funding to support an effective maintenance program. All components of a distribution system have a service life beyond which failures in water supply transmission occur at an accelerated rate that is neither cost-effective to endure nor of acceptable risk for contamination.

WATER STORAGE TANK STRUCTURES

Water storage tank structures are subject to corrosion and wind damage that might destroy the screen barriers to bird passage at air vent locations. Corrosion was a serious problem with the private tank structure owned by Cotton Compress, Gideon, MO. Air vents at this tank and those of the Gideon water system were not properly protected by screen enclosures, so in time, pigeons found access to the inside of the tanks. Not surprising was the report that these tanks had bird feathers and droppings floating on the surface of the water supply. Elevated storage tanks should be inspected every year for needed repairs, barrier screen replacements, and painting. Depending on the nature of the water supply chemistry, tanks need to be drained every 3 to 5 years, sediments removed and appropriate rust-proofing done to the metal surfaces. Preventing access to the tank interior by wildlife and sediment removal are important deterrents to possible pathogen contamination and coliform colonization.

SYSTEM FLUSHING

System flushing is important not only for reduction in taste and odor complaints and reestablishing a chlorine residual in static water zones, but also for removing corrosion particles, sediments, and biofilm. Flushing within a given pressure zone may be effective if the problem is localized; however, this practice is less effective during a biofilm event. In this situation, coliform occurrences are most often scattered in the biofilm fragments released to the bulk flow of water. A total system flushing is then essential. This should be initiated at sites where water enters the distribution system, then each hydrant in line should be flushed as water moves outward toward the ends of the system.

The intense flushing event (750 gal release for 15 min) over a 12.5-h period at Gideon was started in areas around the tanks, not in the immediate vicinity of the wells. Because the water system is small, the impact of this action was to release much of the stored, contaminated water into the northern and western part of the pipe network over a 6-h period and then to diffuse it to the other parts of the system nearer the well source. Starting the flushing program at the well sites might have restricted the area of greatest exposure to *Salmonella typhimurium*.

DISINFECTION RESIDUALS

Disinfection residuals in the distribution system do provide several protective benefits and some indication of immediate deterioration of water supply due to incursion of contaminants. Achieving a consistent disinfectant residual of at least 0.1 mg/l free chlorine is often associated with heterotrophic bacterial densities below 500 organisms per milliliter. Where this is not achieved as in the case with biofilm releases, the problem is often found to be excessive accumulation of protective sediments and tubercles in the pipe environment. Residuals above 0.2 mg/l can be effective in the inactivation of pathogenic bacteria and viruses, provided that the disinfectant is free chlorine. The degree of effectiveness is related to water temperature, contact time, and the disinfectant demand in the contaminant passage. Disinfectant residuals will not provide absolute protection against massive sewage incursion. Fortunately, immediate contaminant dilution with high-quality water greatly reduces the concentration of organics, turbidity, and microbes so that disinfectant residuals may still be an effective deterrent.

Untreated groundwater systems place the total burden of protected delivery of safe water supply upon the distribution system. There can be no margin for error. Both Cabool and Gideon were supplying untreated groundwater until the outbreaks occurred. The intrusion of *E. coli* O157:H7 in Cabool water was thought to have been through two major line breaks near the center of the pipe network. The breaks occurred at night and were not repaired until 6 to 8 h later, during which time there was a significant water pressure drop in the neighboring lines. Line repairs were completed without added disinfectant to the pipeline, although water was flushed through the pipe section by releasing water supply at the nearest downstream fire hydrant. While flushing may have dislodged soil particles from the pipe repair, the water contained no disinfectant residual to inactivate any bacterial contaminants that would mix with water supply before the lines were repaired.

At Gideon, there was no disinfectant applied to the water supply because the source water quality had a good bacteriological record over the years. Again, it was assumed that water quality would not become contaminated going through the system. Unfortunately, the stratified water supply in two storage tanks was contaminated with pigeon droppings. As a consequence of thermal destratification of the contaminated layer of stored water and a vigorous flushing initiated near the storage tanks, *Salmonella typhimurium* was introduced into the pipe network. How effective a disinfectant residual would have been in this situation can only be speculated. Surely the area of risk in the pipe network would have been reduced through inactivation of the pathogen by disinfectant residual rather than by natural die-off in the water supply.

WATER PRESSURES

Water pressures must be maintained continuously in the pipe network to avoid backflow of contaminants from surface-water drainage and possible sewer line leakages. In residential areas, low-pressure zones may be 40 to 60 psi, which are adequate for domestic needs and emergency fire fighting demands. High-pressure zones above 60 psi are often found in commercial and industrial districts and residential areas at higher elevations. When a line break occurs or during high-demand periods, water pressures may drop below 20 psi so that only a trickle of water is released at the faucets. At these times it is advisable to issue a boil water order because of the unknown threat of cross-connections.

At Cabool, the two major water line breaks did not create a system-wide drop in water pressure (110 psi), but localized low water pressure created opportunities for back-siphonage near the break locations. Based on a computerized hydraulic model of the distribution system, it was concluded these low-pressure areas provided opportunities for sewage infiltration over several hours before the breaks were repaired. One bacteriological sample taken at a home across the street from the first main break had 22 coliforms per 100 ml. Two repeat samples taken 3 d later at the site were negative for coliforms, suggesting the slug of contamination had passed beyond that location.

The daily loss of water pressure due to pump shut-downs to conserve electrical energy or reduce electricity costs is a serious problem in many cities in Third World countries. Often the water treatment plant produces a safe water supply only to have it contaminated in a pipe network that is not continuously under a positive pressure. Infiltration of contaminants occur and there is increased chlorine demand. For example, in Lima, Peru, an attempt was made to apply 0.5 mg/l free chlorine in different areas of the system, but free chlorine residuals were infrequently detected. Contaminant incursions during periods of no water pressure had reacted with all available free chlorine. The loss of water pressure was a daily scheduled event due to pump stoppage to reduce the high cost of electrical power in the water treatment budget. This situation also creates many opportunities for reversals of flow and suction that bring contaminated surface water into the water supply pipe network.

REFERENCES

1. Craun, G., D. Swerdlow, R. Tauxe, R. Clark, K. Fox, E. Geldreich, D. Reasoner, and E. Rice. 1991. Prevention of waterborne cholera in the United States. *Jour. Amer. Water Works Assoc.,* 83(11):40-45.
2. Kowal, N.E. 1982. Health effects of land treatment: microbiological. U.S. Environmental Protection Agency, EPA-600/1-82-007. Health Effects Research Laboratory, Cincinnati, OH.
3. Geldreich, E.E. 1972. Waterborne pathogens. In: *Water Pollution Microbiology.* R. Mitchell, Ed., John Wiley & Sons, Inc., New York. pp. 207-241.
4. Craun, G.F. 1988. Surface water supplies and health. *Jour. Amer. Water Works Assoc.,* 80(2):40-52.
5. Rose, J.B. 1990. Emerging issues for the microbiology of drinking water. *Water Eng. Manage.,* 137:23-29.
6. Klein, P.D., D.Y. Graham, A. Gaillour, A.R. Opekun, and E. O'Brian Smith. 1991. Water source as risk factors for *Helicobacter pylori* infection in Peruvian children. *Lancet,* 337:1503-1506.

7. West, A.P., M.R. Millar, and D.S. Tompkins. 1992. Effect of physical environment on survival of *Helicobacter pylori*. *Jour. Clin. Pathol.*, 45:228-231.
8. Chang, S.L. 1970. Unpublished paper presented at the 10th International Congress of Microbiologists, Mexico City.
9. Thornsberry, C., A. Balows, J.C. Feeley, and W. Jakubowski. 1984. *Legionella: Proceedings of the 2nd International Symposium*. Amer. Soc. Microbiol., Washington, D.C.
10. Walter, W.G. and R.P. Bottman. 1967. Microbiological and chemical studies of an open and closed watershed. *Jour. Environ. Health*, 30:157-163.
11. Alter, A.J. 1954. Appearance of intestinal wastes in surface water supplies at Ketchikan, Alaska. In: Proc. Fifth Alaska Sci. Conf. AAAS, Sept 1954, Anchorage, AK, 81-84.
12. Fennel, H., D.B. James, and J. Morris. 1974. Pollution of a storage reservoir by roosting gulls. *Water Treat. Exam.*, 23:5-24.
13. Anon. 1954. Ketchikan laboratory studies disclose gulls are implicated in disease spread. *Alaska's Health*, 11:1-2.
14. Vogt, R.L., H.E. Sours, T. Barrett, R.S. Feldman, R.J. Dickinson, and L. Witherall. 1982. *Campylobacter enteritis* associated with contaminated water. *Ann. Intern. Med.*, 96:292-296.
15. Health and Welfare, Canada. 1981. Possible waterborne *Campylobacter* outbreak — British Columbia. *Can. Dis. Week. Rep.*, 7:223, 226-227.
16. Taylor, D.N., K.T. McDermott, and J.R. Little. 1983. *Campylobacter enteritis* associated with drinking water in back country areas of the Rocky Mountains. *Ann. Intern. Med.*, 99:38-40.
17. Van Donsel, D.J. and E.E. Geldreich. 1970. Unpublished data on recreational waters.
18. Danielsson, D. and G. Laurell. 1964. Detection of enteropathogenic *Escherichia coli* in a Swedish watercourse (the river Fyris) by means of fluorescent antibodies and by conventional methods. *Acta Paediat.*, 53:49-54.
19. Glantz, P.J. and T.M. Jacks. 1968. An evaluation of the use of *Escherichia coli* serogroups as a means of tracing microbial pollution in water. *Water Resources Res.*, 4:625-638.
20. Gustofson, A.A. and J.B. Hundley. 1969. Enteropathogenic *Escherichia coli* serotypes found in sewage lagoons (waste stabilization ponds) in North Dakota. *Health Lab. Sci.*, 6:18-21.
21. Geldreich, E.E. 1967. Fecal coliform concepts in stream pollution. *Water Sew. Works*, 114:R-98-R-110.
22. Rogers, T.O. and H.A. Wilson. 1966. pH as a selecting mechanism of a microbial flora in wastewater-polluted acid mine drainage. *Jour. Water Poll. Contr. Fed.*, 38:990-995.
23. Malaney, G.W., W.D. Sheets, and R. Quillin. 1959. Toxic effects of metallic ions on sewage microorganisms. *Sew. Ind. Wastes*, 31:1309- 1315.
24. Kittrell, F.W. and S.A. Furfari. 1963. Observations of coliform bacteria in streams. *Jour. Water Poll. Contr. Fed.*, 35:1361-1385.
25. Hanes, N.B., G.A. Rohlich, and W.B. Sarles. 1966. Effect of temperature on the survival of indicator bacteria in water. *Jour. New Engl. Water Works Assoc.*, 80:6-18.

26. Gameson, A.L.H. and J.R. Saxon. 1967. Field Studies on effect of daylight on mortality of coliform bacteria. *Water Res.,* 1:279-295.
27. Weiss, C.M. 1951. Adsorption of *E. coli* on river and estuarine silts. *Sew. Ind. Wastes,* 23:227-237.
28. Varon, M. and M. Shilo. 1969. Interaction of *Bellovibrio bacteriovorus* and host bacteria. II. Intercellular growth and development of *Bellovibrio bacteriovorus* in liquid cultures. *Jour. Bact.,* 99:136-141.
29. Kampelmacher, E.H. and L.M. Van Noorle Jansen. 1976. *Salmonella* effluent from sewage treatment plants, wastepipes of butcher's shops and surface water in Walcheren. *Zentralbl. Bakteriol. Hyg. Abt. I Orig. B,* 162:307-319.
30. Claudon, D.G., D.I. Thompson, E.H. Christensen, G.W. Lawton, and E.C. Dick. 1971. Prolonged *Salmonella* contamination of a recreational lake by runoff waters. *Appl. Microbiol.,* 21:875-877.
31. U.S. Department of Health, Education and Welfare. 1965. Report on pollution of interstate waters of the Red River of the North (Minnesota, North Dakota). Public Health Service, Field Investigation Branch, Cincinnati, OH.
32. Hendricks, C.W. 1971. Increased recovery rate of salmonellae from stream bottom sediments vs. surface waters. *Appl. Microbiol.,* 21:379-380.
33. Andre, D.A., H.H. Weiser, and G.W. Maloney. 1967. Survival of bacterial enteric pathogens in farm pond water. *Jour. Amer. Water Works Assoc.,* 59:503-508.
34. Pesigan, T.P. 1965. Studies on the viability of El Tor vibrios in contaminated foodstuffs, fomites, and water. In: Proceedings of the Cholera Research Symposium, PHS Pub #1328, 317-321.
35. Shrewsbury, J.F.D. and G.J. Barson. 1957. On the absolute viability of certain pathogenic bacteria in a synthetic well water. *Jour. Pathol. Bacteriol.,* 74:215-220.
36. Pillai, S.C., M.I. Gurbaxani, and K.P. Menon. 1952. Influence of activated sludge on certain pathogenic bacteria. *Indian Med. Gaz.,* 87:117-119.
37. Abou-Gareeb, A.H. 1960. The detection of cholera vibrios in Calcutta waters: the river Hooghly and canals. *Jour. Hyg. Camb.,* 58:21-33.
38. Rice, E.W., C.H. Johnson, R.M. Clark, K.R. Fox, D.J. Reasoner, M.E. Dunnigan, P. Panigrahi, J.A. Johnson, and J.G. Morris. 1993. *Vibrio cholerae* 01 can assume a 'rugose' survival form that resists killing by chlorine, yet retains virulence. *Intr. Jour. Environ. Health Res.,* 3, 89-98.
39. Grabow, W.O.K. and E.M. Nupen. 1972. The load of infectious microorganisms in the water of two South African hospitals. *Water Res.,* 6:1557-1563.
40. Callagan, P. and J. Brodie. 1969. Laboratory investigation of sewer swabs following the Aberdeen typhoid outbreak of 1964. *J. Hyg.,* 66:489-497.
41. Harvey, R.W.S., T.H. Price, D.W. Foster, and W.C. Griffith. 1969. *Salmonella* in sewage. A study in latent human infections. *J. Hyg. (Brit.),* 67:517-523.
42. Popp, L. 1974. *Salmonella* and natural purification of polluted waters. *Zentralbl. Bakteriol. Ab. I.,* 158(5):432-445.
43. Kampelmacher, E.H. and L.M. Van Noorle Jansen. 1973. *Legionella* and thermotolerant *E. coli* in the Rhine and Meuse at their point of entry into the Netherlands. H_2O, 6:199-200.
44. World Resources Institute. 1986. *World Resources 1986: An Assessment of the Resource Base that Supports the Global Economy.* Basic Books, Inc., New York, NY.

45. Geldreich, E.E. 1972. Buffalo Lake recreational water quality: a study of bacteriological data interpretation. *Water Res.*, 6:913-934.

46. Oliveri, V.P., C.W. Kruse, and K. Kawata. 1977. Microorganisms in urban stormwater. Environ. Protect. Technol. Series, EPA-600/2-77-087. MERL, U.S. Environmental Protection Agency, Cincinnati, OH.

47. McMullen, L.D. 1994. Surviving the flood: teamwork pays off in Des Moines. *Jour. Amer. Water Works Assoc.*, 86(1):68-72.

48. Reid, J. 1994. Overcoming the flood: How midwestern utilities managed disaster. *Jour. Amer. Water Works Assoc.*, 86(1):58-67.

49. Greenberg, A.E. and J. Jongerth. 1966. Salmonellosis in Riverside, California. *Jour. Amer. Water Works Assoc.*, 58:1145-1150.

50. Schroeder, S.A., J.R. Caldwell, T.M. Vernon, P.C. White, S.I. Granger, and J.V. Bennett. 1968. A waterborne outbreak of gastroenteritis in adults associated with *Escherichia coli. Lancet*, 6:737-740.

51. Lassen, J. 1972. *Yersinia enterocolitica* in drinking water. *Scand. Jour. Infect. Dis.*, 4:125-127.

52. Centers for Disease Control. 1973. Typhoid fever — Florida. *Morbid. Mortal. Week. Rep.*, 22:77,78,85.

53. Centers for Disease Control. 1974. Acute gastrointestinal illness — Florida. *Morbid. Mortal. Week. Rep.*, 23:134-135.

54. Centers for Disease Control. 1980. Waterborne disease outbreaks in the United States — 1978. *Morbid. Mortal. Week Rep.*, 29:46-48.

55. Evison, L.M. and A. James. 1973. A comparison of the distribution of intestinal bacteria in British and East African water sources. *Jour. Appl. Bacteriol.*, 36:109-118.

56. Lindel, S.S. and P. Quinn. 1973. *Shigella sonnei* isolated from well water. *Appl. Microbiol.*, 26:424-425.

57. Woodward, W.E., N. Hirschhorn, R.B. Sack, R.A. Cash, I. Brownlee, G.H. Chickadonz, L.K. Evans, R.N. Shephard, and R.C. Woodward. 1974. Acute diarrhea in an Apache Indian reservation. *Amer. Jour. Epidemiol.*, 99:281-290.

58. Dragas, A-Z and M. Tradnik. 1975. Is the examination of drinkable water and swimming pools on presence of entero-pathogenic *E. coli* necessary? *Zentral. Bakt. Hyg. I. Abt. Orig. B*, 160:60-64.

59. Schieman, D.A. 1978. Isolation of *Yersinia enterocolitica* from surface and well waters in Ontario. *Can. J. Microbiol.*, 24:1048-1052.

60. Mentzing, L.O. 1981. Waterborne outbreaks of *Campylobacter enteritis* in central Sweden. *Lancet*, 8:352-354.

61. Vander Velde, T.L. and W.M. Mack. 1973. Poliovirus in water supply. *Jour. Amer. Water Works Assoc.*, 65:345-348.

62. Geldreich, E.E. 1978. Bacterial populations and indicator concepts in feces, sewage, stormwater and solid wastes. In: *Indicators of Viruses in Water and Food*. G. Berg, Ed., Ann Arbor Science Publishers, Inc., Ann Arbor, MI. pp. 51-97.

63. Grimes, J. 1980. Bacteriological water quality effects of hydraulically dredging contaminated upper Mississippi River bottom sediment. *Appl. Environ. Microbiol.*, 39:782-789.

64. Craun, G. 1985. An overview of statistics on acute and chronic water contamination problems. In: *Fourth Domestic Water Quality Symposium: Point-of-Use Treatment and Its Implications*. Water Quality Assoc., Lisle, IL. pp. 5-15.

65. Snead, M.C., V.P. Oliveri, K. Kawata, and C.W. Kruse. 1980. Biological evaluation of benefits of maintaining a chlorine residual in water supply systems. *Water Res.*, 14:403-408.

66. Craun, G.F. 1986. Waterborne giardiasis in the United States, 1975-84. *Lancet*, 2(8505):513-514.

67. Geldreich, E.E., K.R. Fox, J.A. Goodrich, E.W. Rice, R.M. Clark, and D.L. Swerdlow. 1992. Searching for a water supply connection in the Cabool, Missouri disease outbreak of *Escherichia coli* O157:H7. *Water Res.*, 26:1127-1137.

68. Martin, M.L., L.D. Shipman, and J.G. Wells. 1986. Isolation of *Escherichia coli* O157:H7 from dairy cattle associated with two cases of haemolytic uraemic syndrome. *Lancet*, ii: 1043.

69. Borczyk, A.A., M.A. Karmali, H. Lior, and L.M.C. Duncan. 1987. Bovine reservoir for verotoxin producing *Escherichia coli* O157:H7. *Lancet*, i: 98.

70. Brenner, D.J., B.R. Davis, A.G. Strigerwalt, C.F. Riddle, A.C. McWhorter, et al. 1982. Atypical biogroups of *Escherichia coli* found in clinical specimens and description of *Escherichia hermanii* nov. *Jour. Clin. Microbiol.*, 15:703-713.

71. Rice, E.W., E.G. Sowers, C.H. Johnson, M.E. Dunnigan, et al. 1992. Serological cross-reactions between *Escherichia coli* O157 and other species of the genus *Escherichia*. *Jour. Clin. Microbiol.*, 30:1315-1316.

72. Lior, H. and A.A. Borczyk. 1987. False-positive identifications of *Escherichia coli* O157. *Lancet*, i, (8528):333.

73. Clark, R.M. W.M. Grayman, R.M. Males, and J.A. Coyle. 1988. Modeling contaminant propagation in drinking water distribution system. *Aqua*, No. 3, 137-151.

74. Clark, R.M. and J.A. Coyle. 1990. Measuring and modeling variations in distribution system water quality. *Jour. Amer. Water Works Assoc.*, 82(8):46-53.

75. Clark, R.M., E.E. Geldreich, K.R. Fox, E.W. Rice, C.H. Johnson, J.A. Goodrich, et al. 1995. A waterborne *Salmonella* serovar *typhimurium* outbreak in Gideon, Missouri. *Jour. Hyg. (Lond.)*, in preparation.

76. Angulo, F.J., S. Tippin, D.J. Sharp, B.J. Payne, J.E. Hill, T.J. Barrett, et al. 1995. A community waterborne outbreak of salmonellosis and the effectiveness of a boil water order, in preparation.

77. Bourne, W.R.P. 1975. Birds and hazards to health. *The Practitioner*, 215(1286):165-171.

78. Jones, F., P. Smith, and D.C. Watson. 1978. Pollution of a water supply catchment by breeding gulls and the potential environmental health implications. *Jour. Inst. Water Eng. Sci.*, 32:469-482.

79. Koplan, J.P., R.D. Deen, W.H. Swanston, and B. Tota. 1968. Contaminated roof-collected rainwater as a possible cause of an outbreak of salmonellosis. *J. Hyg. (Lond.)*, 81:303-309.

80. Penfold, J.B., H.C. Amery, and P.J. Peet. 1979. Gastroenteritis associated with wild birds in a hospital kitchen. *Br. Med. Jour.*, 2(6193):802.

81. Mutter, D.F. 1990. The pharmacokinetics of dihydrostreptomycin sulfate in domestic pigeons. *Tierarztl. Praf.*, 18:377-381.

82. Rossman, L.A. 1994. EPANET User's Manual, Drinking Water Research Division, U.S. Environmental Protection Agency, Cincinnati, OH.

83. Wachsmuth, I.K., P.A. Blake, and O. Olsvik. 1994. *Vibrio cholera and Cholera: Molecular and Global Perspectives*. Amer. Soc. Microbiol., Washington, D.C.

84. Swerdlow, D.L., E. Mintz, M. Rodriguez, et al. 1992. Waterborne Transmission of epidemic cholera in Trujillo, Peru: Lessons for a continent at risk. *Lancet,* 340:28-32.

85. Geldreich, E.E. 1992. Waterborne pathogen invasions: a case for water quality protection in distribution. Water. Qual. Technol. Conf. Toronto, Canada. pp. 599-616.

86. Mintz, E.D., F.M. Reiff, R.V. Tauxe. 1995. Safe Water Treatment and Storage in the Home, *Jour. Amer. Med. Assoc.* 273:948-953.

87. Lippy, E. and S. Waltrip. 1984. Waterborne Disease Outbreaks, 1946–1980: A Thirty-five Year Perspective. *Jour. Amer. Water Works Assoc.,* 76:60–67.

88. Hayes, E.B. et al. 1989. Large Community Outbreak of Cryptosporidiosis due to Contamination of a Filtered Public Water Supply. *New Eng. Jour. Med.,* 320:1372–1376, pp. 3–31.

89. Geldreich, E.E., 1990. Microbiological Quality of Source Waters for Water Supply. In: *Drinking Water Microbiology,* G.A. McFetens, Ed., Springer-Verlag, New York.

90. Burk, J.A. 1977. The Clinical and Laboratory Diagnosis of Giardiasis. *CRC Critical Reviews in Clin. Lab. Sci.,* 7:373–391.

91. Davies, R.B. and Hibler, C.P. 1979. Animal Reservoirs and Cross-Species Transmission of Giardia. In: *Waterborne Transmission of Giardiasis,* W. Jakublowksi and J.C. Hoff, Eds. EPA-600/9-79-001. U.S. Environmental Protection Agency, Cincinnati, OH. pp 104–126.

92. Grabow, W.O.K. and E.M. Nupen. 1972. The Load of Infectious Microorganisms in the Waste Water of Two South African Hospitals. *Water Research,* 6:1556–1557.

93. U.S. Environmental Protection Agency. 1971. Report on Missouri River Water Quality Studies. Regional Office, Kansas City, MO.

94. Burns, R.J. and R.D. Vaughan. 1966. Bacteriological Comparison Between Combined and Separate Sewer Discharges in Southeastern Michigan. *Jour. Water Poll. Fed.* 38:400-409.

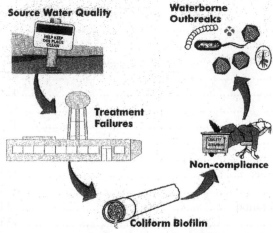

Technology transfer in water supply.

CHAPTER 9

Responses to Microbial Quality Changes in Water Supply Distribution

CONTENTS

INTRODUCTION

Detection of adverse microbial quality changes in distribution water must be followed by an appropriate response to correct the problem promptly. Particularly urgent is the sudden appearance of coliforms because these occurrences represent a potential health hazard and are identified with criteria limits established by federal regulations. Furthermore, it is well established that water treatment should consistently produce a water supply that has no detectable coliform bacteria in 100-ml samples.[1,2] If not, this information suggests a coliform breakthrough in the treatment barriers or a contamination event either in finished water storage or in the pipe network. While these coliform events may, in themselves, not be paralleled by disease organisms, the same pathway could be used by waterborne pathogens.

Direction and intensity of the response are determined from a study of laboratory data and engineering review of system operations. These aspects must be evaluated whenever there is a sudden abrupt change in water quality. More subtle are the trends toward gradual reduction in microbial quality as reflected in loss of disinfectant residual, coliform biofilm development, or regrowth in heterotrophic plate count organisms.

LABORATORY ASSISTANCE

Routine monitoring data that reveals coliform bacteria should promptly lead to a sequence of repeat sampling at the same sites and several locations above and below that location so that the extent of contamination can be determined. Analysis of these special samples should include analyses for fecal coliform or *E. coli* bacteria.[3] This follow-up sampling and analysis may indicate one of three possible patterns: (1) a brief contamination event that is no longer detectable; (2) a zone of contamination near the initial site of coliform detection; or (3) coliform occurrences that are scattered instances throughout the distribution system.

SAMPLE INTEGRITY

Variations in sample collection technique[4-7] are more likely to be the cause of sample contamination than the collector carelessly sticking his finger in the sterile bottle or sample (see Chapter 6, Sample Collections). At the same time, it must be recognized that a variety of personnel are responsible for sample collection, i.e., utility employees, county or state sanitarians, and laboratory personnel (Table 9.3). Because the professional background of sample collectors is varied, there may be different attitudes developed over time on interpreting the concept of sterile technique. For this reason, some state laboratories conduct a mandatory refresher course for all sample collectors in the program on an annual basis. These courses could also provide opportunities to introduce updated information on changes in monitoring policies, provide new instructions on recording sample information and sample transport problems, and encourage feedback from field personnel.

DATA RELIABILITY AND COSTS

Laboratory error is also perceived by some water plant operators as the cause whenever a nonrepeatable coliform detection event occurs. In reality, erroneous reports of coliform occurrence through poor laboratory operations is very unlikely for several reasons. All laboratories committed to public water supply monitoring must be certified periodically (1- to 3-year intervals) in terms of proper test procedures, essential equipment, staffing expertise, and facility constraints.[8-10] In general, more problems were found with small municipal and commercial laboratories because of inadequate technician training and equipment maintenance problems (see Chapter 5, Concepts in Lab Certification).[11]

The cost of performing a bacteriological test is a subject of major concern to laboratory administrators in all state and federal monitoring programs.[12] In an attempt to prepare a realistic cost analysis, non-expendable items must be identified and their life expectancy, annual depreciation, and percent of use per specific test established. Time studies suggest total time for processing a confirmed multiple tube examination of water supply may require only 1.6 min while a similar test performed on polluted surface waters and sewage/stormwater effluents will involve 10 min of processing time. When the membrane filter procedure is used, this total time is further reduced to 1.0 min for analyzing a distribution sample containing no coliforms or 4.7 min for processing a raw source water for coliform bacteria. Thus it can be concluded that facility space and large equipment items factored into cost per test, contribute only a minor additional charge to total test cost estimates.

Expendable items are generally limited to media requirements in the multiple tube procedure, or media, membrane filters and culture dishes in the membrane filter procedure. Both procedures require office supplies and certain standard laboratory items such as pipettes, test tubes, flasks, beakers, dilution water, reagents, marking pens, etc. Extensive use may be made of disposable plastic items (pipettes, culture dishes, etc.), prepackaged sterile media and membrane filters. Some laboratories choose this approach as a means of circumventing the need for permanent laboratory support operations that involve glassware cleaning, preparation of media, and sterilization of equipment.

Laboratory personnel salary requirements that include fringe benefits (health insurance, retirement fund) are the biggest cost items. While the analyst performing routine total coliform analyses at the small water plant laboratory may be a high school graduate with on-the-job training, supervision must be available from either an experienced professional scientist or through an outside consultant on a continuous retainer agreement. A professional microbiologist on staff is a necessity in the utilities serving 10,000 customers or more and all public health laboratories because of involvement in the analysis of a variety of waters and various testing procedures plus defending these procedures and results in legal proceedings. Our study of manpower requirements in 18 state laboratories during the period 1965 to 1970 showed that for each staff microbiologist in the water laboratory, the full time support of 1.4 persons was needed to assist in laboratory support operations and clerical work. Based on data collected from operations in 10 state laboratories, a staff of this size could process an average 5,400 water samples per year using the membrane filter procedure. This estimation of work load capacity was about 10 percent above the number of MPN samples that could be examined in a year by a similar staff.

With these facts as background, tests costs were calculated for the three basic quantitative procedures, i.e., pour plate method, membrane filter technique, and multiple tube procedure. As anticipated, the standard plate count is the more economical test to perform, and contradicts claims that this procedure, recommended as a supplemental check of distribution water quality, is an excessive cost burden.

While coliforms are the primary bacteriological criteria used in monitoring water supply quality, there is increasing interest in identifying biofilm occurrence, checking for fecal contamination and searching for pathogenic agents. These objectives are not achieved without increasing laboratory involvement at a price. Cost per test rapidly escalates in the speciation of coliforms and the search for *Salmonella, E. coli 0157, V. cholera,* viruses, and pathogenic protozoans. The cost of obtaining a coliform profile or speciation from a positive coliform sample is greater than the routine test because of the need to examine a large number of coliform isolates to obtain statistically valid distribution patterns in the microbial population. Virus detection is expensive because of costs involved in obtaining and processing large sample volumes, the use of several cell lines for enterovirus recovery and incubation time required for plaque development in tissue culture. The search for protozoan pathogens involves not only the problems of concentration from large samples but also labor intense microscopic scans of the concentrates for cysts and oocysts among the sediment particulates.

A study of individual quality assurance costs (Table 9.1) reveals that the largest expenses are associated with the autoclave, pH meter, and balance. Their frequent usage on a daily basis accelerates the need for occasional repairs, parts replacement, and semi-annual maintenance to insure satisfactory performance and operational safety. Daily checks of media quality and sterility controls on sample bottles, dilution water and MF funnel sterilization are other labor intensive requirements that are major expense items in the quality assurance effort. The net result of this specified minimal quality assurance program for certified laboratories is that approximately 15 percent of work activities must be devoted to this aspect of essential laboratory activity. Since these requirements are the same for all laboratories, regardless of the volume of tests performed, the greatest financial burden will fall on small to medium sized laboratories and be reflected in their overall cost per test. Additional quality assurance requirements must be met by those laboratories approved to do virus or protozoan monitoring and will be reflected in the cost per sample.

Several water utilities serving cities with populations ranging from 60,000 to 900,000 people were asked for information on cost of bacteriological monitoring of the distribution network. These figures were derived by water plant management in different ways, however, all calculations included salaries, expendable supplies, capital equipment, and indirect costs for both sample collection and bacteriological analysis. Analysis of these data revealed the cost of monitoring the distribution system may approximate $15.00 per sample. In contrast, cost figures for monitoring distribution systems from larger cities (>100,000) becomes more cost effective since depreciation cost of nonexpendable items and facility remain unchanged.

Table 9.1   Costs for a Minimal Quality Assurance Program in Water Microbiology

Quality control procedures	Frequency	Total time (h) per year	% of time	Labor cost ($)	Material cost ($)	Total cost per year ($)
Lab pure-water quality						
Bacterial quality	Monthly	6	0.3[a]	60.90	.24	61.14
Suitability tests	Yearly	6	0.3[a]	60.90	16.52	77.42
Storage reservoir	Yearly	2	0.1[a]	20.30	1.75	22.05
Glassware cleaning performance						
Acidity–alkalinity test	Daily	1	0.05[b]	6.06	0.04	6.10
Detergent toxicity	Yearly	1	0.05[a]	10.15	1.44	11.59
Media quality						
pH check	Daily	44	2.3[b]	266.64	49.50	316.14
Sterility	Daily	20	1.1[b]	121.20	156.00	277.20
Lot quality	Yearly	4	0.2[a]	40.60	2.00	42.60
Temperature monitoring						
Recording	Daily	8.7	0.5[b]	52.72	0.20	52.92
Changing charts	Weekly	3.5	0.02[b]	21.21	7.00	28.21
Thermo standardization	Biannually	6	0.3[a]	60.90	28.70	89.60

Verification of sheeny colonies						
Positive results	Monthly	6	0.3[a]	60.90	1.20	62.10
Technician performance						
Counting colonies	Monthly	6	0.3[a]	60.90	—	60.90
Sterility controls						
Sample bottles	Daily	20	1.1[b]	121.20	36.00	157.20
Dilution (rinse) water	Daily	20	1.1[a]	203.00	78.60	281.60
MF funnel cross-contamination	Daily	20	1.1[a]	203.00	—	203.00
Preventive maintenance/monitoring						
Autoclave	Quarterly	34.6	1.8[a]	351.19	495.00	846.19
Incubator	Quarterly	12.2	0.6[a]	123.83	61.50	185.33
pH meter	Quarterly	44	2.3[a]	446.60	69.50	516.10
Refrigerator	Quarterly	12.2	0.6[b]	73.93	10.00	83.93
Oven	Quarterly	9	0.4[b]	54.54	5.00	59.54
Glassware washer	Quarterly	13	0.6[b]	78.78	7.00	85.78
Water still	Quarterly	11	0.5[b]	66.66	57.00	123.66

Note: Cost estimates based on 1980 prices.

[a] Microbiologist time.
[b] Technician time.
Data from Geldreich and Kennedy.[12]

CHARACTERIZING COLIFORM OCCURRENCES

Total coliform analyses in water supply are used primarily to determine either treatment effectiveness for polluted raw source waters or adequacy of the soil barrier over untreated groundwater aquifers. Within the distribution system, the search for coliforms becomes a verification of pipe network integrity. Such interpretation is made somewhat difficult where there are coliform biofilm events and in the rare instances of a contaminating event in which coliforms fail to be detected because of heterotrophic bacterial population interference.

Much of the confusion over the significance of a coliform occurrence centers on the origins of this broad group of organisms. Many total coliform bacteria originate from soil, vegetation, and aquatic environments unrelated to fecal pollution.[13-15] *Klebsiella, Enterobacter,* and *Citrobacter* are the predominant environmental coliforms; however, some *Klebsiella* strains can be of fecal origin.[16-20] However, the vast majority of *Klebsiella* encountered in contaminated water are nonpathogenic environmental strains that originate from vegetation,[21-24] agricultural products,[25-27] wood pulp[28-30] and textile industry waste (see Table 2.6, Chapter 2).[31]

E. coli is the one coliform that is known to be fecal in origin and generally the predominant coliform in most fecal wastes. Inspection of the bacterial profiles in a variety of treated wastes (Table 2.6, Chapter 2) reveals *E. coli* to be the predominant coliform in municipal wastes and the discharges from meat and other food-processing operations.[32] These observations support the argument that if there is any *E. coli* detected in repeat samples from the distribution system, a serious breakthrough of fecal contamination has occurred. The occurrence of other coliform species should not be ignored. In this instance there is a need for more supporting laboratory information to verify the existence of coliform biofilm rather than a recent contamination breakthrough of sanitary significance.

Recent advances in the development of multiple-test kits have made coliform speciation easier. As a consequence, most water plant laboratories have begun to profile the coliform species present in the distribution network. Speciation often reveals *Klebsiella* to be the predominant coliform, but some systems have reported *Enterobacter* or *Citrobacter*. There is also some evidence that the dominant coliform species in a biofilm may shift during the biofilm episode. A study of data from the speciation of coliforms recovered in New Haven distribution water suggests *Citrobacter* and *Enterobacter* were predominant in springtime occurrences while *Klebsiella* was the dominant coliform found in late July and August.[33] Perhaps the shift in coliform species reflects changes in the profile of organisms present in the source water or is a reflection of subtle differences in water treatment conditions during a seasonal episode. In a compilation of coliform species (Table 3.2, Chapter 3) reported from 111 water distribution systems located in the United States and Canada,[34] most coliform events revealed only one or two coliform species in any given episode. None of the systems had a diverse coliform population, implying these events were more likely to be biofilm related rather than a recent contaminating event. *E. coli* was infrequently reported in these water supplies. This observation suggests *E. coli* is not a permanent resident of pipe biofilms, but is associated with recent fecal contamination breakthrough.[35,36]

INITIAL RESPONSES

Coliform occurrences in the distribution system should always be interpreted as evidence of a contaminating event.[13] What needs to be determined first is whether such an occurrence represents fecal contamination, stressed coliform passage,[37] or a release of coliforms that have colonized the distribution system. Any evidence of fecal coliforms or *E. coli* in the repeat samples calls for an immediate public notification. If the fecal contamination is confined to one zone, then public notification can be restricted to only those customers receiving water in the affected area. When fecal contamination is system-wide, there must be an immediate alert to all customers in the community to boil water for drinking purposes until further notice. During this period all resources must be utilized to find the cause and apply appropriate measures to restore water quality.

Nonfecal contamination events should also be vigorously pursued. If ignored, their repeated occurrence may quickly place the system out of compliance with the federal regulations[38] that limit coliform occurrences to 5% during the month. Appropriate remedial response to these sudden changes must be carefully orchestrated to achieve a quick correction.

FIRST ACTION LEVEL

As an immediate response to the initial discovery of a coliform-positive sample, another sample must be collected from the same site and adjacent sites above and below to verify the contaminating event and location. At this time, quantitative analyses of these samples is important so that a determination can be made of the magnitude of the coliform contamination. It is not uncommon to find that the repeat sampling of water from the same site may not reveal detectable coliforms in 100 ml. In this situation, it is desirable to follow up with collections at 2-h intervals at the site over an 8-h period using liter samples. This approach provides an opportunity to check for contaminant leakages in an intermittent cross-connection.

SECOND ACTION LEVEL

If analyses of the repeat sample reveal coliform bacteria are still present, and the extent of the contamination event in the area is revealed, more drastic action is necessary. Immediately begin localized flushing to remove the contaminant and hopefully restore a disinfectant residual in that area. Emergency booster chlorination in the neighborhood is highly recommended.

PERSISTENT NONCOMPLIANCE RESPONSE

If the incipient coliform occurrence is not quickly suppressed, the situation may then quickly lead to a frequency of occurrence that exceeds the limits established by federal regulation. Obviously, noncompliance calls for drastic measures by the utility to flush out the contaminant, remove sediment accumulations in the pipeline, and restore an effective chlorine residual. This often means the creation of a task force to review the problem, investigate community illnesses, and approve a remedial action strategy.

TASK FORCE ACTIVITIES

Formation of a task force comprised of utility personnel, and state and federal water authorities is essential when occurrence reaches 5% for two consecutive months. Their immediate task is to examine plant records, interview water utility operators on treatment protocols, review laboratory data, and discuss the integrity of the distribution system with water quality management. Local health officials should survey pharmacies for increased sales of antidiarrheal medication and private medical practices, clinics, and hospitals for increased number of patients suffering from intestinal illness that might be water related.

Information supplied from the utility should determine if these coliform occurrences are a result of a major break in the treatment barrier, a cross-connection event in distribution, or a release of environmental coliforms growing in a pipe biofilm. Provided no evidence of fecal contamination exists, frequency of coliform occurrence might be useful in the establishment of a stepwise action plan to avoid sudden increases in taste and odor complaints and increased instability of pipe network sediments. For example, most systems have an annual frequency of coliform occurrence ranging from 1 to 3%, with most individual positive samples containing one to five coliforms per 100 ml. Response to these occurrences would focus on isolating the zone of contamination, flushing those lines, and elevating the disinfectant concentration by use of booster chlorination. When the frequency of coliform occurrences escalates beyond 5%, coliform densities frequently range from 1 to 125 organisms per 100 ml. At that time, the water quality is out of control and the task force must activate an appropriate action plan.

A sanitary survey of source water characteristics, treatment operations, and distribution integrity should be performed promptly. Simultaneously, action should be taken to establish either 0.5 mg/l free chlorine or 1.0 mg/l chloramine residuals in the pipe network. These disinfectant residuals should be achieved at 95% of all sampling sites in the pipe network. Furthermore, coliform monitoring of the system must be expanded to cover more sites and include fecal coliform testing. All major task force decisions should also be made public to assure consumers that the problem is being addressed and that there is no cause for loss of confidence in water plant management to resolve the problem. At the same time, the public should be told that an ongoing health watch is being maintained for any indications of waterborne illness in the community. Using this approach, the need for a boil water order could be minimized unless the laboratory detects fecal coliforms in the water or a waterborne outbreak occurs.

SANITARY SURVEY APPROACH

SURVEY COVERAGE

Water quality as delivered to the consumer's tap is a reflection of source water quality, treatment provided, and the protective capacity of the pipe network to deliver safe water to each individual in the community. Water systems can and do undergo significant changes with time. New treatment plants may be added to treat new source waters, treatment train configurations revised, new operators employed, distribution networks of pipes expanded, reservoirs and standpipes constructed, interconnections made to neighboring utilities to satisfy seasonal water demand, and possible consolidations of water districts, treatment facilities, and distribution networks. Any of these expansions in water supply production can impact distribution water quality. As a consequence, sanitary surveys must cover the adequacy of source, treatment effectiveness, and practices in operating the distribution system.[39] From these on-site inspections and a review of plant and laboratory data, an informed judgment can often be made and a strategy devised to quickly correct emerging problems.

Frequency of these surveys will vary (1 to 5 years), being less frequent on systems having a long history of continuous compliance. For a number of reasons, an annual review is justified for those systems that are marginally satisfactory. In these latter cases, the most serious concerns are related to adequate treatment to cope with surface water fluctuations, failures to comply with drinking water regulations, occurrence of a waterborne outbreak, and impact of natural disasters (floods, droughts, and earthquakes) on facility operations and plant structures. Other conditions that determine the survey frequency include modifications in applied treatment, blending waters of different qualities, variations in operator skills or experience, and changes in ownership.

PLANNING STRATEGY

There are three aspects to every sanitary survey: the presurvey planning, on-site visit, and preparation of the written report based on the fact-finding review of information gathered in the visit. Each of these steps must be carefully developed to obtain maximum benefit from the survey.[40]

Presurvey Planning

Presurvey planning by the state water authority starts by contacting the individual in overall charge of the water system (i.e., the superintendent, company president, manager, or director of public works). At this time, agreement is reached on the dates for the on-site visit. Arrangements are also initiated to make available key staff members and all pertinent plant and laboratory records during the proposed visit. Also requested is that past documentation from the utility on source water characteristics, treatment processes, and distribution configuration be sent in advance for study. Review of such material in presurvey preparation will be useful in conducting a comprehensive and in-depth evaluation of all aspects in an effective manner. Rarely is the sanitary survey planned or executed as a surprise visit. In these instances, the purpose is to develop evidence of malpractice for court action by the state agency. Such measures are only taken as a last resort. It is more desirable to create a cooperative spirit through conferences involving review of engineering and laboratory records that lead to an effective joint agreement on the appropriate response to resolve the persisting noncompliance issue of public health concern.

On-Site Visits

On-site visits involve interviews with individuals in charge of the treatment facilities, water quality (watershed management and monitoring laboratory support), and distribution system operations. Obviously, in small system operations, many of these responsibilities are entrusted to one individual with monitoring requirements delegated to the state laboratory.

Systematic coverage of the many aspects involved in water supply production and protected delivery to the consumer can best be reviewed and documented through the use of a survey form. Rather than the survey form being considered as a checklist of operational practices, it should serve as guidance to the creation of a specific description of source water quality management, treatment practices, protected distribution, and monitoring activities. In addition, this survey form should also provide information on deficiencies observed during visual inspection of treatment plant processes and through review of operation records and laboratory reports.

All the information recorded on the survey form should be discussed in a "wrap-up" conference held at the conclusion of the visit. If deficiencies are identified, they should be brought to the attention of the system manager, as well as with the person directly involved, i.e., treatment plant operator, water quality engineer, etc. Remember, the intent of the survey form is to serve only as a guideline for a thorough on-site review of the utility operations and verification of safe water production and its protected transmission to the consumer. The survey form is not intended to be a grading exercise for answers supplied by utility personnel.

Documentation

Documentation of the on-site visit, personnel contacted, system character-istics, deficiencies observed, and recommended specific improvements with timetables for compliance must be prepared and transmitted to the appropriate water authorities and the utility. This report can also request further informa-tion, identify need for consultants to study particular treatment facility issues or design improvements, and suggest training opportunities for enhancement of staff capabilities. Where the sanitary survey report forms the basis for remedial actions, care should be taken to document precisely what the problem is and this should be supported by specific facts or observations. It is also essential to provide a recommended course of action to correct the deficiencies within a specified time frame. The tracking mechanism, along with appropriate technical assistance follow-up, is essential for continuance of pressure for remedial action and for verifying that improvements are, in fact, being made on schedule.

SEARCHING FOR RISKS IN WATER SUPPLY

Risk reduction is the heart of the sanitary survey. This objective applies to protecting public health from poor-quality water supply and to safeguarding utility employees at the work place. On-site inspections of the many aspects involved in water supply production can often reveal the cause of a reported coliform occurrence in the distribution system or explain the sudden, unac-ceptable passage of a waterborne pathogen in drinking water. In addition, these field visits may also discover other deficiencies that may become a pathway for future contamination events. Achieved benefits of the survey process can be demonstrated from a study of 25 sanitary survey reports prepared after site visits to utilities and water district operations in New England during 1976 to 1984.[41] In this review, attention was focused on deficiencies that relate most strongly to coliform occurrences in monitoring reports or to *Giardia* outbreaks in the community that were caused by contaminated water supply. Other important concerns such as chemical contamination, personnel safety, or administrative responsibility were also noted.

WATERSHED IMPACT ON RAW WATER QUALITY

Nationwide, natural lakes and impoundments (Figure 9.1) form a very significant resource of raw water supply for public utilities. One statistical survey[42] reported that approximately 1700 municipal supplies serving more than 55 million people, or 30.6% of the 180 million total population served, derived all or a portion of their raw water from lakes and impoundments. These natural lakes and impoundments of small streams have become attractive aquatic resources that often create demand for shared uses that may include recreation, resort development, flood control, hydroelectric power generation, or irrigation.[43] Without careful management of these water resources, the desirable features may be quickly lost, the public health risk increased, and lake or impoundment transformed into an unsightly eutrophic body of water.[44]

Figure 9.1   Source water impoundment in a protected watershed. (Courtesy of the Portland, Oregon Water Bureau.)

Historical Information

Historical information indicates that in an effort to provide continued good-quality raw source water from lakes and impoundments, many water utilities acquired large tracts of land on the watershed. Fences were built and the grounds patrolled to limit access to these nature preserves. In theory, the objective was to restrict human activity over the drainage basin and thereby minimize the magnitude of fecal waste released that might contain pathogens. In practice, this vegetation buffer zone around lake shores was found to be effective in minimizing turbidity in stormwater runoff and to act as a deterrent to the rapid movement of fecal contaminants over the land surfaces during rainfall events. The net effect was to mute water quality fluctuations, thereby providing water characteristics that were more amendable to less complex water treatment schemes. In many instances, treatment by disinfection was the only microbial barrier employed.

As we know now, *Giardia* outbreaks over the past 25 years have changed forever this time-honored concept.[45-48] Thousands of people became ill with giardiasis, and minimal treatment barrier concepts were reevaluated for improved public health protection. The source of the pathogen was quickly found to be in the wildlife population living on the protected watershed. Beavers were most often considered to be the primary animal reservoir because of the high incidence of infection in beaver colonies and their habitat association with near-shore water environments. However, coyotes, muskrats, voles, cattle, and pets may also be involved in the perpetuation of this pathogen in the natural environment.[49-51] Many of these animals are either part of the natural fauna or are introduced onto the watershed as domestic farm animals.

Birds and waterfowl are also sources of pollution. Sea gulls are scavengers that frequent open garbage dumps and can contribute *Salmonella* in their fecal droppings to freshwater impoundments and lakes in coastal areas.[52,53] In one instance, sea gulls were the contributors of *Salmonella* to an untreated water supply reservoirs in an Alaskan community and the cause of several cases of salmonellosis.[54] Pigeons inhabiting a water supply storage tank were considered the source of *Salmonella typhimurium* and cause of illness in Gideon, MO.[55]

Groundwaters are used extensively as a source of public water in arid regions and in areas where surface supplies are insufficient in volume or so poor in quality as to require extensive purification through treatment. Of approximately 60,000 public water supplies in the United States, 75% use groundwater as the sole raw water source, and 7% of these public systems use a mixture of ground- and surface water.[56]

Water from deep aquifers is generally of excellent quality because vertical percolation of surface water through soils results in the removal of much of the microbial and organic pollutants. As a consequence, treatment is frequently nonexistent or limited to water hardness reduction or taste and odor removal. By contrast, waters from shallow wells (less than 100 ft deep) are frequently grossly polluted with a variety of wastes. Water derived from these shallow wells are often seriously compromised by surface-water contaminants that enter in stormwater runoff or through periodic flooding. Driven wells provide better protection from surface pollution because of casing construction and better wellhead protection by small enclosures over these sites.

Migration of pollutants through the soil barrier can take many pathways, such as direct injection of wastes through wells, percolation of treated effluents sprayed over the land, leaking or broken sewer lines, seepage from waste lagoons, infiltration of polluted surface streams, interaquifer leakage, downward movement of irrigation waters, leachates from landfills, septic tank effluents, and upwelling of salt water into a freshwater lens. As a result of improper source protection, it is not surprising that many small community and noncommunity water systems have the poorest records for compliance with the drinking water standards.

Investigative Areas

Investigative areas on raw source water should explore not only the quantity and quality of the source, but also adequacy of the watershed management program to protect future use of this resource. On-site inquiries, visual inspections and study of records should focus on the following issues:

- Is source supply being taken from the highest-quality raw water available?
- Is the quantity of available water adequate to meet presentand projected future growth in the community?
- Are the watershed protection activities effective in the protection and enhancement of source waters?
- Are intakes properly located to utilize the highest quality of water available, and are they protected from fouling debris?
- Are major structures (dams, spillways, diversion ditches, and embankments) in good condition, and are they inspected regularly?
- Is soil erosion controlled effectively by vegetation cover?
- Is raw water quality monitored appropriately, using critical criteria that impact treatability?

Field Observations

Field observations obtained from sanitary surveys of 25 utilities revealed that a significant number of small systems with coliform problems are not actively pursuing a watershed management program to protect their raw source water quality.[41] The following combined excerpts of observations and recommendations illustrate the concerns with raw (surface and ground) source water from a microbiological viewpoint.

Watersheds

No comprehensive watershed inspection program or inventory of malfunctioning septic tanks.

Need a watershed map (USGS map) to delineate watershed area and complete watershed survey to: determine if there are controllable areas of erodible soil through vegetation development; locate all septic tanks and those not functioning properly; arrange for toilet facilities at public swimming area and fishing sites; conduct periodic survey for beaver and their removal from watershed; urge the installation of a new force main to transport sewage from failing septic systems in trailer park to an approved disposal field.

Conduct cleanup of refuse dumping on watershed.

Chronic problem with surface water shortages that could be solved by creating surface water impoundments or developing a well field.

Initiate legal action to stop raw sewage discharge from small town on watershed.

Reservoirs/Impoundments

Replace deciduous trees along perimeter of reservoir with evergreens to reduce leaf fall and associated organic inputs to water.

Remove tree growth on edge of dam which could compromise the integrity of the dam.

Divert stormwater runoff and threat of hazardous spills from adjacent road.

Inability to control water level in reservoir has caused serious loss of impounded water.

Establish an optimum $CuSO_4$ dose for controlling algal blooms.

Elevate adjustable intake to receive highest quality of raw water.

Install screen on intake to remove coarse debris.

Initiate a raw water quality monitoring program to continually characterize changes in contaminate concentrations.

Well Fields

No well vent — a vent covered with a 24 mesh corrosion resistant screen should be installed with opening turned downward for protection from dust and rain; faulty valve causes water to constantly discharge into basement of well house; bearing cooling water appears to be going directly into the wells.

Dug wells are notoriously unreliable in terms of quantity and quality. No grouting of tiles to prevent entrance of surface water drainage; located close to stream and may be subject to flooding and surface water contamination; highway only a few hundred feet away subjects water to roadway salt contamination and hazardous material spills from passing truck traffic; well covers not fitted tightly and not locked; frequent bacteriological problem relates to storm events on watershed; need to chlorinate supply continually but a new source of water supply using a drilled well would be more reliable and must be explored.

Area in vicinity of well should be cleaned up, fenced and posted.

Attempt to define direction of groundwater flow from junkyard to determine if well source will become contaminated through leaching.

Drilled well is located in the basement of a building surrounded by a cow pasture.

Direct connection with an animal water tub located near well should have a hose bib vacuum breaker.

Surface water could pond around well casing.

Surface contour the land to divert surface drainage away from well field.

Need an observation well to monitor for contamination migration from nearby closed landfill.

Well head needs to be flood-proofed because it is in a flood plain.

Building housing well is not fenced and is close to main street in town.

Space between pump flooring and well casing must be sealed to prevent surface water from entering groundwater supply.

Electric box for wires to well should be watertight to prevent surface water migration.

Cooling water from auxiliary engine should not be returned to well, should instead be run to waste.

Well not properly vented, need pipe vent with 180° turn down and opening screened to prevent small animal entry.

Dug well not protected from surface runoff; concrete block has holes, not water tight; no fence around well; overflow of well could permit backflow of surface water which drains in direction of well; dug well must be checked frequently for cracks; there are openings in side walls for at least 20 feet.

Spring Source

Spring source is of poor quality, should develop a reliable groundwater source.

Stream with a direct sewage discharge less than 50 feet away from spring source which could be contaminated during periods of high flow or flooding.

Screen near roof of spring house torn away providing access to birds, open space around flooring permits access of small animals.

Vegetation encroachment on spring outlet.

TREATMENT PRACTICES

Freshwater resources used for water supply are often polluted by fecal discharges from man and other animals. These wastes may also introduce a variety of human pathogens: pathogenic bacteria, viruses, protozoans, and several more complex multicellular organisms that can cause gastrointestinal illness. When properly applied, water treatment has had a major impact on the prevention of waterborne disease in the community. Where waterborne outbreaks have occurred, deficiencies in treatment (particularly filter breakthrough and inadequate or interrupted disinfection) have been major causes of the problem.[57]

Historical Records

Historical records suggest to some water authorities that they have successfully achieved adequate source water protection, and the authorities cite evidence of no recorded waterborne outbreaks in their communities. Other water utilities have not been so fortunate in preventing outbreaks of giardiasis[47-50,58] or illness associated with campylobacter enteritis.[59-63] While these outbreaks have occurred in surface waters with minimal treatment, the threat to groundwater cannot be excluded because some waterborne disease has occurred from drinking contaminated well water.[64-77] Included in this group are intestinal diseases caused by *Salmonella, Shigella*, enteropathogenic *E. coli, Yersinia, Campylobacter*, and enteroviruses.

Among the characteristics that play an important part in defining a water that can be safely treated by disinfection as the only controlled engineering process are turbidity, bacterial load, chlorine demand, and color (Table 9.2). Turbidity in raw source water must be minimized because this material may protect microorganisms present (Table 9.3) and facilitate their safe passage into the distribution system. Particulates in surface water are introduced through soil erosion in stormwater runoff, the destratification of raw water impoundments, seasonal algal blooms, chemical spills upstream, and natural siltation in the water course. Groundwater may acquire turbidity from intrusions of leachates originating in nearby landfills, pollutional waste injections that penetrate the aquifer, dissolution of minerals by aeration, and microbial slime releases from colonization of the well casing or water-bearing sand.

Coliform indicators have long been used to measure the probable magnitude of fecal contamination and a relatively clean surface water will often have less than 100 total coliforms per 100 ml.[78,79] At this level of contamination, fecal coliforms can be expected to approximate 10 to 30% of the less-definitive total coliform population (Table 9.4). The reason for such differences in total coliform and fecal coliform densities lies in the more rapid inactivation of *E. coli* as compared with the longer survival of environmental coliform strains of no sanitary significance. Some measure of the potential occurrence of *Salmonella* in surface waters at various average densities for fecal coliforms are given in Table 9.5. The data support the argument that the greater the magnitude of fecal pollution, the greater the chance that some pathogens may also be present.[80,81] Absence of any detectable fecal coliforms in 100 ml does not mean that there is no pathogen risk from *Giardia, Crytosporidium,* or virus. These pathogenic agents frequently occur in small densities requiring the testing of large sample volumes (10 l or more). For this reason, a surrogate indicator (fecal coliform) examination at only 100 ml rather than 1 liter or more is not going to provide sufficient precision to ascertain the potential for extremely low-density occurrences of protozoan pathogens and virus releases.

Table 9.2   Essential Criteria and Limit Amenable
to Disinfection as the only Controlling
Treatment Process

Criteria	Limit
Fecal coliform	20 organisms per 100 ml
Turbidity	1.0 NTU
Color	15 ACU
Chlorine	Demand 2 mg/l

Data from Geldreich, Goodrich, and Clark.[79]

Table 9.3   Comparison of Turbidity and Coliform Densities
from Lake Whitney

Date (February 1985)	Temp (°C)	Color (ACU)	Turbidity (NTU)	Total coliform per 100 ml
1	3.0	10	1.0	8
2	3.0	10	0.7	1
3	3.0	10	0.8	3
4	2.0	10	0.7	2
5	3.0	15	1.1	1
6	3.0	7	0.9	1
7	3.0	10	0.7	3
8	3.0	10	0.8	<1
9	3.0	10	0.7	1
10	3.0	10	1.0	1
11	4.0	12	0.9	2
12	3.0	10	2.0	2
13	3.0	10	0.65	4
14	4.0	15	2.50	66
15	3.0	45	2.00	148
16	5.0	25	16.00	190
17	2.0	30	9.50	158
18	2.0	40	15.00	239
19	3.0	25	13.00	341
20	3.0	30	7.50	134
21	4.0	35	10.00	71
22	4.0	35	6.50	51
23	4.0	30	6.50	70
24	5.0	25	6.50	18
25	3.0	20	3.40	32

Data from Geldreich, Goodrich, and Clark.[79]

Investigative Areas

A safe, untreated, groundwater supply should contain less than ten total coliforms (environmental strains) and no fecal coliforms per 100 ml.[82-83] In this situation, soil is the only barrier to surface contamination. Such reliance

Table 9.4 Total Coliform to Fecal Coliform Relationships in Reservoirs

Reservoir	No. of samples	Coliforms per 100 ml[a] Total	Fecal	Fecal coliforms (%)
Lake Seymour	49	7.0	0.2	2.9
Hemlock Lake	363	64	4.1	6.4
Lake Coquintan	50	3.9	0.5	12.8
Lake Gaillard	58	16.8	2.2	13.1
Casitas	113	8.7	2.3	26.4
Bull Run	78	29	8.5	29.3

[a] Mean values.
Data from Geldreich, Goodrich, and Clark.[79]

Table 9.5 Indicator Correlations with *Salmonella* Occurrences in Ambient Waters

Clostridium per 100 ml	Total coliform per 100 ml	Fecal coliform per 100 ml	Fecal strep per 100 ml	*Salmonella* (% occurrences)
0	0	0	0	0
13	100	10	5	0
25	500	50	50	11
50	1,000	100	100	21
125	10,000	1,000	300	33
200	50,000	5,000	1,500	66
250	100,000	10,000	3,000	99
2,500	1 million	100,000	30,000	100
6,250	2.5 million	850,000	250,000	100

Data modified from Nemedi et al.[81]

on a natural barrier may be satisfactory in some situations, but there is growing evidence that disinfection can rarely be avoided because the water supply may become contaminated passing through a distribution system. Minimal barriers, be they soil protection of the aquifer or disinfection only for a high-quality surface water, will require careful review during the on-site sanitary survey. The major concern is a possible breakthrough of contaminants because the sole barrier to waterborne pathogens may fail under environmental stress. Treatment concerning both groundwater and surface water are addressed in part by the following issues in every system evaluation:

- Is the degree of soil protection or treatment adequate to continuously produce safe drinking water from the groundwater now and in the future?
- If there is a pretreatment of impounded raw source (copper sulfate, powdered carbon, or alum), is it being done effectively without creating other problems (toxicity, sludge build-up, or disinfection interference)?

- If only disinfection is used for a supply treating surface water, are the raw water characteristics (coliform, turbidity, chlorine demand, and color) within a continuously treatable range; will the disinfection C·T values employed provide protection from pathogenic protozoan cysts?
- Is there redundancy of equipment in all treatment processes to assure continuing treatment under all operating conditions (equipment malfunction and power outage)?
- Are treatment facilities protected against flooding, earthquakes, and sabotage?
- Are all treatment facilities and appurtenances in working order?
- Are there common walls between raw, partially treated, and finished water process basins? (If so, are the walls free of structural faults?)
- Are the appropriate, approved chemicals being stored and applied properly?
- Is there appropriate monitoring on all treatment processes to assure their optimal efficiency?
- Does the water utility employ an adequate number of personnel who are trained, certified, and capable of operating the system?
- Are there contingency plans to temporarily suspend raw water operations during a chemical spill into the source water?

Field Observations

Field observations from a study of the 25 sanitary survey reports on public water systems serving populations of 10,000 or less demonstrated the importance of a periodic outside review of treatment operations.[41] Many of these deficiencies were the result of inadequate treatment processes because of operational budget constraints and frequent turnover of water plant operators (often with little interest, training, or experience in water supply treatment). Any combination of these factors can contribute to producing of a water of unpredictable microbial quality and lead to increased occurrence of coliforms in the distribution system. The following statements are excerpts from these reports on deviations noted and corrective measures recommended for an effective response.

Treatment Conditions

Simple chlorination as applied, cannot be relied upon to provide a safe drinking water which is protected from *Giardia* cyst and *Cryptosporidium* oocyst occurrences. Recommend additional treatment (filtration) or the use of a groundwater source.

Chlorination as the only treatment of surface water may be inadequate to provide a safe water at all times because of algae problems and raw water quality deterioration during drought periods.

Install either automatic chlorine changeover devices or additional valves and piping to preclude interruption of chlorine and raw water passage; no standby chlorinator; water meter at chlorination station not registering — impossible to determine total water demand on system related to safe yield of two surface water sources; disinfection interruption when water is by-passed to make repairs on meters because the system is using a hydraulic chlorinator; need automatic chlorine residual recorder with alarm at police or fire station to alert attention since water treatment plant is visited only briefly each day; automatic chlorine residual recorder not working; cylinders were not changed, water leaked into one rotameter, chlorine leak detector not working; first customer no more than 200 ft from point of chlorination, providing insufficient contact time; chlorine feed is not proportional to flow and water is not metered; several chlorine interruptions in the past due to power failure and air binding; increase chlorine feed to achieve 0.2 mg/l free residual at ends of distribution system; should chlorinate continuously; apply chlorine continuously rather than time-pulsed at 15 second intervals.

Polyphosphate used at plant and no chlorine residual in solution feed tank. Phosphate is a nutrient and tank could become a breeding area for algae and bacteria. Need 10 mg/l chlorine residual in feed tank at all times. Chlorine residual must be maintained in all batches of sodium hexametaphosphate to preclude bacterial growth. Batch chlorinate with either liquid or granular hypochlorite.

Activated carbon slurry at rapid sand filter has been in storage tanks for 5 years. Bacterial colonization of carbon slurry may become source of biofilm released to distribution pipe network.

A sheen and water movement on the rapid sand filter is caused by an infestation of nematodes, cycopods and other aquatic organisms.

Raw water permitted to enter distribution system when storage tank is full which in turn shuts off treatment plant. Water continues to flow by gravity because there are a number of customer service lines between plant and tank — install small chlorinator at plant to operate when system shuts down.

Twenty families receive raw water from raw water impoundment — either provide them with safe water or individual deep well supplies.

Hatch to clear well mounted flush with floor would permit a spill of any liquid to enter clear-well — needs a brim around clear-well hatch.

Flow meter not working.

Several cross-connections to various chemical feed systems — use water feed through an air gap.

Underground filter similar to infiltration gallery and has been used for several years, now beginning to clog. Filter cannot be backwashed without excessive reconstruction.

Plant Safety

Are there adequate safety measures in all aspects of water treatment to protect employees?

No emergency power source — could result in air binding and loss of chlorine.

Chlorinator facility should be a designated secure area and accessible only to certified operators; chlorinator unplugged by some unknown person causing air binding of chlorine feed line and loss of chlorine solution.

No full or part-time certified operator. System is visited intermittently which does not provide assurance that treatment is providing a continuous safe supply.

Operator is part-time, not certified and also serves as town dog catcher.

Need new self-contained breathing apparatus to deal with chlorine leaks. Obsolete gas mask does not provide adequate protection.

Chlorination facility needs ventilation to exhaust chlorine gas from building; cylinders stored underground with inadequate ventilation; remove hay from chlorination facility; chlorine cylinders not secured by chain; chlorine leak detector not working; chlorine supply not in enclosed room.

DISTRIBUTION SYSTEM OPERATION

The greatest challenge to protecting supplies of drinking water can be found in the network of pipes that comprise the distribution system. It is in this aspect of public water supplies that there are the most opportunities for microbial contamination to occur. Protocols for pipe disinfection after repairs or for new water lines must be effective to insure external microbial contaminants of sanitary significance are inactivated before there is a passage of drinking water. Cross-connection prevention is complicated by numerous attachments associated with service lines, low water pressure problems in the pipe network, and interconnections to other water systems that could also have water quality problems.

Finished water storage is necessary because water treatment takes time and a reserve of safe water supply is necessary to satisfy peak consumer demand and fire suppression around the clock. Years ago it was considered cost-effective to build open reservoirs for finished water supply and protect the water quality by limiting human access to these water impoundments. Overt contamination of any open reservoir cannot be underestimated because there have been reports of vandalism and threats of sabotage from time to time. In service areas that are hilly, elevated storage tanks and standpipes were constructed to assist in resolving problems of hydraulics involving surges of pressure in pipelines in addition to their function as a depository of drinking water. Many of these elevated water storage supplies were not adequately covered either because of cost-saving measures or the general belief that they were generally inaccessible to the public. Experience now shows that water stored in elevated tanks is often exposed to contamination by birds, dust fallout, and other airborne contaminants through poorly maintained screen vents necessary for air pressure exchange. For these reasons, all finished water reservoirs should be covered as an important barrier to water quality degradation.

Perhaps the most poorly understood activity of distribution system management for some utilities is the selection of appropriate sampling sites to characterize water quality reaching the population served. Often the selection of sampling sites was influenced more by convenience than by the desirability to characterize water quality differences in the system. In this regard, the sanitary survey can provide an important service that identifies weaknesses in a monitoring plan for water supply distribution.

Investigative Areas

Investigative inquiries as part of the distribution system survey should be directed to the following areas:

- Are storage facilities for finished water constructed, covered, and maintained so that no sources of contamination can enter via vents, drain lines, overflows, or flooding?
- Are storage facilities protected from vandalism, and are they frequently inspected for evidence of such?
- Are all instrumentation and appurtenances in the system in good operating condition?
- Is the water system capable of providing an adequate supply of water at all times with adequate pressure in all parts of the distribution system?
- Is the direct connection of blowoffs to sewer lines permitted?
- Are all distribution system components (valves, hydrants, etc.) and pumping stations constructed, maintained, exercised periodically, and operated properly?

Field Observations

Field observations from a review of sanitary and engineering reports reveal that some small utilities have minimized management of the distribution system to a point of careless neglect. As a consequence, it is not surprising that many of these systems have frequent coliform occurrences in samples collected in the distribution system.

Finished Water Storage Deficiencies

Uncovered reservoir, clay and rock lined, not fenced, vegetation around shore line resulting in floating debris, storm runoff from adjacent hill can flow in (needs a better diversion ditch) and no chlorination for water leaving reservoir.

Standpipe is an uncovered distribution storage reservoir with a ladder easily accessible to vandals. No fence around the tank.

Reservoir vent on top of tank dome has no screen to prevent birds from entering. Vent on top of tank dome covered by a wooden box easily removed, leaving tank subject to vandalism and sabotage. Tank walls have holes punched in them to serve as an air vent, allowing rain water to enter. Overflow pipe from underground tank is connected directly to a storm drain. Overflow pipes should be brought to an elevation of 12 to 24 inches above ground level (air break) providing above ground discharge to sewer to avoid this type of cross-connection. Hole in one of the storage tanks permits airborne contaminants, birds and other small animals to enter.

No distribution storage reservoir to serve as a water supply reserve in event of a treatment failure.

Low water levels in the storage reservoir during periods of high demand.

Mice on top of side walls of reservoir covered by a wooden, framed roof which has several small openings.

Above ground steel storage tank easily accessible by vandals.

Steel tanks for water storage have no overflow vent for discharge.

Vent design containing activated carbon filter should be redesigned with 180° downward opening and screen in place.

Vandals could enter underground concrete storage tank through hatch which has broken lock; no air exchange vents on tank, requiring tank to be vented through overflow pipe.

Gaskets removed on two hatches over two compartments to serve as air vents provides a pathway for small animals to enter.

Manholes are not locked and improperly constructed on underground storage facility, permitting surface runoff to enter.

Standpipe has a ladder which is very accessible to vandals as evidenced by graffiti on standpipe.

Small concrete storage reservoir has an opening around the inlet pipe that could permit small animals to enter. Spaces between roof and concrete walls should be sealed properly. Overflow outlet should be screened and any vegetation encroaching on reservoir removed. No emerging power for small booster pumping station that brings water to reservoir. Decaying stairs present a safety hazard.

Pipe Network Integrity

Physical separation of an abandoned reservoir desirable rather than a valve turn off because single valves have been known to leak.

Areas of low pressure known to go to zero on hills around city creates a concern for back siphonage into the system through cross-connection. Low pressure due to leak in distribution system, tuberculation or higher demands for water by users.

No chlorine residual in certain areas of the distribution system.

Heavy tuberculation from corrosion in a number of water mains that need cleaning and lining.

No continuing cross-connection program. Priority should be given to areas of high density coliform occurrences. Establish a program in conjunction with town plumbing or health department to eliminate cross-connections in pipe network. Several outside hose connections found without vacuum breakers. Direct connection from a water main to a catch basin for flushing the main creates a potential cross-connection. Replace the direct connection with a hydrant located near end of line for flushing purposes. Direct connection between water mains and sewer manholes for purposes of flushing should be discontinued or an air gap incorporated. A submerged blowoff of the water mains entering a small brook is a cross-connection. Install a hydrant at this location. The interconnection to another water system is physically broken. Need cross-connection survey, particularly at hospitals and other locations that have alternate water supplies. Cross-connection in potassium permanganate feeder with finished water used at drinking fountain within the plant. Conduct a cross-connection survey to determine if power outages cause back siphonage to contaminate water supply.

Problem with several hydrants freezing in winter to which ethylene glycol is added. This is a toxic chemical used for auto antifreeze. If necessary, use non-toxic propylene glycol or plug drain on hydrants and pump out hydrants periodically to prevent freezing. Flush any hydrants with propylene glycol thoroughly in spring to prevent bacterial regrowth. Hydrants protected from freezing in winter by the addition of methanol to barrel of hydrant. Methanol is toxic, suggested use of glycerine or propylene glycol which must be flushed out in spring to prevent bacterial growth. Transmission line should be covered and insulated to prevent freezing.

No regular flushing program to remove corrosion products, sediment and static water from deadend areas. Should also flush hydrants in central part of the system. Numerous deadends in system. Not all deadends are flushed because of a lack of hydrants at these points. Either eliminate deadends at these locations or install hydrants in these areas. Need to develop a semi-annual flushing program with particular emphasis on deadends.

Need to utilize an effective disinfectant policy on line repairs and new mains. Suggest the protocol recommended by the AWWA Standards. Chlorine tablets not glued to top of pipe nor any measurement made on chlorine residual after 24 hour contact time; need a satisfactory bacteriological sample report before placing new mains in service. New water mains are disinfected with granular hypochlorite placed in each pipe section. This is an unacceptable procedure.

No comprehensive map of sources, transmission line or distribution network. Water system lost their only service record book on pipelines in the network. Distribution system completely rebuilt within last few years but no accurate map of all lines. Poor records on location of service lines.

REGULATION COMPLIANCE MONITORING

Another essential ingredient of the comprehensive sanitary survey is to search for possible causes of a coliform noncompliance based on monitoring information or to verify that safe water is being produced in small systems that cannot meet all of the routine bacteriological testing requirements. Perhaps the greatest benefit lies in the opportunity afforded the small system operator (through review of operations and technical assistance) to communicate with specialists in water supply on ways to improve the water treatment system.

Strategy

The strategy for establishing the microbial quality of a public water supply resides in a the carefully devised monitoring plan. This monitoring plan should include gathering data on changes in source water quality and treatment effectiveness and protection of the water quality during distribution. Furthermore, appropriate criteria should be selected that adequately characterize long-term trends in water quality changes and others that give early warning of treatment failure or breaches in public health safety measures.

Compliance monitoring is the heart of the public safety issue. These official sampling sites are carefully chosen by the utility to be representative of normal water quality conditions throughout the pipe network. The minimum number of samples to be examined each month is determined by federal regulations related to the population served. For example, New York City is required to monitor 480 samples based on the population served (8,080,000) while Gideon, MO (population 1104) must submit a minimum of only two samples per month. Because sample site location becomes critical, input is often provided by the state water authority. Ultimately, the state water authority reviews the plan and the federal regional water supply representative may also critique the monitoring strategy during periodic state program reviews.

As a part of the compliance requirement, any total coliform occurrence must be immediately followed within 24 h by a repeat sampling at the same site. In addition, samples must also be taken within five service connections above and below the site of positive coliform occurrence. Repeat sampling and analysis for total coliforms continues on a daily basis until two consecutive sets of repeat samples are negative. When fecal coliform (E. coli) are detected at a compliance site, an intensive check sampling should be undertaken immediately from the site of a positive sample in an effort to determine the existence of an intermittent cross-connection with wastewater in the immediate vicinity.

Surveillance sample sites are not permanent and are not included in the official compliance monitoring requirement. These sites are chosen by the utility to explore a potential trouble spot due to a faulty valve, defective regulator, or line repairs or to identify zones of stagnate water that need to be flushed. Such special investigative sampling is more flexible than compliance monitoring and provides additional information that could predict an undesirable trend in water quality unless prompt corrective action is taken. These data must be promptly analyzed, and an appropriate response initiated whenever the values obtained are beyond the limits of acceptability.

Investigative Areas

Investigative areas to consider in the evaluation of a monitoring program involve the sampling site selection, frequency of sample collection, criteria examined, and response taken when coliforms are reported. Some questions to be explored include:

- Are all drinking water bacteriological data and disinfectant residuals reported being gathered from representative sites on the distribution system, including a site near the first customer?
- Is there a daily database on raw water turbidity for systems using minimal treatment of surface water, infiltration galleries, and springs?
- Are turbidity samples being taken at representative entry points to the distribution system?
- If the water system has been allowed to use 1 to 5 NTU as an MCL, is the system continuing to demonstrate that the higher level of turbidity does not interfere with disinfection, prevent maintenance of an effective disinfectant throughout the system, or mask coliform detection?
- Are the sampling sites for bacteriological samples representative of all of the various conditions within the distribution system?
- Is the bacteriological monitoring frequency adequate?
- Does the number of bacteriological samples per month equal or exceed the minimum number specified for population served?
- If chlorine residual monitoring is being substituted for coliform testing on a percent of samples required per month, are the following conditions met?
  - a minimum of 0.2 mg/l free chlorine residual maintained throughout the distribution system
  - sites selected do not have a history of coliform occurrence
- Are turbidimeters being calibrated with the appropriate standard (formazin or styrene divinyl benzene polymer)?
- If the water system has failed to comply with the coliform MCL, has the appropriate public notice been given?
- Has the utility maintained the appropriate records such as laboratory data used to determine compliance with MCLs, sanitary survey reports, and action responses to coliform occurrences and turbidity excesses?
- What is the status of corrosion prevention measures and Langlier index in the water supply?

Field Observations

Field observations recorded from sanitary surveys frequently cite inadequate monitoring or a lack of proper response to a confirmed violation with an MCL requirement. Water systems with an inadequate monitoring program are often small water utilities that fail to submit the proper number of bacteriological samples each month. Violations of the maximum contaminant level for coliforms or turbidity are serious because they suggest a possible failure in water treatment or contamination of water supply in distribution. Repeated occurrences of MCL noncompliance triggers the need for a special sanitary survey to discover cause and recommend a prompt action plan to correct the violation. At other times, the scheduled sanitary survey is routine and becomes a verified record of continued success in water treatment and its protection during distribution. A survey done because of noncompliance issues involving microbial quality most often focuses on plant effluent turbidity, chlorine C·T or residuals, monitoring frequency, and coliform frequency of occurrence. The following field reports illustrate the most frequently reported deficiencies:

Turbidity

No turbidity monitoring on raw or finished water. Must measure turbidity of raw source on a daily basis and periodically for other parameters to characterize raw water quality. Turbidity samples are not collected at entry point into the distribution system. Turbidity samples taken at the Town Hall or at the center of the system are not representative of the entry point.

Almost continuously in violation of turbidity MCL. Two surface supplies with turbidity ranging from 0.2 to 0.4 NTU during dry periods but after high intensity rainfall and subsequent runoff over water-shed, NTU values could go beyond 1 NTU for a system using disinfection as only treatment.

Turbidimeter is not working properly.

Chlorine Residual

No chlorine residual detectable in certain areas of the distribution system.

Measure free chlorine residual at all bacteriological sampling points.

New water mains must be disinfected and tested for chlorine residual and coliforms after 24 h. No bacteriological samples are taken to assure the effectiveness of pipeline disinfection.

Bacteriological Monitoring

Location of sampling points on distribution system needs to be reevaluated; monitoring sites are not representative of the distribution system. Bacteriological sampling sites must include a site at the first customer which receives a minimal amount of chlorine contact time. If coliform bacteria are found at first customer location, move chlorination point to provide longer contact time. Need a bacteriological sampling point after the high service reservoir. Another sampling site should be at the end of the pipe network. One sample per month is taken as required for the size population served but because of long residence time for water in the system and large area coverage (17 miles) by the pipe network, this does not adequately characterize water quality in distribution.

Insufficient bacteriological samples taken to meet regulation requirements for population served each month over an eleven month period. Failed to monitor for coliforms for four of first seven months.

Four consecutive coliform MCL violations during 1982 to 1983 (Oct., Nov., Dec., Jan.). No check samples taken when high density coliform counts occurred (2 to over 200 total coliforms per 100 ml).

IN RETROSPECT

A sanitary survey is a critical tool in the investigation of a physical, chemical, or microbial problem with water supply. Key elements of the survey include (1) watershed protection and management, (2) applied treatment, (3) plant operation and review of facilities, equipment, and staff qualifications, (4) maintenance of the distribution system, and (5) monitoring strategy to insure a continuous production and delivery of a safe water supply to all customers. Each public water system is different in some aspect, so it is important to focus on those unique features that have the most potential for degrading water quality.

Integrating laboratory information with the on-site survey of operational practices often provides the clues to the problem (fecal contamination or biofilm growth) and suggests actions to take in restoring water quality. Only through periodic on-site surveys and effective routine monitoring of water quality can there be adequate assurance of the continued availability of a safe public water supply.

REFERENCES

1. Kabler, P. and H.F. Clark. 1960. Coliform Group and Fecal Coliform Organisms as Indicators of Pollution in Drinking Water. *Jour. Amer. Water Works Assoc.*, 53:1577-1579.
2. Walton, G. 1959. Plant Efficiencies May Permit Increased Raw Coliform Loadings. *Water Works Eng.*, 112:289.
3. Office of Drinking Water. 1989. Revised Drinking Water Regulations: Coliform Rule. U.S. Environmental Protection Agency, Washington, D.C.
4. AWWA Committee on Bacteriological Sampling Frequency in Distribution Systems. 1985. Committee Report: Current Practice in Bacteriological Sampling. *Jour. Amer. Water Works Assoc.*, 77:75-81.
5. Thomas, S.B., C.A. Scarlett, W.A. Cuthbert, et al. 1954. The Effect of Flaming Taps before Sampling of the Bacteriological Examination of Farm Water Supplies. *Jour. Appl. Bacteriol.*, 17:175-181.
6. McCabe, L.J. 1969. Trace Metals Content of Drinking Water from a Large System. Presented at Symposium on Water Quality in Distribution Systems. Amer. Chem. Soc. National Meeting, Minneapolis, MN.

7. Walton, G. 1970. Personal communication.
8. Geldreich, E.E. 1975. Handbook for Evaluating Water Bacteriological Laboratories. EPA-670/9-75-006. U.S. Environmental Protection Agency, Cincinnati, OH.
9. Water Supply Quality Assurance Work Group. 1978. Manual for the Interim Certification of Laboratories Involved in Analyzing Public Drinking Water Supplies. EPA 600/8-78-008. U.S. Environmental Protection Agency, Washington, D.C.
10. Geldreich, E.E. 1967. Status of Bacteriological Procedures Used by State and Municipal Laboratories for Potable Water Examination. *Health Lab. Sci.*, 4:9-16.
11. Geldreich, E.E. 1971. Application of Bacteriological Data in Potable Water Surveillance. *Jour. Amer. Water Works Assoc.*, 63:225-229.
12. Geldreich, E.E. and H. Kennedy. 1981. The Cost of Microbiological Monitoring. In: *Bacterial Indicators of Pollution*. W.O. Pipes, Ed. CRC Press, Boca Raton, FL.
13. Geldreich, E.E. and Rice, E.W. 1987. Occurrence, Significance, and Detection of *Klebsiella* in Water Systems. *Jour. Amer. Water Works Assoc.*, 79:74-80.
14. Bagley, S.T. and R.J. Seidler. 1977. Significance of Fecal Coliform Positive *Klebsiella*. *Appl. Environ. Microbiol.*, 33:1141-1148.
15. Bagley, S.T. 1985. Habitat Association of *Klebsiella* Species. *Infect. Contr.*, 6:52-58.
16. Thom, R.T. 1970. *Klebsiella* in Feces. *Lancet*, 2:103-107.
17. Davis, T.J. and J.M. Matsen. 1974. Prevalence and Characteristics of *Klebsiella* species. Relation to Association with a Hospital Environment. *Jour. Infect. Dis.*, 130:402-405.
18. Cooke, E.M., R. Pool, J.C. Brayson, et al. 1979. Further Studies on the Sources of *Klebsiella aerogenes* in Hospital Patients. *Jour. Hyg.*, 83:391-395.
19. Edmondson, A.E., E.M. Cooke, A.P.D. Wilcock, and R. Shinebaum. 1980. A Comparison of the Properties of *Klebsiella* Strains Isolated from Different Sources. *Jour. Med. Microbiol.*, 13:541-550.
20. Jarvis, W.R., Van P. Munn, A.K. Highsmith, D.H. Clulver, and J.M. Hughes. 1985. The Epidemiology of Nosocomial Infections Caused by *Klebsiella pneumoniae*. *Infect. Contr.*, 6:68-74.
21. Duncan, D.W. and W.E. Razzell. 1972. *Klebsiella* Biotypes Among Coliforms Isolated from Forest Environments and Farm Produce. *Appl. Microbiol.*, 24:933-938.
22. Knittel, M.D., R.J. Seidler, and L.M. Cabe. 1977. Colonization of the Botanical Environment by *Klebsiella* Isolates of Pathogenic Origin. *Appl Environ. Microbiol.*, 34:557-563.
23. Bagley, S.T., R.J. Seidler, and H.W. Talbot, Jr. 1978. Isolation of *Klebsiella* from within Living Wood. *Appl. Environ. Microbiol.*, 36:178-185.
24. Talbot, H.W., Jr. and R.J. Seidler. 1979. Cyclitol Utilization Associated with the Presence of *Klebsiella* in Botanical Environments. *Appl. Environ. Microbiol.*, 37:909-915.
25. Brown, C. and R.J. Seidler. 1973. Potential Pathogens in the Environment: *Klebsiella pneumoniae*, a Taxonomic and Ecological Enigma. *Appl. Environ. Microbiol.*, 25:900-904.

26. Miller, P.D., K.E. Ericson, and P.B. Mash. 1982. Bacteria on Closed-Boll and Commercially Harvested Cotton. *Appl. Environ. Microbiol.*, 44:255- 362.
27. Huntley, R.E., A.C. Jones, and V.J. Cabelli. 1966. *Klebsiella* Densities in Waters Receiving Wood Pulp Effluent. *Jour. Water Poll. Contr. Fed.*, 48:1766-1771.
28. Knowles, R., R. Newfield, and S. Simpson. 1974. Acetylene Reduction (Nitrogen Fixation) by Pulp and Paper Mill Effluents and by *Klebsiella* Isolated from Effluents and Environmental Situations. *Appl. Microbiol.*, 28:608-613.
29. Caplenas, N.R., M.S. Kanarek, and A.P. Dufour. 1981. Source and Extent of *Klebsiella pneumoniae* in the Paper Industry. *Appl. Environ. Microbiol.*, 42:779-785.
30. Hendry, G.S., S. Janhurst, and G. Horsnell. 1982. Some Effects of Pulp and Paper Waste Water on Microbiological Water Quality of a River. *Water Res.*, 16:1291-1296.
31. Niemala, W.I. and P. Vaatanen. 1982. Survival in Lake Water of *Klebsiella pneumoniae* Discharged by a Paper Mill. *Appl. Environ. Microbiol.*, 44:264-269.
32. Herman, D.L. 1972. Experiences with Coliform and Enteric Organisms Isolated from Industrial Wastes. In: Proceedings: Seminar on the Significance of Fecal Coliforms in Industrial Wastes. R.H. Bordner and B.J. Carroll, Eds. Office of Enforcement, National Field Investigation Center, U.S. Environmental Protection Agency, Denver, CO.
33. Smith, D. 1988. Personal communication.
34. Geldreich, E.E. 1990. Microbiological Quality in Distribution Systems. In: *Water Quality and Treatment*. 4th ed. American Water Works Association, Denver, CO. pp. 1113-1158.
35. Olson, B.H. 1982. Assessment and Implications of Bacterial Regrowth in Water Distribution Systems. EPA-600/52-82-072. U.S. Environmental Protection Agency, Cincinnati, OH.
36. LeChevallier, M.W., T.M. Babcock, and R.G. Lee. 1987. Examination and Characterization of Distribution System Biofilms. *Appl. Environ. Microbiol.*, 53:2714-2724.
37. McFeters, G.A., J.S. Kippin, and M.W. LeChevallier. 1986. Injured Coliforms in Drinking Water. *Appl. Environ. Microbiol.*, 51:1-5.
38. U.S. Environmental Protection Agency. 1989. Drinking Water: National Primary Drinking Water Regulations; Total Coliforms (Including Fecal Coliform and *E. coli*). Final Rule. 40CFR Parts 141 and 142. *Fed. Reg.*, 54(124):27544-27568.
39. Logsdon, G.S. 1990. Treatment Plant Evaluation During a Waterborne Outbreak. In: Methods for the Investigation and Prevention of Waterborne Disease Outbreaks. G.F. Craun, Ed. U.S. Environmental Protection Agency, EPA/600/1-90/005.
40. Karalekas, P.C., Jr., J.K. Reilly, and J.J. Healey. 1986. Sanitary Survey Techniques. Presented at the Amer. Water Works Assoc. Water Quality Technol. Conf., Portland, OR. Nov. 19, 1986.
41. Karalekas, P.C., Jr. 1984. A Compendium of Sanitary Survey Reports. Water Supply Branch, U.S. Environmental Protection Agency, Boston, MA.
42. Public Health Service. 1963. Statistical Summary of Municipal Water Facilities in United States, PHS Pub 1039, Washington, D.C.

43. Mackenthun, K.M. and W.M. Ingram. 1967. Biologically Associated Problems in Freshwater Environments. Federal Water Pollution Control Administration, U.S. Department of the Interior, Washington, D.C.

44. Geldreich, E.E. 1985. Perception of Lake Water Resource Management. Proc. Water Qual. Tech. Conf., Houston, TX. pp. 677-685

45. Pluntze, J.C. 1984. The Need for Filtration of Surface Water Supplies, Viewpoint. *Jour. Amer. Water Works Assoc.*, 76(12):11, 84.

46. Hibler, C.P., K. MacLead, and D.O. Lyman. 1975. Giardiasis — In Residence of Rome, N.Y. and in U.S. Travelers to the Soviet Union. *Morbid. Mortal. Week. Rep.*, 24:366, 371.

47. Allard, J. et al. 1977. Waterborne Giardiasis Outbreaks — Washington, New Hampshire. *Morbid. Mortal. Week. Rep.*, 26:169-170, 175.

48. Kirner, J.C., J.D. Littler, and L. Angelo. 1978. A Waterborne Outbreak of Giardiasis in Camas, Washington. *Jour. Amer. Water Works Assoc.*, 70:35-40.

49. U.S. Environmental Protection Agency. 1979. Waterborne Transmission of Giardiasis. W. Jakubowski and J.C. Hoff, Eds., EPA-600/9-79-001, Cincinnati, OH.

50. Jakubowski, W., et al. 1985. Waterborne *Giardia*: It's Enough to Make You Sick, Roundtable. *Jour. AWWA*, 77(2):14,19,26, and 84.

51. Lin, S.D. 1985. *Giardia lamblia* and Water Supply. *Jour. Amer. Water Works Assoc.*, 77:40-47.

52. Alter, A.J. 1954. Appearance of Intestinal Wastes in Surface Water Supplies at Ketchikan, Alaska. Proc. 5th Alaska Sci. Conf. AAAS. Anchorage, AK.

53. Fennel, H., D.B. James, and J. Morris. 1974. Pollution of a Storage Reservoir by Roosting Gulls. *Jour. Soc. Water Treat. Exam.*, 23(Pt. 1):5-24.

54. Anon. 1954. Ketchikan Laboratory Studies Disclose Gulls are Implicated in Disease Spread. *Alaska's Health*, 11:1-2.

55. Clark, R.M., E.E. Geldreich, K.R. Fox, E.W. Rice, C.H. Johnson, J.A. Goodrich, J.A. Barnick, F. Abdesaken, J.E. Hill, and F.J. Angulo. In press. A Waterborne *Salmonella* serovar *typhimurium* Outbreak in Gideon, Missouri: Results from a Field Investigation. *Inter. Jour. Envir. Health Res.*

56. McCabe, L.J., J.M. Symons, R.D. Lee, and G.G. Robeck. 1970. Survey of Community Water Supply Systems. *Jour. Amer. Water Works Assoc.*, 62:670-687.

57. Craun, G.F. 1985. An Overview of Statistics on Acute and Chronic Water Contamination Problems. In: *Fourth Domestic Water Quality Symposium: Point-of-Use Treatment and Its Implications*. Water Quality Assoc., Lisle, IL. pp. 5-15.

58. Pluntze, J.C. 1984. The Need for Filtration of Surface Water Supplies, Viewpoint. *Jour. Amer. Water Works Assoc.*, 76:11, 84.

59. Health and Welfare, Canada. 1981. Possible Waterborne *Campylobacter* Outbreak — British Columbia. *Can. Dis. Week. Rep.*, 7:223,226-227.

60. Vogt, R.L., H.E. Sours, T. Barrett, et al. 1981. *Campylobacter enteritis* Associated with Contaminated Water. *Ann. Intern. Med.*, 96:292-296.

61. Blaser, M.J., D.N. Taylor, and R.A. Feldman. 1983. Epidemiology of *Campylobacter jejuni* Infections. *Epidemiol. Rev.*, 5:157-176.

62. Taylor, D.N., K.T. McDermott, J.R. Little, et al. 1983. *Campylobacter enteritis* Associated with Drinking Untreated Water in Back Country Areas of the Rocky Mountains. *Ann. Intern. Med.*, 99:38-40.

63. Blaser, M.M. and H.J. Cody. 1986. Methods for Isolating *Campylobacter jejuni* from Low Turbidity Water. *Appl. Environ. Microbiol.*, 51:312- 315.
64. Greenberg, A.E. and J. Jongerth. 1966. Salmonellosis in riverside, California. *Jour. Amer. Water Works Assoc.*, 58:1145-1150.
65. Schroeder, S.A., J.R. Caldwell, T.M. Vernon, P.C. White, S.I. Granger, and J.V. Bennett. 1968. A Waterborne Outbreak of Gastroenteritis in Adults Associated with *Escherichia coli*. *Lancet*, 6:737-740.
66. Lassen, J. 1972. *Yersinia enterocolitica* in Drinking Water. *Scand. Jour. Infect. Dis.*, 4:125-127.
67. Centers for Disease Control. 1973. Typhoid Fever — Florida. *Morbid, Mortal. Week Rep.* 22:77-78,85.
68. Centers for Disease Control. 1974. Acute Gastrointestinal Illness — Florida. *Morbid. Mortal. Week. Rep.* 23:134.
69. Centers for Disease Control. 1980. Waterborne Disease Outbreaks in the United States — 1978. *Morbid. Mortal. Week. Rep.*, 29:46-48.
70. Evison, L.M. and A. James. 1973. A Comparison of the Distribution of Intestinal Bacteria in British and East African Water Sources. *Jour. Appl. Bacteriol.*, 36:109-118.
71. Lindel, S.S. and P. Quinn. 1973. *Shigella sonnei* Isolated from Well Water. *Appl. Microbiol.*, 26:424-425.
72. Woodward, W.E., N. Hirschhorn, R.B. Sack, R.H. Case, I. Brownlee, G.H. Chickadonz, L.K. Evans, R.N. Shephard, and R.C. Woodward. 1974. Acute Diarrhea on an Apache Indian Reservation, *Amer. Jour. Epidemiol.*, 99:281-290.
73. Dragas, A-Z, and M. Tradnik. 1975. Is the Examination of Drinkable Water and Swimming Pool on Presence of Entero-pathogenic *E. coli* Necessary? *Zentral. Bakt. Hyg. I, Abt. Orig.B*, 160:60-64.
74. Highsmith, A.K., J.D. Feeley, P. Shalig, J.S. Wells, and B.T. Wood. 1977. Isolation of *Yersinia enterocolitica* from Well Water and Growth in Distilled Water. *Appl. Environ. Microbiol.*, 34:745-750.
75. Schieman, D.A. 1978. Isolation of from Surface *Yersinia enterocolitica* and Well Waters in Ontario. *Can. Jour. Microbiol.*, 24:1048-1052.
76. Mentzing, L.A. 1981. Waterborne Outbreaks of *Campylobacter enteritis* in Central Sweden. *Lancet*, 8:352-354.
77. Vander Velde, T.L. and W.M. Mack. 1973. Poliovirus in Water Supply. *Jour. Amer. Water Works Assoc.*, 65:345-348.
78. Water Supply Programs Division. 1971. Manual for Evaluating Public Drinking Water Supplies. PB245-006. U.S. Environmental Protection Agency, Washington, D.C.
79. Geldreich, E.E., J.A. Goodrich, and R.M. Clark. 1988. Characterizing Raw Surface Water Amenable to Minimal Water Supply Treatment. AWWA Annual Conference, Orlando, FL, June 19-23.
80. Geldreich, E.E. 1970. Salmonellae in Fresh Water. *Proc. National Specialty Conference on Disinfection*. Univ. Mass., Amherst. Amer. Soc. Civil Eng., New York. pp. 495-514.
81. Nemedi, L., T. Borbala, P. Gizella, and R. Andrasne 1984. Hydrobiological Evaluation of Water Samples from "Standing Waters" (Lakes, Reservoirs and Ponds) near Budapest. *Budapest Kozegeszsegugy*, (2):40-48.

82. Allen, M.J. and E.E. Geldreich. 1975. Bacteriological Criteria for Groundwater Quality. *Ground Water,* 13:45-52.
83. World Health Organization. 1984. Guidelines for Drinking Water Quality. Volume 1. Recommendations. World Health Organization, Geneva, Switzerland.

CHAPTER 10

Public Awareness of Water Quality Problems

CONTENTS

INTRODUCTION

Everyone must drink water to sustain life so there is a genuine public concern about possible health risks in the public water supply. While it is easy for consumers to recognize poor-quality water that is turbid, off-color, odorous, and unpleasant tasting, it is impossible for them to judge the pathogenic dangers in otherwise acceptable water. This responsibility is delegated to public health authorities and water treatment professionals. Public trust in this mandate can be shaken by news reports on waterborne health hazards or by repeated public notifications of local noncompliance with federal drinking water regulations.

In recent years, regulations have placed greater emphasis on the public's right to know about risks associated with water quality. While most people understand that life is not risk-free, they do not normally associate their drinking water with risk to their well-being, and they do not understand much about water treatment and delivery to the residential tap. They are, therefore, confused and disturbed by sudden boil water notices, and by offhand statements on maximum contaminant levels and on risk assessments. A small but growing segment of the public is convinced that many health risks in community water supplies are nonreversible so that the only option available is bottled water or point-of-use water supply treatment in the home. To counter this growing concern, water utilities professionals need to sharpen their skills at informing the public on water quality issues in a way that instills support and confidence.

PUBLIC PERCEPTIONS OF HEALTH RISKS
IN WATER SUPPLY

Various polls taken in recent years reveal some insights into public thinking on water quality issues, and, most importantly, on the sources of their information. In general, Americans place a higher priority on environmental protection than on economic growth. The public relies overwhelmingly on the mass media as the source of their environmental information from which are formed attitudes on water supply and health risks.

PUBLIC ATTITUDES

Based on a consumer attitude survey dealing with water quality issues conducted by the American Water Works Association Research Foundation,[1] most people believe their local tap water meets or exceeds federal drinking water standards. When asked how this opinion was determined, respondents cited information found in newspaper stories, TV news segments, and radio interviews. At the same time, respondents felt that they had little influence or control over the treatment of their water supply.

Environmental activists in this survey appear to be inherently less trusting than the general public on all environmental issues, including that of local water quality. These special-interest groups perceive the existing standards as not strict enough; rather, they think they should be greatly expanded to include other aspects that could pose an unacceptable health risk.

The water utility industry in this survey thinks that the public sees water quality in a direct relationship to their personal state of health. This concern is best seen among young mothers with newborns and young children because of newspaper and television reports of toxic discharges into surface waters used for public water supply source. Water utility operators also share the same concern but from the perspective of achieving continual effective treatment of raw source waters that are gradually deteriorating over time from new pollutants entering the watershed.

Public officials in this survey optimistically perceive an improvement in the quality of the environment over the last decade, perhaps best represented by the period when they were in office. Their focus on public water supplies places more emphasis on keeping the price of water low and improving the aesthetic considerations of taste and odor. This attitude appears to reflect their sensitivity to their voting constituents.

It is apparent that different segments of the public have differing perspectives of water quality problems and of the regulatory and enforcement process. However, surveys show that nearly all Americans are influenced significantly by mass media coverage of these issues, and an examination of some of that coverage demonstrates that it is often incomplete and unrealistic.

REALITY CHECK ON REGULATION DEVELOPMENT

What are some of the realities about drinking water quality that the public should be aware of? First, the public should be aware that the selection of criteria and standards applied to drinking water quality is not done in a capricious manner. The regulation process begins with early alerts to some new contaminant in water supply. These may come from several directions:

industrial disasters involving new chemical exposures, accidental spills of exotic materials from truck or rail transport, waterborne disease spread through international travel, or discovery of new waterborne agents and awareness of pathogen adaptation to the aquatic ecosystem. Such information may be gathered from incidents in this country or other nations, or through the World Health Organization and other international agencies participating in a global watch over environmental issues.

Investigation of these events and the exploratory study of anticipated threats to water treatment barriers and distribution systems are important first responses in applied research. Alarming discoveries may then lead to exploratory research within the U.S. Environmental Protection Agency or, through federal research grants, by universities, foundations, and private contractors. Once there is evidence of a health threat to water supplies, the best available technology is sought through field evaluation of treatment processes either in pilot-plant or full-scale operation.

Consumers also need to know that parallel to these investigations is a need for a risk assessment to ascertain who in the population may be compromised. Most often those at greatest risk include infants, senior citizens, and special medical care groups (AIDS victims, patients with skin graphs, and individuals on kidney dialysis). Most of these illnesses involve intestinal diseases, respiratory complications, and, to a lesser extent, skin infections. Infection may be acquired from short-term exposures to contaminated water supply — less than 1 d to as long as 30 d — before colonization of the pathogen results in disease. Adverse health effects from chemical contaminants (THMs, lead, arsenic, pesticides, herbicides, radioactivity, etc.) may require continued exposures over 10 to 40 years.

Consumers also need to know that following completion of a health risk assessment of the agent, the best available treatment is selected on the basis of cost of treatment compared to the benefit in avoiding one death among 10,000 or 100,000 consumers. At the same time, appropriate monitoring criteria are chosen and limits in water established within the sensitivity of detection methods. At that point, draft regulations are prepared with the assistance of agency scientists, university researchers, and water supply professionals. The "straw man" rule (initial draft of a proposed rule) is then discussed in regional public hearings, revised, and released for a public comment period. After the public comment period is closed, input is carefully reviewed and negotiated changes are made to finalize the rules. The new regulation is then published in the *Federal Register* with a date for national implementation. Thereafter, local utilities may obtain technical assistance related to rule changes and compliance issues from the federal and state agencies involved, and from various consulting engineering firms. Technical assistance areas of microbial interest include watershed management, filtration, C·T values for disinfection effectiveness, coliform biofilm occurrences, fecal coliform contamination, and waterborne outbreak investigations.

MASS MEDIA AND PUBLIC AWARENESS

There are a variety of reasons why some members of the public have lost confidence in their municipal water supply and mass media play a role in most of them. News stories on boil water orders, waterborne outbreaks in the community, deficiencies in water treatment, infrastructural problems in distribution systems, and reported spills of pollutants upstream of the intake, create an uneasiness about the safety of tap water. Persistent media complaints about the taste, odor, color, and hardness of local water have caused some people to seek alternate solutions through the installation of point-of-use or point-of-entry treatment devices. Water softeners have a long history of use, especially in communities that have a hard water problem that aggravates the laundering of clothes. Water supply problems will occur and the media will continue to make the most of them; however, much of the anxiety and distrust of public water supply could be avoided by better communication between water authorities, the press and public — another reason for a trained public relations person on staff.

PRINT MEDIA

Newspapers and other periodicals record in print those events that directly impact people's lives, or those that may be of interest to others who relate somehow to these events.[2] News is occurring all the time and each reported event competes for time and space in every edition. This public information mechanism has a tremendous impact, good or bad, on the actions of citizens, industry, and government.

From the editors viewpoint, top priority in the paper will be given to news that sells newspapers. In the case of a major story, the editor may place a reporter or a team of reporters on special assignment to investigate a particular issue in depth. Such was the case for the *Milwaukee Journal* during the 1993 *Cryptosporidium* outbreak in the water supply of that city.[3] By contrast, stories that tell of improvements in water treatment or distribution of water supply, water quality upgrade, future expansion of water system coverage, and staff accomplishments will receive less attention. To place such positive information requires a good relationship with local editors and an understanding of their perspective of community needs.

Newspapers may have differing agendas, depending on their role and their readership. For example, the *New York Times,* the *Washington Post,* or the *Christian Science Monitor* may give more coverage to national issues involving water safety than do local newspapers. Similarly, big-city daily newspapers will give front-page coverage to emergency news such as the need to boil water or to noncompliance issues, while community weekly newspapers are more likely to address long-term water problems that are not time-dated for immediate attention. Their mostly suburban readers want information on efforts made to minimize traffic slowdowns from water line construction, low water pressure problems, and taste and odor complaints. Weeklies are also more likely to carry stories that justify cost increases for water supply or progress being made to extend system coverage in their community.

Tabloid newspapers found at the supermarket checkout stand often create distortions in news reporting. These stories are designed to arouse a quick emotional response and are often written with little regard for the facts and poor record of story verification. Utilities managers must exercise caution in talking to such journalists because of concern about distorted quotes.

Magazines devoted to family health and personal fitness have provided occasional space to general discussions on water quality issues and health effects. Reporters for these publications often have more time to develop a story and frequently call their sources back to verify information accuracy. Outdoor camping magazines are more interested in doing articles on the preparation of a safe outdoor water supply. Issues of interest are waterborne pathogens, treating surface waters for a safe supply, and evaluating treatment options such as boiling, tablets, and filters. Travel magazines will focus on health risks associated with water supplies in foreign countries and the acceptance of bottled water and soft drinks as an alternative source of drinking water. Cooperation with many of these media can result in positive coverage of water supply issues.

ELECTRONIC MEDIA

The invention of radio, television, and satellites has brought information to the public far faster than is possible through print media. Radio station newsbreaks and scheduled news programs throughout the day have become a major avenue for getting information to the public in emergency situations. Boil water notifications because of contaminated water, information on major line breaks that affect traffic flow, and temporary shutdown of water service are just a few examples of public service messages speedily communicated through broadcast media.

National Public Radio news and information programs provide extended commentary on current topics, including environmental issues and health. Some of these topics have included changes in drinking water regulations, new treatment approaches, waterborne outbreaks, alternative water supplies (bottled water and point-of-use devices), water reuse, and water conservation.

Talk show radio has grown in popularity in recent years. While these programs are often outlets for some citizens to vent their personal displeasure on a wide range of subjects, station management may structure the agenda to permit water authority personnel to rebut the arguments from their perspectives with facts to squelch distortions. Such interfacing with the public must be skillfully done by water professionals trained in public relations techniques, to avoid further verbal antagonisms.

Television adds the dimension of visualization to the public understanding of water issues. Viewers can see some of the problems: engineering treatment for water supply, distribution network repairs and pipeline extensions, laboratory processing of samples, and flushing of lines. Through television, audiences can view the physical facility and some of the plant operations, and watch personnel at the plant or in the field. These brief snapshots can do much to bring a better understanding of the treatment and delivery of a safe water supply.

Television news programs can take the viewer to the scene, close up to the problem through on-site reporting. News sound "bites" are typically short, but can convey much information when the job is done properly. As in print media, special investigative reports are an additional feature of television programming. Some are broadcast in tabloid style characterized by sensationalism, while others are carefully researched and formatted into a documentary. These latter reports often involve discussions with water plant operators, consumer advocates, state water supply engineers, university researchers, and U.S. Environmental Protection Agency officials. Areas of interest have been raw water quality, treatment effectiveness, and health concerns over pathogens, lead, and disinfection by-products, all topics that gave utilities an opportunity to tell their story.

Cable news networks provide both brief reports of waterborne outbreaks in major cities and panel discussions by experts on public water supply health issues. Government access channels are another source of information on committee reports and legislative debates on funding demonstration grants, regulation of clean water resources, and safe public water supplies. Community cable service provides the opportunity for citizens to view public hearings on proposed water rate increases and allow utilities to gain support for plant improvements and infrastructure repairs to the distribution pipe and reservoir network.

Of course, print and electronic communication links have been known to falter in small communities without their own radio station or daily newspaper. Such was the case in Gideon, MO during the salmonellosis outbreak in January 1994. The mayor notified a small radio station with coverage in the Gideon area and several announcements were made. But few local people heard the news and some were not convinced it could affect their personal health. A hand-written notice to boil water was also placed on the door at City Hall and another at the town supermarket. These, too, had little impact on reducing the growing number of outbreak cases. Finally, only after the field epidemiologist from the National Center for Infectious Diseases prepared an information sheet clearly explaining the need to boil water and had it hand-delivered to every residence was there a significant improvement in the situation.[4]

This event was, however, an exception to the normal rule that mass media often plays in alerting and informing the public on water supply issues. Through media attention, the public often gains a better understanding of the science of water treatment, and provides the support for future bond issues and water rate increases that will eventually occur.

NEWS DEVELOPMENT

The link to media exposure is the reporter. This contact is most often in response to an announcement of a water supply noncompliance issue, a major water main break interrupting traffic flow, a pollution spill upstream of the water intake, or consumer complaints about water pressure, taste, or odor. Often, the newspaper or television news deadline for a developing story is only a few hours away. It is important to remember that the reporter is trying to develop a story that is as accurate as possible within the time constraints. Rarely will there be time for a reporter to call back for review and comment on the story; occasionally, reporters call to confirm their understanding of some technical aspect. This is an opportunity to be cooperative and provide the technical assistance needed to improve accuracy. Precise input may prove helpful in providing the public with a better understanding of the news event.

It is important that all phone inquiries from the media be answered by the designated individuals in the utility. Maintaining a positive image of cooperation is important. If an individual is not immediately available, every effort should be made to call back within the hour. If the inquiry requires more specific information to answer the question, the reporter should be told he will get the information promptly by a return the call within 30 min. The phone response should not be avoided; it is an opportunity to put a positive light on story development.

Truthfulness is essential. If the answer is known, the facts should be provided in a concise manner. Elaboration is not required; it is not desirable to extrapolate or volunteer an opinion. Above all, false information should not be given. If the answer to the question is not known, it should be stated and then a request made for time to get the facts. During the phone call, suggestions can be made for other potential contacts that might provide the reporter with more information or a different perspective, but no attempt to manipulate story development should occur. The reporter may become suspicious of motives and be antagonistic in future encounters.

Once the story has been prepared in the newsroom by the reporter, it is reviewed by the city editor for changes or additional details.[5] The news editor is next to review the story. At this point, the story must compete for space with other news events of the day and for placement on the front page or regional or metro news sections of the newspaper. The competition for space with other news stories is often acute. Some stories may be delayed or minimized because of other important, fast-breaking news events or because of the news judgement of the editor. Once selected, the material is sent to the copy editor to check story content and is edited to fit available space and given a headline. Headline writing is an art in itself. The purpose of the headline is to encapsulate the story in a way that will draw reader interest. At times, the headline can be an overstatement in the attempt to gain more reader attention. Having verified that the story is trimmed to meet its alloted space and that the headline is suitable, the copydesk chief then transmits the story to be typeset for the next edition.

At the small newspaper, the position of news editor and city editor may be performed by the same person, reducing the benefits of additional review. However, it must be noted that even with a small editorial staff, some small newspapers have consistently achieved remarkably high standards, as measured by their numerous awards in journalism.

News reporting on television is routinely condensed to short bursts of information over a 30- to 60-s segment in the program. These sound "bites" are linked to video shots related to the event. Where possible the utility public relations officer may be able to set up a good photo opportunity that will be appreciated by the assignment editor. At other times, file tape of scenes at the water plant, collecting water samples at the tap for water testing, flushing lines at fire hydrants, or repairing line breaks may be used. These stories must also compete with other developing news for air time. If successful, the stories will air on both the early and late news programs to reach a majority of the viewing audience. Occasionally, television investigative reporting may devote longer stories on drinking water quality in special news programs at the local level or over national networks. These stories are developed over periods of several days to several weeks and provide more in-depth information to viewers.

Interviews on camera introduce another dimension to communicating — the use of body language. Voice inflection, facial expressions, and body movements can project a sense of honesty, sincerity, and credibility if done properly. If not, all statements cast doubts in the viewers mind about the speaker's credibility. Body language is used to reinforce the response to interviewers' questions. In speaking to the public, the findings of the investigation should be translated into language free of technical expressions so the viewer can clearly understand.

EDITORIALS AND CARTOONS

Articles and cartoons on the editorial page of the newspaper provide the press and public with a forum to express opinion on current events. The editorial group functions independently of other editors and reporters. In editorial staff meetings, current events are discussed and ideas selected for editorial comment; an individual editor is assigned the task of preparing the article. The writer may review material developed by reporters, but because of seniority, remains independent to provide opinion on the subject. There is no review of this article by the beat reporter for technical accuracy and rarely is there any review by other members of the editorial staff.

Not every reader is interested in editorials. Those who are, are often better educated and have a wider interest in local and national affairs involving politics or environmental subjects. Many of these readers have predetermined opinions on a subject and prefer those editorials that reaffirm their views. As a consequence, editorials often do little to change attitudes or perceptions.

Editorial cartoons are prepared with the intent of providing humor, caricature, or satire on current events. They can be a powerful tool that appeals to many readers and they have a long history. Perhaps the most influential cartoon during the pre-Revolutionary War period in this country was one circulated in leaflets that made the point, "Don't Tread on Me." In England, political cartoons predate the era of Dickens and the Pickwick Papers.

The political cartoonist is generally given a theme that relates to current events. Other than selection of the subject, there is complete freedom to illustrate the message and little review of the artwork is done before publication. The cartoonist's assessment of drinking water issues may fall into one of three categories: visualization of consequences, exaggerated situations to create whimsical impressions, or sarcastic reproach to a problem or solution. The latter category creates distrust of the public water supply and often stimulates the consumer's search for alternative water supplies. Three examples from newspapers in New York, Florida, and Massachusetts illustrate different approaches used by cartoonists.

The public water supply of Utica, NY received serious criticism in several cartoons published during 1991 to 1992, a period when the utility experienced several water quality problems: coliform releases from pipe biofilm and releases of excess levels of lead and trihalomethanes in some areas of the distribution system. Randall Kimberly, cartoonist for the *Sunday Observer-Dispatch* conceived several cartoons suggesting that exposure to the Utica water supply might lead to being physically altered or subjected to a life-threatening event (Figures 10.1 to 10.4). It was also intimated that the water had a bad smell from elevated chlorine dosages and contained many "coliform bugs." One panel suggested that restaurants serve two kinds of drinking water: "regular or unleaded." What impact these opinions had on a permanent loss of confidence in the public water supply is unknown. However, as a result of the news stories and cartoons, bottled water sales rose dramatically during the noncompliance period. However, not all readers responded positively to the cartoons. A regular reader of the newspaper for 50 years wrote to the editor calling the cartoon "a cheap shot." He added:

Whether your placement of the cartoon next to the explanatory letter from the technical people at the Water Board was by accident or design, I cannot guess. However, ... I believe your responsibility to your readers is to present objective, factual information concerning the problem, not to sensationalize it with massive headlines and corny, misleading artwork. (David L. Mahoney, Little Falls, NY, Letter to the Editor, *Sunday Observer-Dispatch [Utica]*, December 20, 1991.)

Crumbs Along the Mohawk/Randall Kimberly

Figure 10.1 From the *Sunday Observer-Dispatch (Utica)*, 1991–2. With permission.

More humorous were cartoons published in several Florida cities with public water supply quality problems. The water system at Boca Raton, FL had a brief episode of coliforms in 1991 that required public notification. The cartoonist for the *Palm Beach Post* suggested that the contaminated water supply should be packaged in bombs for use in the Gulf War (Figure 10.5). Another cartoon in the Ft. Lauderdale *Sun-Sentinel* cartoon (Figure 10.6) proposed using "Plan B" if the coliform problem persisted. In this plan, the water utility would disconnect the water treatment plant and hook up a large supply of bottled water to the distribution system.

The Worcester, MA water supply had a long history of freshwater shrimp (copepods) infesting the old pipe network. When increased chlorine dosage, flushing, and mechanical scraping proved unsuccessful, new pipe was laid to replace the 50- to 100-year-old pipe sections. In the meantime, most homeowners installed screen attachments to their faucets as a barrier to shrimp passage into their drinking water. Bob Sullivan, cartoonist for the *Worcester Telegram and Gazette,* proposed all home plumbing systems be modified with an additional faucet to dispense cocktail sauce (Figure 10.7).

Crumbs Along the Mohawk/Randall Kimberly

Figure 10.2   From the *Sunday Observer-Dispatch (Utica)*, December 15, 1991. With permission.

A STUDY OF MEDIA COVERAGE VS. TECHNICAL REALITY

Public awareness of microbial quality deterioration in the municipal water supply is most often derived from newspaper stories, radio talk shows, or television news programs. How accurate are these information alerts? Do stories developed from investigative reports uncover hidden facts or merely enhance the story to sell newspapers or capture a larger viewing audience? Perhaps a study of how faithfully newspaper headlines reflected the technical reality of several water supply problems might provide answers to these questions.

SPRINGFIELD, IL (1982 EPISODE)

In June 1982, Springfield water officials observed that coliform bacteria in distribution system samples were appearing at an alarming rate.[6] Eventually the average coliform density from all samples for the month exceeded the 1.0 per 100 ml limit acceptable to the Illinois Environmental Protection Agency.

Crumbs Along the Mohawk/Randall Kimberly

Figure 10.3 From the *Sunday Observer-Dispatch (Utica),* December 29, 1992. With permission.

This was occurring even though no coliforms were detected in finished water leaving the treatment plant. After consultation with state and federal water authorities, a boil water order was issued through the news media on July 19; it remained in effect for 10 d.

Review of the laboratory data revealed evidence of a growing occurrence of coliform throughout the system. Over a 4-week period (June 20 to July 17) coliforms suddenly began to occur with dramatic frequency. Of 336 samples analyzed, 43% had coliforms and the maximum density reached 80 coliforms per 100 ml. The background count of noncoliform bacteria growing on m-Endo medium ranged from <1 to 490 per 100 ml. Considering the Endo medium was designed to suppress more than 95% of noncoliform bacteria in the heterotrophic population, these findings suggested a regrowth event was occurring in the pipeline environment and had developed into a significant biofilm.

Study of plant records revealed the chlorine application rate was gradually increased from 4.5 to 11.0 mg/l with corresponding proportional increases in ammonia feed. The chloramine residual increased from 1.1 to 2.0 mg/l, indicating there was a significant disinfectant demand in the process water. While contact time was estimated to be seldom less than 6 h at pH 9.4 to 9.6, the

Crumbs Along the Mohawk/Randall Kimberly

Figure 10.4 From the *Sunday Observer-Dispatch (Utica)*, July 26, 1992. With permission.

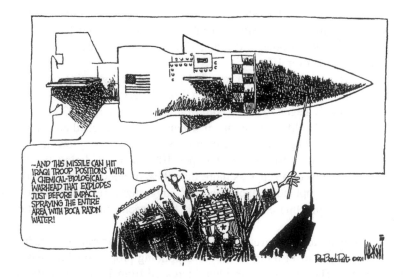

Figure 10.5 From the *Palm Beach Post*, February 15, 1992. With permission.

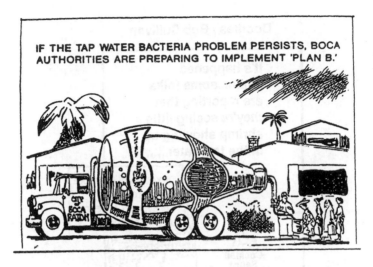

IF THE TAP WATER BACTERIA PROBLEM PERSISTS, BOCA AUTHORITIES ARE PREPARING TO IMPLEMENT 'PLAN B.'

Figure 10.6 From the *Sun-Sentinel (Ft. Lauderdale)*, February 12, 1991. Reprinted with permission of the Tribune Media Services.

effectiveness of chloramines for quick inactivation of microorganisms did not match that of free chlorine under more favorable pH conditions and less organic demand. The organic demand not only reduced disinfectant power, but also provided accumulated supply of nutrients for biofilm development.

Remedial actions to combat the coliform regrowth problem took several directions. Most significant among the treatment modifications were changes in chemical addition to the mixing basin to reduce turbidity and bacteria, adjustment of lime dosage to control pH, and alternate disinfectant application (free chlorine to replace chloramination). To clear out the pipe sediments and some of the habitat sites for microbial growth, systematic flushing of the distribution system was instituted, something that had never been done in the past. Previous practice was to use a 30-min flush of fire hydrants near areas of "red water complaints." Otherwise "flushing" was done by the fire department personnel to check line pressure and operation characteristics of each hydrant. Analysis of the laboratory data indicated the maintenance checks by fire department personnel temporarily stirred up sediments in the immediate vicinity and increased the chances of detecting coliforms for a few hours to a few days later. Obviously, the evidence suggested nutrients, organisms, and loss of disinfectant at these sites were creating a serious problem over the long term. Flushing the entire distribution system was done systematically over an 8-d period. Following these actions the numbers of both coliform and non-coliform organisms in the system were gradually reduced and by March 1983 coliforms in the system had subsided to within acceptable baseline values.

Doodles / Bob Sullivan

Figure 10.7   From the *Worcester Telegram and Gazette*, 1981. With permission.

Press Response

Press response to the public notification has been encapsulated from a study of newspaper headlines to stories filed in chronological order (Table 1). Analysis of these headlines reveal concerns of the public and investigative tacks that the reporter used to discover news beyond the public announcement from the utility to boil water.

The reporter began his coverage by developing public interest stories on the impact of a boil water announcement by covering supermarket reports of increased bottled water sales, the annoyance of boiling drinking water supply, and the effect of highly chlorinated water on house plants. Follow-up articles appeared on status of the boil water order and public reaction when the order was lifted. Reports of utility actions to find the source of contamination and progress in resolving the problem were part of a daily inquiry. Some of the subjects covered were the need to flush the entire system, search for cross-connections, and the request for technical help from state and federal experts in drinking water engineering and research. Investigative reporting produced

Table 10.1   Newspaper Headlines on Springfield, IL Water Quality Problem

"CWLP Seeks Source of Bacteria in Water"
  July 19, 1982 — *State Journal-Register*
"Water Dumped in CWLP's Attempts to Find Source of Bacteria"
  July 20, 1982 — *State Journal-Register*
"Bottled Water Firms Rush to Fill Orders From Public"
  July 20, 1982 — *State Journal-Register*
"Rumors Flowed Before Water Problem Surfaced"
  July 20, 1982 — *State Journal-Register*
"Don't Drink the Bath Water Either"
  July 21, 1982 — *State Journal-Register*
"Source of Bacteria May be Under I-55"
  July 21, 1982 — *State Journal-Register*
"City's Water Woes Nearly A Repeat of Muncie Experience"
  July 22, 1982 — *State Journal-Register*
"Cross-Connections Often Contaminate Water"
  July 23, 1982 — *State Journal-Register*
"Keep Boiling: Water Flunks Latest Tests"
  July 23, 1982 — *State Journal-Register*
"Extra Chlorine in City's Water Should Not Harm Your Plants"
  July 23, 1982 — *State Journal-Register*
"More Contamination Found in Water"
  July 24, 1982 — *State Journal-Register*
"Week-Old Boil Order Still On"
  July 25, 1982 — *State Journal-Register*
"Contamination Not Present In Latest Tests"
  July 26, 1982 — *State Journal-Register*
"Contamination Returns To Samples: City Alters Plan"
  July 27, 1982 — *State Journal-Register*
"Boiling The Water — 'Annoying' Job Still Carried Out By Most"
  July 28, 1982 — *State Journal-Register*
"Contamination Pattern Puzzles Officials"
  July 28, 1982 — *State Journal-Register*
"Boil Order Lifted: Test Continue"
  July 29, 1982 — *State Journal-Register*
"Water Was OK: Samples Were Contaminated"
  July 30, 1982 — *State Journal-Register*
"CWLP Stops Taking Own Water Samples, Will Depend On Other Laboratories"
  July 31, 1982 — *State Journal-Register*
"Contaminated Water Samples Still Puzzle CWLP"
  August 8, 1982 — *State Journal-Register*
"CWLP Chlorine Levels Stay Higher Than Normal"
  August 20, 1982 — *State Journal-Register*

information that the problem had begun to develop earlier than announced to the public and that Muncie, IN had experienced a similar coliform noncompliance issue. Confusion and puzzlement of utility officials surfaced when there was a cycling of coliform occurrence that suggested to the utility management that the laboratory was the cause of sample contamination. As a consequence, management elected to have all official samples examined by other laboratories. When this was demonstrated not to be a factor, the utility decided to maintain elevated chlorine levels until the coliform problem faded away. So did further interest in the story by the newspaper. None of the articles were uncomplimentary to the utility and the public remained mostly stoic with the inconvenience. This was good job of reporting without distortion or sensationalism.

FLORISSANT, MO (1984 EPISODE)

Florissant, MO owns only the public water supply distribution system. Potable water transmitted through these lines is purchased from the St. Louis County Water Authority. A storage reservoir in the center of the pipe network stores water for use during peak periods of water demand in the summer. The tank is refilled during nighttime hours of low water use to take advantage of a lower service charge. At other times during the year, the stored water is slowly released over 2-week periods into the distribution system.

With a change in operators running the system, stored water was bypassed after the summer high-demand period and held in storage from October 1983 to February 1984. During early February, the new water plant operator released a portion of this static water into the distribution system. Suddenly there was a large density increase in heterotrophic bacteria reported from routine monitoring samples followed in a few days by coliform occurrences in the distribution system. No coliform bacteria were detected at the entry point for treated water purchased from the St. Louis County Water Authority. The source of the coliform contamination in the tank was thought to be either the result of new line installations or a water main repair that was not disinfected properly before acceptance by the water system manager. The contaminated water was in the pathway to the storage tank where the organisms adjusted to survival in the static water environment and, as a consequence of partial release of tank water, were passed on into the distribution network. Elevating the chloramine residual did not immediately eliminate the coliform occurrences. Flushing lines and refilling the storage tank with fresh water that received additional disinfection (free chlorine) at the entry point were eventually successful in restoring the water quality. (For more details see pages 366–367, Chapter 7).

Press Response

The press was at first understanding of efforts being made to restore water quality (Table 10.2). Stories at first discussed the possibility of a contamination leak in the distribution system barrier and efforts being made to find the source. Soon articles began to appear on public displeasure with the inconvenience of boiling water. One reporter concluded the problem was caused by a "super bacteria" because disinfection was not effective in inactivating these coliform organisms. Consumer displeasure continued to rise largely because the boil water order was lifted then reissued two more times over a 46-d period. It was not surprising that sales of soft drinks were noticeably strong in the community

Table 10.2   Newspaper Headlines on Florissant, MO Water Quality Problem

"Residents Told to Boil Water in Florissant"
    March 3, 1984 — *St. Louis Post-Dispatch*
"Officials Still Seeking Leak in Florissant Water System"
    March 4, 1984 — *St. Louis Globe-Democrat*
"Bacteria Put at 20 Times Safe Level"
    March 4, 1984 — *St. Louis Post-Dispatch*
"Florissant Residents Boiling At Delay in Action on Bacteria in Water"
    March 4, 1984 — *St. Louis Post-Dispatch*
"Florissant Continues to Monitor Water"
    March 4, 1984 — *Northwest County Journal*
"Florissant Still Boiling Water While Bacteria Source is Sought"
    March 5, 1984 — *St. Louis Post-Dispatch*
"Residents Stomaching Boil Water Order"
    March 5, 1984 — *St. Louis Post-Dispatch*
"Residents Advised to Boil Water"
    March 5, 1984 — *St. Joseph News-Press*
"Florissant Water Back to Normal"
    March 5, 1984 — *St. Louis Post-Dispatch*
"Florissant Hard-Boil Spurs Soft-Drink Sale"
    March 6, 1984 — *St. Louis Post-Dispatch*
Editorial, "Public Safety Comes First"
    March 6, 1984 — *St. Louis Post-Dispatch*
"Boil Order for Florissant Water To Remain At Least Until Friday"
    March 7, 1984 — *St. Louis Post-Dispatch*
"Water Contamination Source May Be Found"
    March 18, 1984 — *St. Louis Post-Dispatch*
"Florissant Ordered to Boil Its Drinking Water"
    March 24, 1984 — *St. Louis Post-Dispatch*
"Florissant At A Boil: Water, Tempers"
    March 24, 1984 — *St. Louis Post-Dispatch*
"Task Force Seeks Source of Water Woes"
    March 25, 1984 — *St. Louis Post-Dispatch*
"Aid Sought on Tainted Water"
    March 26, 1984 — *Kansas City Star*
"Florissant Flushes Water System in 2nd Attempt to Rout Bacteria"
    March 26, 1984 — *St. Louis Post-Dispatch*
"Recent Water Problem in Florissant"
    March 27, 1984 — Florissant Valley Reporter
"Super Bacteria May Be What Keeps Florissant Boiling"
    March 28, 1984 — *St. Louis Post-Dispatch*
"Florissant Water Problem Laid To Lack of Cleansing"
    April 8, 1984 — *St. Louis Post-Dispatch*
(Third boil water order issued April 9, 1984 by water supply authorities to all news media)
"Florissant — Coffee Won't Be Good Till The Last Drop"
    April 10, 1984 — *USA Today*
(Boil water order lifted April 17, 1984 by water supply authorities to all news media)

during this period. Finally, the task force recommendations were made public as the city water department began a renewed effort to clean the pipe network and restore fresh water to the storage tank.

In retrospect, the repeated reissuance of boil water orders caused loss of confidence with statements issued by the water authority. The *St. Louis Post-Dispatch* reported a serious flaw in the monitoring program and management reaction in a March 6 editorial:

Public Safety Comes First

In a shocking dereliction of their official duties and public responsibilities, city officials in Florissant made an unconscionable delay in issuing an order to boil all public drinking water.

The first indication that the city's water system was tainted occurred on Feb. 8, when water inspectors discovered that six of 15 test sites had unusually high levels of coliform bacteria, which is an indirect measure of fecal contamination. The only defense offered by officials for why a boil order was not issued at that time was that the test results were erratic and officials wanted to recheck their own water testing procedures.

Still, the 55,000 citizens of the largest city in St. Louis County should have been warned and water should have been boiled, if only to have been on the safe side. But what is more outrageous was the failure to issue a public warning after a series of water tests on Feb. 22, which found 10 of 15 sites contaminated with bacteria that were in some cases 20 times the amount considered safe. It was 10 d after this series of tests before a boil order was finally issued.

The first and most important responsibility of any city official is the public safety. Even though there may have been legitimate doubts as to the exact extent of the contamination, a boil order should have been the first order of business. such an order causes inconvenience; it may spread public alarm. But not to issue the order is to sanction a totally unacceptable risk with the public's health. (From an editorial, *St. Louis Post-Dispatch,* March 6, 1984. With permission.)*

Public health protection is a responsibility that should not be taken lightly in any decision on the need to go public.[7] What is difficult to reconcile is **public pressure** to lift a boil water order and **time needed to confirm** water quality has been restored permanently.

ROCHESTER, NY (1986 EPISODE)

Most of the raw source water for this utility is obtained from two lakes (Hemlock and Canadice) with a lesser amount derived from Lake Ontario.[1]

* The following editorial correction was issued March 8, 1994.

Tuesday's editorial, 'Public Safety Comes First,' wrongly criticized the city officials of Florissant for dereliction of their responsibilities in failing to issue a prompt order to boil public drinking water after it was discovered to be tainted. That failure properly should have been charged to St. Louis County health officials, who are in charge of the water testing and did not officially inform Florissant officials of the need to boil water until Friday, Mar. 2, weeks after the contamination was first discovered. At that point, Florissant officials immediately moved to inform the public and correct the problem. The Post-Dispatch regrets the error.

The water is chlorinated, fluoridated, and gravity-fed into the city's storage and distribution system. Of the monthly turbidity averages for a recent 5-year period, 90% were less than 2 NTU, with a maximum monthly average of 3 NTU. However, in the late winter of 1986 there were periods of exceptionally heavy rains, resulting in elevated turbidity and increased chlorine demand, which at times exceeded 1 mg/l. Compensating increases in chlorine applied were not promptly made to these changing source water conditions. By February, coliform occurrences started to increase substantially to the point that the coliform MCL was exceeded for two consecutive months. Coliforms and other heterotrophic bacteria had penetrated the unfiltered treatment process and within the next few months biofilm became established in the nutrient-rich pipe sediments. With onset of the warm-water conditions of late spring and summer, accelerated growth of the biofilm began to fragment, releasing coliforms throughout the distribution network. (For more details see pages 327–328, Chapter 7).

Press Response

Press response to the boil water order was at first slow to develop, probably because the utility had previously announced through the press that there was a sudden return of the puzzling coliform problem from the previous summer. However, within a few days, several reporters from the same newspaper began to investigate (Table 10.3). They reported the increased utility sampling that led to more coliform discoveries, which in turn escalated the sampling over the entire distribution system. Another article reported on the search for leaks and cross-connections in the system. Recognition was also given to a large brewery that offered the public free bottles of their safe water supply as a goodwill gesture to the community. As the boil water order continued, reporters began to dig for theories on the source of the contamination. A search of newspaper files soon revealed that a waterborne outbreak had occurred 46 years earlier, resulting in six cases of typhoid and 34,000 cases of diarrhea disturbances. Reports also revealed that there were fecal coliforms in water from the Cobbs Hill finished water reservoir and that there had been internal disagreement on the need for a filtration plant.

A decision to lift the boil water order came on July 31, based on the argument that because the boil water order was issued due to the presence of *E. coli* and because *E. coli* was no longer found in the water samples, the boil water order could be lifted. The continued presence of fecal *Klebsiella* was discounted as a matter of convenience. This rationale never appeared in the news stories. Rather, the water bureau elected to place all blame on a huge dropping of fecal wastes (frass) from gypsy moth caterpillars.

The caterpillar theory was introduced by an inquisitive, retired forest pest-control official from Pennsylvania. He happened to be visiting Rochester on the day of the boil water advisory and decided to take a look around the city-owned Hemlock Lake. Following visual observation of caterpillar infestation among the trees surrounding Hemlock Lake, he wrote a letter on July 9 to the Monroe County health director expressing his concern. The letter drew little

**Table 10.3   Newspaper Headlines on Rochester, NY
Water Quality Problem**

"A Puzzling Bacteria Invasion in Water Supply"
April 20, 1986 — *Democrat and Chronicle*
"Ryan Says Filter Plant 'Desirable'"
June 20, 1986 — *Democrat and Chronicle*
Boil water order issued to the public
July 7, 1986 — Utility Press Conference
"City Looks For Cracks In Pipes"
July 10, 1986 — *Democrat and Chronicle*
"Samples, After Samples, After Samples"
July 11, 1986 — *Democrat and Chronicle*
"'Genny' Giving Away 75,000 Gallons of Water"
July 11, 1986 — *Democrat and Chronicle*
"Theories Flow on Bacterial Source"
July 13, 1986 — *Democrat and Chronicle*
"A System of Innovation"
July 13, 1986 — *Democrat and Chronicle*
"A City Water Crisis 46 Years Ago Resulted in Sickness, Scandal"
July 13, 1986 — *Democrat and Chronicle*
"Old Memo Muddies Waters Over Filtration Plant Idea"
July 16, 1986 — *Democrat and Chronicle*
"City to Still Test The Waters — But Quietly From Now On"
July 16, 1986 — *Democrat and Chronicle*
"A Hint of Trouble in Reservoir"
July 16, 1986 — *Democrat and Chronicle*
"Extra Chlorine Adding to The Problem: Officials Wary"
July 16, 1986 — *Democrat and Chronicle*
"Relief From Water Advisory?"
July 17, 1986 — *Democrat and Chronicle*
"Great Bear Firm Uses Disinfected Water From City"
July 18, 1986 — *Democrat and Chronicle*
"You Can Drink The Water"
July 31, 1986 — *Democrat and Chronicle*
"Some Rochester Residents Still a Little Leery of Tap-Water"
July 31, 1986 — *Democrat and Chronicle*
"In Rochester, Caterpillars Get The Blame"
July 31, 1986 — *New York Times*
"Caterpillar Theory Bugs Some"
August 1, 1986 — *Democrat and Chronicle*
"City Water Fouled By Tone of Caterpillar Dung"
August 1, 1986 — *Louisville Courier-Journal*
"Alert News To End This Week Anyway"
August 1, 1986 — *Democrat and Chronicle*
"Rochester Not Only City With Water Problems"
August 1, 1986 — *New York Times*
"Rochester Water Problems Not Over"
August 1, 1986 — *New York Times*
"Gypsy Moth Theory A Year From Proof"
August 2, 1986 — *Democrat and Chronicle*

interest until other possible causes were rejected. After careful review of the theory, city officials told the press that this type of bacterial contamination might have begun with gypsy moth infestation and defoliation of a $1^{1}/_{2}$-mi-long stretch of Hemlock Lake shoreline in May and early June. Then just after the caterpillar population reached a peak in early June, heavy rain washed several tons of frass and dead caterpillars into Hemlock Lake. It was presumed that this fecal waste contained large amounts of bacteria.

Several microbiology experts disagreed with this scenario. While it was possible for an environmental coliform to be found in caterpillar fecal waste because of ingestion of leaves, it was thought improbable that these bacteria would occur in high densities. More likely, coliform bacteria would be transient in this and other insects, passing through the digestive tract without colonizing the insect. No further evidence was found to associate the caterpillar infestation with the coliforms in the Rochester water supply. Several reporters and some of the public could not accept this theory and thought something else made it necessary to boil water for almost three weeks.

Reader comments on August 1 flowed freely after the *Democrat and Chronicle* asked for response on the caterpillars and the water alert. A total of 75 readers submitted opinions, many like the following samples:

I Don't Believe That

Traced to caterpillars? I don't believe any part of that. I think that theres more to the water contamination than Rochester water works is telling us. There probably has been raw sewage and stuff like that, rather than blaming it on these little caterpillars.

Clarence Gerstner, Rochester

A Monster of a Problem

It took them $2^{1}/_{2}$ weeks to decide whether to let us know about the water. Now they're telling us that its gypsy moths? Why don't they tell us that it's feces from the monster of the black lagoon.

Lori Amering, Rochester

A Big Hoax

I think the water was a big hoax all the time. There had been nothing wrong with the water all this while.

Anonymous

I Can't Buy Moth Theory

I don't think the gypsy moth caused the water problem because we have had gypsy moths in Rochester as long as Rochester has been here and all of a sudden this problem surfaces?

Anonymous

BRADENTON, FL (1990 EPISODE)

The city built a new surface-water treatment facility and brought it on-line in February 1990. Treated water was passed through the old plant with the approval of state officials. In April, the city unilaterally began putting the water directly into the distribution system. At this point, the turbidity increased from 1 NTU to peaks over 6 NTU. At the same time, chloraminated residuals disappeared and coliforms began to be detected. The state regional laboratory at Tampa identified these coliforms as *Enterobacter cloacae* and *Klebsiella pneumoniae*. By July 24, a boil water order was issued and remained in effect until July 31. During this boil water advisory period, the entire system was flushed and chloramine concentrations were elevated. This action brought a halt to further coliform occurrences. High densities of heterotrophic bacteria as reflected in excessive noncoliform colonies on the m-Endo test cultures also declined. A subsequent review of water system operations revealed several problems: a flow of raw source water from a branch line bypassing the treatment plant; inadequate chlorination since the new plant opened; and irregularities in blending with the county water system during the summer to meet demand.

The City of Bradenton filed legal action against the Florida State Department of Health and Rehabilitative Services (HRS) and the County Health Department for "false order to boil water." Reasons cited were bad sample collection by county sanitarians, wrong test used by the health department to establish water quality, and the alleged willful contamination of a sample by a county employee. The state agency countersued the city for its numerous violations under the old regulations. As a consequence, public confidence in the public water supply was shaken.

Press Response

Press response to the developing story provided readers with in-depth reporting of all the events, which eventually culminated in the temporary closing of the water plant laboratory, a rebate to consumers, and a legal battle between the city and state officials. Reporters for the *Sarasota Herald-Tribune* contacted numerous local and state health officials, utility personnel, state water supply authorities, hospitals, and private laboratories to gather information about the problem. The public was informed of city locations where to obtain a safe water supply and provided instructions on boiling water, applying

Table 10.4   Newspaper Headlines on Bradenton, FL Water Quality Problem

"State: Water Worse Than Feared, Tests Find Fecal Coliform in City's Water Supply"
July 26, 1990 — *Sarasota Herald-Tribune*
"Purifying Your Water"
July 26, 1990 — *Sarasota Herald-Tribune*
"Callers Swamp Health Unit, Hospitals With Questions"
July 26, 1990 —*Sarasota Herald-Tribune*
"Developments in Bradenton's Water Crisis"
July 31, 1990 — *Sarasota Herald-Tribune*
"Officials Criticize State Tests, Say Criminal Probe Is Needed"
July 31, 1990 — *Sarasota Herald-Tribune*
"Water Samples Pass, But More Tests Ordered"
July 31, 1990 — *Sarasota Herald-Tribune*
"Water Crisis Panel Says Bradenton Water Is Fit To Drink"
August 1, 1990 — *Sarasota Herald-Tribune*
"Ex-City Employee Was First to Find Problem With Water"
August 1, 1990 — *Sarasota Herald-Tribune*
"State Faults City's Water Lab: Mayor Evers Says State Is All Wet"
August 2, 1990 — *Sarasota Herald-Tribune*
"Chronology Of Water Crisis"
August 5, 1990 — *Sarasota Herald-Tribune*
"HRS Tells City: Shut Lab, Train Workers"
August 7, 1990 — *Sarasota Herald-Tribune*
"Bradenton Forced To Close Water Lab"
August 8, 1990 — *Sarasota Herald-Tribune*
"Water Customers To Get Rebate"
August 9, 1990 — *Sarasota Herald-Tribune*

common household bleach as a disinfectant, or purchasing bottled water. These instructions were given in the paper on July 26 and repeated on July 31 to insure maximum coverage among consumers.

Inquiries revealed that water samples often were positive when analyzed by the state laboratory, but negative when done on samples collected by the city and two private laboratories. A review of the city laboratory by the state certification officer indicated there were five deviations from acceptable practices. Most serious was the use of methanol as a substitute for ethanol in the preparation of the medium used to detect coliform bacteria. Methanol is very toxic to coliform bacteria and its use explains why there was no agreement on coliform detection between the state and city laboratories. The city laboratory was ordered temporarily closed for a month while technicians were retrained at the state lab.

Perhaps the most novel reporting of the events was the chronology of the water crisis (see below) prepared by the *Sarasota Herald-Tribune* on August 5, 1990. In an effort to compensate water customers for their "fear and inconvenience" during the 8-d water crisis, the Bradenton City Council decided to authorize a refund of $6.95 to the average customer. Customers with service lines larger than 3/4 in. received a larger rebate. Other costs incurred included legal fees, tests performed by private laboratories, laboratory personnel training,

Chronology of Water Crisis

Here's a chronology of Bradenton's water crisis and events leading up to it:

■ April 16: A new water plant goes into operation next to the Bill Evers Reservoir, off State Road 70. The old plant, on First Street West, is still operating.

■ May 21: After approving operations at the new plant and testing water samples, state health officials allow the city to take its old plant off line.

■ May 22: The city informs the state Department of Health and Rehabilitative Services that Bill Green, the city's only licensed Class A water-testing operator, is stepping down.

■ May 30: Acting on a complaint from a citizen about poor drinking water, the HRS takes water samples at the resident's home and finds low pH and chlorine levels. Health officials notify the city, which takes steps to raise water chlorine levels.

■ June 4-7: Additional water samples are taken at the residence, and results again indicate low chlorine levels. City staff members are again told of the problem.

■ June 12: Three samples are collected by the HRS from the city's distribution system. Bacteriological results are satisfactory, but one sample shows low chlorine levels. The city staff is again notified. The city begins flushing, draining and cleaning both ground and elevated storage tanks.

■ June 18: Five water samples collected by the HRS from the east side of Bradenton show low levels of chlorine.

■ June 20: An additional sample is collected from the original complaint site. The sample is found to be satisfactory.

■ Early July: The city requests exemption from state Department of Environmental Regulation requirements for a Class A operator at any plant as large as Bradenton's.

■ July 5: Five more water samples are collected by the HRS. All test satisfactory for bacteria, and four test satisfactory for chlorine.

■ July 10: The DER denies request for an exemption for the city, saying a competent operator is needed to ensure good water.

■ July 18: The HRS conducts an inspection of the city's water-treatment plant. Samples are collected from water lines at the plant and from a service pump at the main storage tank.

■ July 19: The HRS takes 35 water samples throughout the city's distribution system.

■ July 20: HRS officials obtain initial test results but inspectors do not read them for another three days.

■ July 23: After reading the test results, HRS officials call Bradenton city officials at home and tell them that there are problems with city water.

■ July 24: Information about the finding of contamination is leaked to a local radio station, which begins broadcasting warnings. HRS officials declare the city water unsafe to drink and announce that tests conducted on July 18 found high levels of coliform bacteria in the water. City officials say their own tests show low chlorine levels but no bacteria.

■ July 25: HRS officials declare that the water situation is worse than originally thought. They say their tests of city water samples show contamination by fecal coliform, a type of bacteria found in animal wastes. The presence of fecal coliform indicates that the water could be contaminated by other, more dangerous bacteria that can cause diseases such as cholera and typhoid. County hospitals and the HRS receive more than 1,000 calls from people with questions

about the water. The HRS reports several city residents seeking treatment for vomiting and stomach cramps.

■ July 26: The Manatee County Commission votes unanimously to offer county water to city residents. HRS tests show contamination in only three of 12 city water samples. But tests of city water conducted by two independent private laboratories still show no contamination. Free water is made available to city residents.

■ July 27: HRS officials speculate that water contamination may have been caused by the absence of a fully qualified operator at the city's water plant; by the lack of chlorination before the water is filtered; and/or by a failure to operate the plant for enough hours each day to maintain a constant flow of treated water through the system. Teams from the HRS, the city's laboratory and two private laboratories begin 560 additional tests.

■ July 28: HRS officials discover that two of the tests samples that showed contamination were taken from condominiums using county water. County water officials perform their own tests at those sites, and results show the water to be clean. HRS officials refuse to release results of the tests.

■ July 29: The head of a local environmental group criticizes the HRS officials for a five-day delay in announcing that the city's water could be contaminated. A state official admits that city officials did not find out about the contaminated water sooner because the HRS officials did not read the test results for several days.

■ July 30: After a 3½-hour meeting, officials say the 560 water samples taken earlier show that the city water is good enough to drink. But they also say that the water restrictions will remain in effect for another day to allow further testing. Bradenton Mayor Bill Evers calls for a criminal investigation of the HRS Public Health Unit's handling of the water crisis.

■ July 31: HRS officials declare Bradenton water safe to drink and list several recommendations to the city's Public Works Department. The recommendations include hiring a Class A operator within 48 hours, doubling testing efforts, staffing the water plant 24 hours a day and having an independent engineering firm analyze the plant's operations. The county's leading epidemiologist says he cannot confirm that there were any unusual illnesses during the water scare. The same day, Evers calls the governor's office and asks for an investigation into the HRS handling of the water crisis.

■ Aug. 1: After inspecting the city's laboratory at the water plant and finding several problems, the state threatens to close the lab. Evers instructs City Attorney Bill Lisch to prepare a memorandum to the governor's office outlining the city's position.

■ Aug. 2: The city says it has found evidence that helps prove that the state's tests were wrong and that the water was never contaminated. The evidence includes rusty lids on state sampling bottles and what city officials call poor sampling methods. Also, city Public Works officials send a letter to the HRS in Tallahassee, saying they plan to correct all deficiencies at the lab. Lisch sends his letter to the governor's office requesting an investigation.

■ Aug. 3: The state delays a decision on whether to revoke the city's certification to run a laboratory. A report sits on HRS Secretary Greg Coler's desk waiting for his review.

From *Sarasota Herald - Tribune*, August 5, 1990.

and the consultant fee to critique the water treatment operations and recommend improvements.

PUBLIC RELATIONS PROGRAM

Public understanding of community water supply operations should be a very important utility policy. How favorably a utility is perceived by its community may ultimately be determined by the effectiveness of a public information program.[2] Unfortunately, a 1993 survey of members of the American Water Works Association suggests that only 19% of 275 water treatment systems had either a staff person or department dedicated to public relations.[9] While increasing requirements for public notification of noncompliance issues will force this situation to change, it is far more effective to have a voluntary, proactive public relations program in place before problems arise.

Dissemination of community water supply information should be done in a professional manner and by a mechanism that reaches the majority of consumers. A public relations specialist can provide positive contributions through improving the utility's image and building consumer trust. For large water utilities, these duties are the function of a public affairs officer; in a medium-sized utility, the plant manager or senior engineer is the designated individual. In communities serving fewer than 3000 customers, the responsibility often resides jointly with the mayor and the water plant operator. Whoever is responsible, the task of communicating water quality problems to the community requires a dedication to professional ethics that is above local politics and partisan attempts to cover up reality. Investigative news reporters can readily penetrate these smoke screens and turn a manageable problem into a startling exposé.

COMMUNITY OUTREACH

Public relations can also be invaluable in many ways beyond working with the press. Movement of information can become a two-way process. Not only do public information specialists inform the community about utility matters, they can also monitor public reaction through interaction with citizen advisory groups. This feedback on attitudes can be very useful in making management decisions related to water rates, water quality, and bond issues. Other valuable activities include development of fact sheets on utility policy, operations, and future expansion; information packets and facility tours for residents; employee newsletters; and educational programs and speakers for schools and local club groups.

A good public relations program is carefully designed to keep the public aware of utility efforts to provide and protect a high-quality product at the tap, satisfy industry needs for adequate water supply, and maintain water pressure to fight fires in an emergency and to water lawns and gardens. The program should provide frequent press releases on all activities such as planned interruptions in service for construction of larger service lines, announcements of service expansion to keep up with suburban development, scheduled line flushing in the distribution system, and comparison of local water quality with limits established in new regulations. More direct communication can be achieved through information included with water bills. These inserts or bill stuffers can provide useful information on the cleaning and protection of highrise building water storage tanks, service line flushing, water supply attachment devices, point-of-use treatment units, and water conservation measures.

Other examples of good public relations activities include the participation of utility employees in community festivals, Junior Achievement programs, high school science fairs, school field trips to the water treatment facility and laboratory, job fairs, and local government proclamations for National Drinking Water Week. Invitations to local environmental groups like the Sierra Club or the National Wildlife Association to take guided tours of the watershed, and restrictively permitting local fishing clubs to fish the reservoir are also very positive steps to attract public attention to water quality preservation.

CRISIS COMMUNICATION

Communicating with the public about problems in water quality requires well-timed messages that transform the technical language of the industry into laymen's terms so that the problem, its significance, and remedial action strategy are clearly understood. In fast-developing news events such as boil water orders and waterborne outbreaks, the press will be in frantic search for new information to answer the traditional questions of who, what, when, where, why, and how. Press releases and interviews should provide the first details of a serious problem.

Written Word

The written word released to the press must be carefully composed, presenting only the facts as known, explaining activities underway to correct the problem, and identifying water and health authorities involved in the working group. A schedule of updates on the water quality status should also be provided.

The news release should be confined to one or two pages and composed of short but informative sentences. If the information is sensitive, the public relations or senior staff member should be sure to get appropriate clearance from management before release to the press.

Spoken Word

The spoken word in the form of press conferences or interviews is an important form of public communication. At this point, the public relations officer, utility management, or the expert are in the spotlight. Prior to the press conference, there should be agreement as to the facts to be released and the strategy to be followed during the interviews. It is vital to anticipate possible lines of inquiry. Time permitting, proposed responses should be rehearsed with the senior staff to develop accurate, consensus statements.

Once the interview has started, answers should be honest and try to stress the positive resolution of the problem.[10] In replying to questions, subjects should speak clearly, avoiding ambiguous statements or fuzzy answers. It is important that a noncontentious attitude be maintained while talking to reporters, even under intensive questioning. If a reporter asks several questions, the subject should answer the most important question first, then ask for a repeat of the other questions. Asking a reporter to rephrase the question may provide more time to think out the answer. The subject should be on the alert for tricky questions such as, "why are you poisoning the water?" In reply, the negative statement should not be repeated. Instead, a positive response should be given, such as, "chemicals are added in water treatment to clarify the water of turbidity (or cloudiness)," or, "chlorine is applied in water treatment to kill harmful bacteria and viruses that cause waterborne outbreaks of disease in a community."

It is undesirable to make statements "off the record" because they may be used anyway. Similarly, a response of "no comment" suggests the subject has something to hide from the public. The subject should always redefine technical terms (such as coliform bacteria, indicators, water treatment, compliance criteria, risk assessment, and routes of exposure) in language that the press and the public will understand. Experts may be asked a variety of questions by reporters, including those beyond their area of experience. It is imperative to limit their answers to their specialties. Finally, it is important for those interviewed not to get emotional during interviews even though a reporter may become abrasive. Maintaining composure is a great aid in combating attempts to extract words of anger or frustration and, especially during filming, helps win audience respect.

CRISIS MANAGEMENT

Orders to boil water, waterborne outbreaks, and natural disasters such as floods, hurricanes, and earthquake disruption of the public water supply, lead to crises in the community. The farsighted utility management team will have rehearsed crisis scenarios and developed contingency plans to deal with each type of disaster. Having a crisis plan to put into operation provides a head start on restoring a safe water supply at the earliest moment. An important part of the unfolding emergency is to keep the public informed on the magnitude of the crisis, the progress being made in response to it, and options for an alternative water supply. This is where public relations plays an important part in crisis management.

Ideally, the strategy should include, in addition to an operations command center to direct restoration of treatment processes, emergency assignment of personnel, restarting of pressure to move the water in distribution, and acquisition of emergency financial support, a public information center. This center will be involved with the continuous updating of information in the form of carefully developed press releases as frequently as new information warrants. Like all staff members, the public relations officer must be accessible over extended working hours. In between press conferences there will be a deluge of phone calls from concerned citizens and investigative reporters seeking new information and a possible news scoop. The same information should be given to all press representatives without playing favorites among the media. News coverage should be monitored and any erroneous information corrected promptly.

Flood Case Study

Flooding over nine Midwestern states in the summer of 1993 created numerous water supply crises. This natural disaster caused an estimated $15 billion in damage to farm crops, animals, homes, businesses, public buildings, roads, bridges, and municipal utilities. More than 250 public water supplies in Missouri, Kansas, Illinois, and Iowa were contaminated by flooding and over 500,000 people dealt with the lack of safe water for periods of a few days to a month.[11] Among the larger public water systems that were forced to suspend operations was Des Moines, IA, serving 250,000 people.

As Des Moines can attest, coping with flood disaster required not only professional dedication, disaster preparedness, and team effort, but significant public relations skills as well. At the onset, there was intense media scrutiny from local and national newspapers and television stations, and network news anchorpersons.[12] Reporters were given early access to the flooded water supply facility so their stories, photos, and videos would show the community the extent of damage and the activities in progress to restore service. At first, some reporters tried to picture the residents as frustrated and angry over the loss of water. However, once the community realized the magnitude of the problem and saw that round-the-clock efforts were being made to restore water supply, it became almost sacrilegious to complain about the water situation. This change in public perception of the problem required an intensive public relations commitment. The utility general manager devoted at least four hours a day to answering questions from the media, industry, and the public. The 235 water plant employees were also kept informed of progress through a daily, one-page newsletter. This communication boosted morale by stopping misinformation from being circulated as rumors, and by recording progress in restoring the water supply.

Keeping the public informed of daily progress provided several additional payoffs. During the refilling of the distribution system by pressure zones, the utility communicated the importance of public cooperation in closing off all open water taps in households. This information gave the community a better appreciation of the work done in restoring a safe water supply. Business and industry leaders became more receptive to the need to build a second water treatment facility that could provide future plant redundancy.

Overcoming this disaster was an enormous problem because of power and communication failures, river sediment over the filter beds and in the clear-well, and heaving of some water service lines below flooded streets. These problems called for an all-out, round-the-clock effort in restoring the system. Teams of plant employees were organized for cleaning the treatment facility, restarting the chemical feed to process basins, reinstate electrical service, rehabilitate service pumps, and reestablish the distribution of a safe water supply.

The combination of an all-out, round-the-clock effort in restarting the system and good communication practices helped Des Moines to declare a safe water supply within an incredible 19 d. All water-use restrictions were removed by day 29 after the shutdown.

Hurricane Case Study

Hurricane damage to the distribution system can be enormous. In this situation, it is easy for the public to see the magnitude of the problem of water supply: broken service lines, damaged fire hydrants, distribution pipes disconnected or broken by the uprooting of trees, and water flowing, uncontrolled, from destroyed buildings. These images help the public to be more accepting of disrupted water service. Hurricane Andrew, which struck the southeastern coast of the U.S. in 1992, is a prime example of such a crisis. Total damage caused by Hurricane Andrew has been estimated at $20 billion. Of this, $100 million in damage was done to public water systems.

Over the years, many utilities have learned to be prepared for the impact of hurricanes, constructing concrete structures around wellheads and stockpiling treatment chemicals, emergency generators, pipe supplies, fuel, and food in anticipation. Unfortunately, nothing could have been done to lessen the terrible force of wind generated by Hurricane Andrew. This supercharged hurricane leveled entire communities and disrupted all utility services, (electricity, telephone, water supply, sewage treatment, and gas lines) to more than 3 million homes and businesses. In just one housing project (Florida City), nearly 3000 homes were obliterated, exposing free-flowing water from all broken service lines.[13] At other locations, hoses left attached to outside taps were ripped away along with the connecting faucet, creating fountains of water at the site. One fire hydrant was broken off in the street, but much of the water loss in this case was from numerous water leaks in an adjacent building that reduced water pressure in the area.

Major media coverage was generated by widespread need for bottled water and other basics such as shelter, food, and clothing. Many water utility employees lost their homes and possessions, but once assured that their families were safe, reported for work to help restore the system. The Florida City utility also provided shelter to many of its employees at the water plant during the recovery period. In little more than 1 week, work crews had repaired some 500 service line leaks and 175 main breaks. Much effort was spent in locating shut-off valves to abandoned and damaged homes with severe water leaks, and to repairing secondary line breaks resulting from bulldozers clearing away collapsed homes. Volunteers came from many utilities not affected by the storm. One utility (Sumter, SC) supplied a complete team, equipment, service truck, and their food supply to help close out the leaks and line breaks and cap the service lines to destroyed homes. This emergency response team was very effective in providing quick assistance in combating massive water losses in the pipe network.

Earthquake Case Study

Earthquake motion can be catastrophic to both water treatment plants and associated distribution systems. Ground movement of this magnitude often causes line breaks, collapse of corroded pipe, separation of pipe sections at joints, fire hydrant disconnects, service line separations, cracks in cement water storage basins, and seam separations in steel water storage tanks. If the water plant is near the quake epicenter, serious damage may be done to control room monitors, treatment basins, flocculators, and chemical feeds. Electric service outage alone can disrupt the movement of water to the distribution system. Source water intake towers may suffer structural damage, clogging the intake pipe or physically separating the intake pipe from the tower and anchorage.

On January 17, 1994, southern California was shaken by an extensive earthquake. Aqueducts lost raw water, treatment plants shut down because of power outages, storage tanks ruptured, and water mains snapped.[14] Two treatment plants near the quake epicenter suffered some damage from shockwaves. Equipment was scattered at the Metropolitan Water District, Jensen Filtration Plant. Minor damage at the L.A. Aqueduct Filtration Plant caused a shutdown for 2 d, then operation resumed at half capacity until repairs were completed. Loss of electrical power was the major factor in treatment plant shutdowns (loss of ozone generation at the Sylmar plant), rather than physical damage to treatment facilities. Pumping stations serving the higher elevations of the valley and the Santa Monica Mountains were also out of service until electrical service was restored 3 d later. The biggest problem in restoring service in the Los Angeles service area was caused by damage to the four main trunk lines. Utility field crews were supplemented by help from eight other water departments in the area, working 24 h a day to repair dozens of main breaks and more than 1000 leaks discovered after the lines were repressurized. Eight of the twenty storage tanks on the system were also damaged. To offset the loss of aqueduct water, the Los Angeles Department of Water had to rely on existing supplies of treated water in the reservoirs and the emergency connections to the Metropolitan Water District of Southern California, some of which also sustained damage. Many outlying area communities who buy their water from the Metropolitan Water District were also forced to draw upon their stored water reserves. All were affected by lack of power to operate pumps. An estimated two million customers in southern California found their water service disrupted. Initially more than 100,000 of the 660,000 customers in the Los Angeles service area were without water. About 2000 of these had no water for several weeks.

The immensity of the affected area and the number of local and regional jurisdictions involved posed unique communication problems in the California case. Water authorities tried to keep an exhausted and frightened public informed, and potable water was distributed at schools and other public places during the emergency. But there was public confusion on warnings to boil water, along with a run on bottled water supplies in the area. Perhaps the problem of disorganized responses from various local and state agencies would best be solved by the creation of a statewide emergency response command center with authority to mobilize resources from all government groups mandated to respond to water supply issues. To support this effort, a public information center staffed by water professionals with public relations expertise must be included in the team to provide a focal point for release of information that is accurate and responsive to public concerns.

PUBLIC AWARENESS: WHAT WE HAVE LEARNED

The federal regulatory requirement to inform customers when the quality of their water supply quality becomes unsatisfactory has increased public concern and often resulted in a loss of confidence, particularly when there are repeated notifications. As a consequence, the public now demands more information and, at the same time, doubts that statements made by water authorities or the press are always completely factual.

Much public information is obtained from the press, which can be a powerful force in matters of public interest, especially when it takes on the role of a watchdog. The media cannot be ignored, misled, or manipulated to try to reduce public pressure. Any attempt to do so will quickly lead to distrust and stimulate the investigative reporter to search further for hidden facts. This may also lead to dangerous speculations of no value to anyone.

From the utility position, reporters should be viewed as "friendly adversaries". By providing accurate information on a regular basis to local reporters, utilities have the opportunity to educate the press on the technical aspects of source water management, water treatment, and quality protection through the distribution system. Thus, when a crisis does occur, the informed reporter might provide a positive spin on the problem through better understanding of the process of producing and supplying a safe water supply. It is essential to provide a single voice to respond to press inquires. This individual must be kept informed of all developments in resolving a problem so that he(she) can communicate the perception of confidence and expertise during press conferences and television interviews. For the larger utilities, this interfacing with the press and public is a responsibility of a public relations specialist, whose daily job is to build the utility's image. Finally, the progressive utility should develop strategies for combating disasters and programs of preventive maintenance to avoid unexpected system failures. Contingency plans, personnel duty rehearsals, and standby emergency equipment will do much to lessen the impact of the unexpected and build confidence in utility operations.

REFERENCES

1. Hurd, R.E. 1993. *Consumers Attitude Survey on Water Quality Issues*. American Water Works Association Research Foundation, Denver, CO.
2. Public Information Committee. 1989. *Public Information*. American Water Works Association, Denver, CO.
3. Rowen, J. and D. Behm. 1993. Fatal Neglect: A Special Report. *Milwaukee Journal*. Special Reprint, September 19-26.
4. Angulo, F.J., S. Tippin, D. Sharp, B.J. Payne, J. Hill, et al. In press. A Community Waterborne Outbreak of Salmonellosis and the Effectiveness of a Boil Water Order, *Jour. Infect. Dis.*
5. Brooks, B.S., G. Kennedy, D.R. Moen, and D. Ranly. 1992. *News Reporting and Writing*. 4th ed. St. Martin's Press, New York.
6. Hudson, L.D., J.W. Hankins, and M. Battaglia. 1983. Coliforms in a Water Distribution System: A Remedial Approach. *Jour. Amer. Water Works Assoc.*, 75:564-568.
7. Sandman, P.M. 1994. Mass Media and Environmental Risk: Seven Principles. *Health Safety Environ.*, 5(3):251-260.
8. Kriewall, D.F. and L.G. Schantz. 1985. The Laboratory Role in Watershed Management, Rochester, New York. Water Qual. Tech. Conf., Dec. 8-11, Houston, TX. pp. 319-338; *Jour. Amer. Water Works Assoc.*, 75:564-568.
9. American Water Works Association. 1994. The Public Relations Advantage. *Reservoir 1, 3*. April/May Issue.
10. Elwell, F.H. 1991. Dealing with the Media. Amer. Water Works Assoc., Water Quality Tech. Conf., Orlando, FL.
11. Reid, J. 1994. Overcoming the Flood: How Midwestern Utilities Managed Disaster. *Jour. Amer. Water Works Assoc.*, 86:58-67.
12. McMullen, L.D. 1994. Surviving the Flood: Teamwork Pays Off in Des Moines. *Jour. Amer. Water Works Assoc.*, 86:68-72.
13. Murphy, M. 1994. Weathering the Storm: Water Systems versus Hurricanes. *Jour. Amer. Water Works Assoc.*, 86:74-83.
14. Waterweek. 1994. Los Angeles Quake Crushes City's Water Distribution System. *Amer. Water Works Assoc.*, 3:3, January 31.

Appendix: Field Investigation Checklist

WATERSHED MANAGEMENT

☐ Agricultural activity
☐ Residential development
☐ Recreation uses
☐ Utility ownership or control on land use
☐ Sewage treatment (septic tanks, primary or secondary effluent)
☐ _____
_____
_____

RAW WATER SOURCE

☐ Surface water (lake, impoundment, river)
☐ Groundwater (depth, soil structure, well protection)
☐ Blended sources
☐ Characterize raw water quality (coliforms, turbidity, pH, chlorine demand)
☐ _____
_____
_____

WATER TREATMENT

☐ Untreated
☐ Disinfection only
☐ Conventional (describe treatment train)

☐ Continuous processing of raw water 24 h/d
☐ Disinfection (>0.2 mg/l) applied continuously (free chlorine, ozone, chlorine dioxide, chloramine); C.T. value adequate
☐ Plant effluent turbidities <0.5 NTU
☐ Plant effluent <1 coliform per 100 or 1000 ml
☐ _____
_____
_____

WATER DISTRIBUTION SYSTEM

☐ Coliform percent occurrence per month
☐ Heterotrophic bacterial densities per milliliter
☐ Disinfectant residual detected in 95% of the monitoring sites
☐ Measurable chlorine residual at dead ends (0.5 to 1.0 mg/l for free chlorine or 1.0 to 2.0 mg/l for chloramines)
☐ System flushing procedure
☐ Flushing systematically from plant outwards through each fire plug for 15 min
☐ Static water zones flushed frequently
☐ Pulse flushing
☐ Foam plug passage
☐ Flushing frequency (spring and autumn)
☐ Minimize flow reversals and reduce water hammer effects
☐ Program to eliminated dead-end areas
☐ Finished water reservoir annual inspection, repair, and cleaning scheduled
☐ Corrosion inhibitor additives (concentration, protocol, pH adjustment)
☐ Bimetallic glassy phosphate
☐ Zinc polyphosphate
☐ Sodium hexametaphosphate
☐ Zinc orthophosphate
☐ High-molecular weight polyphosphate
☐ Pipe materials and age
☐ Program for long-term cleaning, relining or replacement of unlined cast-iron pipes
☐ Pipeline breaks, percent frequency from records
☐ Water pressure and number of pressure zones
☐ Water supply retention time in distribution system
☐ Cross-connection program
☐ Monitoring sites include first customer(s), all pressure zones, dead ends
☐ Seasonal interconnection to neighboring utility water supply
☐ Utility operates distribution system, purchased water enters system
☐ Seasonal (Summerfest) distribution line
☐ Pipeline depth
☐ Configuration

☐ Cross-connection potential (washing facilities, concession stand tap-ins, irrigation connection)

☐ Winterizing pipeline (antifreeze type, springtime flushing, and testing for coliforms, HPC, and disinfectant residue)

☐ Water quality monitoring (sampling sites, frequency, number per month for coliform, HPC, and disinfectant residual)

☐ _____

_____

_____

BIOFILM DETERMINATION

☐ Coliform occurrence pattern (localized or random in the distribution system)

☐ Coliform record during episode (densities, percent frequency per month)

☐ Coliform speciation (*Klebsiella, Enterobacter, Aerobacter, Citrobacter,* and *E. coli*)

☐ Chlorine residual pattern in distribution system

☐ Water temperature above 15°C.

☐ Flushing impact on coliform densities

☐ Use of alternative disinfectants

☐ Evaluate conversion from free chlorine to chloramines (alert hospitals and clinics before change)

☐ Flush ends of system more often or when HPC growth exceeds 1-log increase

☐ If chloramines are used, switch to free chlorine for several weeks each year to avoid excessive bacterial regrowth, taste, and odor

☐ _____

_____

_____

LABORATORY INFORMATION

☐ Utility or city health department laboratory certified by state for drinking water microbiological testing

☐ Frequency of recertification

☐ Deviation problems

☐ Sample collection practices

☐ Faucets selected are flushed for 1 min or flamed

☐ Transported to the lab promptly (within ___ hours)

☐ Sample collectors periodically recertified

☐ Sample bottles sterilized and QC record

☐ Laboratory involvement in positive coliform alerts

☐ Samples processed same day collected

☐ MPN, MF, or presence–absence technique (positive results verified)

☐ Coliform speciation method
☐ Fecal coliform or *E.coli* testing
☐ Prompt notification to water authority
☐ _____
   _____
   _____

RESPONSE

☐ Repeat sampling including samples taken above and below site of positive result
☐ Booster chlorination
☐ Flushing program activated (localized or system-wide)
☐ Expanded monitoring
☐ Use R-2A agar (7 d at 28°C) for early indication of regrowth (1-log increase or more) and need to flush out sediments
☐ Switch from P–A test to MF for quantitation of coliform results (verify and speciate coliform colonies)
☐ Activate a search for fecal coliforms/*E. coli* in all samples until the biofilm is suppressed
☐ State and local health departments monitor hospital and clinic admissions for illness cases possibly attributable to public water supply
☐ Issue boil water order if fecal coliform or *E. coli* are confirmed in repeat sample, loss of water pressure, or outbreak is waterborne
☐ _____
   _____
   _____

RECOMMENDATIONS

Index

Milton Keynes UK
Ingram Content Group UK Ltd.
UKHW021915071024
449327UK00022B/1667